Springer Tracts in Natural Philosophy

Volume 37

Springer
New York
Berlin
Heidelberg
Barcelona
Budapest
Hong Kong
London
Milan
Paris
Santa Clara
Singapore
Tokyo

Edited by C. Truesdell

Springer Tracts in Natural Philosophy

Vol. 1 Gundersen: Linearized Analysis of One-Dimensional Magnetohydrodynamic Flows (1964).

Vol. 2 Walter: Differential- und Integral-Ungleichungen und ihre Anwendung bei Abschatzungs- und Eindeutigkeitsproblemen (1964).

Vol. 3 Gaier: Konstruktive Methoden der Konformen Abbildung (1964).

Vol. 4 Meinardus: Approximation von Funktionen und Ihre Numerische Behandlung (1964).

Vol. 5 Coleman/Markovitz/Noll: Viscometric Flows of Non-Newtonian Fluids: Theory and Experiment (1966).

Vol. 6 Eckhaus: Studies in Non-Linear Stability Theory (1965).

Vol. 7 Leimanis: The General Problem of the Motion of Coupled Rigid Bodies About a Fixed Point (1965).

Vol. 8 Roseau: Vibrations Non Linéaires et Théorie de la Stabilité (1966).

Vol. 9 Brown: Magnetoelastic Interactions (1966).

Vol. 10 Bunge: Foundations of Physics (1967).

Vol. 11 Lavrentiev: Some Improperly Posed Problems of Mathematical Physics (1967).

Vol. 12 Kronmuller: Nachwirkung in Ferromagnetika (1968).

Vol. 13 Meinardus: Approximation of Functions: Theory and Numerical Methods (1967).

Vol. 14 Bell: The Physics of Large Deformation of Crystalline Solids (1968).

Vol. 15 Buchholz: The Confluent Hypergeometric Function with Special Emphasis on Its Applications (1969).

Vol. 16 Slepian: Mathematical Foundations of Network Analysis (1968).

Vol. 17 Gavalas: Nonlinear Differential Equations of Chemically Reacting Systems (1968).

Vol. 18 Martin: Introduction to the Theory of Bases (1969).

Vol. 19 Knops/Payne: Uniqueness Theorems in Linear Elasticity

Vol. 20 Edelen/Wilson: Relativity and the Question of Discretization in Astonomy (1970).

Vol. 21 McBride: Obtaining Generating Functions (1971).

Vol. 22 Day: The Thermodynamics of Simple Materials with Fading Memory (1972).

(continued after index)

Ingo Müller Tommaso Ruggeri

Rational Extended Thermodynamics

Second Edition

With 84 Illustrations

Springer

Ingo Müller
Technische Universität, Berlin
FB 6 - Institut für Verfahrenstechnik
Sekr. HF2 - Thermodynamik
Strasse des 17. Juni 135
D-10623 Berlin
Germany

Tomasso Ruggeri
Universita degli Studi di Bologna
Departimento de Matematica
C.I.R.A.M.
Centro Interdipartimentale di Ricerca
per le Applicazioni della Matematica
Via Satagozza, 8
I-40123 Bologna
Italy

Mathematics Subject Classification (1991): 82C35, 80A10, 73B30, 76A02

Library of Congress Cataloging-in-Publication Data
Müller, Ingo, 1936-
Rational extended thermodynamics / Ingo Müller, Tommaso Ruggeri, — 2nd ed.
p. cm. — (Springer tracts in natural philosophy : 37)
Rev. ed. of: Extended thermodynamics. c 1993.
Includes bibliographical references.
ISBN 978-1-4612-7460-5 (alk. paper)
1. Second sound. 2. Thermodynamics. 3. Gases—Thermal properties. 4. Fluids—Thermal properties. 5. Second sound. I. Müller, Ingo, 1936- Extended thermodynamics. II. Ruggeri, Tommaso. III. Title. IV. Series: Springer tracts in natural philosphy ; v. 37.
QC311.M784 1998
536'.7—dc21 97-31936

Printed on acid-free paper.

ISBN-13: 978-1-4612-7460-5 e-ISBN-13: 978-1-4612-2210-1
DOI: 10.1007/978-1-4612-2210-1

Softcover reprint of the hardcover 2nd edition 1998

Production managed by Timothy Taylor; manufacturing supervised by Jacqui Ashri.
Photocomposed copy prepared using LaTeX.

9 8 7 6 5 4 3 2 1

ISBN 978-1-4612-7460-5 Springer-Verlag New York Berlin Heidelberg SPIN 10654306

Preface

Ordinary thermodynamics provides reliable results when the thermodynamic fields are smooth, in the sense that there are no steep gradients and no rapid changes. In fluids and gases this is the domain of the equations of Navier-Stokes and Fourier. Extended thermodynamics becomes relevant for rapidly varying and strongly inhomogeneous processes. Thus the propagation of high-frequency waves, and the shape of shock waves, and the regression of small-scale fluctuation are governed by extended thermodynamics.

The field equations of ordinary thermodynamics are parabolic while extended thermodynamics is governed by hyperbolic systems. The main ingredients of extended thermodynamics are

- field equations of balance type,
- constitutive quantities depending on the present local state and
- entropy as a concave function of the state variables.

This set of assumptions leads to first order quasi-linear symmetric hyperbolic systems of field equations; it guarantees the well-posedness of initial value problems and finite speeds of propagation.

Several tenets of irreversible thermodynamics had to be changed in subtle ways to make extended thermodynamics work. Thus, the entropy is allowed to depend on nonequilibrium variables, the entropy flux is a general constitutive quantity, and the equations for stress and heat flux contain inertial terms. New insight is therefore provided into the principle of material frame indifference.

With these modifications an elegant formal structure can be set up in which, just as in classical thermostatics, all restrictive conditions—derived from the entropy principle—take the form of integrability conditions.

Also the modifications made by extended thermodynamics render the theory fully consistent with the kinetic theory of gases, in particular Grad's 13-moment version of the kinetic theory of gases. In fact, extended thermodynamics is most restrictive for gases or, more generally, for bodies whose constituent particles have large mean free paths. Most of this book, therefore, deals with gases: classical ideal gases, degenerate gases, relativistic gases, and mixtures of gases. It puts into perspective the various phenomena called *second sound*, viz. heat propagation, propagation of shear stress, and the second sound in superfluid helium.

Phonons and photons may have large mean free paths as well, and therefore they are amenable to a treatment by extended thermodynamics. Two chapters describe the present status of the systematic theory in this field, which is still progressing.

A certain disappointment with extended thermodynamics of 13 or 14 fields is created by the observation that it describes resonance experiments and light scattering data only slightly better than the conventional theory. These data require further extensions to many moments. Also the shock wave structure calculated in extended thermodynamics of 13 fields is worse than the shock wave structure in ordinary thermodynamics; and again: many moments are needed to put things right. All this will be demonstrated in the second half of this book.

When enough moments are used to describe the state, extended thermodynamics leads to perfect agreement of theory and experiment. Actually—even without reference to experiments—extended thermodynamics carries its own evaluation of the range of validity: As soon as more moments do not change the predictions, the extant number of moments provides a proper description of the state.

The present book is a new edition; at least half of the material is new and the rest is revised and streamlined to a considerable degree. Also the title is changed: *Rational* Extended Thermodynamics. The literature is full of papers referring to extended thermodynamics which, however, are devoid of rational methodology and mathematical cohesion. The epithet *rational* in the present title is chosen so as to emphasize the systematic procedure which the book espouses—a procedure typical for a deductive science.

Three chapters, Chapters 12 through 14, carry the names of Drs. Struchtrup and Weiss, because the material presented there is the work of these authors. However, the chapters are entirely embedded—in contents, format, and style—into the scope of this book. We are grateful to Drs. Struchtrup and Weiss for their contribution.

We wish to thank Mrs. Marlies Hentschel, who spent long hours on the word processor with unfailing enthusiasm for this work.

Ingo Müller — Berlin, Germany
Tommaso Ruggeri — Bologna, Italy

Contents

Chapter 1

Tour d'Horizon

Rational extended thermodynamics solves some problems that arise in ordinary thermodynamics and puts others into the right perspective. Originally motivated by the desire to resolve the paradox of heat conduction – a fairly esoteric problem – extended thermodynamics has grown into a predictive theory of light scattering, sound dispersion, shock wave structure and non-equilibrium radiation. It now appears concomitant to the kinetic theory of gases and closely related to the mathematical theory of hyperbolic systems.

Extended thermodynamics embraces the various phenomena called second sound, viz. heat propagation, propagation of shear stress and concentration, and the second sound in liquid helium and low-temperature semiconductors. All of these are natural consequences of the symmetric hyperbolic character of the field equations.

The matter of the principle of material frame indifference under Euclidean transformations is decided by extended thermodynamics. This principle is recognized as approximate, – valid only close to equilibrium, where the phenomenological equations of Navier-Stokes and Fourier may be relied on. Inertial terms are explicit in the field equations, and the constitutive equations of extended thermodynamics are local and instantaneous so that the question of frame indifference never arises, – except for Galilean transformations where frame indifference was never called in question.

Also extended thermodynamics allows us to identify the Euler equations and the Navier-Stokes-Fourier theory as mutilated residues of moment equations when either dissipative terms are ignored or when neither rapid changes nor steep gradients occur. The transition is effected by a procedure which we call the Maxwell iteration, – in analogy to a similar procedure in the kinetic theory of gases.

Rational Extended Thermodynamics

The basic equations of continuum mechanics and thermodynamics are balance laws, e.g. balance laws for a volume V with the fixed surface ∂V

$$\frac{d}{dt}\int_V \mathbf{F}^0 dV + \int_{\partial V} \mathbf{F}^i n_i dA = \int_V \mathbf{\Pi}\, dV.$$

$\mathbf{F}^0$ and $\mathbf{F}^i$ are densities and fluxes of thermodynamic fields and $\mathbf{\Pi}$ are production densities. [1]

In ordinary thermodynamics the vector $\mathbf{F}^0$ has five components, viz. the densities of mass, momentum and energy which define the state of a fluid. In that case the productions $\mathbf{\Pi}$ vanish, so that the balance equations are in fact conservation laws. Even so, the system of ordinary thermodynamics is not closed because of the fluxes $\mathbf{F}^i$; closure is achieved by relating the fluxes to the state—defined by $\mathbf{F}^0$—in a materially dependent manner by constitutive relations. Such relations are also called phenomenological equations, because originally they were empirical laws motivated by close observation of the phenomena.

The simplest constitutive equation is the thermal equation of state, which relates the pressure at one point and time to the densities of mass and energy at that point and time. Thus a hyperbolic system emerges which suffices for the characterization of an Euler fluid. Viscosity and heat conduction are then ignored.

A more realistic description of a fluid is provided by the constitutive equations of Navier-Stokes and Fourier for viscous stresses and heat flux. These empirical laws are *nonlocal*, because the fluxes $\mathbf{F}^i$ at a position do not only depend on the state at that position but also on the state in the immediate neighborhood. In other words $\mathbf{F}^i$ depends on $\mathbf{F}^0$ as well as on the gradients of $\mathbf{F}^0$. This additional dependence gives the Navier-Stokes-Fourier system a parabolic structure. However, the system has been most successful; it is sufficient for *most* fluids and leads to satisfactory results under *most* circumstances.

But the Navier-Stokes-Fourier theory is not sufficient for *all* fluids under *all* circumstances! Even monatomic ideal gases—the simplest of all thermodynamic materials—are not well described by the Navier-Stokes-Fourier constitutive equations, if steep gradients or rapid changes occur. This is where extended thermodynamics has found a fertile field.

In rational extended thermodynamics the state $\mathbf{F}^0$ is not given by the densities of mass, momentum and energy alone; it is extended (sic!) to represent further densities, typically the densities of the fluxes of momentum and energy. Whatever the additional variables are, however, the basic equations all have the structure of balance equations, although not all of them are conservation laws. And the closure is achieved by constitutive equations for the fluxes $\mathbf{F}^i$ and the productions $\mathbf{\Pi}$. The constitutive functions are *local* and *instantaneous* so that $\mathbf{F}^i$ and $\mathbf{\Pi}$ at one point and time depend only on the state at that point and time. Nonlocality and history dependence have no room in extended thermodynamics.

[1] We ignore the possibility of supplies and inertial contributions in these introductory considerations.

Therefore the local forms of the balance equations, viz.

$$\frac{\partial \mathbf{F}^0}{\partial t} + \frac{\partial \mathbf{F}^i}{\partial x^i} = \mathbf{\Pi} \qquad \text{with} \qquad \mathbf{F}^i = \mathbf{F}^i(\mathbf{F}^0) \quad \text{and} \quad \mathbf{\Pi} = \mathbf{\Pi}(\mathbf{F}^0) \tag{1}$$

provide a system of first-order, quasi-linear differential equations for the densities $\mathbf{F}^0$. The form of the constitutive functions is restricted in its generality by just two universal principles

- the entropy principle, and
- the principle of relativity.

In this respect rational extended thermodynamics reflects the procedure of rational mechanics by Truesdell & Noll [1] and rational thermodynamics by Truesdell [2].

The entropy principle requires that the entropy inequality

$$\frac{\partial h^0}{\partial t} + \frac{\partial h^i}{\partial x^i} = \Sigma \geq 0 \qquad \text{with} \qquad h^0 = h^0(\mathbf{F}^0), \;\; h^i = h^i(\mathbf{F}^0), \;\; \Sigma = \Sigma\,(\mathbf{F}^0) \tag{2}$$

hold for all solutions of the field equations (1) and that $h^0(\mathbf{F}^0)$ be a concave function. The principle of relativity states that the field equations (1) have the same form in all Galilean frames. The two principles together go a long way to make the constitutive functions explicit. And they ensure that the system (1) has symmetric hyperbolic structure.

Physicist like hyperbolicity, because it precludes action at a distance and ensures finite speeds. But symmetric hyperbolicity is also a desirable feature mathematically, because it ensures well-posedness of local Cauchy problems, e.g. see Fisher & Marsden [3]. Therefore mathematicians have studied quasi-linear first-order systems for their own sake, e.g. see Godunov [4], Friedrichs & Lax [5] and Boillat [6]. These authors discovered that such a system may be reduced to a symmetric hyperbolic system, if it is compatible with a "convex extension," i.e. an additional equation of the type (2). This requirement tallies conveniently with the physical axiom of the entropy inequality. To be sure, the entropy is concave whereas the Friedrichs & Lax extension is convex, but the difference in sign is unimportant.

Therefore it is fair to say that rational extended thermodynamics and the theory of hyperbolic systems are symbiotic. At least they are *now*. But this is not how extended thermodynamics started.

When extended thermodynamics started with the work of Müller [7], [8] its sole and—from the present point of view—rather naive objective was the resolution of the so-called paradoxes of thermal and shear waves: The Navier-Stokes-Fourier theory has a parabolic structure and it predicts infinite pulse speeds. Working within the then prevailing theory of Thermodynamics of Irreversible Processes (TIP), Müller allowed the entropy to depend on the heat flux and the viscous stress; thus he arrived at a modified Gibbs equation and, eventually, at a hyperbolic system of equations. This work is briefly reviewed in this book—see Chapter 2—for two reasons: First, because it foreshadowed rational extended thermodynamics, and second, because a part of the thermodynamic community has never progressed beyond it. Indeed, the alteration of the Gibbs equation has become a machine to generate literally hundreds of papers on many branches of physics, see Jou, Casas-Vazquez & Lebon [9], and Nettleton & Sobolev [10] for reviews and extensive lists of references.

Even before extended thermodynamics encountered the mathematical theory of hyperbolic systems, it profited from contact with the kinetic theory of gases—still in [7]—particularly with the moment method of Grad [11], [12]. Thus the kinetic theory helped to recognize the fact that the nonconvective entropy flux φ_i is not universally related to the heat flux q_i by $\varphi_i = q_i/T$. This contact between thermodynamics and the kinetic theory became more and more intimate as extended thermodynamics proceeded to more and more variables, and as it concentrated on moments of the kinetic phase density for their identification. A step forward occurred with the closure by entropy maximization which was reinvented and generalized by Dreyer [13]. The application of extended thermodynamics to "gases" of photons and phonons and application to the phenomena of light scattering, shock wave structure, and dispersion of sound would all have been impossible without recognition of the concomitant existence of the kinetic theory. These applications were worked out by Struchtrup [14], [15] and Weiss [16], two authors who have kindly agreed to let their theories be incorporated into this book in Chapters 12, 13, and 14.

TIP was replaced by Rational Thermodynamics in the 1960's when Coleman & Noll [17] recognized the essential role of the entropy principle in restricting the constitutive functions. This theory in its original form was still hampered by specific—unwarranted and/or unneeded—assumptions on the flux and supply of entropy. Müller [18], [19] removed these unnecessary features from Rational Thermodynamics, and Liu [20] introduced Lagrange multipliers to make the entropy balance compatible with the basic balance laws of thermodynamics. Liu & Müller [21] combined their ideas and formulated the first paper of rational extended thermodynamics.

Thus the subject was finally prepared for being joined to the mathematical theory of hyperbolic systems: Ruggeri & Strumia [22] recognized that the Lagrange multipliers—their *main field*—could be chosen as the thermodynamic variables and, if they were, the field equations turned out to be symmetric hyperbolic. With this observation it became possible to reveal the simple formal structure of extended thermodynamics. This formal structure is presented in Chapters 3 and 9 of this book; it was refined and extended by Boillat and Ruggeri in several papers, e.g. [23], [24] and—most recently—these authors [25] reached an anticlimax by proving that the pulse speed in nonrelativistic thermodynamics really *is* infinite. To be sure, as long as the number of state variables is finite, so is the speed; however, the speed tends to infinity when the number of variables does.

But this does not matter! After all, extended thermodynamics has long outgrown the original motive of resolving the paradox of heat conduction. It has now become a predictive theory which is needed when steep gradients and rapid changes occur—as they do in shock waves and high-frequency sound propagation.

Seeing that extended thermodynamics is a theory of symmetric hyperbolic systems, we are not surprised that sound propagation is an integral part of it. And indeed, all the multifarious phenomena called second sound emerge from the equations of extended thermodynamics: The original second sound in liquid helium as well as the second sound in low temperature semiconductors. Moreover, more and more frequently one hears the thermal waves and shear waves in viscous heat-conducting fluids dubbed second sound; and, of course, these are solutions of the equations of extended thermodynamics.

There is also the matter of the principle of material frame indifference under Euclidean transformations. Müller [26] showed that the kinetic theory of gases contradicts that principle, because stress and heat flux are affected by the Coriolis force which—in a noninertial frame—bends the mean free paths of the atoms into circular arcs. Extended thermodynamics confirms that observation. Indeed, in extended thermodynamics the field equations—being balance laws – contain explicit time derivatives and these produce explicit inertial terms in all of them, when written in a noninertial frame. To the extent that the phenomenological equations of ordinary thermodynamics are mutilated forms of the complete balance equations, they *must* depend on frame, unless of course the frame dependent terms fall victim to the process of mutilation.

Instead of Euclidean invariance extended thermodynamics focuses attention on Galilean invariance as a physicist would. This is good enough to identify the role of velocity in the densities and fluxes and to recognize the constitutive functions as isotropic functions of their variables.

$$
\begin{aligned}
&\frac{\partial \varrho}{\partial t} + \frac{\partial v_i}{\partial x_i} = 0,\\
&\frac{\partial v_i}{\partial t} + \frac{\left(\frac{k}{m}\bar{T}\right)}{\bar{\varrho}}\frac{\partial \varrho}{\partial x_i} + \frac{\partial}{\partial x_i}\left(\frac{k}{m}T\right) - \frac{1}{\bar{\varrho}}\frac{\partial t_{(ij)}}{\partial x_j} = 0,\\
&\frac{\partial\left(\frac{k}{m}T\right)}{\partial t} + \frac{2}{3}\left(\frac{k}{m}\bar{T}\right)\frac{\partial v_k}{\partial x_k} + \frac{2}{3}\frac{1}{\bar{\varrho}}\frac{\partial q_k}{\partial x_k} = 0,\\
&\frac{\partial t_{(ij)}}{\partial t} - \frac{4}{5}\frac{\partial q_{(i}}{\partial x_{j)}} - \frac{\partial \varrho_{(ijk)}}{\partial x_k} - 2\bar{\varrho}\left(\frac{k}{m}\bar{T}\right)\frac{\partial v_{(i}}{\partial x_{j)}} = \sigma t_{(ij)},\\
&\frac{\partial q_i}{\partial t} - \left(\frac{k}{m}\bar{T}\right)\frac{\partial t_{(ik)}}{\partial x_k} + \frac{1}{6}\frac{\partial \Delta}{\partial x_i} + \frac{5}{2}\bar{\varrho}\left(\frac{k}{m}\bar{T}\right)\frac{\partial\left(\frac{k}{m}T\right)}{\partial x_i} = \frac{2}{3}\sigma q_i,\\
&\frac{\partial \varrho_{(ijk)}}{\partial t} - 3\frac{k}{m}\bar{T}\left\{\frac{\partial t_{((ij)}}{\partial x_{k)}} - \frac{2}{5}\frac{\partial t_{(r(i)}}{\partial x_r}\delta_{jk)}\right\} = \frac{3}{2}\sigma \varrho_{(ijk)},\\
&\frac{\partial \Delta}{\partial t} + 8\frac{k}{m}\bar{T}\frac{\partial q_i}{\partial x_i} = \frac{2}{3}\sigma\Delta.
\end{aligned}
$$

$$
\begin{aligned}
&\frac{\partial \varrho}{\partial t} + \bar{\varrho}\frac{\partial v_i}{\partial x_i} = 0,\\
&\frac{\partial v_i}{\partial t} + \frac{\left(\frac{k}{m}\bar{T}\right)}{\bar{\varrho}}\frac{\partial \varrho}{\partial x_i} + \frac{\partial}{\partial x_i}\left(\frac{k}{m}T\right) - \frac{1}{\bar{\varrho}}\frac{\partial t_{(ij)}}{\partial x_j} = 0,\\
&\frac{\partial\left(\frac{k}{m}T\right)}{\partial t} + \frac{2}{3}\left(\frac{k}{m}\bar{T}\right)\frac{\partial v_k}{\partial x_k} + \frac{2}{3}\frac{1}{\bar{\varrho}}\frac{\partial q_k}{\partial x_k} = 0,\\
&\frac{\partial t_{(ij)}}{\partial t} - \frac{4}{5}\frac{\partial q_{(i}}{\partial x_{j)}} - \frac{\partial \varrho_{(ijk)}}{\partial x_k} - 2\bar{\varrho}\left(\frac{k}{m}\bar{T}\right)\frac{\partial v_{(i}}{\partial x_{j)}} = \sigma t_{(ij)},\\
&\frac{\partial q_i}{\partial t} - \left(\frac{k}{m}\bar{T}\right)\frac{\partial t_{(ik)}}{\partial x_k} + \frac{1}{6}\frac{\partial \Delta}{\partial x_i} + \frac{5}{2}\bar{\varrho}\left(\frac{k}{m}\bar{T}\right)\frac{\partial\left(\frac{k}{m}T\right)}{\partial x_i} = \frac{2}{3}\sigma q_i,\\
&\frac{\partial \varrho_{(ijk)}}{\partial t} - 3\frac{k}{m}\bar{T}\left\{\frac{\partial t_{((ij)}}{\partial x_{k)}} - \frac{2}{5}\frac{\partial t_{(r(i)}}{\partial x_r}\delta_{jk)}\right\} = \frac{3}{2}\sigma \varrho_{(ijk)},\\
&\frac{\partial \Delta}{\partial t} + 8\frac{k}{m}\bar{T}\frac{\partial q_i}{\partial x_i} = \frac{2}{3}\sigma\Delta.
\end{aligned}
$$

$$
\begin{aligned}
&\frac{\partial \varrho}{\partial t} + \bar{\varrho}\frac{\partial v_i}{\partial x_i} = 0,\\
&\frac{\partial v_i}{\partial t} + \frac{\left(\frac{k}{m}\bar{T}\right)}{\bar{\varrho}}\frac{\partial \varrho}{\partial x_i} + \frac{\partial}{\partial x_i}\left(\frac{k}{m}T\right) - \frac{1}{\bar{\varrho}}\frac{\partial t_{(ij)}}{\partial x_j} = 0,\\
&\frac{\partial\left(\frac{k}{m}T\right)}{\partial t} + \frac{2}{3}\left(\frac{k}{m}\bar{T}\right)\frac{\partial v_k}{\partial x_k} + \frac{2}{3}\frac{1}{\bar{\varrho}}\frac{\partial q_k}{\partial x_k} = 0,\\
&\frac{\partial t_{(ij)}}{\partial t} - \frac{4}{5}\frac{\partial q_{(i}}{\partial x_{j)}} - \frac{\partial \varrho_{(ijk)}}{\partial x_k} - 2\bar{\varrho}\left(\frac{k}{m}\bar{T}\right)\frac{\partial v_{(i}}{\partial x_{j)}} = \sigma t_{(ij)},\\
&\frac{\partial q_i}{\partial t} - \left(\frac{k}{m}\bar{T}\right)\frac{\partial t_{(ik)}}{\partial x_k} + \frac{1}{6}\frac{\partial \Delta}{\partial x_i} + \frac{5}{2}\bar{\varrho}\left(\frac{k}{m}\bar{T}\right)\frac{\partial\left(\frac{k}{m}T\right)}{\partial x_i} = \frac{2}{3}\sigma q_i,\\
&\frac{\partial \varrho_{(ijk)}}{\partial t} - 3\frac{k}{m}\bar{T}\left\{\frac{\partial t_{((ij)}}{\partial x_{k)}} - \frac{2}{5}\frac{\partial t_{(r(i)}}{\partial x_r}\delta_{jk)}\right\} = \frac{3}{2}\sigma \varrho_{(ijk)},\\
&\frac{\partial \Delta}{\partial t} + 8\frac{k}{m}\bar{T}\frac{\partial q_i}{\partial x_i} = \frac{2}{3}\sigma\Delta.
\end{aligned}
$$

$$
\begin{aligned}
&\frac{\partial \varrho}{\partial t} + \bar{\varrho}\frac{\partial v_i}{\partial x_i} = 0,\\
&\frac{\partial v_i}{\partial t} + \frac{\left(\frac{k}{m}\bar{T}\right)}{\bar{\varrho}}\frac{\partial \varrho}{\partial x_i} + \frac{\partial}{\partial x_i}\left(\frac{k}{m}T\right) - \frac{1}{\bar{\varrho}}\frac{\partial t_{(ij)}}{\partial x_j} = 0,\\
&\frac{\partial\left(\frac{k}{m}T\right)}{\partial t} + \frac{2}{3}\left(\frac{k}{m}\bar{T}\right)\frac{\partial v_k}{\partial x_k} + \frac{2}{3}\frac{1}{\bar{\varrho}}\frac{\partial q_k}{\partial x_k} = 0,\\
&\frac{\partial t_{(ij)}}{\partial t} - \frac{4}{5}\frac{\partial q_{(i}}{\partial x_{j)}} - \frac{\partial \varrho_{(ijk)}}{\partial x_k} - 2\bar{\varrho}\left(\frac{k}{m}\bar{T}\right)\frac{\partial v_{(i}}{\partial x_{j)}} = \sigma t_{(ij)},\\
&\frac{\partial q_i}{\partial t} - \left(\frac{k}{m}\bar{T}\right)\frac{\partial t_{(ik)}}{\partial x_k} + \frac{1}{6}\frac{\partial \Delta}{\partial x_i} + \frac{5}{2}\bar{\varrho}\left(\frac{k}{m}\bar{T}\right)\frac{\partial\left(\frac{k}{m}T\right)}{\partial x_i} = \frac{2}{3}\sigma q_i,\\
&\frac{\partial \varrho_{(ijk)}}{\partial t} - 3\frac{k}{m}\bar{T}\left\{\frac{\partial t_{((ij)}}{\partial x_{k)}} - \frac{2}{5}\frac{\partial t_{(r(i)}}{\partial x_r}\delta_{jk)}\right\} = \frac{3}{2}\sigma \varrho_{(ijk)},\\
&\frac{\partial \Delta}{\partial t} + 8\frac{k}{m}\bar{T}\frac{\partial q_i}{\partial x_i} = \frac{2}{3}\sigma\Delta.
\end{aligned}
$$

Figure 1.1 Upper left: Euler. Upper right: Navier-Stokes-Fourier.
Lower left: 13 moments. Lower right: 20 moments.

Of course, everything that is said here will be demonstrated in subsequent chapters of the book,— and sometimes more than once. We have tried to keep the complex formalism of extended thermodynamics out of this introductory chapter but we depart from this effort now, because we wish to manifest the mutilation aspect mentioned above.

Figure 1.1 shows a tableau of four systems, all representing the linearized equations of extended thermodynamics of 21 moments, viz. density ρ, velocity v_i, temperature T, stress deviator $t_{<ij>}$, heat flux q_i, traceless third moment $\rho_{<ijk>}$ and double trace of fourth moment Δ in nonequilibrium. The frames and crosses indicate to what extent that system has to be mutilated for the Euler equations to appear, or the Navier-Stokes-Fourier equations, or the 13- and 20-moment equations.

We note that the left-hand side of the system is completely explicit in the variables. The only unknown is the coefficient σ on the right hand side. $\frac{1}{\sigma}$ represents a relaxation time, which we may take to be of the order of magnitude of the mean time of free flight of the atoms.

The Euler equations result from dropping all dissipative terms from the system. The Navier-Stokes-Fourier equations do recognize the stress deviator and the heat flux, but they ignore their space- and time derivatives. This omission is surely not justified when steep gradients and rapid changes occur, as they do in shock waves for example. Therefore one might have hoped that the 13-moment system, which preserves the time derivatives of stress and heat flux, gives satisfactory results where Navier-Stokes-Fourier fails.

This expectation has not come true, unfortunately. Indeed, our experience with sound dispersion and light scattering indicates that the 13-moment system—or even the 20-moment system—is not much better than Navier-Stokes-Fourier. It takes *many* moments—hundreds or even thousands of them—to achieve good results. That circumstance is satisfactory and disappointing at the same time.

What is satisfactory about it is the fact *that moment methods do work.* It is common among physicists—and somehow fashionable—to sneer at moments, particularly many moments. But the works of Weiss [16], [27] and Weiss & Müller [28] have shown convincingly that moments are effective tools for a quantitative description of light scattering and shock wave structure. Moreover, extended thermodynamics of moments is a "theory of theories," which carries its own proof of validity: As soon as an increase of the number of moments does not improve a result, we know that the extant number is sufficient.

A certain disappointment comes from the fact that so many moments are often needed. This is bad, of course, for the solution of initial and boundary value problems, because nobody can prescribe the values of a high-order moment initially, nor can anyone control it on the boundary. It seems that we shall have to live with that uncertainty. But then, there are problems which can be solved even when there is no suggestive interpretation of the variables, if only equilibrium is well defined. Light scattering is one of those problems, shock wave structure is another one.

Relativists have often shown a keen interest in the development of extended thermodynamics, because the "paradox" of infinite speed of thermal waves disturbed them acutely. Indeed, immediately after the formulation of the early version of extended thermodynamics Müller [7] formulated the theory relativistically; his work was not properly published until Israel [29] presented a very similar theory. That theory was revisited by Liu, Müller & Ruggeri [30], who produced a rational systematic formulation of relativistic thermodynamics. Chapter 6 gives a detailed account of their work, which concentrates on the particle flux vector and the energy-momentum tensor as variables, so that there are 14 of them.

Of course, hyperbolicity and, in particular, finite speeds are built into all nonrelativistic and relativistic theories in this book. And while in the nonrelativistic theory of moments the speed tends to infinity, this is not so in relativity: Indeed, in relativity the speed limit for infinitely many moments is below the speed of light. Nobody is surprised, of course, but the proof was not easy and it was only recently obtained by Boillat & Ruggeri [25].

To a certain extent extended thermodynamics identifies the transport coefficients of irreversible thermodynamics, i.e. shear and bulk viscosity and thermal conductivity. At least it determines these coefficients to within a relaxation time, or a mean time of free flight of atoms or molecules. The shear viscosity in the Navier-Stokes equation of Figure 1.1 provides a simple example; it reads $\mu = \rho \frac{k}{m} T \frac{1}{\sigma}$. We shall often have ulterior knowledge about the relaxation time and its dependence on ρ and T, so that—with such knowledge—μ is determined.

A less simple case was recently discovered by Kremer & Müller [31] in extended thermodynamics of chemically reacting mixtures of gases. They identified the thermal conductivity and the bulk viscosity—always to within a relaxation time—and found these coefficients to depend strongly on the heat of reaction (see Chapter 7). The full consequences of that discovery have not yet been investigated.

It is fair to say that rational extended thermodynamics has been most successful for gases of particles, or quasiparticles—e.g. photons and phonons—in which the mean free path between collisions is large. Under such circumstances a Boltzmann-type transport equation for the phase density can usually be written and that equation may give rise to an extended thermodynamics of moments. Metal electrons may be treated in this way (see Chapter 15); and Anile [32] has thus formulated an extended thermodynamics of electrons and holes in semiconductors. This work has made a promising start. If it is to proceed successfully, the balance equations of electromagnetism must be firmly tied to the balance equations of extended thermodynamics.

Since rational extended thermodynamics has local and instanteneous constitutive relations, it might seem that viscoelastic fluids cannot be part of it. After all, viscoelastic fluids are the paradigms of history dependence and fading memory, and the nth-grade fluid is quite successful in describing viscometric flows. But then, it is also true that there are problems with thermodynamics of nth-grade fluids, as witness the *imbroglio* about the sign of the first normal stress coefficient (see Dunn & Fosdick [33] and Joseph [34]). Extended thermodynamics can offer a fresh view of this problem. This is a field which future research may find fertile (see Chapter 16).

A natural question to be asked of extended thermodynamics concerns the relation to ordinary thermodynamics. Usually this means the transition from a hyperbolic system of equations to parabolic equations, and it is clear that some approximation is involved—an approximation that raises havoc with characteristic speeds and stability. In this book we use the Maxwell iteration for the purpose—a purely formal iterative procedure which leads to an expansion in terms of mean free paths or relaxation times. It may be a rewarding task for an analyst to investigate that procedure with the tools at his disposal and thus improve our knowledge on both ordinary thermodynamics and extended thermodynamics.

Another fantastic opportunity for an analyst occurs in the treatment of shock structure by extended thermodynamics; see Chapter 12. It turns out—as if by a miracle—that all characteristic speeds give rise to *regular* singular points, except one: the largest characteristic speed. This is a result of the numerical calculation of the shock structure, but no less convincing for being numerical. Therefore the question arises: What is the intrinsic—so far unidentified—property of the system of symmetric hyperbolic equations that brings this situation about? Note that without that property no smooth shock structure could be calculated.

Chapter 2

Early Version of Extended Thermodynamics and Kinetic Theory of Gases

The laws of Fourier, Fick and Navier-Stokes were originally proposed by ad hoc arguments to describe the phenomena of heat conduction, diffusion and shear stresses in fluids. This was long before there was any non-equilibrium thermodynamics. It was much later – essentially when Eckart's papers [35], [36], [37] appeared – that these laws became incorporated into thermodynamics of irreversible processes. That theory, while not entirely systematic, is simple and has great suggestive value. We start out with a brief description of how thermodynamics of irreversible processes arrives at the laws of Fourier and Navier-Stokes.

Once these laws are applied to the equations of balance of mechanics and thermodynamics, a system of parabolic equations appears which predicts infinite speeds for the propagation of heat and shear stress. That result has been called a paradox and upon reflection it is clear that the laws of Fourier and Navier-Stokes are to blame.

Cattaneo [38] was the first to propose a remedy. He was concerned with heat conduction and – basing his arguments on molecular kinetics – he formulated a modification of Fourier's law which is now called the Cattaneo equation. Thus Cattaneo arrived at a hyperbolic differential equation for the propagation of temperature. Much later Müller [7], [8] made an attempt to incorporate the Cattaneo equation – *and* a corresponding modification of the Navier-Stokes equations – into thermodynamics of irreversible processes. He was thus able to predict finite speeds for heat waves and shear waves. The theory that emerged has been called *extended* thermodynamics of irreversible processes.

That extended theory is characterized by non-equilibrium terms in the entropy and by a modified entropy flux, both ad hoc assumptions. Those assumptions can be shown, however, to have specific counterparts in the kinetic theory of gases at the level of Grad's 13 moment approximation [11]. And while the extended theory is not properly invariant under a change of frame, the kinetic theory of gases provides the arguments by which that flaw may be remedied.

1 Paradoxes of Heat Conduction and Shear Diffusion

1.1 Heuristic Derivation of the Laws of Fourier and Navier-Stokes

One objective of irreversible thermodynamics of fluids is the determination of the 5 fields of

$$\begin{array}{ll} \text{mass density} & \varrho(\mathbf{x},t) \\ \text{velocity} & v_i(\mathbf{x},t) \\ \text{temperature} & T(\mathbf{x},t) \end{array} \tag{1.1}$$

in all points of the fluid and at all times.

For this purpose we need field equations, and these are based upon the equations of balance of mechanics and thermodynamics, viz. the conservation laws of mass and momentum and the equation of balance of internal energy

$$\begin{aligned} \dot{\varrho} &+ \varrho\frac{\partial v_j}{\partial x_j} = 0 \\ \varrho\dot{v}_i &- \frac{\partial t_{ij}}{\partial x_j} = 0 \\ \varrho\dot{\varepsilon} &+ \frac{\partial q_i}{\partial x_j} = t_{ij}\frac{\partial v_i}{\partial x_j}. \end{aligned} \tag{1.2}$$

While these are 5 equations, they are not field equations for the fields (1.1). Indeed, T does not even appear in the equations (1.2), and instead they contain new quantities, viz. the

$$\begin{array}{ll} \text{(symmetric) stress tensor} & t_{ij} \\ \text{heat flux} & q_i \\ \text{specific internal energy} & \varepsilon. \end{array} \tag{1.3}$$

In order to close the system we must find relations between t_{ij}, q_i, ε and the fields (1.1), so-called constitutive relations.

In thermodynamics of irreversible processes, a theory universally known as TIP, such relations are derived in a heuristic manner from an entropy inequality that is based upon the Gibbs equation of equilibrium thermodynamics

$$\dot{s} = \frac{1}{T}\left(\dot{\varepsilon} - \frac{p}{\varrho^2}\dot{\varrho}\right) \tag{1.4}$$

s is the specific entropy. ε and the pressure p are considered to be functions of ϱ and T as prescribed by the caloric and thermal equations of state of equilibrium thermodynamics.[1] Elimination of $\dot{\varepsilon}$ and $\dot{\varrho}$ between (1.4) and $(1.2)_{1,3}$ and some rearrangement leads to the equation

$$\varrho\dot{s} + \frac{\partial}{\partial x_i}\left(\frac{q_i}{T}\right) = q_i\frac{\partial\frac{1}{T}}{\partial x_i} + \frac{1}{T}t_{\langle ij\rangle}\frac{\partial v_{\langle i}}{\partial x_{j\rangle}} + \frac{1}{T}\left(\frac{1}{3}t_{ii} + p\right)\frac{\partial v_i}{\partial x_i} \tag{1.5}$$

which may be interpreted as an equation of balance of entropy. That interpretation implies that

[1]This reference to equilibrium thermodynamics is sometimes called "principle of local equilibrium."

$$\begin{aligned} \varphi_i &= \frac{q_i}{T} && \text{is the entropy flux,} \\ \Sigma &= -\frac{q_i}{T^2}\frac{\partial T}{\partial x_i} + t_{ij}\frac{\partial v_{<i}}{\partial x_{j>}} + \left(\frac{1}{3}t_{ii} + p\right)\frac{\partial v_n}{\partial x_n} && \text{is the entropy production.} \end{aligned} \tag{1.6}$$

Angular brackets characterize symmetric traceless tensors. Inspection shows that the entropy production is a sum of products of

thermodynamic fluxes	and	thermodynamic forces
heat flux q_i		temperature gradient $\frac{\partial T}{\partial x_i}$
stress deviator $t_{<ij>}$		deviatoric velocity gradient $\frac{\partial v_{<i}}{\partial x_{j>}}$
dynamic pressure $\pi = -\frac{1}{3}t_{ii} - p$		divergence of velocity $\frac{\partial v_n}{\partial x_n}$.

(1.7)

The entropy production must be nonnegative. Assuming only linear relations between forces and fluxes TIP ensures the nonnegative entropy production by constitutive equations—phenomenological equations in the jargon of TIP—of the type

$$\begin{aligned} q_i &= -\kappa\frac{\partial T}{\partial x_i} && \kappa \geq 0 \\ t_{<ij>} &= 2\mu\frac{\partial v_{<i}}{\partial x_{j>}} && \mu \geq 0 \\ \pi &= -\lambda\frac{\partial v_n}{\partial x_n} && \lambda \geq 0. \end{aligned} \tag{1.8}$$

Along with the thermal and caloric equations of state $p = p(\varrho, T), \varepsilon = \varepsilon(\varrho, T)$ the equations (1.8) are the constitutive equations of TIP. They are known as the laws of Fourier and Navier-Stokes, with κ being the thermal conductivity and μ and λ the shear and bulk viscosity respectively. All of these coefficients may be functions of ϱ and T.

1.2 Parabolic Laws of Heat Conduction and Shear Diffusion

Insertion of (1.8) and of the equations of state into the balance equations (1.2) leads to field equations for the determination of the fields (1.1). The field equations are parabolic in character. In order to emphasize that fact we investigate two special cases:

1. the case of a fluid at rest with constant density, and
2. the case of a fluid with constant density and temperature and with a velocity field of the form $\mathbf{v} = (0, v(x^1, t), 0)$.

In these cases the system of equations (1.2) reduces to a single partial differential equation for T and v respectively. Neglecting nonlinear terms in gradients and time derivatives we obtain

$$\frac{\partial T}{\partial t} = \frac{\kappa}{\varrho\varepsilon_T}\Delta T \quad \text{and} \quad \frac{\partial v}{\partial t} = \frac{\mu}{\varrho}\frac{\partial^2 v}{\partial x_1^2}, \tag{1.9}$$

where $\varepsilon_T = \left(\frac{\partial \varepsilon}{\partial T}\right)_\varrho$ is the specific heat at constant volume, a positive quantity. Both equations are of the same type; they are known as ***diffusion equations.***

We may write the solution of $(1.9)_1$ for an initial value problem in infinite space in the form

$$T(\mathbf{x},t) = \frac{1}{(4\pi Dt)^{3/2}} \int_{-\infty}^{\infty} T(\mathbf{y},0) e^{-\frac{(\mathbf{y}-\mathbf{x})^2}{4Dt}} d\mathbf{y}, \tag{1.10}$$

where D stands for $\frac{\kappa}{\varrho\varepsilon_T}$. Equation $(1.9)_2$ has the same general solution if we make allowance for its one-dimensional character. This solution implies that $T(\mathbf{x},t)$ is unequal to zero for all $\mathbf{x}$ and $t > 0$ even though $T(\mathbf{x},0)$ may be nonzero in a finite volume only. Thus the temperature and the shear velocity spread throughout the whole space infinitely fast, a phenomenon that has been called a paradox.

It is fair to say that few people cared about the paradox, because the equations (1.9) are perfectly well suited for the problems of the engineers and the physicists. Only a few mathematicians and theoretical physicists were concerned. However, the desire to have hyperbolic equations in thermodynamics, and hence finite speeds, was the main motivation for the development of extended thermodynamics. It began with Cattaneo.

2 Paradox Removed

2.1 The Cattaneo Equation

In 1948 Cattaneo wrote a paper entitled "Sulla Conduzione del Calore" [38] in which he addressed the question of the paradox of heat conduction. He modified Fourier's law and based his argument on the elementary kinetic theory of gases. The reasoning proceeds as follows:

Cattaneo first reviews an old argument of Maxwell's which gives a molecular interpretation of heat conduction. To understand the gist of that argument we focus attention upon a small volume element of linear dimensions of the mean free path of the molecules in which a temperature gradient prevails; see Figure 2.1.

A particle moving upwards will, in the mean, carry more energy than a particle moving downwards. Therefore, the pair of particles carry a net energy upwards, i.e. opposite to the temperature gradient. This amounts to a heat flux across the middle layer, and we obtain the

$$\textit{stationary}\text{ Fourier law} \qquad q_i = -\kappa \frac{\partial T}{\partial x_i}. \tag{2.1}$$

Cattaneo changed this argument slightly. He argued that there is a time lag between the start of the particles at their point of departure and the time of passage through the middle layer. If the temperature changes in time, it is thus clear that the heat flux at a certain time depends on the temperature gradient at an earlier time. Therefore it seemed reasonable to formulate the

$$\textit{instationary}\text{ Fourier law} \qquad q_i = -\kappa\left(\frac{\partial T}{\partial x_i} - \tau\left(\frac{\partial T}{\partial x_i}\right)^{\bullet}\right), \tag{2.2}$$

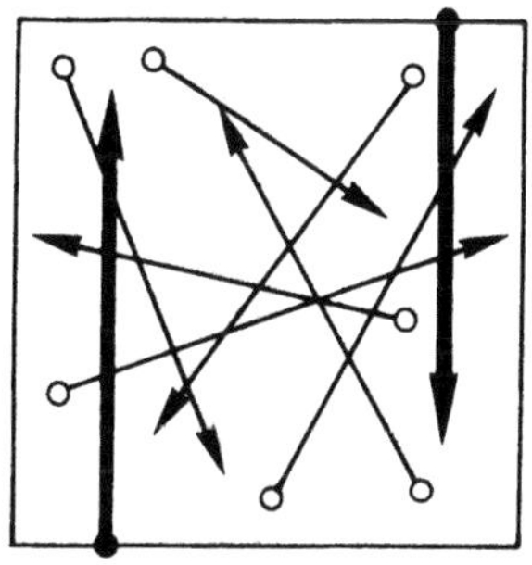

Figure 2.1 On Fourier's Law.

where the right-hand side is a Taylor expansion of the temperature gradient at the present time. If the equations (2.1) and (2.2) are inserted into the energy balance $(1.2)_3$ for a body at rest and with constant density, viz.

$$\varrho\varepsilon_T \dot{T} + \frac{\partial q_i}{\partial x_i} = 0, \tag{2.3}$$

we obtain differential equations for heat condition, namely

$$\dot{T} = \frac{\kappa}{\varrho\varepsilon_T}\Delta T \qquad \text{from (2.1)} \tag{2.4}$$

$$\dot{T} = \frac{\kappa}{\varrho\varepsilon_T}(\Delta T - \tau(\Delta T)^{\bullet}) \qquad \text{from (2.2).} \tag{2.5}$$

Neither of these two equations are hyperbolic and both predict infinite speeds for the propagation of thermal disturbances. Cattaneo must have noticed this, because he proceeds to modify equations (2.2) assuming that the operator $\tau\frac{d}{dt}$ is small such that

$$\left(1 - \tau\frac{d}{dt}\right)^{-1} \approx 1 + \tau\frac{d}{dt}.$$

In this approximation the equation assumes the form

$$q_i + \tau\dot{q}_i = -\kappa\frac{\partial T}{\partial x_i}. \tag{2.6}$$

The last equation is often called the *Cattaneo equation*; it is again a modification of Fourier's law and, if we combine it with the energy equation (2.3), we obtain

$$\tau\ddot{T} + \dot{T} = \frac{\kappa}{\varrho\varepsilon_T}\Delta T. \tag{2.7}$$

This is the telegraph equation. It is hyperbolic if $\tau > 0$ holds, and it predicts the propagation of heat pulses at the finite speed

$$V = \pm\sqrt{\frac{\kappa}{\varrho\varepsilon_T\tau}}.$$

We conclude that the transition from (2.2) to (2.6) is all-important for finite speeds and that it changes a parabolic equation into a hyperbolic one. And yet that transition is impossible to justify rigorously. Cattaneo had this to say about it:

> *Nel risultato ottenuto approfitteremo della piccolezza del parametro τ per trascurare il termine che contiene a fattore il suo quadrato, conservando per altro il termine in cui τ compare a primo grado. Naturalmente, per delimitare la portata delle conseguenze che stiamo per trarre converrà precisare un pò meglio le condizioni in cui tale approssimazione è lecita. Allo scopo ammetteremo esplicitamente che il fenomeno di conduzione calorifica avvenga nell'intorno di uno stato stazionario o, in altri termini, che durante il suo svolgersi si mantengano abbastanza piccole le derivate temporali delle varie grandezze in gioco.*

Thus he argues that his derivation is correct, if both τ and d/dt are small. But lack of rigor in the derivation is not the only criticism to be leveled at Cattaneo's equation, because it is also not invariant under a change of frame. We shall come back to that point later; see Paragraph 2.4 and Section 3.

2.2 Extended TIP

Nevertheless, Cattaneo's argument points in the right direction, and his work motivated Müller [7], [8] to try and derive modifications of the laws of Fourier and Navier-Stokes from the principles of TIP. Müller noticed that the arguments of TIP—as presented above in Section 1.1—can be used to derive a Cattaneo-like equation, if only the principle of local equilibrium were abandoned. All one needs is the assumption that the specific entropy s, which heretofore was considered to be a function of ε and ϱ only, may depend on q_i as well. Actually, however, if q_i is a possible variable in s, there is no reason why s should not also depend on $t_{<ij>}$ and π. Of course, these dependencies must be such that s is a scalar and that s has a maximum in equilibrium, where $q_i, t_{<ij>}$ and π all vanish. Therefore, restricting attention to quadratic terms, Müller allowed s to depend on the scalars $q_iq_i, t_{<ij>}t_{<ij>}$ and π^2 with negative coefficients a, b, c, and thus he generalized the Gibbs equation (1.4) to read

$$\dot{s} = \frac{1}{T}\left(\dot{\varepsilon} - \frac{p}{\varrho^2}\dot{\varrho} + 2aTq_i\dot{q}_i + 2bTt_{<ij>}\dot{t}_{<ij>} + 2cT\pi\dot{\pi}\right). \tag{2.8}$$

Now again, $(1.2)_{1,3}$ are used to eliminate $\dot{\varrho}$ and $\dot{\varepsilon}$ and, after some rearrangement, one obtains an entropy balance of the form [2]

$$\varrho\dot{s} + \frac{\partial}{\partial x_i}\left(\frac{q_i}{T}\right) = q_i\left(\frac{\partial\frac{1}{T}}{\partial x_i} + 2\varrho a\dot{q}_i\right) + t_{<ij>}\left(\frac{1}{T}\frac{\partial v_{<i}}{\partial x_{j>}} + 2\varrho b\dot{t}_{<ij>}\right) - \pi\left(\frac{1}{T}\frac{\partial v_i}{\partial x_i} + 2\varrho c\dot{\pi}\right). \tag{2.9}$$

We may again interpret q_i/T as the entropy flux and the right side of (2.9) as the entropy production, which is a sum of products of the thermodynamic fluxes $q_i, t_{<ij>}, \pi$ and *modified* thermodynamic forces, viz.

[2] Terms of third order in the entropy balance are ignored, as they should be in TIP.

$$\frac{\partial T}{\partial x_i} - 2\varrho T^2 a\dot{q}_i, \quad \frac{\partial v_{<i}}{\partial x_{j>}} + 2\varrho T b\dot{t}_{<ij>}, \quad \frac{\partial v_i}{\partial x_i} + 2\varrho T c\dot{\pi}. \tag{2.10}$$

But this is not all: There is an inherently weak point in TIP as it proceeds from the Gibbs equation to an entropy balance, because the interpretation of the entropy flux is ambiguous. Indeed, it is possible to add the term $Kt_{<ij>}q_j + L\pi q_i$—with arbitrary coefficients K and L—to both sides of the entropy balance (2.9). Thus $Kt_{<ij>}q_j + L\pi q_i$ is added to the usual entropy flux q_i/T and the entropy production retains its structure as a sum of products of forces and fluxes. If this is done, the entropy flux assumes the form

$$\varphi_i = \frac{q_i}{T} + Kt_{<ij>}q_j + L\pi q_i, \tag{2.11}$$

and the entropy balance reads

$$\begin{aligned} \varrho\dot{s} + \frac{\partial \varphi_i}{\partial x_i} &= q_i\left(\frac{\partial \frac{1}{T}}{\partial x_i} + 2\varrho a\dot{q}_i + K\frac{\partial t_{ij}}{\partial x_j} + L\frac{\partial \pi}{\partial x_i}\right) + \\ &+ t_{<ij>}\left(\frac{1}{T}\frac{\partial v_{<i}}{\partial x_{j>}} + 2\varrho b\dot{t}_{<ij>} + K\frac{\partial q_{<j}}{\partial x_{i>}}\right) - \pi\left(\frac{1}{T}\frac{\partial v_i}{\partial x_i} + 2\varrho c\dot{\pi} + L\frac{\partial q_i}{\partial x_i}\right). \end{aligned} \tag{2.12}$$

The fluxes are the same ones as always, namely $q_i, t_{<ij>}, \pi$, but the thermodynamic forces have changed again, from (2.10), because we have to compensate for the changed entropy flux by also changing the entropy production. However, we still assume linear phenomenological relations between the fluxes and the forces and obtain

$$\begin{aligned} q_i &= -\kappa\left\{\frac{\partial T}{\partial x_i} - 2\varrho T^2 a\dot{q}_i \quad \underline{- KT^2\frac{\partial t_{<ij>}}{\partial x_j} - LT^2\frac{\partial \pi}{\partial x_i}}\right\} \\ t_{<ij>} &= 2\mu\left\{\frac{\partial v_{<i}}{\partial x_{j>}} + 2\varrho T b\dot{t}_{<ij>} + \underline{KT\frac{\partial q_{<i}}{\partial x_{j>}}}\right\} \\ \pi &= -\lambda\left\{\frac{\partial v_i}{\partial x_i} + 2\varrho T c\dot{\pi} \quad + \underline{LT\frac{\partial q_i}{\partial x_i}}\right\}. \end{aligned} \tag{2.13}$$

Thus the Fourier law $(1.8)_1$ is now replaced by $(2.13)_1$ which—without the underlined terms that were introduced by the modification of the entropy flux—is equivalent to the Cattaneo equation (2.6). Also a corresponding change has appeared in the Navier-Stokes equations, because we now have $(2.13)_{2,3}$ instead of $(1.8)_{2,3}$. Thus, as we shall see, the theory provides finite speeds not only for heat conduction but also for shear diffusion.

This theory is called extended TIP, because it has extended the list of variables on which the specific entropy and the entropy flux depend.

For rigid heat conductors a different procedure was outlined. In fact since the Cattaneo equation can be expressed as an integral over the history of the temperature gradient, viz.

$$q_i = -\frac{\kappa}{\tau}\int_{-\infty}^{t} \exp\left(-\frac{t-t'}{\tau}\right)\frac{\partial T(\mathbf{x},t')}{\partial x_i}\,dt',$$

some authors developed the idea that the Cattaneo equation is a particular case of materials with fading memory. See in particular the papers of Gurtin & Pipkin [39], of Green & Lindsay [40], and of Grioli [41]. A review paper of this approach has recently been given by Joseph & Preziosi [42]. A limitation of this *functional approach* lies in the fact that, when the material is not rigid, we have interaction between the thermal and mechanical fields. In such cases it is difficult to interpret the Cattaneo equation and its counterparts in (2.13) as integral equations.

2.3 Finite Pulse Speeds in Extended TIP

Comparing the phenomenological equations (1.8) and (2.13), we realize that extended TIP has replaced the simple algebraic equation between q_i, t_{ij} and $\frac{\partial T}{\partial x_i}, \frac{\partial v_i}{\partial x_j}$ by a complex set of partial differential equations. The complete set of equations is represented by the balance laws (1.2), by the phenomenological equations (2.13), and by the thermal and caloric equations of state $p = p(\varrho, T)$ and $\varepsilon = \varepsilon(\varrho, T)$. It is impossible now to characterize a reasonably realistic special case in which these equation decouple to give simple equations like (1.9) for T, or for the shear velocity v. Therefore it is not immediately obvious, whether the speeds of propagation are finite.

In order to find that out, we consider plane harmonic waves of small amplitude assuming two types of solutions, namely

$$\text{I: } \varrho = \tilde{\varrho} + \bar{\varrho}E,\ v_i = \begin{bmatrix} \bar{v} \\ 0 \\ 0 \end{bmatrix} E,\ T = \tilde{T} + \bar{T}E,\ t_{<ij>} = \begin{bmatrix} \bar{P} & 0 & 0 \\ 0 & -\frac{1}{2}\bar{P} & 0 \\ 0 & 0 & -\frac{1}{2}\bar{P} \end{bmatrix} E,\ \pi = \bar{\pi}E,\ q_i = \begin{bmatrix} \bar{q} \\ 0 \\ 0 \end{bmatrix} E; \tag{2.14}$$

$$\text{II: } \varrho = \tilde{\varrho}, \quad v_i = \begin{bmatrix} 0 \\ \bar{v} \\ 0 \end{bmatrix} E, \quad T = \tilde{T}, \quad t_{<ij>} = \begin{bmatrix} 0 & \bar{P} & 0 \\ \bar{P} & 0 & 0 \\ 0 & 0 & 0 \end{bmatrix} E, \quad \pi = 0, \quad q_i = \begin{bmatrix} 0 \\ \bar{q} \\ 0 \end{bmatrix} E. \tag{2.15}$$

For a generic field g we have $\tilde{g}$ as a constant field, while $\bar{g}$ denotes small complex amplitudes of which products are neglected. E in (2.14), (2.15) stands for

$$E = e^{i(\omega t - kx)}, \tag{2.16}$$

where ω is the frequency and k is the wave number. The latter is generally complex.

The wave (2.14) is longitudinal in velocity and heat flux, and it propagates disturbances in density and temperature, hence also in pressure. The wave (2.15) on the other hand is transversal in velocity and heat flux, and we may say that it is a shear wave.

If (2.14) is introduced into the system of equations (1.2) and (2.13), all but 6 of these equations are identically satisfied. The remaining 6 form a linear homogeneous algebraic system for the determination of $\bar{\varrho}$ through $\bar{q}$, viz.

$$\begin{bmatrix} 1 & -\varrho\frac{k}{\omega} & 0 & 0 & 0 & 0 \\ -p_\varrho\frac{k}{\omega} & \varrho & -p_T\frac{k}{\omega} & \frac{k}{\omega} & \frac{k}{\omega} & 0 \\ -\frac{T}{\varrho}p_T & 0 & \varrho\varepsilon_T & 0 & 0 & -\frac{k}{\omega} \\ 0 & 0 & \kappa\frac{k}{\omega} & -\kappa KT^2\frac{k}{\omega} & -\kappa LT^2\frac{k}{\omega} & \left[2\kappa\varrho T^2 a + i\frac{1}{\omega}\right] \\ 0 & -\frac{4}{3}\mu\frac{k}{\omega} & 0 & \left[4\mu\varrho Tb + i\frac{1}{\omega}\right] & 0 & -\frac{4}{3}\mu KT\frac{k}{\omega} \\ 0 & -\lambda\frac{k}{\omega} & 0 & 0 & \left[2\lambda\varrho Tc + i\frac{1}{\omega}\right] & -\lambda LT\frac{k}{\omega} \end{bmatrix} \begin{bmatrix} \bar{\varrho} \\ \bar{v} \\ \bar{T} \\ \bar{P} \\ \bar{\pi} \\ \bar{q} \end{bmatrix} = 0. \qquad (2.17)$$

p_T, p_ϱ, and ε_T stand for the partial derivatives of p and ε with respect to T and ϱ. The determinant of this matrix must vanish. This condition gives rise to the *dispersion relation* from which we may calculate the permissible wave numbers for a given frequency ω. The phase speed v_{ph} and the attenuation α may be calculated from

$$v_{ph} = \frac{\omega}{\mathcal{R}e(k)} \qquad \text{and} \qquad \alpha = -\mathcal{I}m(k). \qquad (2.18)$$

In the special case in which $\lambda = 0$ holds we have also $\pi = 0$; see $(2.13)_3$. Therefore (2.17) reduces to the slightly simpler system

$$\begin{bmatrix} 1 & -\varrho\frac{k}{\omega} & 0 & 0 & 0 \\ -p_\varrho\frac{k}{\omega} & \varrho & -p_T\frac{k}{\omega} & \frac{k}{\omega} & 0 \\ -\frac{T}{\varrho}p_T & 0 & \varrho\varepsilon_T & 0 & -\frac{k}{\omega} \\ 0 & 0 & \kappa\frac{k}{\omega} & -\kappa KT^2\frac{k}{\omega} & \left[2\kappa\varrho T^2 a + i\frac{1}{\omega}\right] \\ 0 & -\frac{4}{3}\mu\frac{k}{\omega} & 0 & \left[4\mu\varrho Tb + i\frac{1}{\omega}\right] & -\frac{4}{3}\mu KT\frac{k}{\omega} \end{bmatrix} \begin{bmatrix} \bar{\varrho} \\ \bar{v} \\ \bar{T} \\ \bar{P} \\ \bar{q} \end{bmatrix} = 0. \qquad (2.19)$$

Later in Chapter 11, when we treat monatomic gases, we shall calculate the phase speed as a function of ω. Here, however, we concentrate on the maximum speed of propagation allowed by the system of equations (1.2), (2.13). We ask whether that speed of propagation, the *pulse speed* V, is finite or infinite. It can be proved that V is the limiting value of the phase speed for $\omega \to \infty$; see Chapter 8, where this proof will be presented.

For $\omega \to \infty$ the dispersion relation simplifies considerably, because the imaginary terms drop out. We calculate the pulse speed implied by the system (2.19) and obtain

$$\left(\frac{V}{v_{ph}(0)}\right)^2 = \frac{1}{2}\left\{1 - \frac{1}{12\varrho^2 T v_{ph}^2(0)} \frac{4a + \frac{6}{T\varepsilon_T}b - 2TK^2}{ab}\right\} + \\ + \sqrt{\frac{1}{4}\{\ldots\}^2 - \frac{1}{6}\frac{(1 - T^2 p_T K)^2 - p_\rho \rho^2 T(3b - K^2 T^2 \varepsilon_T)}{\rho^4 \varepsilon_T T^3 ab\, v_{ph}^4(0)}}, \qquad (2.20)$$

where, for abbreviation, we have introduced

$$v_{ph}(0) = \sqrt{p_\rho + \frac{T}{\rho^2 \varepsilon_T} p_T^2}, \tag{2.21}$$

the *adiabatic sound speed.* The pulse speed $\overset{I}{V}$ will depend on the values of a, b, and K, and it is *finite.*

Introducing (2.15) into the system of equations (1.2) and (2.13), we note that all but 3 equations are satisfied and that the remaining 3 form a linear homogeneous system of equations for the determination of $\bar{v}, \bar{P}$, and $\bar{q}$, viz.

$$\begin{bmatrix} \varrho & \frac{k}{\omega} & 0 \\ 0 & -\kappa K T^2 \frac{k}{\omega} & \left[2\kappa\varrho T^2 a + i\frac{1}{\omega}\right] \\ -2\mu\frac{k}{\omega} & \left[4\mu\varrho T b + i\frac{1}{\omega}\right] & -2\mu K T\frac{k}{\omega} \end{bmatrix} \begin{bmatrix} \bar{v} \\ \bar{P} \\ \bar{q} \end{bmatrix} = 0. \tag{2.22}$$

We obtain the dispersion relation by setting the determinant of the matrix equal to zero. Again we determine the pulse speed V as ω/k for $\omega \to \infty$ and we obtain

$$\overset{II}{V} = \sqrt{\frac{1}{8\varrho^2 T} \frac{-4a + 2K^2 T}{ab}} \tag{2.23}$$

which is a finite speed as long as a, b, and K are nonzero; recall that a and b are negative.

2.4 Conclusion and Criticism

Extended TIP has provided finite speeds, but in order to achieve that, it had to make the unconventional assumption that the entropy depends on t_{ij} and q_i. Also the entropy flux was assumed to be given by the complex expression (2.11) rather then simply by q_i/T. Clearly it is desirable to have some support for these assumptions. Such support can be derived from the kinetic theory of monatomic gases, as we shall soon see.

There is a severe criticism to be leveled both at the Cattaneo equation (2.6) and at the phenomenological equations (2.13) of extended TIP. Indeed, neither of these can be true physical laws, because they are not properly invariant. The time derivatives $\dot{q}_i$ and $\dot{t}_{<ij>}$ in (2.6) and (2.13) are not objective tensors, while all other terms in those equations are objective tensors. Therefore, even if the equations hold in an inertial frame, they do not hold in a noninertial frame. The kinetic theory will furnish an interpretation of this phenomenon and, in so doing, it will provide the basis for a new systematic—rational—version of extended thermodynamics.

Also the system of equations (1.2), (2.13) is awkward in a mathematical sense. Although we have seen that small disturbances of a constant equilibrium state propagate with a finite speed, it is not sure at this time that the nonlinear system is hyperbolic so that the speeds of nonlinear waves are finite as well. [3]

[3] see Chapter 8 for the properties of hyperbolic systems.

It seems desirable—both from the mathematical and the physical points of view—to deal with field equations which guarantee existence, uniqueness and continuous dependence for the solutions of Cauchy problems. That is to say, we want equations for which Cauchy problems with reasonable data are well posed. The present system (1.2), (2.13), however, is complicated to a degree that there exists no theorem guaranteeing well-posedness of Cauchy problems.

Finally we know that in quasilinear systems even smooth data may in time lead to the formation of shocks, thus calling for jump conditions across the shock front. Such conditions can only be had if the field equations are of balance type. And while the equations (1.2) are indeed balance equations, this seems not to be the case for the phenomenological equations (2.13). Here again, the kinetic theory will help us to find the correct interpretation.

3 Kinetic Theory of Monatomic Gases

3.1 Boltzmann Equation and Moments

The kinetic theory describes the state of a gas by the phase density $f(\mathbf{x}, \mathbf{c}, t)$ such that

$$f(\mathbf{x}, \mathbf{c}, t)d\mathbf{c} \tag{3.1}$$

is the number density of atoms at the point $\mathbf{x}$ and at time t that have velocities between $\mathbf{c}$ and $\mathbf{c} + d\mathbf{c}$. The phase density obeys the Boltzmann equation

$$\frac{\partial f}{\partial t} + c_i \frac{\partial f}{\partial x_i} + (f_i + \underset{ci}{i}) \frac{\partial f}{\partial c_i} = \int (f' f^{1'} - f f^1) r\, g \sin\theta\, d\theta\, d\varepsilon d\mathbf{c}^1. \tag{3.2}$$

f_i is the specific external body force, assumed independent of c_i and $\underset{ci}{i}$ is the inertial acceleration of an atom, viz.

$$\underset{ci}{i} = 2W_{ik}(c_k - \dot{b}_k) - W^2_{ik}(x_k - b_k) + \dot{W}_{ik}(x_k - b_k) + \ddot{b}_i, \tag{3.3}$$

where $\mathbf{W}$ is the matrix of the angular velocity of the frame with respect to an inertial frame and b_i is the distance vector between the origins of the two frames. The right-hand side of (3.2) is due to collisions between the atoms, and for its derivation we refer the reader to books on the kinetic theory of gases, e.g. [43], [44], or [45]. f, f^1, f', and $f^{1'}$ are the values of the phase density for the velocities $\mathbf{c}, \mathbf{c}^1, c'$, and $c^{1'}$ for two atoms before and after the collision.

Most macroscopic thermodynamic quantities are formed by taking moments of the phase density. The generic moment has the form

$$F_{i_1 i_2 \dots i_N} = \int m c_{i_1} c_{i_2} \dots c_{i_N} f\, d\mathbf{c}. \tag{3.4}$$

Thus we have in particular

$$\begin{array}{ll} \text{mass density} & F = \varrho \\ \text{momentum density} & F_i = \varrho v_i. \end{array} \tag{3.5}$$

With the velocity v_i of the gas we form the *peculiar velocity* $C_i = c_i - v_i$ of an atom, and we use these to define *the internal moments*

$$\varrho_{i_1 i_2 \ldots i_N} = \int m C_{i_1} C_{i_2} \ldots C_{i_N} f d\mathbf{x}. \tag{3.6}$$

The first few moments and internal moments have their own canonical notation and names that are suggestive of their interpretation, viz.

$$\begin{array}{llll} \frac{1}{2}F_{ii} = \varrho e & \text{- energy density} & \frac{1}{2}\varrho_{ii} = \varrho\varepsilon & \text{- internal energy density} \\ F_{ij} & \text{- momentum flux} & -\varrho_{ij} = t_{ij} & \text{- stress tensor} \\ \frac{1}{2}F_{ijj} = J_i & \text{- energy flux} & \frac{1}{2}\varrho_{ijj} = q_i & \text{- heat flux.} \end{array} \tag{3.7}$$

Since $c_i = C_i + v_i$ holds, there are simple relations between the moments and the internal moments, namely

$$\begin{aligned} F &= \varrho \\ F_i &= \varrho v_i \\ F_{ij} &= \varrho_{ij} + \varrho v_i v_j \\ F_{ijk} &= \varrho_{ijk} + 3\varrho_{(ij}v_{k)} + \varrho v_i v_j v_k \\ F_{ijk\ell} &= \varrho_{ijk\ell} + 4\varrho_{(ijk}v_{l)} + 6\varrho_{(ij}v_k v_{\ell)} + \varrho v_i v_j v_k v_\ell. \end{aligned} \tag{3.8}$$

Round brackets indicate symmetrization, i.e. the sum of the permutations of all N indices divided by $N!$. Prototypes of the relations (3.8) are the well-known decompositions of momentum flux, energy density, and energy flux into velocity-dependent and internal parts. We have

$$\begin{aligned} F_{ij} &= -t_{ij} + \varrho v_i v_j, \\ \varrho e &= \varrho\varepsilon + \frac{\varrho}{2}v^2, \\ J_i &= q_i + \varrho\left(\varepsilon + \frac{v^2}{2}\right)v_i - t_{ij}v_j. \end{aligned} \tag{3.9}$$

In a monatomic ideal gas there is no dynamic pressure. Therefore $t_{ii} = -3p$ holds, and from (3.7) we conclude

$$p = \frac{2}{3}\varrho\varepsilon. \tag{3.10}$$

This is a well-known relation between the thermal and caloric equations of state in a monatomic gas. The kinetic theory confirms this relation, which shows that we have chosen the definitions (3.7) properly.

We know that in monatomic ideal gases we have $\varepsilon = \frac{3}{2}\frac{k}{m}T$, and this relation may be used to define temperature in the kinetic theory by

$$\frac{3}{2}kT = \frac{1}{\varrho}\int \frac{m}{2}C^2 f d\mathbf{c}. \tag{3.11}$$

3.2 Equations of Balance for Moments

Multiplication of the Boltzmann equation by a generic function $\psi(\mathbf{x}, \mathbf{c}, t)$ and integration over all $\mathbf{c}$ gives an equation of balance for the mean value $\overline{\psi}$, viz.

$$\rho\bar{\psi} = \int \psi f \, d\mathbf{c}. \tag{3.12}$$

The equation is derived under the assumption that the phase density falls off rapidly for large values of $\mathbf{c}$, and it reads

$$\begin{aligned}\frac{\partial \varrho\bar{\psi}}{\partial t} + \frac{\partial \overline{\varrho\psi c_i}}{\partial x_i} - \overline{\varrho(f_i + \underset{ci}{i})\frac{\partial \psi}{\partial c_i}} = \\ = \overline{\varrho\left(\frac{\partial \psi}{\partial t} + c_i\frac{\partial \psi}{\partial x_i}\right)} + \frac{1}{4}\int(\psi + \psi^1 - \psi' - \psi^{1'})(f'f^{1'} - ff^1)rg\sin\theta \, d\theta \, d\mathbf{c}_1 \, d\mathbf{c}.\end{aligned} \tag{3.13}$$

This equation is an equation of balance in which the third term represents the supply due to external body forces and inertial forces, while the right-hand side represents the production.

Of particular interest in thermodynamics are the equations of balance for moments. We set

$$\psi = m\, c_{i_1} c_{i_2} \ldots c_{i_N} \tag{3.14}$$

and obtain

$$\frac{\partial F_{i_1\ldots i_N}}{\partial t} + \frac{\partial F_{i_1\ldots i_N j}}{\partial x_j} - NF_{(i_1\ldots i_{N-1}}\left(f_{i_N)} + i^0_{i_N)}\right) - NF_{k(i_1\ldots i_{N-1}}2W_{i_N)k} = S_{i_1\ldots i_N}, \tag{3.15}$$

where i^0_j is the part of the inertial acceleration that is independent of the velocity. The right-hand side represents the collision production. Comparison with (3.13) shows that S, S_i, and S_{ii} must vanish, because mass, momentum, and energy of a particle are conserved in a collision. More explicitly we write the balance equations for the first thirteen moments $F = \varrho, F_i = \varrho v_i, F_{ij}$ and F_{ijj}. They read

$$\begin{aligned}
&\frac{\partial \varrho}{\partial t} + \frac{\partial \varrho v_i}{\partial x_i} && = 0\\
&\frac{\partial \varrho v_i}{\partial t} + \frac{\partial F_{ij}}{\partial x_j} - \varrho(f_i + i^0_i) - 2\varrho v_k W_{ik} && = 0\\
&\frac{\partial F_{ij}}{\partial t} + \frac{\partial F_{ijk}}{\partial x_k} - 2\varrho v_{(i}(f_{j)} + i^0_{j)}) - 4F_{k(i}W_{j)k} && = S_{<ij>}\\
&\frac{\partial F_{ijj}}{\partial t} + \frac{\partial F_{ijjk}}{\partial x_k} - 3F_{(ij}(f_{j)} + i^0_{j)}) - 6F_{k(ij}W_{j)k} && = S_{ijj}.
\end{aligned} \tag{3.16}$$

In this explicit form it is immediately obvious that the first two of the equations (3.15) represent the conservation laws of mass and momentum, while the trace of the third equation is the conservation law of energy. We note that the contributions from body- and inertial forces have the familiar form, well known in mechanics. The remaining equations in (3.16) are balance laws for the deviatoric momentum flux and for the energy flux. These balance equations also contain contributions due to external forces and inertia, but above all they are not conservation laws, because they contain productions due to atomic collisions. For some atomic interaction forces the collision terms on the right-hand sides of $(3.16)_{3,4}$ may be calculated easily. Thus for instance for Maxwellian atoms we have

$$
\begin{aligned}
S_{<ij>} &= -\sigma t_{<ij>} = -\sigma(F_{<ij>} - \varrho v_{<i}v_{j>}) \\
S_{ijj} &= -\frac{4}{3}\sigma q_i \quad = -\frac{2}{3}\sigma(F_{ijj} - 3F_{(ij}v_{j)} - \varrho v_i v^2),
\end{aligned}
\tag{3.17}
$$

where $1/\sigma$ is a relaxation time whose value depends on the strength of the atomic interaction. We may consider $1/\sigma$ as the mean time of free flight of the atoms; certainly $1/\sigma$ is of the same *order of magnitude* as the mean time of free flight.

Note that upon first sight the system (3.16) with (3.17) appears to be a system of equations for the thirteen moments $\varrho, v_i, F_{ij}, F_{ijj}$. But the system is not closed, because of the appearance of the fluxes $F_{<ijk>}$ and F_{ijjk} in $(3.16)_{3,4}$. Nor does it help to proceed to an equation for the higher moments, because the divergence term in (3.15) will always bring a moment of the next order into the system. This constitutes the closure problem of the kinetic theory, and we shall come back to it.

3.3 Balance of Entropy and Possible Equilibria

If we insert

$$
\psi = -k\left(\ln\frac{f}{y} - 1\right)
\tag{3.18}
$$

into the general equation of balance (3.13), we obtain the equation

$$
\frac{\partial\left(-k\int \ln f f d\mathbf{c}\right)}{\partial t} + \frac{\partial\left(-k\int c_i \ln f f d\mathbf{c}\right)}{\partial x_i} = \frac{m}{4}\int \ln\frac{f' f^{1\prime}}{f f^1}(f' f^{1\prime} - f f^1) s \sin\theta\, d\theta\, d\varepsilon\, d\mathbf{c}^1\, d\mathbf{c}.
\tag{3.19}
$$

$1/y$ is the smallest element in $(\mathbf{x}, \mathbf{c})$-space that can accommodate a position $\mathbf{x}$ and a velocity $\mathbf{c}$. It is equal to $\frac{h^3}{m^3}\frac{1}{2s+1}$, where h is Planck's constant and $s\frac{h}{2\pi}$ is the spin of an atom. The occurence of y is unimportant at this stage and so is the additive term k in (3.18); both drop out of the subsequent relations, or rather they affect only the entropy constant.

The collision production on the right-hand side of (3.19) is obviously nonnegative, and that fact suggests that (3.19) represents the balance of entropy with

$$
\begin{aligned}
s &= -\tfrac{k}{\varrho}\textstyle\int\left(\ln\tfrac{f}{y} - 1\right) f d\mathbf{c} && \text{- as the specific entropy} \\
h_i &= -k\textstyle\int c_i\left(\ln\tfrac{f}{y} - 1\right) f d\mathbf{c} && \text{- as the entropy flux} \\
\Sigma &= \tfrac{k}{4}\textstyle\int \ln\tfrac{f' f^{1\prime}}{f f^1}(f' f^{1\prime} - f f^1) s g \sin\theta\, d\theta\, d\varepsilon\, d\mathbf{c}_1\, d\mathbf{c} && \text{- as the entropy production.}
\end{aligned}
\tag{3.20}
$$

We expect the entropy production to vanish in equilibrium and, by $(3.20)_3$, this requires that the equilibrium phase density f_{E} satisfy the relation

$$
\ln f_{\mathrm{E}} + \ln f_{\mathrm{E}}^1 = \ln f_{\mathrm{E}}' + \ln f_{\mathrm{E}}^{1\prime},
\tag{3.21}
$$

that is to say $\ln f_{\mathrm{E}}$ is a summational invariant in the collision. There are, however only 5 independent summational invariants in a collision, namely m, mc_i and $m/2c^2$, the atomic masses, momenta, and energies. Therefore $\ln f_{\mathrm{E}}$ must be the linear combination

$$\ln f_{\mathrm{E}} = \alpha + \beta_i c_i + \gamma c^2, \tag{3.22}$$

where α, β_i, and γ are coefficients that may be determined from (see (3.5) and (3.11))

$$\varrho = \int mf\, d\mathbf{c}, \qquad \varrho v_i = \int mc_i f\, d\mathbf{c}, \qquad \varrho \frac{3}{2}kT = \int \frac{m}{2} C^2 f\, d\mathbf{c}.$$

A simple calculation shows that f_{E} has the form

$$f_{\mathrm{E}} = \frac{\varrho}{m}\sqrt{\frac{m}{2\pi kT}}^{3} e^{-\frac{m(\mathbf{c}-\mathbf{v})^2}{2kT}} \tag{3.23}$$

which is called the Maxwell distribution.

The Maxwell distribution makes the entropy production $(3.20)_3$ vanish irrespective of the form of the fields $\varrho(\mathbf{x},t), v_i(\mathbf{x},t)$ and $T(\mathbf{x},t)$. But not all such fields satisfy the Boltzmann equation. In fact those which do must be solutions of the equations

$$\begin{aligned}
&\frac{\partial \varrho}{\partial t} + \frac{\partial \varrho v_i}{\partial x_i} = 0 \\
&\varrho \dot{v}_i + \frac{\partial \varrho \frac{k}{m} T}{\partial x_i} = \varrho(f_i + i_i), \qquad \text{where} \quad i_i = i_i^0 + 2W_{ik}v_k \\
&\frac{\partial T}{\partial t}\delta_{ij} + 2T\frac{\partial v_{(i}}{\partial x_{j)}} = 0, \\
&\frac{\partial T}{\partial x_i} = 0.
\end{aligned} \tag{3.24}$$

Thus equilibria in a monatomic gas need not be characterized by constant and uniform fields of ϱ, v_i, and T. Rather the equilibrium fields contain rigid rotations, isotropic expansions, a wide variety of density gradients, and even a time-dependent temperature. However, the temperature gradient must vanish.

3.4 The Grad Distribution

Without essential loss of generality the phase density f may be written as

$$f = \left(a + a_i\frac{\partial}{\partial c_i} + a_{ij}\frac{\partial^2}{\partial c_i \partial c_j} + a_{ijk}\frac{\partial^3}{\partial c_i \partial c_j \partial c_k} + \ldots\right) f_{\mathrm{E}}. \tag{3.25}$$

Indeed, by (3.23) this equation is equivalent to

$$\begin{aligned}
f = f_{\mathrm{E}}\Big\{a - a_i\frac{m}{kT}C_i \;&+\; a_{ij}\left(\frac{m}{kT}\right)^2\left(C_iC_j - \frac{kT}{m}\delta_{ij}\right) - \\
&-\; a_{ijk}\left(\frac{m}{kT}\right)^3\left(C_iC_jC_k - \frac{kT}{m}(\delta_{ij}C_k + \delta_{jk}C_i + \delta_{ki}C_j)\right) + \ldots\Big\}.
\end{aligned} \tag{3.26}$$

and this is an expansion in terms of Hermite polynomials which form a complete set of functions. The coefficients $a, a_i, \ldots$ in this expansion must be determined by inserting (3.26) into (3.4). The best-known application of this expansion is Grad's 13-moment distribution f_G, in which only the first 13 polynomials are considered. After a little calculation one obtains

$$f_G = f_E \left\{ 1 - \frac{1}{2p} t_{<ij>} \left(\frac{m}{kT} C_i C_j - \delta_{ij} \right) - \frac{1}{p} \frac{m}{kT} q_i C_i \left(1 - \frac{1}{5} \frac{m}{kT} C^2 \right) \right\}. \tag{3.27}$$

This is called the Grad distribution.

3.5 Entropy and Entropy Flux in Grad's 13-Moment Theory

We recall from the considerations of Paragraph 3.3 that the entropy balance in the kinetic theory has the form

$$\varrho \dot{s} + \frac{\partial \varphi_i}{\partial x_i} \geq 0 \quad \text{with} \quad \begin{aligned} \varrho s &= -k \int \left(\ln \tfrac{f}{y} - 1 \right) f dc \\ \varphi_i &= -k \int C_i \left(\ln \tfrac{f}{y} - 1 \right) f dc. \end{aligned} \tag{3.28}$$

If we replace f in $\ln f$ by the Grad distribution f_G given in (3.27), we obtain—for processes close to equilibrium, where f_G differs little from f_E—

$$s = \frac{k}{m} \ln \frac{T^{3/2}}{\varrho} + \varsigma - \frac{t_{<ij>} t_{<ij>}}{4 p \varrho T} - \frac{q_i q_i}{5 p^2 T} \tag{3.29}$$

$$\varphi_i = \frac{q_i}{T} + \frac{2}{5pT} t_{<ij>} q_j. \tag{3.30}$$

Thus indeed, the specific entropy of the kinetic theory in the Grad approximation depends on $t_{<ij>}$ and q_i. Also the entropy flux is not simply equal to q_i/T but it contains a term proportional to $t_{<ij>} q_j$. These facts confirm the conjectures of extended TIP. Moreover, the kinetic theory identifies the coefficients a, b, c and K, L of extended TIP appropriate for a monatomic ideal gas. Comparison of (3.29), (3.30) with (2.8), (2.11) shows that we have

$$a = -\frac{1}{5p^2T}, \qquad b = -\frac{1}{4p\varrho T}, \qquad c = 0, \qquad K = \frac{2}{5pT}, \qquad L = 0. \tag{3.31}$$

3.6 Phenomenological Equations derived from the Kinetic Theory

We now ask whether the phenomenological equations (2.13) of extended TIP have a counterpart in the kinetic theory. Upon reflection it is clear that such equations in the kinetic theory can only come from the equations of balance $(3.16)_{3,4}$ which, by (3.17), already have $t_{<ij>}$ and q_i on their right-hand sides. However, those equations contain the higher moments $F_{<ij>n}$ and F_{ijjk} so that the system of equations (3.16) is not closed. We propose to close the system by calculating $F_{<ij>n}$ and F_{ijjk}, or the corresponding internal quantities

$$\varrho_{<ij>k} = \int m C_{<i} C_{j>} C_k f dc \qquad \text{and} \qquad \varrho_{ijjk} = \int m C_i C^2 C_k f dc \tag{3.32}$$

from Grad's 13-moment distribution (3.27). Thus we obtain

$$\varrho_{<ij>k} = \frac{2}{5}\left(q_i\delta_{jk} + q_j\delta_{ik} - \frac{2}{3}q_k\delta_{ij}\right), \quad \varrho_{ijjk} = 5p\frac{kT}{m}\delta_{ij} - \frac{7}{2}t_{<pq>}\frac{kT}{m}(\delta_{ip}\delta_{jq} + \delta_{iq}\delta_{jp}). \quad (3.33)$$

When these expressions are introduced into $(3.16)_{3,4}$ and when all the other moments $\mathbf{F}$ occurring in those equations are replaced by the internal moments $\boldsymbol{\varrho}$ as dictated by (3.8), we obtain after some trivial calculation

$$t_{<ij>} = 2\frac{p}{\sigma}\left\{\frac{\partial v_{<i}}{\partial x_{j>}} - \frac{1}{2p}\left[\dot{t}_{<ij>} + 2t_{<n<i>}\frac{\partial v_{j>}}{\partial x_n} - 4t_{<n<i>}W_{j>n}\right] + \frac{2}{5p}\frac{\partial q_{<i}}{\partial x_{j>}}\right\} - \boxed{-\sigma t_{<ij>}\frac{\partial v_n}{\partial x_n}}$$

$$(3.34)$$

$$q_i = -\frac{15}{4}\frac{k}{m}\frac{p}{\sigma}\left\{\frac{\partial T}{\partial x_i} + \frac{2}{5}\frac{m}{k}\frac{1}{p}\left[\dot{q}_i + q_n\frac{\partial v_i}{\partial x_n} - 2q_nW_{in}\right] + \frac{2}{5}\frac{T}{p}\frac{\partial t_{<in>}}{\partial x_n}\right\} -$$

$$\boxed{-\frac{3}{2}\frac{1}{\sigma}\left(\frac{7}{5}q_i\frac{\partial v_n}{\partial x_n} + \frac{4}{5}q_n\frac{\partial v_{(i}}{\partial x_{n)}} - \frac{7}{2}\frac{k}{m}t_{<iq>}\frac{\partial T}{\partial x_q} - \frac{1}{\varrho}t_{<ik>}\frac{\partial t_{<kn>}}{\partial x_n} + \frac{1}{\varrho}t_{<ik>}\frac{\partial p}{\partial x_k}\right)}\,.$$

Comparison of these equations with the equations $(2.13)_{1,2}$ reveals enough similarity to justify the suggestion that the balance laws $(3.16)_{3,4}$, from which we have derived (3.34), correspond to the phenomenological relations of extended TIP. Of course, there is no counterpart in (2.13) to the framed terms in (3.34); TIP would have neglected such terms as nonlinear anyhow, so let us ignore them in the comparison. This leaves only the curly brackets on the right-hand sides of (3.34).

We introduce

$$\mu = \left(\varrho\frac{k}{m}T\right)\frac{1}{\sigma} \quad \text{and} \quad \kappa = \frac{15}{4}\frac{k}{m}\left(\varrho\frac{k}{m}T\right)\frac{1}{\sigma} \quad (3.35)$$

into (3.34) and the special values (3.31) into (2.13) to obtain a significant similarity between the two sets of equations—viz. $(2.13)_{1,2}$ and (3.34). Indeed the only difference is that $\dot{q}_i$ and $\dot{t}_{<ij>}$ in (2.13) are now replaced by

$$\dot{q}_i + \underbrace{q_n\frac{\partial v_i}{\partial x_n} - 2q_nW_{in}} \quad \text{and} \quad \dot{t}_{<ij>} + \underbrace{2t_{<n<i>}\frac{\partial v_{j>}}{\partial x_n} - 4t_{<n<i>}W_{j>n}}\,. \quad (3.36)$$

Thus the kinetic theory suggests that the noninvariance of the equations (2.13), which was discussed in Paragraph 2.4, must be mended by joining the braced terms in (3.36) to $\dot{q}_i$ and $\dot{t}_{<ij>}$. It is indeed true that the sum of the three nonobjective terms in $(3.36)_{1,2}$ form an objective vector and an objective tensor respectively.

There is no way in extended TIP that would lead us to conjecture the presence of the expressions (3.36) in the phenomenological relations. Indeed, had the noninvariance been noticed in extended TIP, it would probably have been mended by replacing $\dot{q}_i$ and $\dot{t}_{<ij>}$ by the objective derivatives

$$\dot{q}_i - q_n \frac{\partial v_i}{\partial x_n} \quad \text{and} \quad \dot{t}_{<ij>} - 2t_{<n<i>} \frac{\partial v_{j>}}{\partial x_n}, \tag{3.37}$$

the so-called Jaumann derivatives. But the kinetic theory insists on inertial terms in (3.34), namely the terms with $\mathbf{W}$. Later, in Chapter 4, we shall discuss those terms and find them eminently sensible.

3.7 Pulse Speeds

If we introduce the specific values (3.31) for a, b, and K into the general expressions for the pulse speeds $\overset{I}{V}$ and $\overset{II}{V}$ in (2.20) and (2.23), we obtain

$$\begin{array}{ll} \text{for the longitudinal sound wave} & \overset{I}{V} = 1.65\, v_{ph}(0) \\ \text{for the transversal shear wave} & \overset{II}{V} = 1.30\, v_{ph}(0). \end{array} \tag{3.38}$$

3.8 Conclusions

We have learned one thing above all from the kinetic theory of gases, namely that the phenomenological relations of extended TIP are in fact equations of balance. Indeed, the system (1.2), (2.13) of extended TIP assumes the elegant form (3.16) in the kinetic theory in which all equations are equations of balance with

$$\begin{array}{ccc} \text{densities} & \text{fluxes} & \text{productions} \\ \varrho, \varrho v_i, F_{ij}, F_{ijj} & \varrho v_i, F_{ij}, F_{ijk}, F_{ijjk} & S_{<ij>}, S_{ijj}. \end{array} \tag{3.39}$$

Also in the kinetic theory the densities, fluxes, and productions depend locally on the variables

$$\varrho,\ v_i,\ \varepsilon,\ t_{<ij>},\ q_i \tag{3.40}$$

which we may consider as the basic fields to be determined by thermodynamics of viscous heat conducting fluids.

Moreover, there is an entropy inequality of balance form, viz. by (3.19), (3.20)

$$\frac{\partial \varrho s}{\partial t} + \frac{\partial h_i}{\partial x_i} = \Sigma \geq 0 \tag{3.41}$$

in which the density ϱs and the flux h_i also depend locally on the variables (3.40).

These observations will guide us in Chapter 4 in the formulation of the systematic rational theory of extended thermodynamics of viscous heat-conducting gases.

Chapter 3

Formal Structure of Extended Thermodynamics

Extended Thermodynamics extends the number of fields beyond the ones of ordinary thermodynamics which are the densities of mass, momentum and energy. Typical extensions are the fluxes of momentum and energy. Essential ingredients of the extended theory are

- basic equations are of balance type
  hence possibility of weak solutions and shocks.
- constitutive equations are local in space-time
  hence quasilinear first-order field equations
- entropy inequality with a concave entropy density
  hence symmetric hyperbolic field equations.

The exploitation of the entropy principle introduces Lagrange multipliers and a four-vector potential from which all constitutive relations may be derived. The Lagrange multipliers may be considered as privileged fields in extended thermodynamics, because the four-vector potential depends on them and because the system of field equations becomes symmetric when written in terms of Lagrange multipliers.

In all thermodynamic theories the velocity belongs to the basic fields and, infact, the velocity plays a prominent role among thermodynamic variables, because the constitutive quantities depend on it in a specific manner. This dependence can be determined by exploiting the requirement of Galilei invariance of the field equations.

The formulation of extended thermodynamics is facilitated by the assumption that the basic fields are components of symmetric tensors of increasing rank. A simple – nearly trivial – case of that type is the Euler fluid which, despite its simplicity, may serve to illustrate most ingredients of the formal structure of rational extended thermodynamics

1 Field Equations

n fields are to be determined, which in the present chapter are synthetically denoted by an n-vector $\mathbf{u}$. An appropriate number of balance equations are postulated from which field equations result by the formulation of constitutive relations. The form of the constitutive functions is restricted by universal physical principles, viz. the entropy principle and the principle of relativity.

1.1 Thermodynamic Processes and Principles of the Constitutive Theory

The objective of extended thermodynamics is the determination of an n-vector field

$$\mathbf{u}(x^D), \tag{1.1}$$

which defines the state of a body at the event x^D.[1]

The field equations necessary for the determination of the n-vector field $\mathbf{u}(x^D)$ are based upon n equations of balance of the form

$$\mathbf{F}^A{}_{,A} = \mathbf{\Pi}. \tag{1.2}$$

The comma denotes partial differentiation. The components $\mathbf{F}^0$ are called densities and the components $\mathbf{F}^i$ are called fluxes. We refer to $\mathbf{F}^A$ as four-dimensional fluxes. Their divergence in space-time is equal to the production vector $\mathbf{\Pi}$.

In order to provide field equations for the fields $\mathbf{u}$, the balance equations must be supplemented by constitutive equations. Such constitutive equations relate the fluxes $\mathbf{F}^A$ and the productions $\mathbf{\Pi}$ to the fields $\mathbf{u}$ in a materially dependent manner. In extended thermodynamics the constitutive equations are assumed to be of the forms

$$\mathbf{F}^A \equiv \mathbf{F}^A(\mathbf{u}) \qquad \text{and} \qquad \mathbf{\Pi} \equiv \mathbf{\Pi}(\mathbf{u}) \tag{1.3}$$

so that the fluxes $\mathbf{F}^A$ and productions $\mathbf{\Pi}$ at one event depend only on the values of the fields $\mathbf{u}$ at the same event. We may thus say that the constitutive equations are *local* in space-time.

If the constitutive functions $\mathbf{F}^A(\mathbf{u})$ and $\mathbf{\Pi}(\mathbf{u})$ were known explicitly, we could eliminate $\mathbf{F}^A$ and $\mathbf{\Pi}$ between (1.2) and (1.3) and obtain explicit field equations for $\mathbf{u}$. They form a quasilinear system of partial differential equation of first order. Every solution of this system is called a *thermodynamic process*.

1.2 Universal Principles of the Constitutive Theory

In reality, of course, the constitutive functions are not known, and it is the task of the constitutive theory to determine these functions or, at least, to reduce their generality. The tools of the constitutive theory are certain universal physical principles which represent expectations whose validity we accept from long experience. The main principles are

i.) the entropy inequality
ii.) the requirement of convexity
iii.) the principle of relativity.

[1] An event consists of the time t and the spatial position with coordinates $x^d = (x^1, x^2, x^3)$. Capital Latin indices run from 0 to 3, Latin minuscules from 1 to 3. In the nonrelativistic theory these minuscules are up or down, whatever is more convenient; however, the summation convention is used on all repeated indices.

The first two of these combined represent the entropy principle. In particular item (ii) guarantees thermodynamic stability.

The entropy inequality is an additional balance law—additional to the equations (1.2). We write

$$h^A{}_{,A} = \Sigma \geq 0 \tag{1.4}$$

and require this inequality to hold for all thermodynamic processes. h^A is the four-dimensional entropy flux and Σ is the density of entropy production. Both are constitutive quantities such that in extended thermodynamics we have

$$h^A \equiv h^A(\mathbf{u}), \qquad \Sigma = \Sigma(\mathbf{u}). \tag{1.5}$$

h^o is the density of entropy and h^a is its flux. It is now generally accepted that the entropy flux should be given by a constitutive relation rather than universally related to the heat flux. That idea was first proposed by Müller [18].

The requirement of convexity is also part of the entropy principle. It demands that the entropy density $h^o(\mathbf{u})$ be a concave function[2] of $\mathbf{u}$, i.e.

$$\frac{\partial^2 h^o}{\partial \mathbf{u} \partial \mathbf{u}} \sim \text{ negative definite.} \tag{1.6}$$

The principle of relativity states that the field equations (1.2), (1.3), and the entropy inequality (1.4), (1.5) have the same forms in all Galilean frames or in all Lorentz frames, depending on whether the theory is nonrelativistic or relativistic.

2 Entropy Inequality and Symmetric Hyperbolic Systems

The exploitation of the entropy inequality introduces an n-vector $\mathbf{\Lambda}$ of Lagrange multipliers. This vector may replace the original field vector $\mathbf{u}$: if it does, the field equations form a symmetric system. The system is also hyperbolic and as such it has convenient and desirable properties, viz. well-posedness of Cauchy problems and finite speeds.

2.1 Exploitation of the Entropy Inequality

We recall that the entropy inequality must hold for all thermodynamic processes, i.e. solutions of the field equations. In a manner of speaking the field equations provide constraints for the fields that must satisfy the entropy inequality. A lemma by Liu [20] proves that it is possible to use Lagrange multipliers $\mathbf{\Lambda}$ to eliminate such constraints. Indeed, the new inequality

$$h^A{}_{,A} - \mathbf{\Lambda} \cdot (\mathbf{F}^A{}_{,A} - \mathbf{\Pi}) \geq 0 \tag{2.1}$$

must hold for *all* continuous differentiable fields $\mathbf{u}$. The Lagrange multipliers are functions of $\mathbf{u}$.

By use of the chain rule the inequality (2.1) may be written in the form

$$\left(\frac{\partial h^A}{\partial \mathbf{u}} - \mathbf{\Lambda} \cdot \frac{\partial \mathbf{F}^A}{\partial \mathbf{u}}\right) \mathbf{u}_{,A} + \mathbf{\Lambda} \cdot \mathbf{\Pi} \geq 0 \tag{2.2}$$

[2]Often in the mathematical literature $-h^A$ replaces the entropy flux. In that case $h^o(\mathbf{u})$ is a convex function and the requirement of convexity is then properly named.

and, as this inequality must hold for all fields $\mathbf{u}$, it must hold in particular for all $\mathbf{u}_{,A}$ at one event. The factor of $\mathbf{u}_{,A}$ must therefore vanish, lest the inequality be violated for some choice of $\mathbf{u}_{,A}$. Thus we conclude

$$\frac{\partial h^A}{\partial \mathbf{u}} = \mathbf{\Lambda}\cdot\frac{\partial \mathbf{F}^A}{\partial \mathbf{u}} \quad \text{and} \quad \mathbf{\Lambda}\cdot\mathbf{\Pi} \geq 0. \tag{2.3}$$

The condition $(2.3)_1$ may be written in the short form

$$dh^A = \mathbf{\Lambda}\cdot d\mathbf{F}^A \tag{2.4}$$

which makes it clear that $\mathbf{\Lambda}$ is independent of the choice of the variables $\mathbf{u}$. In particular, if we choose $\mathbf{u} = \mathbf{F}^o$, which may be done without essential loss of generality, we obtain from (2.3) and (2.4)

$$\frac{\partial h^0}{\partial \mathbf{u}} = \mathbf{\Lambda} \quad \text{or, by differentiation with respect to } \mathbf{u}: \quad \frac{\partial^2 h^0}{\partial \mathbf{u}\partial \mathbf{u}} = \frac{\partial \mathbf{\Lambda}}{\partial \mathbf{u}}. \tag{2.5}$$

We see that $\frac{\partial \mathbf{\Lambda}}{\partial \mathbf{u}}$ is symmetric and—by the requirement of convexity (1.6)—that it is negative definite. From a theorem of nonlinear analysis (see e.g. Berger & Berger [46]) we therefore conclude that the map from $\mathbf{u}$ to $\mathbf{\Lambda}$ is not only locally but globally invertible. Therefore we may write the constitutive quantities in the form

$$\mathbf{F}^A \equiv \mathbf{F}^A(\mathbf{\Lambda}), \quad \mathbf{\Pi} \equiv \mathbf{\Pi}(\mathbf{\Lambda}), \quad h^A \equiv h^A(\mathbf{\Lambda}). \tag{2.6}$$

With the new variables $\mathbf{\Lambda}$ it is convenient to introduce the four-vector potential

$$h'^A = \mathbf{\Lambda}\cdot\mathbf{F}^A - h^A \tag{2.7}$$

for which (2.4) assumes the form

$$dh'^A = \mathbf{F}^A\cdot d\mathbf{\Lambda}. \tag{2.8}$$

Therefore we have

$$\mathbf{F}^A = \frac{\partial h'^A}{\partial \mathbf{\Lambda}}, \quad \text{hence by (2.7)} \quad h^A = -h'^A + \mathbf{\Lambda}\cdot\frac{\partial h'^A}{\partial \mathbf{\Lambda}} \tag{2.9}$$

such that the constitutive functions $\mathbf{F}^A(\mathbf{\Lambda})$ and $h^A(\mathbf{\Lambda})$ may be derived from a single vector function, viz. $h'^A(\mathbf{\Lambda})$. This fact justifies the name "vector potential" for h'^A. The components of this vector field have also been called "generators"; see Ruggeri & Strumia [22] and Boillat [47].

There is a residual inequality, namely

$$\Sigma = \mathbf{\Lambda}\cdot\mathbf{\Pi}(\mathbf{\Lambda}) \geq 0 \tag{2.10}$$

which ensures that the density of entropy production Σ is nonnegative.

The conditions (2.9), (2.10) have been derived as necessary conditions for the entropy inequality (1.4) to hold. They can be shown to be sufficient as well. Indeed, if we define

$$h^A = -h'^A + \mathbf{\Lambda}\cdot\mathbf{F}^A,$$

we obtain by differentiation

$$\begin{aligned} h^A{}_{,A} &= -h'^A{}_{,A} + \mathbf{\Lambda}_{,A}\cdot\mathbf{F}^A + \mathbf{\Lambda}\cdot\mathbf{F}^A{}_{,A} \\ &= -\left(\frac{\partial h'^A}{\partial \mathbf{\Lambda}} - \mathbf{F}^A\right)\cdot\mathbf{\Lambda}_{,A} + \mathbf{\Lambda}\cdot\mathbf{\Pi} \quad \text{or, by (2.9), (2.10)} \\ &\geq 0. \end{aligned}$$

Therefore (2.9), (2.10) exhaust the consequences of the entropy inequality. Note that the integrability condition for h'^A implied by $(2.9)_1$ states that the four $n \times n$ matrices

$$\frac{\partial \mathbf{F}^A}{\partial \mathbf{\Lambda}} \quad \text{must be symmetric.} \tag{2.11}$$

These conditions provide severe restrictions on the functions $\mathbf{F}^A(\mathbf{\Lambda})$.

2.2 Symmetric Hyperbolic Field Equations

We proceed to investigate the character of the system of field equations (1.2), (1.3), or (1.2), $(2.6)_{1,2}$. If we combine (1.2) and $(2.6)_1$ we obtain for the field equations

$$\frac{\partial \mathbf{F}^A}{\partial \mathbf{\Lambda}} \cdot \mathbf{\Lambda}_{,A} = \mathbf{\Pi}, \quad \text{or by } (2.9)_1 \quad \frac{\partial^2 h'^A}{\partial \mathbf{\Lambda} \partial \mathbf{\Lambda}} \cdot \mathbf{\Lambda}_{,A} = \mathbf{\Pi}. \tag{2.12}$$

Inspection shows that we may call the system *symmetric*, since the matrices multiplying $\mathbf{\Lambda}_{,A}$ are Hessian matrices.

It is important to observe that the system of field equations (1.2) is symmetric *only* if the fields are chosen as the n-vector of Lagrange multipliers $\mathbf{\Lambda}$ (see $(2.12)_2$). This special role of $\mathbf{\Lambda}$ has motivated Ruggeri & Strumia [22] to call this field the *main field.*

Symmetric hyperbolicity for the system (2.12) requires that $\frac{\partial^2 h'^0}{\partial \mathbf{\Lambda} \partial \mathbf{\Lambda}}$ be definite, and indeed, that matrix is *negative* definite, if $\mathbf{u}$—without essential loss of generality—is chosen to be the density vector $\mathbf{F}^0$. We proceed to show that this is true. From (2.7) we have for $A = 0$

$$h'^0 = \mathbf{\Lambda} \cdot \mathbf{u} - h^0, \tag{2.13}$$

so that h'^0 is the Legendre transform of h^0. This implies that h'^0 is a concave function of $\mathbf{\Lambda}$, if h^0 is a concave function of $\mathbf{u}$. The proof is simple: We consider the quadratic form $Q = \delta\mathbf{\Lambda}\delta\mathbf{u}$ and write

i.) by $\mathbf{\Lambda} = \frac{\partial h^0}{\partial \mathbf{u}}$ (see (2.4)) $Q = \delta\left(\frac{\partial h^0}{\partial \mathbf{u}}\right)\delta\mathbf{u} = \frac{\partial^2 h^0}{\partial \mathbf{u}\partial \mathbf{u}}\delta\mathbf{u}\delta\mathbf{u}$ and

ii.) by $\mathbf{u} = \frac{\partial h'^0}{\partial \mathbf{\Lambda}}$ (see (2.8)) $Q = \delta\mathbf{\Lambda}\delta\left(\frac{\partial h'^0}{\partial \mathbf{\Lambda}}\right) = \frac{\partial^2 h'^0}{\partial \mathbf{\Lambda}\partial \mathbf{\Lambda}}\delta\mathbf{\Lambda}\delta\mathbf{\Lambda}$.

We conclude from this and (1.6) that

$$\frac{\partial^2 h'^0}{\partial \mathbf{\Lambda} \partial \mathbf{\Lambda}} \delta\mathbf{\Lambda}\delta\mathbf{\Lambda} = \frac{\partial^2 h^0}{\partial \mathbf{u} \partial \mathbf{u}} \delta\mathbf{u}\delta\mathbf{u} < \mathbf{0} \tag{2.14}$$

holds, and therefore the system (2.12) is *symmetric hyperbolic*; it represents a particular case of symmetric hyperbolic systems defined by Friedrichs [48].

2.3 Discussion

Hyperbolicity guarantees finite speeds of propagation. The reason why we insist in having *symmetric* hyperbolic systems lies in their convenient and desirable mathematical properties, namely, the well-posedness of Cauchy problems, i.e. existence, uniqueness, and continuous dependence of the solutions on the data; see Fisher & Marsden [3].

The fact that a system of balance-type field equations is symmetric hyperbolic if it is compatible with the entropy inequality and the concavity of the entropy density was discovered by Godunov

[4] in a special case. In general this was proved by Boillat [6] and Ruggeri & Strumia [22]; the latter authors were motivated by a paper by Friedrichs & Lax [5]. Ruggeri [49] first recognized the applicability of these ideas to the theory of viscous, heat-conducting fluids.

The ingredients of the theory explained in this section are not only mathematically desirable, they also please the physicist. The attractive features include

- field equations of balance type,
- local constitutive relations,
- existence of an entropy density,
- thermodynamic stability,
- well-posedness of Cauchy problems, and
- finite speeds.

We suggest, therefore, that thermodynamic theories ought to fit into the scheme explained in this section. Rational extended thermodynamics does that.

2.4 Characteristic Speeds

We refer to the system (2.12), which we rewrite as

$$\frac{\partial^2 h'^0}{\partial \mathbf{\Lambda} \partial \mathbf{\Lambda}} \cdot \mathbf{\Lambda}_{,t} + \frac{\partial^2 h'^a}{\partial \mathbf{\Lambda} \partial \mathbf{\Lambda}} \cdot \mathbf{\Lambda}_{,a} = \mathbf{\Pi}, \tag{2.15}$$

displaying the time derivative $\mathbf{\Lambda}_{,t}$ and spatial derivatives $\mathbf{\Lambda}_{,a}$ separately. Inspection shows that all matrices are symmetric Hessians, and we recall that the matrix multiplying $\mathbf{\Lambda}_{,t}$ is negative definite. This makes the system symmetric hyperbolic, a property which has followed from the entropy principle. Such systems exhibit finite speeds of propagation of weak discontinuities, the so-called acceleration waves.

The speeds of acceleration waves are called *characteristic speeds* and we proceed to show how they are calculated. In an acceleration wave the fields $\mathbf{u}$, or $\mathbf{\Lambda}$ are continuous across the wave front, but their derivatives are not. Infact, denoting the jumps by $[\![\bullet]\!]$, we have

$$[\![\mathbf{\Lambda}_{,a}]\!] = \delta\mathbf{\Lambda}\, n_a \quad \text{and} \quad [\![\mathbf{\Lambda}_{,t}]\!] = -V\delta\mathbf{\Lambda}\ ; \tag{2.16}$$

n_a is a unit normal of the front and V is its speed. Thus $\delta\mathbf{\Lambda}$ is the jump of the gradient of $\mathbf{\Lambda}$ normal to the front. We substract the equation (2.15) on the two sides of the front and obtain with (2.16) the eigenvalue problem

$$\left(\frac{\partial^2 h'^a}{\partial \mathbf{\Lambda} \partial \mathbf{\Lambda}} n_a - V \frac{\partial^2 h'^0}{\partial \mathbf{\Lambda} \partial \mathbf{\Lambda}}\right) \cdot \delta\mathbf{\Lambda} = 0. \tag{2.17}$$

This homogeneous linear system must satisfy the characteristic equation

$$\det\left(\frac{\partial^2 h'^a}{\partial \mathbf{\Lambda} \partial \mathbf{\Lambda}} n_a - V \frac{\partial^2 h'^0}{\partial \mathbf{\Lambda} \partial \mathbf{\Lambda}}\right) = 0. \tag{2.18}$$

from which the possible speeds V may be calculated.

The symmetric hyperbolic character of the system (2.15) ensures that all V's are real and finite; otherwise, of course, they could not be interpreted as speeds. Also the eigenvalues $\delta\mathbf{\Lambda}$ are linearly independent.

3 Main Subsystems

We may need a large number of components of the field vector $\mathbf{u}$, – or indeed the main field vector $\boldsymbol{\Lambda}$, – where steep gradients and fast changes occur. In other parts of the body where conditions are smoother, a lesser number of fields may describe the state sufficiently well, so that less equations are needed. The reduced system of field equations is again symmetric hyperbolic and its characteristic speeds are nested within the interval of the speeds of the large system. These results were recently proved by Boillat & Ruggeri [24].

3.1 Constraints on the Main Field

Let some of the n components of the main field—the field of Lagrange multipliers $\boldsymbol{\Lambda}(x^D)$—be constrained to be constants. In that case a corresponding number of field equations is redundant, and we are left with a simpler theory. We proceed to investigate the system of field equations of such restricted theories.

We recall from (1.2) and $(2.9)_1$ that the complete system has the form

$$\left(\frac{\partial h'^A}{\partial \boldsymbol{\Lambda}}\right)_{,A} = \boldsymbol{\Pi}. \tag{3.1}$$

In preparation for the imposition of the constraints we split the n-vector $\boldsymbol{\Lambda}$ into the m-vector $\mathbf{L}$ and the $(n-m)$-vector $\mathbf{l}$. Accordingly the n-vector $\boldsymbol{\Pi}$ of productions is split into the m-vector $\mathbf{P}$ and the $(n-m)$-vector $\mathbf{p}$. The system (3.1) may thus be split into two partial systems, viz.

$$\left(\frac{\partial h'^A(\mathbf{L},\mathbf{l})}{\partial \mathbf{L}}\right)_{,A} = \mathbf{P}(\mathbf{L},\mathbf{l}) \quad \text{and} \quad \left(\frac{\partial h'^A(\mathbf{L},\mathbf{l})}{\partial \mathbf{l}}\right)_{,A} = \mathbf{p}(\mathbf{L},\mathbf{l}). \tag{3.2}$$

Obviously there are 2^n-2 possibilities for such a decomposition if we exclude the trivial possibilities $m=0$ and $m=n$.

Now we think of the fields $\mathbf{l}$ as constrained to be constant, and we write $\mathbf{l} = \mathbf{l}^* = const.$ In that case the system $(3.2)_2$ is ignored and we rely on $(3.2)_1$ to determine $\mathbf{L}$. We write $(3.2)_1$ in the form

$$\frac{\partial^2 h'^A(\mathbf{L},\mathbf{l}^*)}{\partial \mathbf{L}\partial \mathbf{L}} \cdot \mathbf{L}_{,A} = \mathbf{P}(\mathbf{L},\mathbf{l}^*) \tag{3.3}$$

and call this system a *main* subsystem, since the constraints—and the remaining fields—are components of the *main* field. The right-hand side is a function of $\mathbf{L}$.[3]

3.2 A Main Subsystem Implies an Entropy Inequality

The solutions of the main subsystem (3.3) satisfy an entropy balance of the form

$$\overline{h}^A{}_{,A} = \overline{\Sigma} \quad \text{with} \quad \begin{aligned} &\overline{h}^A(\mathbf{L}) = h^A(\mathbf{L},\mathbf{l}^*) - \mathbf{l}^* \cdot \tfrac{\partial h'^A(\mathbf{L},\mathbf{l})}{\partial \mathbf{l}}\big|_{\mathbf{l}=\mathbf{l}^*} \\ &\overline{\Sigma}(\mathbf{L}) = \mathbf{L}\cdot \mathbf{P}(\mathbf{L},\mathbf{l}^*). \end{aligned} \tag{3.4}$$

For obvious reasons we call $\overline{h}^A$ and $\overline{\Sigma}$ the subentropy flux and the subentropy production respectively. It seems reasonable to postulate that the latter be nonnegative, so that we have an entropy *inequality* for the subsystem.

[3]The definition of main subsystem can be generalized to the case in which we assign a *function* $\mathbf{l}^*(x^D)$, rather than a constant. In this case the main subsytem depends also explicitly of x^D; see Boillat & Ruggeri [24].

For proof of (3.4) we multiply the subsystem (3.3) by $\mathbf{L}$ and perform some easy manipulations. Thus we may define

$$\overline{h}^A(\mathbf{L}, \mathbf{l}^*) = -h'^A(\mathbf{L}, \mathbf{l}^*) + \mathbf{L} \cdot \frac{\partial h'^A(\mathbf{L}, \mathbf{l}^*)}{\partial \mathbf{L}} \tag{3.5}$$

which is—for the subsystem—equivalent to the relation $(2.9)_2$ between $h^A(\mathbf{\Lambda})$ and $h'^A(\mathbf{\Lambda})$ for the full system. The equations (3.4) follow from this by simple analysis.

We note that $\overline{h}^A(\mathbf{L}, \mathbf{l}^*)$, the subentropy flux, is not simply the restriction of $h^A(\mathbf{L}, \mathbf{l})$ to the constrained vector $\mathbf{l}^*$; rather there is an additional term. The entropy production $\overline{\Sigma}$ of the subsystem consists of the term $\mathbf{L} \cdot \mathbf{P}(\mathbf{L}, \mathbf{l}^*)$ which is equivalent—for the subsystem—to the expression $\Sigma = \mathbf{\Lambda} \cdot \mathbf{\Pi}(\mathbf{\Lambda})$ for the full system.

3.3 A Main Subsystem Is Symmetric Hyperbolic

The main subsystem (3.3) is symmetric hyperbolic. Indeed, in the form (3.3) we see that the symmetry of the coefficient matrices is guaranteed by their Hessian character. The concavity of $h'^0(\mathbf{L}, \mathbf{l}^*)$ with respect to $\mathbf{L}$ is implied by the concavity of $h'^0(\mathbf{\Lambda})$ with respect to $\mathbf{\Lambda}$; see (2.14). And, since $h'^0(\mathbf{L}, \mathbf{l}^*)$ is concave in $\mathbf{L}$, so is the subentropy $\overline{h}^0$ with respect to $\partial h'^0(\mathbf{L}, \mathbf{l}^*)/\partial \mathbf{L}$; this follows from the Legendre transfrom (3.5) for $A = 0$.

3.4 Characteristic Speeds of the Subsystems

We recall the full system (2.15), its eigenvalue problem (2.17), and its characteristic equation (2.18). The corresponding equations for the subsystem read

$$\begin{aligned} &\frac{\partial^2 h'^0(\mathbf{L}, \mathbf{l}^*)}{\partial \mathbf{L} \partial \mathbf{L}} \cdot \mathbf{L}_{,t} + \frac{\partial^2 h'^a(\mathbf{L}, \mathbf{l}^*)}{\partial \mathbf{L} \partial \mathbf{L}} \cdot \mathbf{L}_{,a} = \mathbf{P}(\mathbf{L}, \mathbf{l}^*) \quad \text{and} \\ &\left(\frac{\partial^2 h'^a(\mathbf{L}, \mathbf{l}^*)}{\partial \mathbf{L} \partial \mathbf{L}} n_a - \overline{V} \frac{\partial^2 h'^0(\mathbf{L}, \mathbf{l}^*)}{\partial \mathbf{L} \partial \mathbf{L}} \right) \cdot \delta \mathbf{L} = 0 \\ &\det \left(\frac{\partial^2 h'^a(\mathbf{L}, \mathbf{l}^*)}{\partial \mathbf{L} \partial \mathbf{L}} n_a - \overline{V} \frac{\partial^2 h'^0(\mathbf{L}, \mathbf{l}^*)}{\partial \mathbf{L} \partial \mathbf{L}} \right) = 0 \end{aligned} \tag{3.6}$$

$\overline{V}$, the roots of the characteristic equation $(3.6)_3$, are the characteristic speeds of the subsystem.

We proceed to prove that the speeds $\overline{V}(\mathbf{L}, n)$ of the subsystem lie within the interval defined by the speeds $V(\mathbf{L}, \mathbf{l}^*, n)$ of the original system. We express this by writing

$$\begin{aligned} V_{\max}(\mathbf{L}, \mathbf{l}^*, n) &\geq \overline{V}_{\max}(\mathbf{L}, n) \\ V_{\min}(\mathbf{L}, \mathbf{l}^*, n) &\leq \overline{V}_{\min}(\mathbf{L}, n). \end{aligned} \tag{3.7}$$

Thus the extremal speeds of the susbsystems are neatly "nested" within the interval of the extremal speeds of bigger systems.

For proof we recall from linear algebra that

$$-\frac{\partial^2 h'^a(\mathbf{\Lambda})}{\partial \mathbf{\Lambda} \partial \mathbf{\Lambda}} n_a + V_{\substack{\max \\ \min}}(\mathbf{\Lambda}, \mathbf{n}) \frac{\partial^2 h'^0(\mathbf{\Lambda})}{\partial \mathbf{\Lambda} \partial \mathbf{\Lambda}} \sim \begin{matrix} \text{negative} \\ \text{positive} \end{matrix} \text{ semidefinite} \tag{3.8}$$

holds, since both matrices are symmetric and the latter is negative definite. In particular therefore we have

$$\left(-\frac{\partial^2 h'^a(\mathbf{L},\mathbf{l}^*)}{\partial\mathbf{L}\partial\mathbf{L}}\, n_a + V_{\substack{\max\\ \min}}(\mathbf{L},\mathbf{l}^*,\mathbf{n})\frac{\partial^2 h'^0(\mathbf{L},\mathbf{l}^*)}{\partial\mathbf{L}\partial\mathbf{L}}\right)\delta\mathbf{L}\delta\mathbf{L} \lesseqgtr 0. \tag{3.9}$$

But for the subsystem (3.6) we have

$$\left(-\frac{\partial^2 h'^a(\mathbf{L},\mathbf{l}^*)}{\partial\mathbf{L}\partial\mathbf{L}}\, n_a + \overline{V}_{\substack{\max\\ \min}}(\mathbf{L},\mathbf{n})\frac{\partial^2 h'^0(\mathbf{L})}{\partial\mathbf{L}\partial\mathbf{L}}\right)\delta\mathbf{L}\delta\mathbf{L} = 0, \tag{3.10}$$

and hence follows (3.7) by comparison of (3.9), (3.10). Boillat & Ruggeri [24] have called (3.7) the ***subcharacteristic conditions***, because this name has been applied in the literature to similar conditions where a subsystem was created by the vanishing of a relaxation time: First by Witham [50] in the context of a linear system and then by Chen, Levermore & Liu for the nonlinear case [51].

3.5 Other Subsystems

For the symmetric hyperbolicity and for the subcharacteristic condition it is important that the subsystem is a *main* subsystem. In other words the constraints must be on the Lagrange multipliers—the *main* field. Boillat & Ruggeri [24] have given an example where hyperbolicity was lost when a component of the field vector $\mathbf{u}$—rather than $\mathbf{\Lambda}$—was constrained.

4 Galilean Invariance

The requirement of Galilean invariance of the field equations goes a long way in determining the velocity dependence of the constitutive quantities of extended thermodynamics. It permits the identification of convective parts, non-convective parts and internal parts of densities, fluxes and productions, the latter as the Galilei invariant residue appropriate to the rest frame of the body. If the field vector consists of blocks of tensors, the velocity dependence of the constitutive quantities can be calculated explicitly. The present systematic exploitation of Galilei invariance is due to Ruggeri [23].

4.1 Tensors, Galilean Tensors, and Euclidean Tensors

In classical, i.e. nonrelativistic thermodynamics we encounter three relevant types of transformations between frames, viz.

coordinate rotation	Galileian transformation	Euclidean transformation
$x_i^* = O_{ij}x_j$	$x_i^* = O_{ij}x_j + c_i t$	$x_i^* = O_{ij}(t)x_j + b_i(t)$
$t^* = t$	$t^* = t$	$t^* = t.$

(4.1)

A quantity is called a tensor of rank A if its components in the two frames are related by the equation

$$T_{i_1 i_2 \ldots i_A} = O_{i_1 j_1} \ldots O_{i_A j_A} T_{j_1 \ldots j_A} \tag{4.2}$$

and, of course, scalars and vectors are included as tensors of rank 0 and 1. In case of the Euclidean and Galilean transformations we call the tensors *objective* and *Galilean* respectively, and in the case of coordinate rotation we speak of tensors, without further qualification. The velocity v_i transforms according to the rules

$$v_i^* = O_{ij}v_j \qquad v_i^* = O_{ij}v_j + c_i \qquad v_i^* = O_{ij}(t)v_j + \dot{O}_{ij}(t)x_j + \dot{b}_i(t) \tag{4.3}$$

under the transformations (4.1); therefore in our jargon the velocity is a vector, but neither a Galilean nor an objective vector.

4.2 Principle of Relativity

In most specific cases the generic field $\mathbf{u}$ contains, as some of its components, the velocity components v^i of the body. The remaining $(n-3)$ variables are denoted by $\mathbf{w}$; they are assumed to be components of Galilean tensors.

The most prominent occurence of the velocity v^j is in the fluxes $\mathbf{F}^i$ and h^i, which can be split into convective and nonconvective fluxes, viz.

$$\mathbf{F}^i = \mathbf{F}^0 v^i + \mathbf{G}^i \qquad \text{and} \qquad h^i = h^0 v^i + \varphi^i. \tag{4.4}$$

$\mathbf{G}^i$ and φ^i are nonconvective fluxes.

This does not mean, however, that $\mathbf{F}^0, \mathbf{G}^i, \mathbf{\Pi}$, or h^0, φ^i, Σ are necessarily independent of v^i. In fact we may write the constitutive equations (1.3) and (1.5) in the form

$$\begin{array}{lll} \mathbf{F}^0 = \mathbf{F}^0(\mathbf{v}, \mathbf{w}) & \mathbf{G}^i = \mathbf{G}^i(\mathbf{v}, \mathbf{w}) & \mathbf{\Pi} = \mathbf{\Pi}(\mathbf{v}, \mathbf{w}) \\ h^0 = h^0(\mathbf{v}, \mathbf{w}) & \varphi^i = \varphi^i(\mathbf{v}, \mathbf{w}) & \Sigma = \Sigma(\mathbf{v}, \mathbf{w}) \end{array} \tag{4.5}$$

where $\mathbf{w}$ represents those $(n-3)$ components of the generic field $\mathbf{u}$ which are not the velocities v^i. Let (4.5) be the constitutive equations in one Galileian frame; in another frame they are assumed to have the forms

$$\begin{array}{lll} \mathbf{F}^{*0} = \mathbf{F}^0(\mathbf{v}^*, \mathbf{w}^*) & \mathbf{G}^{*i} = \mathbf{G}^i(\mathbf{v}^*, \mathbf{w}^*) & \mathbf{\Pi}^* = \mathbf{\Pi}(\mathbf{v}^*, \mathbf{w}^*) \\ h^{*0} = h^0(\mathbf{v}^*, \mathbf{w}^*) & \varphi^{*i} = \varphi^i(\mathbf{v}^*, \mathbf{w}^*) & \Sigma^* = \Sigma(\mathbf{v}^*, \mathbf{w}^*). \end{array} \tag{4.6}$$

Note that the constitutive *functions* are unchanged while both their values and the values of their variables do change. This assumption is part of the principle of relativity, because the field equations cannot have the same forms in all Galilean frames unless that is first true for the constitutive functions.

The principle of relativity further requires that the field equations and entropy balance equations in the two frames are equivalent, viz.

$$\begin{array}{ll} \dfrac{\partial \mathbf{F}^0(\mathbf{v}, \mathbf{w})}{\partial t} + \dfrac{\partial \mathbf{F}^0(\mathbf{v}, \mathbf{w})\, v^i + \mathbf{G}^i(\mathbf{v}, \mathbf{w})}{\partial x^i} & = \mathbf{\Pi}(\mathbf{v}, \mathbf{w}), \\[2ex] \dfrac{\partial \mathbf{F}^0(\mathbf{v}^*, \mathbf{w}^*)}{\partial t^*} + \dfrac{\partial \mathbf{F}^0(\mathbf{v}^*, \mathbf{w}^*) v^{i*} + \mathbf{G}^i(\mathbf{v}^*, \mathbf{w}^*)}{\partial x^*_i} & = \mathbf{\Pi}(\mathbf{v}^*, \mathbf{w}^*) \end{array} \tag{4.7}$$

$$\begin{array}{ll} \dfrac{\partial h^0(\mathbf{v}, \mathbf{w})}{\partial t} + \dfrac{\partial h^0(\mathbf{v}, \mathbf{w})\, v^i + \varphi^i(\mathbf{v}, \mathbf{w})}{\partial x^i} & = \Sigma(\mathbf{v}, \mathbf{w}) \\[2ex] \dfrac{\partial h^0(\mathbf{v}^*, \mathbf{w}^*)}{\partial t^*} + \dfrac{\partial h^0(\mathbf{v}^*, \mathbf{w}^*) v^{i*} + \varphi^i(\mathbf{v}^*, \mathbf{w}^*)}{\partial x^*_i} & = \Sigma(\mathbf{v}^*, \mathbf{w}^*). \end{array} \tag{4.8}$$

4.3 Exploitation of the Principle of Relativity for the Entropy Balance

We have assumed that the $(n-3)$ fields $\mathbf{w}$ are components of Galilean tensors. With $O_{ij} = \delta_{ij}$ this means $\mathbf{w}^* = \mathbf{w}$ and

$$\frac{\partial}{\partial x_i^*} = \frac{\partial}{\partial x_i} \quad \text{and} \quad \frac{\partial}{\partial t^*} = \frac{\partial}{\partial t} - c_i \frac{\partial}{\partial x_i} \tag{4.9}$$

We may thus write $(4.8)_2$ in the form

$$\frac{\partial h^0(\mathbf{v}+\mathbf{c},\mathbf{w})}{\partial t} + \frac{\partial h^0(\mathbf{v}+\mathbf{c},\mathbf{w})v^i + \varphi^i(\mathbf{v}+\mathbf{c},\mathbf{w})}{\partial x^i} = \Sigma(\mathbf{v}+\mathbf{c},\mathbf{w}) \tag{4.10}$$

and this must be equivalent to $(4.8)_1$. Hence follows by comparison

$$h^0(\mathbf{v}+\mathbf{c},\mathbf{w}) = h^0(\mathbf{v},\mathbf{w}), \quad \varphi^i(\mathbf{v}+\mathbf{c},\mathbf{w}) = \varphi^i(\mathbf{v},\mathbf{w}), \quad \Sigma(\mathbf{v}+\mathbf{c},\mathbf{w}) = \Sigma(\mathbf{v},\mathbf{w}) \tag{4.11}$$

which must hold for all $\mathbf{v}$ and $\mathbf{c}$. Inspection shows that all three quantities, i.e. h^0, φ^i, and Σ must be independent of $\mathbf{v}$. We have

$$h^0 = h^0(\mathbf{w}), \quad \varphi^i = \varphi^i(\mathbf{w}), \quad \Sigma = \Sigma(\mathbf{w}). \tag{4.12}$$

4.4 Exploitation of the Principle of Relativity for the Field Equations

We continue with $O_{ij} = \delta_{ij}$, so that $\mathbf{w}^* = \mathbf{w}$ and (4.9) holds. Therefore the system $(4.7)_2$ may be written in the form

$$\frac{\partial \mathbf{F}^0(\mathbf{v}+\mathbf{c},\mathbf{w})}{\partial t} + \frac{\partial \mathbf{F}^0(\mathbf{v}+\mathbf{c},\mathbf{w})v^i + \mathbf{G}^i(\mathbf{v}+\mathbf{c},\mathbf{w})}{\partial x^i} = \mathbf{\Pi}(\mathbf{v}+\mathbf{c},\mathbf{w}) \tag{4.13}$$

and this must be equivalent to $(4.7)_1$. At first sight it would seem that this is just like comparing the entropy equations $(4.8)_1$ and (4.10), and yet the result is quite different, because the entropy equation is a single equation, while the field equations form a *system* of n equations. This system may be Galilei invariant even though the densities, fluxes, and productions are velocity-dependent, albeit in a restricted form.

Indeed, the equivalence of (4.13) and $(4.7)_1$ requires that the former system is a linear combination of the latter one. And this means that there exists an $n \times n$ nonsingular matrix $\mathbf{X}(\mathbf{c})$ such that

$$\mathbf{F}^0(\mathbf{v}+\mathbf{c},\mathbf{w}) = \mathbf{X}(\mathbf{c})\mathbf{F}^0(\mathbf{v},\mathbf{w}), \mathbf{G}^i(\mathbf{v}+\mathbf{c},\mathbf{w}) = \mathbf{X}(\mathbf{c})\mathbf{G}^i(\mathbf{v},\mathbf{w}), \mathbf{\Pi}(\mathbf{v}+\mathbf{c},\mathbf{w}) = \mathbf{X}(\mathbf{c})\mathbf{\Pi}(\mathbf{v},\mathbf{w}) \tag{4.14}$$

holds for all $\mathbf{v}$ and $\mathbf{c}$. In particular for $\mathbf{v} = 0$ we have for all $\mathbf{c}$

$$\mathbf{F}^0(\mathbf{c},\mathbf{w}) = \mathbf{X}(\mathbf{c})\mathbf{F}^0(0,\mathbf{w}), \quad \mathbf{G}^i(\mathbf{c},\mathbf{w}) = \mathbf{X}(\mathbf{c})\mathbf{G}^i(0,\mathbf{w}), \quad \mathbf{\Pi}(\mathbf{c},\mathbf{w}) = \mathbf{X}(\mathbf{c})\mathbf{\Pi}(0,\mathbf{w}). \tag{4.15}$$

$\mathbf{F}^0(0,\mathbf{w})$, $\mathbf{G}^i(0,\mathbf{w})$, and $\mathbf{\Pi}(0,\mathbf{w})$ represent the densities, fluxes, and productions in the rest frame of the body. We call these quantities *internal* and denote them by $\hat{\mathbf{F}}^0, \hat{\mathbf{G}}^i, \hat{\mathbf{\Pi}}$. Thus (4.15) assumes the form

$$\mathbf{F}^0(\mathbf{v},\mathbf{w}) = \mathbf{X}(\mathbf{v})\hat{\mathbf{F}}^0(\mathbf{w}), \quad \mathbf{G}^i(\mathbf{v},\mathbf{w}) = \mathbf{X}(\mathbf{v})\hat{\mathbf{G}}^i(\mathbf{w}), \quad \mathbf{\Pi}(\mathbf{v},\mathbf{w}) = \mathbf{X}(\mathbf{v})\hat{\mathbf{\Pi}}(\mathbf{w}). \tag{4.16}$$

The matrix $\mathbf{X}$ has the properties

$$\mathbf{X}(\mathbf{v}^{(1)} + \mathbf{v}^{(2)}) = \mathbf{X}(\mathbf{v}^{(1)})\,\mathbf{X}(\mathbf{v}^{(2)}) = \mathbf{X}(\mathbf{v}^{(2)})\mathbf{X}(\mathbf{v}^{(1)}), \quad \mathbf{X}(0) = \mathbf{1} \tag{4.17}$$

which follow from (4.14) through (4.16). And (4.17) implies the differential relations

$$\frac{\partial \mathbf{X}}{\partial v_r}(\mathbf{v})=\mathbf{A}^r\mathbf{X}(\mathbf{v})=\mathbf{X}(\mathbf{v})\mathbf{A}^r, \mathbf{X}(0)=\mathbf{1},\ \mathbf{A}^r=\frac{\partial \mathbf{X}}{\partial v_r}(0), \text{ and } \mathbf{A}^r\mathbf{A}^s=\mathbf{A}^s\mathbf{A}^r. \tag{4.18}$$

Thus $\mathbf{X}(\mathbf{v})$ is an exponential function of the components v_r with constant matrices $\mathbf{A}^r$ as coefficients, viz.

$$\mathbf{X}(\mathbf{v}) = e^{\mathbf{A}^r v_r} = \mathbf{1} + \mathbf{A}^r v_r + \frac{1}{2}\mathbf{A}^r\mathbf{A}^s v_r v_s + \ldots \tag{4.19}$$

We summarize (4.16) and (4.19) by saying that the v-dependence of the constitutive quantities $\mathbf{F}^0$, $\mathbf{G}^i$ and $\mathbf{\Pi}$ is not dependent on material; in fact it is explicit if only the three constant matrices $\mathbf{A}^r$ are known. The form of these matrices depends on the tensorial character of the densities $\mathbf{F}^0$ about which—so far—we have not made a commitment.

4.5 Field Equations for Internal Quantities

Since the velocity dependence of $\mathbf{F}^0, \mathbf{G}^i$, and $\mathbf{\Pi}$ is explicit there is a lot of redundance in the field equations

$$\frac{\partial \mathbf{F}^0}{\partial t} + \frac{\partial \mathbf{F}^0 v^i + \mathbf{G}^i}{\partial x^i} = \mathbf{\Pi}. \tag{4.20}$$

This redundance may be removed by insertion of the decompositions (4.16), viz.

$$\mathbf{F}^0 = \mathbf{X}\hat{\mathbf{F}}^0, \qquad \mathbf{G}^i = \mathbf{X}\hat{\mathbf{G}}^i \quad \text{and} \quad \mathbf{\Pi} = \mathbf{X}\hat{\mathbf{\Pi}} \tag{4.21}$$

into (4.20). A short calculation with reference to (4.18) provides a new hierarchy of equations, which contains only the internal densities, fluxes, and productions and the *derivatives* of the velocity v. We obtain with $\frac{d}{dt} = \frac{\partial}{\partial t} + v_i \frac{\partial}{\partial x_i}$ as material time derivative

$$\mathbf{X}\left\{\frac{d\hat{\mathbf{F}}^0}{dt} + \hat{\mathbf{F}}^0\frac{\partial v_i}{\partial x_i} + \frac{\partial \hat{\mathbf{G}}^i}{\partial x_i} + \mathbf{A}^r\left\{\hat{\mathbf{F}}^0\frac{dv_r}{dt} + \hat{\mathbf{G}}^i\frac{\partial v_r}{\partial x_i}\right\} - \hat{\mathbf{\Pi}}\right\} = 0. \tag{4.22}$$

Thus we obtain an alternative form of the equations of balance, viz.

$$\frac{d\hat{\mathbf{F}}^0}{dt} + \hat{\mathbf{F}}^0\frac{\partial v_i}{\partial x_i} + \frac{\partial \hat{\mathbf{G}}^i}{\partial x_i} + \mathbf{A}^r\left\{\hat{\mathbf{F}}^0\frac{dv_r}{dt} + \hat{\mathbf{G}}^i\frac{\partial v_r}{\partial x_i}\right\} = \hat{\mathbf{\Pi}}. \tag{4.23}$$

These equations are much more useful for practical purposes than the seemingly simpler set (4.30). However, it is difficult to recognize the equations (4.22) as equations of balance type. And of course, they contain the so far unknown constant matrices $\mathbf{A}^r$.

4.6 Galilei Invariance for Subsystems

We recall that it may sometimes be useful to restrict the attention to subsystems and, obviously, we shall then not wish to lose the Galilean invariance. The desire to preserve Galilean invariance in all relevant subsystems will provide a natural order for the n field equation of the original system, and a simplification of the matrix $\mathbf{X}(\mathbf{v})$.

Let us imagine that the n-vectors $\mathbf{F}^0, \mathbf{G}^i, \mathbf{\Pi}$ are split according to

$$\mathbf{F}^0 = \begin{bmatrix} \mathbf{\Phi} \\ \varphi \end{bmatrix}, \quad \mathbf{G}^i = \begin{bmatrix} \mathbf{\Gamma}^i \\ \gamma^i \end{bmatrix}, \quad \mathbf{\Pi} = \begin{bmatrix} \mathbf{P} \\ \mathbf{p} \end{bmatrix}, \tag{4.24}$$

into $(n-m)$-vectors $\mathbf{\Phi}, \mathbf{\Gamma}^i, \mathbf{P}$ and the remaining m components: $\varphi, \gamma^i, \mathbf{p}$. Thus the balance equations (4.20) read

$$\frac{\partial \mathbf{\Phi}}{\partial t} + \frac{\partial \mathbf{\Phi} v^i + \mathbf{\Gamma}^i}{\partial x^i} = \mathbf{P} \quad \text{and} \quad \frac{\partial \varphi}{\partial t} + \frac{\partial \varphi v^i + \gamma^i}{\partial x^i} = \mathbf{p}. \tag{4.25}$$

Now we restrict the attention to the subsystem $(4.25)_1$, ignoring the rest, i.e. $(4.25)_2$. We require that the old system (4.20) and the new subsystem $(4.25)_1$ both exhibit Galilean invariance, and therefore we must have

$$\mathbf{F}^0 = \mathbf{X}(\mathbf{v})\hat{\mathbf{F}}^0 \quad \text{and} \quad \mathbf{\Phi} = \mathbf{Y}(\mathbf{v})\hat{\mathbf{\Phi}} \tag{4.26}$$

and analogous equations for $\mathbf{G}^i, \mathbf{\Gamma}^i$ and $\mathbf{\Pi}, \mathbf{P}$. The internal part $\hat{\mathbf{\Phi}}$ of $\mathbf{\Phi}$ is related to $\mathbf{\Phi}$ by a matrix $\mathbf{Y}(\mathbf{v})$ which has all the general properties of the matrix $\mathbf{X}(\mathbf{v})$, except that it is of lower order, of course. We write the first $(n-m)$ components of $(4.26)_1$ in the form

$$\mathbf{\Phi} = \mathbf{X}^{\Phi\Phi}\hat{\mathbf{\Phi}} + \mathbf{X}^{\Phi\varphi}\hat{\varphi} \tag{4.27}$$

and, comparing $(4.26)_2$ and (4.27), we obtain

$$\mathbf{X}^{\Phi\Phi}(\mathbf{v}) = \mathbf{Y}(\mathbf{v}) \quad \text{and} \quad \mathbf{X}^{\Phi\varphi}(\mathbf{v}) = 0. \tag{4.28}$$

It follows that $\mathbf{X}(\mathbf{v})$ has the generic form

$$\mathbf{X} = \begin{bmatrix} \mathbf{X}^{\Phi\Phi} & 0 \\ \mathbf{X}^{\varphi\Phi} & \mathbf{X}^{\varphi\varphi} \end{bmatrix}. \tag{4.29}$$

In fact, if we drop one equation after the other starting at the bottom of the system—and always requiring the residual system to have Galilean invariance—a simple extrapolation of the above argument will furnish a matrix $\mathbf{X}$ which is subtriangular, i.e. whose entries in the upper half are all equal to zero.

These arguments seem to furnish a good criterion for ordering the system: Put the equations in such an order that the last equation—or last equations—can be removed without losing Galilean invariance for the remaining subsystems.

4.7 Galilean Invariance and Entropy Principle

Once the matrices $\mathbf{A}^r$—and hence $\mathbf{X}(\mathbf{v})$—are known, we conclude from (4.16) that only the internal parts $\hat{\mathbf{F}}^0, \hat{\mathbf{G}}^i$, and $\hat{\mathbf{\Pi}}$ of densities, fluxes, and productions remain as constitutive quantities and that $\mathbf{v}$ has dropped out as a variable. We shall now investigate how these remaining constitutive relations are restricted by the entropy principle. In other words we wish to find out how the general restrictions (cf. (2.3), (2.4))

$$dh^A = \mathbf{\Lambda} \cdot d\mathbf{F}^A \quad \text{and} \quad \mathbf{\Lambda} \cdot \mathbf{\Pi} \geq 0 \tag{4.30}$$

are reflected in restrictions on the internal quantities.

By (4.12) and (4.16) we may write $(4.30)_1$ as

$$A=0: dh^0 = \mathbf{\Lambda} d(\mathbf{X}\hat{\mathbf{F}}^0) \quad \Big| \quad A=i: d(h^0 v^i + \varphi^i) = \mathbf{\Lambda} d[\mathbf{X}(\hat{\mathbf{F}}^0 v^i + \hat{\mathbf{G}}^i)]. \tag{4.31}$$

With $d\mathbf{X} = \mathbf{X}\mathbf{A}^r dv_r$ these two relations assume the forms

$$dh^0 - \mathbf{\Lambda X} d\hat{\mathbf{F}}^0 = \mathbf{\Lambda X A}^r \hat{\mathbf{F}}^0 dv_r \quad \Big| \quad d\varphi^i - \mathbf{\Lambda X} d\hat{\mathbf{G}}^i = \left(\mathbf{\Lambda X A}^r \hat{\mathbf{G}}^i - [\hat{h}_0 - \mathbf{\Lambda X}\hat{\mathbf{F}}^0]\delta^{ri}\right) dv_r.$$

Since this must hold for all $d\mathbf{v}$ we have

$$dh^0 = \mathbf{\Lambda X} d\hat{\mathbf{F}}^0, \mathbf{\Lambda X A}^r \hat{\mathbf{F}}^0 = 0 \quad \Big| \quad d\varphi^i = \mathbf{\Lambda X} d\hat{\mathbf{G}}^i, \mathbf{\Lambda X A}^r \hat{\mathbf{G}}^i = [\hat{h}_0 - \mathbf{\Lambda X}\hat{\mathbf{F}}^0]\delta^{ri}. \tag{4.32}$$

Moreover $h^0, \hat{\mathbf{F}}^0$ and $\varphi_i, \hat{\mathbf{G}}_i$ are independent of $\mathbf{v}$. Therefore $\mathbf{\Lambda X}$ is also independent of $\mathbf{v}$, and we write[4]

$$\hat{\mathbf{\Lambda}} = \hat{\mathbf{\Lambda}}(\mathbf{w}) = \mathbf{\Lambda X}. \tag{4.33}$$

The $\mathbf{\Lambda}$'s may be called the *internal* Lagrange multipliers. Thus (4.32) reads

$$dh^0 = \hat{\mathbf{\Lambda}} d\hat{\mathbf{F}}^0, \quad \hat{\mathbf{\Lambda}} \mathbf{A}^r \hat{\mathbf{F}}^0 = 0, \quad d\varphi^i = \hat{\mathbf{\Lambda}} d\hat{\mathbf{G}}^i, \quad \hat{\mathbf{\Lambda}} \mathbf{A}^r \hat{\mathbf{G}}^i = \left[\hat{h}_0 - \hat{\mathbf{\Lambda}}\hat{\mathbf{F}}^0\right]\delta^{ri}. \tag{4.34}$$

By $(4.16)_3$ and (4.33) the residual inequality $(4.29)_2$ may be written in the form

$$\hat{\mathbf{\Lambda}}\hat{\mathbf{\Pi}} \geq 0. \tag{4.35}$$

The relations (4.34) and (4.35) represent the restrictions that were drawn from the entropy inequality *and* the requirement of Galilean invariance. The present form contains only internal quantities, i.e. the velocity does not appear anymore. The differential form $(4.34)_1$, giving the differential of the entropy density, may be called the Gibbs equation of extended thermodynamics. Its equilibrium counterpart can identify the Lagrange mulitipliers $\mathbf{\Lambda}$ in equilibrium. And there is an analogous relation, viz. $(4.34)_3$ for the nonconvective entropy flux.

4.8 Explicit Velocity Dependence of Constitutive Quantities. The Determination of $\mathbf{A}^r$.

We consider the case where the densities $\mathbf{F}^0$, (nonconvective) fluxes $\mathbf{G}^i$, and productions $\mathbf{\Pi}$ have the forms

$$\mathbf{F}^0 = \begin{bmatrix} F \\ F_{k_1} \\ F_{k_1k_2} \\ \vdots \\ F_{k_1k_2\dots k_N} \end{bmatrix}, \quad \mathbf{G}^i = \begin{bmatrix} G_i \\ G_{ik_1} \\ G_{ik_1k_2} \\ \vdots \\ G_{ik_1k_2\dots k_N} \end{bmatrix}, \quad \mathbf{\Pi} = \begin{bmatrix} \Pi \\ \Pi_{k_1} \\ \Pi_{k_1k_2} \\ \vdots \\ \Pi_{k_1k_2\dots k_N} \end{bmatrix} \tag{4.36}$$

with $F_{k_1\dots k_j}, G_{ik_1\dots k_j}, \Pi_{k_1\dots k_j}$ as symmetric tensors. Thus the system of balance equations (1.2), or (4.20) has a natural order induced by the increasing tensorial character of the equations of the system. Each block of tensorial equations of rank j governs the evolution of a new quantity $F_{k_1k_2\dots k_j}$.

[4]The conclusion (4.33) can obviously be drawn from $(4.32)_{1,3}$, if none of the components of $\hat{\mathbf{F}}^0$ and $\hat{\mathbf{G}}^i$ vanish. Later, however, we shall see that some of these components are zero, *but never the same ones* and therefore the conclusion remains valid. In fact, if the same components of $\hat{\mathbf{F}}^0$ and $\hat{\mathbf{G}}^i$ were to vanish, the corresponding balance equations and the system of n equations would be underdetermined. We exlude that possibility.

The field equations have the form

$$\frac{\partial F_{i_1\dots i_l}}{\partial t}+\frac{\partial F_{i_1\dots i_l}v_i+G_{ii_1\dots i_l}}{\partial x^i}=\Pi_{i_1\dots i_l}\quad (l=0,1,2,\dots,N). \tag{4.37}$$

We assume that all subsystems that result from (4.36) by ignoring the tensor equation of rank N, or N and $N-1$, or N and $N-1$ and $N-2$, etc. have Galilean invariance. Thus by the arguments of Paragraph 4.6 the matrix $\mathbf{X}(\mathbf{v})$ in (4.16) must be a subtriangular block matrix of the form

$$\mathbf{X}(\mathbf{v})=\begin{bmatrix} x & 0 & 0 & & 0\\ x_{k_1} & x^{h_1}_{k_1} & 0 & & 0\\ x_{k_1k_2} & x^{h_1}_{k_1k_2} & x^{h_1h_2}_{k_1k_2} & \cdots & 0\\ \vdots & \vdots & \vdots & & \vdots\\ x_{k_1k_2\dots k_N} & x^{h_1}_{k_1k_2\dots k_N} & x^{h_1h_2}_{k_1k_2\dots k_N} & \cdots & x^{h_1h_2\dots h_N}_{k_1k_2\dots k_N}\end{bmatrix}. \tag{4.38}$$

The blocks $x_{jl}=x^{h_1\dots h_j}_{k_1\dots k_l}$ are fully symmetric with respect to the indices $h_1\dots h_j$ and $k_1\dots k_l$.

Since $\mathbf{X}(\mathbf{v})$ is a subtriangular block matrix, so are the constant matrices $\mathbf{A}^r$, by (4.18); i.e. we have

$$\mathbf{A}^r=\begin{bmatrix} a^r & 0 & 0 & & 0\\ a^r_{k_1} & a^{rh_1}_{k_1} & 0 & & 0\\ a^r_{k_1k_2} & a^{rh_1}_{k_1k_2} & a^{rh_1h_2}_{k_1k_2} & \cdots & 0\\ \vdots & \vdots & \vdots & & \vdots\\ a^r_{k_1k_2\dots k_N} & a^{rh_1}_{k_1k_2\dots k_N} & a^{rh_1h_2}_{k_1k_2\dots k_N} & \cdots & a^{rh_1h_2\dots h_N}_{k_1k_2\dots k_N}\end{bmatrix}. \tag{4.39}$$

The block elements $a^{rh_1\dots h_j}_{k_1\dots k_l}$ must be constant tensors, and this means that they must be combinations of Kronecker symbols. Therefore the blocks with an even sum $j+l$ must be zero and, in particular, the diagonal blocks vanish, in addition to the elements in the upper triangle. But this is not all. It is possible to show that $\mathbf{A}^r$ without loss of generality can be made to contain only terms in the subdiagonal, i.e. the one immediately below the main diagonal. Moreover the entries are all pure Kronecker symbols—with coefficients equal to one. Thus the matrices $\mathbf{A}^r$ have the forms [5]

$$\mathbf{A}^r=\begin{bmatrix} 0 & 0 & 0 & & & 0\\ \delta^r_{k_1} & 0 & 0 & & & 0\\ 0 & 2\delta^{(rh_1)}_{(k_1k_2)} & 0 & \cdots & & 0\\ \vdots & \vdots & \vdots & & & \vdots\\ 0 & 0 & 0 & \cdots & N\delta^{(rh_1h_2\dots h_{N-1})}_{(k_1k_2\dots k_N)} & 0\end{bmatrix}. \tag{4.40}$$

The proof of this form of $\mathbf{A}^r$ is indicated by Ruggeri [23]. It proceeds by several steps of rescaling the quantities $\mathbf{F}^0,\mathbf{G}^i$, and $\mathbf{\Pi}$: First the entries below the subdiagonal are absorbed into terms on the subdiagonal and then the terms on the subdiagonal are all given the factor 1. The rescaling procedure does not change the only property which so far characterizes $\mathbf{F}^0,\mathbf{G}^i$, and $\mathbf{\Pi}$, namely their Galilean invariance.

[5] This is a special case of a nilpotent matrix. We have $\mathbf{A}^{k_1}\mathbf{A}^{k_2}\dots\mathbf{A}^{k_{N+1}}=0$ for all $k_1,k_2,\dots,k_{N+1}$ equal to 1, 2 or 3.

Accordingly the matrix $\mathbf{X}(\mathbf{v})$ is a polynomial matrix in $\mathbf{v}$ of order N whose diagonal elements are blocks of Kronecker symbols. It reads

$$\begin{bmatrix} 1 & & & & & & \\ v_{k_1} & \delta^{h_1}_{k_1} & & & & & \\ v_{k_1}v_{k_2} & 2\delta^{h_1}_{(k_1}v_{k_2)} & \cdots & & & & \\ v_{k_1}v_{k_2}v_{k_3} & 3\delta^{h_1}_{(k_1}v_{k_2}v_{k_3)} & \cdots & \delta^{h_1}_{k_1}\delta^{h_2}_{k_2}\delta^{h_3}_{k_3} & & & \\ \vdots & \vdots & \vdots & \vdots & & & \\ v_{k_1}v_{k_2}\cdots v_{k_n} & \binom{n}{1}\delta^{h_1}_{(k_1}v_{k_2}\cdots v_{k_n)} & \cdots & \vdots & & & \\ \vdots & \vdots & \vdots & \vdots & & \vdots & \vdots \\ v_{k_1}v_{k_2}\cdots v_{k_N} & \binom{N}{1}\delta^{h_1}_{(k_1}v_{k_2}\cdots v_{k_N)} & \cdots & \binom{N}{3}\delta^{h_1}_{(k_1}\delta^{h_2}_{k_2}\delta^{h_3}_{k_3}v_{k_4}\cdots v_{k_N)} & & \cdots & \delta^{h_1}_{k_1}\delta^{h_2}_{k_2}\cdots\delta^{h_N}_{k_N} \end{bmatrix}. \tag{4.41}$$

Thus we are able to decompose tensors of arbitrary rank into velocity-dependent and internal parts. In particular the decomposition of an arbitrary tensorial density is given by

$$\begin{aligned} F_{i_1\ldots i_l} = \hat{F}_{i_1\ldots i_l} + \binom{l}{1}\hat{F}_{(i_1\ldots i_{l-1}}v_{i_l)} & + \binom{l}{2}\hat{F}_{(i_1\ldots i_{l-2}}v_{i_{l-1}}v_{i_l)} + \cdots \\ & + \binom{l}{l-1}\hat{F}_{(i_1}v_{i_2\ldots}v_{i_l)} + \hat{F}v_{i_1\ldots}v_{i_l}. \end{aligned} \tag{4.42}$$

The same decompositions can be made for $\mathbf{G}^i$ and $\mathbf{\Pi}$.

In all physically relevant cases F and F_k are the densities of mass and momentum, while $\frac{1}{2}F_{kk}$ is the energy density. [6] Accordingly the nonconvective flux G_i and the internal part of F_k vanish: there is no nonconvective flux of mass nor does the momentum density have a velocity-independent part. Also, since mass, momentum, and energy are conserved, we have $\Pi = 0, \Pi_k = 0$, and $\Pi_{kk} = 0$.

In fact, now we proceed, for illustrative purposes, to consider a case where F, F_k, and F_{kk} are the *only* variables.

5 An Illustrative Trivial Case. Thermodynamics of an Euler Fluid

One of the simplest examples of extended thermodynamics – a case where the theory barely deserves its name – is the Euler fluid. We study it, because it serves to illustrate many aspects of the formal structure explained heretofore.

Thus here we have balance laws, convective and internal densities and fluxes, an entropy equality, Lagrange multipliers, a vector potential, explicit conditions of convexity, characteristic speeds, non-trivial subsystems and subcharacteristic conditions. In short, the theory has many of the ingredients of the general structure, – and it has the advantage that we recognize their status from ordinary equilibrium thermodynamics: kinetic and internal energy, absolute temperature, Gibbs and Helmholtz free energies, adiabatic and isothermal speeds of sound.

What the Euler fluid does not have is a nonconvective entropy flux and an entropy production. Nor does it reveal any result that we have not known before. The Euler fluid is too simple to provide new results. It merely shines a new light upon what we did know already.

---

[6]In later chapters the tensors F will be moments of the phase density. We call F_{kk} *twice* the energy density to be consistent in notation with those later chapters.

5.1 The Euler Fluid

We shall assume an even simpler case than (4.26), namely the case

$$\mathbf{F}^0 = \begin{bmatrix} F \\ F_k \\ \frac{1}{2}F_{kk} \end{bmatrix} \quad \mathbf{G}^i = \begin{bmatrix} G_i \\ G_{ik} \\ \frac{1}{2}G_{ikk} \end{bmatrix} \quad \mathbf{\Pi} = \begin{bmatrix} 0 \\ 0 \\ 0 \end{bmatrix}. \tag{5.1}$$

The five densities represent the densities of the conserved quantities mass, momentum, and energy; hence the productions are all zero.

The matrices $\mathbf{A}^r$ and $\mathbf{X}(\mathbf{v})$ read (cf. (4.39), (4.40))

$$\mathbf{A}^r = \begin{bmatrix} 0 & 0 & 0 \\ \delta^r_{k_1} & 0 & 0 \\ 0 & \delta^r_{h_1} & 0 \end{bmatrix} \quad \text{and} \quad \mathbf{X}(\mathbf{v}) = \begin{bmatrix} 1 & 0 & 0 \\ v_{k_1} & \delta^k_{k_1} & 0 \\ \frac{1}{2}v^2 & v_{h_1} & 0 \end{bmatrix} \tag{5.2}$$

so that the decompositions of $\mathbf{F}^0$ and $\mathbf{G}^i$ are given by

$$\begin{aligned} F &= \hat{F} & G_i &= \hat{G}_i \\ F_k &= \hat{F}v_k + \hat{F}_k & G_{ik} &= \hat{G}_i v_k + \hat{G}_{ik} \\ \frac{1}{2}F_{kk} &= \frac{1}{2}\hat{F}v^2 + \hat{F}_k v_k + \frac{1}{2}\hat{F}_{kk} & \frac{1}{2}G_{ikk} &= \frac{1}{2}\hat{G}_i v^2 + \hat{G}_{ik}v_k + \frac{1}{2}\hat{G}_{ikk} \end{aligned} \tag{5.3}$$

Of course, $\hat{F}_k$ and $\hat{G}_i$ are zero, because there is no nonconvective mass flux nor internal momentum density.

For easy recognition we introduce the canonical notation of thermodynamics, which may be read off from the table (5.4)

	Density	Nonconvective Flux
Mass	ρ	0
Momentum	ρv_k	$-t_{ik}$
Energy	$e + \frac{\rho}{2}v^2$	$-t_{ik}v_k + q_i$

(5.4)

ρ is the mass density, and e is the internal energy density. t_{ik}, the internal momentum flux—to within sign—is called the stress, and q_i is the internal energy flux, or heat flux.

The balance laws read

$$\begin{aligned} &\frac{\partial \rho}{\partial t} + \frac{\partial \rho v_j}{\partial x_j} = 0 \\ &\frac{\partial \rho v_i}{\partial t} + \frac{\partial \rho v_i v_j - t_{ij}}{\partial x_j} = 0 \\ &\frac{\partial\left(e + \frac{\rho}{2}v^2\right)}{\partial t} + \frac{\partial\left(e + \frac{\rho}{2}v^2\right)v_j - t_{ij}v_i + q_j}{\partial x_j} = 0. \end{aligned} \tag{5.5}$$

Constitutive equations must be written for the internal fluxes. We recall that these may only depend—in extended thermodynamics – on the present and local values of the fields, and that the internal quantities are independent of the velocity. Thus the constitutive relations read

$$t_{ik} = t_{ik}(\rho, e), \qquad q_i = q_i(\rho, e).$$

Since t_{ik} and q_i are Galilean tensors, dependent only on the Galilean scalars ρ, e, we obtain from this

$$t_{ik} = -p(\rho, e)\, \delta_{ik}, \qquad q_i = 0, \tag{5.6}$$

where $p(\rho, e)$ is the pressure, which is now seen to be the only remaining constitutive quantity in (5.5). The constitutive equations (5.6) are those of an Euler fluid, a fluid incapable of supporting shear stresses or conducting heat.

The entropy inequality has its usual form

$$\frac{\partial h^0}{\partial t} + \frac{\partial h^0 v^i + \varphi^i}{\partial x^i} = \Sigma \geq 0 \tag{5.7}$$

and it introduces two more constitutive equations

$$h^0 = h^0(\rho, e) \qquad \text{and} \qquad \varphi^i = 0; \tag{5.8}$$

the internal flux φ^i vanishes, just as q_i does; see $(5.6)_2$.

5.2 Lagrange Multipliers

In order to exploit the entropy inequality we introduce Lagrange multipliers, as explained in Paragraph 2.1, and write

$$\begin{aligned} \frac{\partial h^0}{\partial t} + \frac{\partial h^0 v_j}{\partial x_j} &-\Lambda^\rho \left(\frac{\partial \rho}{\partial t} + \frac{\partial \rho v_j}{\partial x_j}\right) - \\ &-\Lambda^{v_i} \left(\frac{\partial \rho v_i}{\partial t} + \frac{\partial \rho v_i v_j - t_{ij}}{\partial x_j}\right) - \\ &-\Lambda^\varepsilon \left(\frac{\partial (e + \frac{\rho}{2} v^2)}{\partial t} + \frac{\partial (e + \frac{\rho}{2} v^2) v_j - t_{ij} v_i + q_j}{\partial x_j}\right) = \Sigma \geq 0. \end{aligned} \tag{5.9}$$

The left-hand side is linear in the derivatives

$$\frac{\partial \rho}{\partial t}, \frac{\partial v_i}{\partial t}, \frac{\partial e}{\partial t}, \frac{\partial \rho}{\partial x_j}, \frac{\partial v_i}{\partial x_j}, \frac{\partial e}{\partial x_j}$$

and these may be chosen arbitrarily. It follows that the factors of these derivatives must vanish, lest the inequality be violated. Hence follows after a brief calculation

$$\Lambda^{v_i} = -\Lambda^\varepsilon v_i, \qquad \text{since } \frac{\partial h^0}{\partial v_i} = 0, \tag{5.10}$$

$$dh^0 = \left(\Lambda^\rho - \Lambda^\varepsilon \frac{1}{2} v^2\right) d\rho + \Lambda^\varepsilon\, de \tag{5.11}$$

$$d(h^0 v_j) = \left(\Lambda^\rho - \Lambda^\varepsilon \frac{1}{2} v^2\right) v_j d\rho + \left(\rho \Lambda^\rho + \Lambda^\varepsilon \left(e + p - \frac{\rho}{2} v^2\right)\right) dv_j + \Lambda^\varepsilon v_j de \tag{5.12}$$

which is the explicit form of (2.4) for the present case.

Nothing remains on the left hand side of (5.9), so that $\Sigma = 0$ holds: There is no entropy production in an Eulerian fluid, since there are no productions in the balance equations of mass, momentum, and energy.

The equations (5.10) through (5.11) exhaust the results that can be obtained from the entropy inequality, and we conclude that nothing is learned about the only constitutive function of this case, viz. $p(\rho, e)$. Indeed, all we can learn from these results are relations between the Lagrange multipliers Λ^{ρ}, Λ^{v_i}, and Λ^{ε}, and a differential equation for h^0 in terms of Λ^{ε}. We have

$$\rho\Lambda^{\rho} = -\Lambda^{\varepsilon}\left(e + p - \frac{1}{\Lambda^{\varepsilon}}\, h^0 - \frac{\varrho}{2}\, v^2\right),\ \Lambda^{v_i} = -\Lambda^{\varepsilon} v_i,\ d\left(\frac{h^0}{\rho}\right) = \Lambda^{\varepsilon}\left(d\left(\frac{e}{\rho}\right) + pd\left(\frac{1}{\rho}\right)\right). \tag{5.13}$$

5.3 Internal Lagrange Multipliers

It is instructive to write the alternative (4.23) of the system (5.5), because this is also well-known in thermodynamics, at least in this special case. We obtain from (5.2) and (5.3) and with the notation (5.4)

$$\begin{aligned} &\frac{d\varrho}{dt} + \varrho\frac{\partial v_i}{\partial x_i} &&= 0 \\ &-\frac{\partial t_{ik}}{\partial x_i} + \left\{\varrho\frac{dv_r}{dt}\right\} &&= 0 \\ &\frac{de}{dt} + e\frac{\partial v_i}{\partial x_i} + \frac{\partial q_i}{\partial x_i} + \left\{-t_{ir}\frac{\partial v_r}{\partial x_i}\right\} &&= 0. \end{aligned} \tag{5.14}$$

The gaps in this arrangement are due to the form of $\mathbf{A}^r$ and to the fact that $\hat{F}_k = 0$ and $\hat{G}_i = 0$ hold. we recognize in (5.14) familiar forms of the balance of mass, momentum and *internal* energy.

Seeing that the system (4.20) is equivalent to (4.22) we may rewrite (5.9) in the form, with $\hat{\mathbf{\Lambda}} = \mathbf{\Lambda X}$, cf. (4.33), as internal Lagrange multipliers.

$$\begin{aligned} \frac{\partial h^0}{\partial t} + \frac{\partial h^0 v_j}{\partial x_j} \quad &-\hat{\Lambda}^{\rho}\left(\frac{d\rho}{dt} + \rho\frac{\partial v_i}{\partial x_i}\right) - \\ &-\hat{\Lambda}^{v_i}\left(\rho\frac{dv_i}{dt} - \frac{t_{ij}}{\partial x_j}\right) - \\ &-\hat{\Lambda}^{\varepsilon}\left(\frac{de}{dt} + e\frac{\partial v_i}{\partial x_i} + \frac{\partial q_i}{\partial x_i} - t_{ir}\frac{\partial v_r}{\partial x_i}\right) \geq 0. \end{aligned} \tag{5.15}$$

By $(5.2)_2$ and (5.13) the internal Lagrange multipliers are

$$\begin{aligned} \hat{\Lambda}^{\rho} &= \Lambda^{\rho} + \Lambda^{v_i} v_i + \Lambda^{\varepsilon}\frac{1}{2}v^2 = \Lambda^{\rho} - \frac{1}{2}\Lambda^{\varepsilon} v^2 \\ \hat{\Lambda}^{v_i} &= \Lambda^{v_i} + \Lambda^{\varepsilon} v_i = 0 \\ \hat{\Lambda}^{\varepsilon} &= \Lambda^{\varepsilon} \end{aligned} \tag{5.16}$$

so that the constraints (5.13) assume the simpler form

$$\rho\hat{\Lambda}^{\rho} = -\hat{\Lambda}^{\varepsilon}\left(e + p - \frac{1}{\Lambda^{\varepsilon}}\, h^0\right),\ \hat{\Lambda}^{v_i} = 0,\ d\left(\frac{h^0}{\rho}\right) = \hat{\Lambda}^{\varepsilon}\left(d\frac{e}{\rho}\right) + pd\left(\frac{1}{\rho}\right). \tag{5.17}$$

These conditions could also be derived from (5.15) by direct exploitation of the inequality (5.15). In some ways this is easier than the exploitation of (5.9), because the velocity does not get in the way so much.

5.4 Absolute Temperature

We conclude from (5.13) that Λ^{ε} is an integrating factor of

$$d\left(\frac{e}{\rho}\right) + pd\left(\frac{1}{\rho}\right).$$

We know from classical thermodynamics (e.g. see Carathéodory [53]) that among all integrating factors, *the one* that has the specific entropy h^0/ρ as its generator function is equal to $1/T$, the reciprocal of the absolute temperature.

Thus $(5.13)_3$ becomes the Gibbs equation

$$d\left(\frac{h^0}{\rho}\right) = \frac{1}{T}\left(d\left(\frac{e}{\rho}\right) + pd\left(\frac{1}{\rho}\right)\right) \quad \text{or} \quad dh^0 = \frac{1}{T}(de - gd\rho) \tag{5.18}$$

and all Lagrange multipliers are related to T, viz.

$$\Lambda^{v^i} = -\frac{1}{T}v_i, \quad \Lambda^{\rho} = -\frac{g - \frac{1}{2}v^2}{T}, \quad \Lambda^{e} = \frac{1}{T}, \tag{5.19}$$

where $\rho g = e - Th^0 + p$ is the density of *free enthalpy*, or Gibbs free energy; g is sometimes called the chemical potential.

In thermodynamics it is convenient and customary to introduce specific quantities, like s for entropy and ε for internal energy. Thus $h^0 = \rho s$ and $e = \rho\varepsilon$ and (5.19) assumes the form

$$ds = \frac{1}{T}\left(d\varepsilon - \frac{p}{\rho^2}d\rho\right).$$

Also it is common to use T as a variable instead of e or ε. In that case (5.19) reads

$$ds = \frac{1}{T}\left(\varepsilon_T dT + \left(\varepsilon_\rho - \frac{p}{\rho^2}\right)d\rho\right). \tag{5.20}$$

This equation implies the important integrability condition

$$\rho^2\varepsilon_\rho = p - Tp_T. \tag{5.21}$$

which permits the calculation of ε as function of ρ, if $p(\rho,T)$ is given. ε_ρ and p_T in (5.21) are derivatives of $\varepsilon(\rho,T)$ and $p(\rho,T)$ with respect to ρ and T.

5.5 Vector Potential

Knowing the Lagrange multipliers, we may now calculate the vector potential $h'^A = -h^A + \mathbf{\Lambda}\cdot\mathbf{F}^A$; see (2.7). A quick calculation shows that we have

$$h'^0 = -\frac{p}{T} \qquad h'^i = -\frac{p}{T}v^i. \tag{5.22}$$

The special form of $(2.9)_1$ for the present case reads

$$\begin{array}{l|l|l}
\rho = \dfrac{\partial\left(-\frac{p}{T}\right)}{\partial\left(-\frac{g-\frac{1}{2}v^2}{T}\right)} & \rho v_j = \dfrac{\partial\left(-\frac{p}{T}\right)}{\partial\left(-\frac{1}{T}v_j\right)} & e + \frac{\rho}{2}v^2 = \dfrac{\partial\left(-\frac{p}{T}\right)}{\partial\,\frac{1}{T}} \\
\rho v^i = \dfrac{\partial\left(-\frac{p}{T}v^i\right)}{\partial\left(-\frac{g-\frac{1}{2}v^2}{T}\right)} & \rho v_i v_j + p\delta_{ij} = \dfrac{\partial\left(-\frac{p}{T}v^i\right)}{\partial\left(-\frac{1}{T}v_j\right)} & \left(e + \frac{\rho}{2}v^2 + p\right)v_i = \dfrac{\partial\left(-\frac{p}{T}v_i\right)}{\partial\,\frac{1}{T}}.
\end{array} \tag{5.23}$$

In all derivatives two of the three main field variables $-\frac{g-\frac{1}{2}v^2}{T}, -\frac{\mathbf{v}}{T}, \frac{1}{T}$ are kept constant.

5.6 Convexity

We recall the requirement of convexity (1.6), which states in the present case that the matrix

$$\begin{vmatrix} \frac{\partial^2 h^0}{\partial \rho^2} & \frac{\partial^2 h^0}{\partial \rho \partial e} \\ \frac{\partial^2 h^0}{\partial e \partial \rho} & \frac{\partial^2 h^0}{\partial e^2} \end{vmatrix} \text{ – be negative definite.}$$

By (5.18) this requirement implies the relations

$$\left(\frac{\partial e}{\partial T}\right)_\rho > 0 \quad \left(\frac{\partial \frac{g}{T}}{\partial \rho}\right)_e > 0 \quad \left(\frac{\partial \frac{g}{T}}{\partial \rho}\right)_e \left(\frac{\partial T}{\partial e}\right)_\rho - \frac{1}{T^2}\left(\frac{\partial T}{\partial \rho}\right)_e^2 > 0 \tag{5.24}$$

which are the "stability conditions" of thermodynamics. In more conventional notation these conditions imply the positivity of specific heat and compressibility, viz.

$$\left(\frac{\partial \varepsilon}{\partial T}\right)_\rho > 0 \quad \left(\frac{\partial p}{\partial \rho}\right)_T > 0. \tag{5.25}$$

5.7 Characteristic Speed

The Euler fluid is governed by the equations (5.5), (5.6), or

$$\begin{aligned} \frac{\partial \rho}{\partial t} &+ v_j \frac{\partial \rho}{\partial x_j} + \rho \frac{\partial v_j}{\partial x_j} = 0 \\ \frac{\partial v_i}{\partial t} &+ v_j \frac{\partial v_i}{\partial x_j} + \frac{1}{\rho}\frac{\partial p}{\partial x_i} = 0 \\ \frac{\partial \varepsilon}{\partial t} &+ v_j \frac{\partial \varepsilon}{\partial x_j} + \frac{p}{\rho}\frac{\partial v_j}{\partial x_j} = 0. \end{aligned} \tag{5.26}$$

We introduce $p(\rho, T)$ and $\varepsilon(\rho, T)$ and calculate the characteristic speeds. According to the number of equations there are five of them, which are easily calculated from the characteristic equation, viz.

$$V_{\min} = v_i n_i - \sqrt{p_\rho + \frac{T}{\rho^2}\frac{p_T^2}{\varepsilon_T}}, \quad V_{-1,0,+1} = v_i n_i, \quad V_{\max} = v_i n_i + \sqrt{p_\rho + \frac{T}{\rho^2}\frac{p_T^2}{\varepsilon_T}}. \tag{5.27}$$

The speeds $(5.27)_{1,3}$ are called the *adiabatic* speeds of sound; this is the sound speed usually observed in low-frequency sound propagation for which heat conduction is negligible.

5.8 Subsystems

Simple and thermodynamically trivial as the Euler fluid may be, its 5 equations imply 30 subsystems, or six of them, if we take $\mathbf{v}$ as *one* variable. We recall that in a subsystem we fix a Lagrange multiplier or Lagrange multipliers to be constants and ignore the corresponding field equation. For fields like (4.36) or (5.1), with a natural order the physicist will be interested in studying the system that results from elimination of the last equation or equations.

Thus if we fix $\Lambda^{\varepsilon} = \frac{1}{T}$, we ignore the last equation, which is the energy equation, and we are left with the conservation laws of mass and momentum appropriate to an *isothermal fluid*, viz.

$$\begin{aligned} \frac{\partial \rho}{\partial t} + v_j \frac{\partial \rho}{\partial x_j} + \rho \frac{\partial v_j}{\partial x_j} &= 0 \\ \frac{\partial v_i}{\partial t} + v_j \frac{\partial v_i}{\partial x_j} + \frac{\partial p(\rho, T^*)}{\partial x_i} &= 0. \end{aligned} \tag{5.28}$$

We use the notation of Section 3 and calculate characteristic speeds, of which there are four now, viz.

$$\overline{V}_{\min} = v_i n_i - \sqrt{p_\rho}, \qquad \overline{V}_{-1,+1} = v_i n_i, \qquad \overline{V}_{\max} = v_i n_i + \sqrt{p_\rho}. \tag{5.29}$$

Obviously the extreme speeds are smaller than the ones for the full system (5.26), which confirms the general result (3.8) for main subsystems. The speed $\overline{V} - v_i n_i$ is sometimes called the *isothermal* sound speed.

According to (3.4) and with (5.19), (5.22), (5.23) we calculate the subentropy and the subentropy flux for the subsystem to have the forms

$$\begin{aligned} \overline{h}^0 &= h^0(\rho, T^*) - \frac{1}{T^*}\left(e(\rho, T^*) + \frac{\rho}{2} v^2\right) \\ \overline{h}^i &= \left\{ h^0(\rho, T^*) - \frac{1}{T^*}\left(e(\rho, T^*) + \frac{\rho}{2} v^2 + p(\rho, T^*)\right)\right\} v_i. \end{aligned} \tag{5.30}$$

Therefore the subentropy law reads in this case

$$\frac{\partial\left(f(\rho, T^*) + \frac{\rho}{2} v^2\right)}{\partial t} + \frac{\partial\left(f(\rho, T^*) + \frac{\rho}{2} v^2\right) v_i + p(\rho, T^*) v_i}{\partial x_i} = 0. \tag{5.31}$$

It represents the equation of balance or, in fact, the conservation law for the *free energy*, or Helmholtz free energy $f = e - T h^0$.

We obtain an even more restricted subsystem if we fix $\Lambda^{\varepsilon} = \frac{1}{T}$ *and* $\Lambda^{v^i} = -\frac{v^i}{T}$ to be constants. In this case the only surviving equation of our system is the mass balance for a fluid in isochoric flow

$$\frac{\partial \rho}{\partial t} + v_j^* \frac{\partial \rho}{\partial x_j} = 0. \tag{5.32}$$

The single characteristic speed is

$$\overline{V} = v_i^* n_i \tag{5.33}$$

and obviously this is smaller than the extreme speeds (5.29) and (5.27) of either the isothermal fluid or the full Euler fluid. That relation again illustrates the general result (3.7).

In this case of the isochoric fluid the subentropy law is a simple restatement of the single surviving equation, viz. (5.32). We do not write it down. Nor are other subsystems investigated. There seems no need for that and, besides, we lose Galilean invariance, unless we arrive at a subsystem by elimination of the last equation, or equations; cf. Paragraph 4.6.

5.9 Discussion

While the Euler fluid illustrates, in conventional terms, much of the formal synthetic structure of rational extended thermodynamics, it does not give new results. Extended thermodynamics comes into its own only when irreversible phenomena are considered, i.e. for dissipative materials. The simplest nontrivial case is a viscous, heat-conducting gas, and we proceed to consider that.

Chapter 4

Extended Thermodynamics of Viscous, Heat-Conducting Monatomic Gases

The new version of extended thermodynamics to be described in this chapter is based on Rational Thermodynamics whereas the old version was based on TIP. There are several ingredients that make rational thermodynamics more powerful than TIP, viz.

i.) the explicit recognition that the entropy inequality must hold for all solutions of the field equations and that the constitutive relations must ensure this validity; this was clearly expressed by Coleman & Noll [17] and Coleman & Mizel [53].

ii.) the realization that the entropy flux is a constitutive quantity in its own right unrelated – a priori – to the heat flux divided by temperature. This was proposed by Müller [18].

iii.) the use of Lagrange multipliers in the exploitation of the entropy inequality as proved by Liu [20].

iv.) the recognition that first order field equations of balance type form a symmetric hyperbolic system, if the Lagrange multipliers are chosen as variables. This was shown by Ruggeri [49].

After all this was known, it was possible to revisit the problem of the formulation of extended thermodynamics in a systematic way. Liu & Müller [21] formulated that theory for monatomic gases. The theory considers the thirteen variables $\varrho, v_i, \varepsilon, t_{<ij>}$ and q_i as basic and derives field equations from the conservation laws of mass, momentum, energy and from balance equations for the fluxes of momentum and energy. It is capable of calculating specific forms for all constitutive coefficients except two and those two can be related to the measurable values of viscosity and heat conductivity.

Full agreement is achieved with the results of the kinetic theory of gases in the Grad 13-moment approximation even though extended thermodynamics is a macroscopic phenomenological theory with no explicit reference to the kinetic theory in its tenets. In particular ***it is not assumed that the variables are moments.*** The entropy and its flux are proved to contain the non-linear terms in $t_{<ij>}$ and q_i postulated by extended TIP and the corresponding coefficients can be calculated explicitly.

Extended thermodynamics also derives the inertial terms in the equations for stress and heat flux which we noted in the kinetic theory.

The theory applies to all mon-atomic gases with a thermal equation of state of the form

$$p = T^{5/2} F\left(\frac{\varrho}{T^{3/2}}\right)$$

where F is some function. This includes the classical ideal gas as well as degenerate gases of Bose and Fermi particles.

1 The Equations of Extended Thermodynamics of Monatomic Gases

1.1 Thermodynamic Processes

Extended thermodynamics of viscous heat conducting fluids is a field theory whose principal objective is the determination of the thirteen fields of densities

$$\begin{array}{rl} \text{mass density} & F = \varrho \\ \text{momentum density} & F_i = \varrho v_i \\ \text{momentum flux density} & F_{ij} \\ \text{energy flux density} & \frac{1}{2} F_{ppi}.^{1} \end{array} \tag{1.1}$$

The necessary field equations are based on the equations of balance of mass, momentum, momentum flux, and energy flux, viz.

$$\begin{aligned} \frac{\partial F}{\partial t} &+ \frac{\partial F_k}{\partial x_k} &= 0 \\ \frac{\partial F_i}{\partial t} &+ \frac{\partial F_{ik}}{\partial x_k} &= 0 \\ \frac{\partial F_{ij}}{\partial t} &+ \frac{\partial F_{ijk}}{\partial x_k} &= S_{<ij>} \\ \frac{\partial F_{ppi}}{\partial t} &+ \frac{\partial F_{ppik}}{\partial x_k} &= S_{ppi}. \end{aligned} \tag{1.2}$$

All tensors are symmetric and $S_{<ij>}$ is traceless, because the trace of $(1.2)_3$ represents the conservation law of energy.

In order to close this system we need constitutive relations for the quantities

$$F_{<ijk>}, F_{ppik}, S_{<ij>}, S_{ppi}. \tag{1.3}$$

$F_{<ijk>}$ is the traceless part of F_{ijk}. In extended thermodynamics the constitutive quantities (1.3) at one point and time depend on the values of the fields (1.1) at that point and time; so that we have

$$\begin{aligned} F_{<ijk>} &= \hat{F}_{<ijk>} & (F, F_i, F_{ij}, F_{ppi}) \\ F_{ppik} &= \hat{F}_{ppik} & (\underline{\qquad\qquad}) \\ S_{<ij>} &= \hat{S}_{<ij>} & (\underline{\qquad\qquad}) \\ S_{ppi} &= \hat{S}_{ppi} & (\underline{\qquad\qquad}). \end{aligned} \tag{1.4}$$

If we know the constitutive functions $\hat{F}_{<ijk>}$ through $\hat{S}_{ppi}$ we can eliminate $F_{<ijk>}, F_{ppik}, S_{<ij>}$ and S_{ppi} between (1.3) and (1.4) and obtain a system of field equations. This is a system of quasi-linear first-order differential equations. Each solution is called a *thermodynamic process.*

[1]We use this synthetic notation in order to illustrate the simple structure of the theory. The notation recalls that in the kinetic theory the fields (1.1) are moments, although that fact is not used explicitly in this chapter.

1.2 Discussion

The balance equations (1.2) are rather special in several ways: First the flux in one equation appears as the density in the subsequent one. In particular this means that the energy density e is related to the trace F_{ii} of the momentum flux by $e = \frac{1}{2}F_{ii}$. This is a relation which we know to be true for monatomic gases, but not otherwise. Secondly the symmetry of all tensors is a restrictive assumption. It may not hold true for real gases and liquids, in particular in regard to F_{ijk}. Body forces and radiation have been ignored in the formulation of the balance equations (1.2) and the equations are written in an inertial frame so that there are no inertial contributions. The form of such inertial contributions can be determined after the velocity dependence of the fluxes—as dictated by Galilean invariance—has been established. We proceed to do this.

1.3 Galilean Invariance. Convective and Nonconvective Fluxes

We require that the balance laws (1.2) hold in every inertial frame so that they must be invariant under a Galilean transformation i.e. a time-independent rotation O_{ij} and a translation with the constant velocity c_i, viz.

$$x_i^* = O_{ij}x_j + c_i t. \tag{1.5}$$

This requirement will permit us to determine the velocity dependence of the $\mathbf{F}'s$ and $\mathbf{S}'s$ explicitly. We shall prove that the following decompositions are valid:

$$\begin{aligned}
F &= \varrho \\
F_i &= \quad +\varrho v_i \\
F_{ij} &= \varrho_{ij} + \quad +\varrho v_i v_j \\
F_{ijk} &= \varrho_{ijk} + 3\varrho_{(ij}v_{k)} + \quad +\varrho v_i v_j v_k \\
F_{ppij} &= \varrho_{ppij} + 4\varrho_{(ijp}v_{p)} + 6\varrho_{(ip}v_p v_{j)} \quad + \varrho v^2 v_i v_j \\
S_{\langle ij\rangle} &= s_{\langle ij\rangle} \\
S_{ppi} &= s_{ppi} + 2s_{\langle ip\rangle}v_p.
\end{aligned} \tag{1.6}$$

The quantities $\boldsymbol{\varrho}$ and $\mathbf{s}$ in (1.6) are Galilean tensors so that their components in the two inertial frames are related by the equations

$$\overset{*}{\varrho}_{i_1 i_2 \ldots i_n} = O_{i_1 j_1} \ldots O_{i_n j_n} \varrho_{j_1 \ldots j_n} \quad \text{and} \quad \overset{*}{s}_{i_1 \ldots i_n} = O_{i_1 j_1} \ldots O_{i_n j_n} s_{j_1 \ldots j_n}. \tag{1.7}$$

The general proof of the decomposition (1.6) was given in Chapter 3 for systems of an arbitrary number of equations. Here we give a short inductive proof appropriate for the simple system of equations (1.2).

The relations $(1.6)_{1,2}$ require no proof; after all, F and F_i are the densities of mass and momentum. For the proof of the other relations (1.6) it is convenient to decompose the $\mathbf{F}'s$ into convective and nonconvective parts according to

$$F_{i_1 \ldots i_n k} = F_{i_1 \ldots i_n} v_k + G_{i_1 \ldots i_n k}. \tag{1.8}$$

For $n = 0$ we have

$$G_k = 0, \tag{1.9$_1$}$$

because there is no nonconvective flux of mass nor is there a momentum density in the rest frame. Therefore the mass balance $(1.2)_1$ may be written in the form

$$\dot{\varrho} + \varrho \frac{\partial v_k}{\partial x_k} = 0. \tag{1.10$_1$}$$

All terms are Galilean scalars so that this equation holds in all inertial frames. By use of $(1.9)_1$, $(1.10)_1$ the momentum balance $(1.2)_2$ assumes the form

$$\varrho \dot{v}_i + \frac{\partial G_{ik}}{\partial x_k} = 0. \tag{1.10$_2$}$$

The first term is a Galilean vector, so that G_{ik} must be a Galilean tensor, if the momentum balance is to be valid in all inertial frames. We emphasize this conclusion in the notation by writing.

$$G_{ik} = \varrho_{ik}. \tag{1.9$_2$}$$

By use of $(1.9)_{1,2}, (1.10)_{1,2}$ we write the momentum flux balance $(1.2)_3$ in the form

$$\varrho \left[\left(\frac{\varrho_{ij}}{\varrho} \right)^{\bullet} + 2 \frac{\varrho_{k(i}}{\varrho} \frac{\partial v_{j)}}{\partial x_k} \right] + \frac{\partial (G_{ijk} - 2v_{(i}\varrho_{j)k})}{\partial x_k} = S_{<ij>}. \tag{1.10$_3$}$$

The square bracket is a Galilean tensor, so that $G_{ijk} - 2v_{(i}\varrho_{j)k}$ and $S_{<ij>}$ must also be Galilean tensors, if Galilean invariance of the balance law is required. We therefore write

$$G_{ijk} = \varrho_{ijk} + 2v_{(i}\varrho_{j)k} \qquad \text{and} \qquad S_{<ij>} = s_{<ij>}. \tag{1.9$_3$}$$

By use of $(1.9)_{1,2,3}$ and $(1.10)_{1,2,3}$ the energy flux balance $(1.2)_4$ may be rewritten as

$$\begin{aligned} \varrho \left[\left(\frac{\varrho_{ppi}}{\varrho} \right)^{\bullet} + \frac{\varrho_{ppk}}{\varrho} \frac{\partial v_i}{\partial x_k} \right] &+ 2\varrho_{ijk} \frac{\partial v_j}{\partial x_k} - 2 \frac{\varrho_{ij}}{\varrho} \frac{\partial \varrho_{jk}}{\partial x_k} - \frac{\varrho_{pp}}{\varrho} \frac{\partial \varrho_{ik}}{\partial x_k} + \\ &+ \frac{\partial \left(G_{ppik} - 2\varrho_{kij}v_j - \varrho_{ppk}v_i - \varrho_{ki}v^2 - \varrho_{kl}v_l v_i \right)}{\partial x_k} = S_{ppi} - 2S_{<ij>}v_j. \end{aligned} \tag{1.10$_4$}$$

Once again, the first line is a Galilean vector. Therefore, Galilean invariance of the balance requires that $G_{ppik} - 2\varrho_{kij}v_j - \varrho_{ppk}v_i - \varrho_{ki}v^2 - \varrho_{kl}v_l v_i$ and $S_{ppi} - 2S_{<ij>}v_j$ be Galilean tensors. We denote them by ϱ_{ppik} and s_{ppi} and have

$$G_{ppik} = \varrho_{ppik} + 2\varrho_{kij}v_j + \varrho_{ppk}v_i + \varrho_{ki}v^2 + \varrho_{kl}v_l v_i \quad \text{and} \quad S_{ppi} = s_{ppi} + 2S_{<ij>}v_j. \tag{1.9$_4$}$$

The combination of (1.8) with the relations $(1.9)_{1-4}$ provides the decomposition (1.6) after an easy rearrangement.

Some of the ϱ's are well-known quantities in continuum mechanics and thermodynamics. For those we shall replace the synthetic notation in much of the sequel and use the canonical letters. Thus

$$t_{ij} = -\varrho_{ij} \qquad \varrho\varepsilon = \frac{1}{2}\varrho_{ii} \qquad q_i = \frac{1}{2}\varrho_{ppi} \tag{1.11}$$

are the stress, the internal energy density, and the heat flux. The pressure p is defined as $-\frac{1}{3}t_{ii}$ and therefore we conclude from $(1.11)_{1,2}$

$$p = \frac{2}{3}\varrho\varepsilon, \tag{1.12}$$

a relation that is typical for monatomic gases.

1.4 Euclidean Invariance. Inertial Effects

We require that the fluxes $\boldsymbol{\varrho}$ and productions $\mathbf{s}$ introduced in (1.6) be objective tensors rather than merely Galilean tensors; i.e. in an Euclidean transformation

$$x_i^* = O_{ij}(t)x_j + b_i(t) \tag{1.13}$$

they transform according to the rule

$$\varrho^*_{i_1 \dots i_n} = O_{i_1 j_1} \dots O_{i_n j_n} \varrho_{j_1 \dots j_n} \quad \text{and} \quad s^*_{i_1 \dots i_n} = O_{i_1 j_1} \dots O_{i_n j_n} s_{j_1 \dots j_n}. \tag{1.14}$$

The Euclidean transformation represents a change of frame from an inertial, or Galilean frame to a noninertial, or Euclidean frame. $\mathbf{O}(t)$ is a time-dependent rotation and $\mathbf{b}(t)$ describes the time-dependent relative translation of the frames.

We shall prove that in a noninertial frame the equations of balance (1.2) assume the following forms

$$\begin{aligned}
&\frac{\partial F^*}{\partial t^*} + \frac{\partial F^*_k}{\partial x^*_k} &&= 0 \\
&\frac{\partial F^*_i}{\partial t^*} + \frac{\partial F^*_{ik}}{\partial x^*_k} - F^* i^0_i - 2F^*_k W_{ik} &&= 0 \\
&\frac{\partial F^*_{ij}}{\partial t^*} + \frac{\partial F^*_{ijk}}{\partial x^*_k} - 2F^*_{(i} i^0_{j)} - 4F^*_{k(i} W_{j)k} &&= S^*_{\langle ij \rangle} \\
&\frac{\partial F^*_{ppi}}{\partial t^*} + \frac{\partial F^*_{ppik}}{\partial x^*_k} - 3F^*_{(ij} i^0_{j)} - 6F^*_{k(ij} W_{j)k} &&= S^*_{ppi}.
\end{aligned} \tag{1.15}$$

$\mathbf{W} = \dot{\mathbf{O}}\mathbf{O}^T$ is the matrix of angular velocity of the noninertial frame with respect to an inertial one and $\mathbf{i}^0$ is the velocity-independent part of the specific inertial force, namely

$$i^0_i = -2W_{ik}\dot{b}_k - W^2_{ik}(x^*_k - b_k) + \dot{W}_{ik}(x^*_k - b_k) + \ddot{b}_i. \tag{1.16}$$

Note that except for the mass balance all balance equations have inertial contributions. The inertial force in $(1.15)_2$ is well known; it consists of the Coriolis, centrifugal, and Euler forces and of the force of relative translation.

For the proof of the relations (1.15) we must realize that by (1.13) the velocity transforms according to the rule

$$v_i = O_{ji}(v^*_j - R_j), \qquad \text{where} \qquad R_j = W_{jk}(x^*_k - b_k) + \dot{b}_j. \tag{1.17}$$

It is then easy to show from (1.6) that the fluxes $\mathbf{F}$ and productions $\mathbf{S}$ transform as follows.

$$\begin{aligned}
F &= F^* \\
F_i &= O_{ri} (F^*_r - F^* R_r) \\
F_{ij} &= O_{ri}O_{sj} (F^*_{rs} - 2F^*_{(r} R_{s)} + F^* R_r R_s) \\
F_{ijk} &= O_{ri}O_{sj}O_{tk} (F^*_{rst} - 3F^*_{(rs} R_{t)} + 3F^*_{(r} R_s R_{t)} - F^* R_r R_s R_t) \\
F_{ppij} &= O_{ri}O_{sj} (F^*_{pprs} - 4F^*_{(rsp} R_{p)} + 6F^*_{(rp} R_p R_{s)} + 6F^*_{(r} R_s R_p R_{p)} + F^* R_r R_s R_p R_p) \\
S_{\langle ij \rangle} &= O_{ri}O_{sj} S^*_{\langle rs \rangle} \\
S_{ppi} &= O_{ri} (S^*_{ppr} - 2S^*_{\langle rp \rangle} R_p).
\end{aligned} \tag{1.18}$$

Space- and time derivatives in the two frames are related by the equations

$$\frac{\partial}{\partial x_k} = O_{nk}\frac{\partial}{\partial x_n^*} \qquad \text{and} \qquad \frac{\partial}{\partial t} = \frac{\partial}{\partial t^*} + R_l\frac{\partial}{\partial x_l^*}. \tag{1.19}$$

We insert the transformation rules (1.18) and (1.19) into the equations of balance (1.2) and obtain the equations of balance (1.15), appropriate to a noninertial frame. The calculation is straightforward; we need only realize that we have to set

$$\frac{\partial R_j}{\partial x_k^*} = W_{jk} \qquad \text{and} \qquad \frac{\partial R_j}{\partial t^*} = i_j^0 + W_{jk}R_k. \tag{1.20}$$

2 Constitutive Theory

2.1 Restrictive Principles

We recall that the system of balance equations (1.2) is closed by the constitutive equations (1.4). Thus, if the constitutive functions $\hat{\mathbf{F}}$ and $\hat{\mathbf{S}}$ were known, we should be able, in principle, to calculate thermodynamic processes, that is solutions of given initial and boundary value problems.

In reality, of course, it is difficult to know the constitutive relations and therefore the largest part of the work of thermodynamicists consists of attempts to restrict the generality of the constitutive functions. Ideally those functions might be reduced to some scalar coefficients which the experimentalist may measure. Such efforts form the subject of the thermodynamic constitutive theory.

The tools of the constitutive theory are universal physical principles which we rely on from long experience. The most important ones of these principles are

i.) the entropy principle,

ii.) the principle of relativity (material frame indifference),

iii.) the requirement of convexity.

We proceed to formulate and exploit these requirements.

The *entropy principle* provides the balance equation

$$\frac{\partial h}{\partial t} + \frac{\partial h_i}{\partial x_i} = \Sigma \geq 0 \tag{2.1}$$

which must be satisfied for all thermodynamic processes, i.e. solutions of the field equations which result from (1.2), (1.4). The entropy density h and the entropy flux h_i are constitutive quantities, and we write generic constitutive relations of the form (1.4), i.e.

$$\begin{aligned} h &= \hat{h}\ (F, F_i, F_{ij}, F_{ppi}) \\ h_i &= \hat{h}_i(\underline{\qquad\qquad}). \end{aligned} \tag{2.2}$$

The entropy density has no convective term. The entropy flux is decomposed into a convective and an internal part as follows:

$$h_i = hv_i + \varphi_i. \tag{2.3}$$

Thus we may write the inequality (2.1) in the form

$$\varrho\left(\frac{h}{\varrho}\right)^{\bullet} + \frac{\partial \varphi_i}{\partial x_i} = \Sigma \geq 0. \tag{2.4}$$

We assume that this inequality holds in every Euclidean frame and conclude that h and Σ are objective scalars while φ_i is an objective vector.

By (1.6) and (2.3) we have a one-to-one correspondence between $F, F_i, F_{ij}, F_{ijk}, F_{ppik}, S_{<ij>}, S_{ppi}, h$ and h_i and the corresponding quantities $\varrho, v_i, \varrho_{ij}, \varrho_{ijk}, \varrho_{ppik}, s_{\langle ij\rangle}, s_{ppi}, h$ and φ_i, which are Galilean tensors—or even objective tensors. Therefore we may write the constitutive relations (1.4), (2.2) in the equivalent form

$$\begin{array}{lcll}
\varrho_{<ijk>} & = & \hat{\varrho}_{<ijk>} & (\varrho, v_i, \varrho_{ij}, \varrho_{ppi}) \\
\varrho_{ppik} & = & \hat{\varrho}_{ppik} & (\underline{\qquad\qquad}) \\
s_{<ij>} & = & \hat{s}_{<ij>} & (\underline{\qquad\qquad}) \\
s_{ppi} & = & \hat{s}_{ppi} & (\underline{\qquad\qquad}) \\
h & = & \hat{h} & (\underline{\qquad\qquad}) \\
\varphi_i & = & \hat{\varphi}_i & (\underline{\qquad\qquad}).
\end{array} \tag{2.5}$$

These are the constitutive relations in the form appropriate to an observer at rest in an inertial frame. An observer in a noninertial frame sees different values of the dependent variables, viz. $\varrho^*_{<ijk>}$ through φ^*_i and of the independent variables, viz. $\varrho, v^*_i, \varrho^*_{ij}, \varrho^*_{ppi}$. But the constitutive *functions* $\hat{\varrho}_{<ijk>}$ through $\hat{\varphi}_i$ are unchanged by the change of frame. This assumption is known as the principle of *material frame indifference*. Formally we express this principle by writing

$$\begin{array}{lll}
\varrho^*_{<ijk>} & = \hat{\varrho}_{<ijk>} & (\varrho, v^*_i, \varrho^*_{ij}, \varrho^*_{ppi}) \\
\varrho^*_{ppik} & = \hat{\varrho}_{ppik} & (\underline{\qquad\qquad}) \\
s^*_{<ij>} & = \hat{s}_{<ij>} & (\underline{\qquad\qquad}) \\
s^*_{ppi} & = \hat{s}_{ppi} & (\underline{\qquad\qquad}) \\
h & = \hat{h} & (\underline{\qquad\qquad}) \\
h^*_i & = \hat{\varphi}_i & (\underline{\qquad\qquad}).
\end{array} \tag{2.6}$$

Note that the constitutive functions $\hat{\varrho}_{<ijk>}$ through $\hat{\varphi}_i$ are the *same ones* in (2.5) and (2.6).

The *requirement of convexity*—sometimes called requirement of convexity and causality—demands that $-h$ be a convex function of the densities F, F_i, F_{ij}, F_{ppi} so that the matrix of second derivatives of h is negative definite. This requirement is eminently sensible both mathematically and physically. Mathematically it ensures well-posedness of Cauchy problems for the field equations and physically it guarantees that all wave speeds are finite and that thermodynamic stability holds.

While the principles of entropy and material frame indifference are common ingredients in thermodynamics and continuum mechanics generally, the requirement of convexity and causality is particular to extended thermodynamics, although there are traces of it in the stability conditions of equilibrium thermodynamics. We have discussed its general consequences in Chapter 3 where we describe the mathematical theory of extended thermodynamics. Here we observe only that for causality, i.e. finite speeds, it would have been sufficient to require hyperbolic field equations. Convexity strengthens this demand, because it makes the field equations *symmetric* hyperbolic, and thus implies well-posedness and thermodynamic stability.

2.2 Exploitation of the Principle of Material Frame-Indifference

We recall that the conditions (2.5), (2.6) refer to Euclidean transformations

$$x_i^* = O_{ij}(t)x_j + b_i(t)$$

between an inertial frame and a noninertial one. For the exploitation we proceed in two steps.

First we consider translations with $O_{ij} = \delta_{ij}$. In that case the conditions (2.5), (2.6) may be combined to read

$$\hat{C}(\varrho, v_i, \varrho_{ij}, \varrho_{ppi}) = \hat{C}(\varrho, v_i + \dot{b}_i, \varrho_{ij}, \varrho_{ppi}), \tag{2.7}$$

where $\hat{C}$ stands for any one of the constitutive functions. Since this relation must hold for all $\dot{b}_i$, we conclude that none of the constitutive quantities may depend on v_i.

Secondly we consider arbitrary Euclidean transformations and make use of the transformation properties of objective tensors to combine the equations (2.5), (2.6), in the form of the following functional relations

$$\begin{array}{llclll}
O_{ir}O_{js}O_{kt} & \hat{\varrho}_{\langle rst\rangle} & (\varrho, \varrho_{ij}, \varrho_{ppi}) & = & \hat{\varrho}_{\langle ijk\rangle} & (\varrho, O_{ir}O_{js}\varrho_{rs}, O_{ir}\varrho_{ppr}) \\
O_{ir}O_{ks} & \hat{\varrho}_{pprs} & (\underline{\qquad}) & = & \hat{\varrho}_{ppik} & (\underline{\qquad\qquad}) \\
O_{ir}O_{ks} & \hat{s}_{\langle rs\rangle} & (\underline{\qquad}) & = & \hat{s}_{\langle ik\rangle} & (\underline{\qquad\qquad}) \\
O_{ir} & \hat{s}_{ppr} & (\underline{\qquad}) & = & \hat{s}_{ppi} & (\underline{\qquad\qquad}) \\
 & \hat{h} & (\underline{\qquad}) & = & \hat{h} & (\underline{\qquad\qquad}) \\
O_{ir} & \hat{\varphi}_r & (\underline{\qquad}) & = & \hat{\varphi}_r & (\underline{\qquad\qquad}).
\end{array} \tag{2.8}$$

We express this in words by saying that the constitutive functions are isotropic functions of their variables.

There are representation theorems for isotropic functions (e.g. see Smith [54]). Here we are not interested in those, because we shall be content with a linear constitutive theory in which the ϱ's and s's are linear functions of the stress deviator $-\varrho_{\langle ij\rangle}$ and of the heat flux $\frac{1}{2}\varrho_{ppi}$. The representations read in this case

$$\begin{aligned}
\varrho_{\langle ijk\rangle} &= 0 \\
\varrho_{ppij} &= \alpha\delta_{ij} + \beta\varrho_{\langle ij\rangle} \\
s_{\langle ij\rangle} &= -\frac{1}{\varsigma_t}\varrho_{\langle ij\rangle} \\
s_{ppi} &= -\frac{1}{\varsigma_q}\varrho_{ppi}.
\end{aligned} \tag{2.9}$$

Note that there is no way to construct an isotropic function for the traceless tensor $\varrho_{<ijk>}$ from $\varrho_{<ij>}$ and ϱ_{ppi} if we require linear dependence on these variables. The coefficients α through ς_q in (2.9) may be functions of ϱ and ϱ_{pp}. The entropy principle will impose restrictions on these functions.

2.3 Exploitation of the Entropy Principle

2.3.1 Six Steps

The exploitation of the entropy principle is the cornerstone of the constitutive theory of extended thermodynamics. It requires careful and cumbersome calculations and a certain ingenuity in finding the best, i.e. most time-saving path. In order to give the argument some structure, we identify seven steps, viz.

i.) Introduction of Lagrange multipliers.

ii.) Removal of velocity-dependent terms.

iii.) Lagrange multipliers as variables. The scalar potential and the vector potential.

iv.) Equilibrium.

v.) Determination of near-equilibrium potentials.

vi.) Residual inequality.

vii.) Summary of results.

For a first reading we suggest that the reader should skip the first steps and appreciate the final results (2.45), (2.47). By these results the coefficients in (2.9) are explicitly related to the thermal equation of state; and this is also true for the entropy density and entropy flux.

2.3.2 Lagrange Multipliers (Step i)

The key to the evaluation of the entropy principle is the statement that the entropy inequality (2.1) must hold for all solutions of the field equations rather than for all fields F, F_i, F_{ij}, F_{ppi}. In a manner of speaking the field equations provide constraints for the fields that satisfy the inequality. Liu has proved in [20] that these constraints may be taken into account by the use of Lagrange multipliers, so that the new inequality

$$\begin{aligned}\frac{\partial h}{\partial t}+\frac{\partial h_i}{\partial x_i} &- \Lambda \left(\frac{\partial F}{\partial t}+\frac{\partial F_k}{\partial x_k}\right)- \\ &- \Lambda_i \left(\frac{\partial F_i}{\partial t}+\frac{\partial F_{ik}}{\partial x_k}\right)- \\ &- \Lambda_{ij} \left(\frac{\partial F_{ij}}{\partial t}+\frac{\partial F_{ijk}}{\partial x_k}-S_{<ij>}\right)- \\ &- \Lambda_{qqi} \left(\frac{\partial F_{ppi}}{\partial t}+\frac{\partial F_{ppik}}{\partial x_k}-S_{ppi}\right)\geq 0\end{aligned} \tag{2.10}$$

must hold for *all* fields F through F_{ppi}. The **Λ**'s are Lagrange multipliers and according to Liu's proof they may be functions of F, F_i, F_{ij}, F_{ppi} whose form depends on material.

We insert the constitutive relations (1.4) and (2.2) for $h, h_i, F_{<ijk>}$, and F_{ppik} and obtain an expression that is explicitly linear in the derivatives

$$\frac{\partial F}{\partial t}, \frac{\partial F_i}{\partial t}, \frac{\partial F_{ij}}{\partial t}, \frac{\partial F_{ppi}}{\partial t}, \frac{\partial F}{\partial x_k}, \frac{\partial F_i}{\partial x_k}, \frac{\partial F_{ij}}{\partial x_k}, \frac{\partial F_{ppi}}{\partial x_k}. \tag{2.11}$$

The inequality must be satisfied for arbitrary values of these derivatives and therefore their contributions to the inequality must vanish. Thus we obtain

$$\begin{aligned} dh &= \Lambda dF + \Lambda_i dF_i + \Lambda_{ij} dF_{ij} + \Lambda_{ppi} dF_{qqi} \\ dh_k &= \Lambda dF_k + \Lambda_i dF_{ik} + \Lambda_{ij} dF_{ijk} + \Lambda_{ppi} dF_{qqik}. \end{aligned} \tag{2.12}$$

There remains the residual inequality

$$\Sigma = \Lambda_{ij} S_{<ij>} + \Lambda_{ppi} S_{ppi} \geq 0, \tag{2.13}$$

which represents the nonnegative entropy production.

2.3.3 Removal of Velocity-Dependent Terms (Step ii)

There is a certain redundance in the conditions (2.12) and (2.13), because by (1.6) and (2.3) the **v**-dependence of the **F**'s and of h_i is explicitly known and that knowledge may be used to simplify (2.12), (2.13). After an easy but fairly long calculation we obtain

$$\begin{aligned} dh &= \lambda d\varrho + \qquad\qquad + \lambda_{ij} d\varrho_{ij} + \lambda_{qqi} d\varrho_{ppi} \\ d\varphi_k &= \qquad \lambda_i d\varrho_{ik} + \lambda_{ij} d\varrho_{ijk} + \lambda_{qqi} d\varrho_{ppik} \end{aligned} \tag{2.14}$$

$$\begin{aligned} \lambda_t \varrho \qquad &+ \lambda_{qqi}(2\varrho_{it} + \varrho_{pp}\delta_{it}) = 0 \\ 2\lambda_{tj}\varrho_{jk} &+ \lambda_{qqi}(2\varrho_{itk} + \varrho_{ppk}\delta_{it}) = [h - \lambda\varrho - \lambda_{ij}\varrho_{ij} - \lambda_{qqi}\varrho_{ppi}]\,\delta_{tk} \end{aligned} \tag{2.15}$$

$$\Sigma = \lambda_{ij} s_{<ij>} + \lambda_{qqi} s_{ppi} \geq 0. \tag{2.16}$$

The λ's introduced here are the velocity-independent—*internal*—parts of the Lagrange multipliers. They are related to the previous $\mathbf{\Lambda}$'s by the equations[1]

$$\begin{aligned}
\lambda &= \Lambda + \Lambda_i v_i + \Lambda_{ij} v_i v_j + \Lambda_{ppi} v^2 v_i \\
\lambda_i &= \Lambda_i + 2\Lambda_{ij} v_j + \Lambda_{ppj}(v^2 \delta_{ji} + 2 v_j v_i) \\
\lambda_{ij} &= \Lambda_{ij} + \Lambda_{ppt} 3 v_{(t} \delta_{ij)} \\
\lambda_{ppi} &= \Lambda_{ppi}.
\end{aligned} \tag{2.17}$$

Since by (2.14), the λ's are derivatives of the objective scalar h and of the objective vector φ_k with respect to the objective quantities ϱ, they are themselves objective tensors. For the same reason the λ's must be isotropic functions of $\varrho, \varrho_{ij}, \varrho_{ppi}$, since h and φ_k are such functions.

The conditions (2.15) result from the fact that h and φ_k are both independent of the velocity. Note that the velocity occurs nowhere in (2.14) through (2.16).

This step of removing the velocity dependence in the equations (2.12) through (2.17) is made in a *systematic manner* in Chapter 3 Section 4.

2.3.4 Lagrange Multipliers as Variables. The Scalar Potential and the Vector Potential. (Step iii)

It is not a priori obvious, but experience has shown that the further calculations are much simplified by replacing $\varrho, \varrho_{ij}, \varrho_{ppi}$ as independent variables by the Lagrange multipliers $\lambda, \lambda_{ij}, \lambda_{ppi}$.[2] We introduce the scalar potential h' and the vector potential h'_k

$$\begin{aligned}
h' &= -h + \lambda\varrho + \lambda_{ij}\varrho_{ij} + \lambda_{qqi}\varrho_{ppi} \\
h'_k &= -\varphi_k + \quad + \lambda_{ij}\varrho_{ijk} + \lambda_{qqi}\varrho_{ppik}
\end{aligned} \tag{2.18}$$

and rewrite (2.14) and $(2.15)_2$ in the alternative forms

$$\begin{aligned}
dh' &= \varrho d\lambda + \varrho_{rs} d\lambda_{rs} + \varrho_{ppn} d\lambda_{qqn} \\
dh'_k &= \left[\lambda_{qqe} a_{ei} \frac{\partial \varrho_{ij}}{\partial \lambda}\right] d\lambda + \left[\varrho_{rsk} + \lambda_{qqe} a_{ei} \frac{\partial \varrho_{ik}}{\partial \lambda_{rs}}\right] d\lambda_{rs} + \left[\varrho_{ppnk} + \lambda_{qqe} a_{ei} \frac{\partial \varrho_{ik}}{\partial \lambda_{qqn}}\right] d\lambda_{qqn}
\end{aligned} \tag{2.19}$$

$$2\lambda_{tj}\varrho_{jk} + \lambda_{qqi}(2\varrho_{itk} + \varrho_{ppk}\delta_{it}) = -h'\delta_{tk}. \tag{2.20}$$

Note that λ_t has been eliminated by use of $(2.15)_1$. This elimination has brought the tensor a_{ei} into $(2.19)_2$, which is defined as

$$a_{ei} = \frac{1}{\varrho}(2\varrho_{ei} + \varrho_{pp}\delta_{ei}). \tag{2.21}$$

[1] See Chapter 3, equation (4.33), where the internal Lagrange multipliers were denoted by $\hat{\mathbf{\Lambda}}$.

[2] In the mathematical theory of extended thermodynamics this change of variables appears much less artificial than here. Also there—in Chapter 3—it is shown that the change is always possible.

We conclude from (2.19) and (2.21) that the knowledge of h' and h'_k as functions of the variables $\lambda, \lambda_{ij}, \lambda_{qqi}$ determines $\varrho, \varrho_{ik}, \varrho_{ppk}, \varrho_{<ijk>}$, and ϱ_{ppik} as functions of these variables. For that reason h' and h'_k are properly called potentials.

However expedient the Lagrange multipliers may be as variables for the purpose of exploiting the entropy principle, they are also nonsuggestive and nonmeasurable except for their equilibrium values. Therefore eventually we shall have to remove them from the results. But first we calculate their equilibrium values.

2.3.5 Equilibrium. (Step iv)

Equilibrium is a process in which the productions $s_{<ij>}$ and s_{ppi} vanish. Therefore, by (2.16), the entropy production has a minimum in equilibrium, namely zero. Hence follow as necessary conditions

$$\lambda_{<ij>}|_{\mathrm{E}} = 0, \qquad \lambda_{qqi}|_{\mathrm{E}} = 0, \tag{2.22}$$

where E denotes equilibrium.

From $(2.14)_1$ we conclude with $\varrho_{ii} = 2\varrho\varepsilon$ (see $(1.11)_2$)

$$dh_{\mathrm{E}} = \frac{2}{3}\lambda_{ii}|_{\mathrm{E}} d(\varrho\varepsilon) + \lambda_{\mathrm{E}} d\varrho. \tag{2.23}$$

Comparison with the Gibbs equation identifies $\lambda_{ii}|_{\mathrm{E}}$ and λ_{E} as follows:

$$\lambda_{ii}|_{\mathrm{E}} = \frac{3}{2}\frac{1}{T} \qquad \text{and} \qquad \lambda_{\mathrm{E}} = -\frac{g}{T}, \tag{2.24}$$

where T is the absolute temperature and $g = \varepsilon - T\frac{h_{\mathrm{E}}}{\varrho} + \frac{p_{\mathrm{E}}}{\varrho}$ is the specific free enthalpy or Gibbs free energy, sometimes called the chemical potential.

Thus by (2.22) and (2.24) we have identified the equilibrium values of the Lagrange multipliers. Before we use this knowledge for the further evaluation of the conditions (2.19), (2.20) we determine the consequences of the Gibbs equation. Combination of (2.23) and (2.24) gives

$$\begin{aligned} dh_{\mathrm{E}} &= \frac{1}{T} d(\varrho\varepsilon) - \frac{g}{T} d\varrho \qquad \text{or} \\ d\left(\frac{h_{\mathrm{E}}}{\varrho}\right) &= \frac{1}{T}\left(d\varepsilon - \frac{p_{\mathrm{E}}}{\varrho^2} d\varrho\right). \end{aligned} \tag{2.25}$$

and this Gibbs equation implies the integrability condition

$$\left(\frac{\partial \varepsilon}{\partial \varrho}\right)_T = \frac{1}{\varrho^2}\left(p_{\mathrm{E}} - T\left(\frac{\partial p_{\mathrm{E}}}{\partial T}\right)_\varrho\right). \tag{2.26}$$

For a monatomic gas with $p_{\mathrm{E}} = \frac{2}{3}\varrho\varepsilon$ the first of these conditions may be integrated to give, with $z = \frac{\varrho}{T^{3/2}}$,

$$p_{\mathrm{E}} = T^{5/2} F(z) \qquad \text{hence} \qquad \frac{h_{\mathrm{E}}}{\varrho} = \frac{3}{2}\int\left(\frac{F'}{z} - \frac{5}{3}\frac{F}{z^2}\right) dz + \varsigma. \tag{2.27}$$

$F(z)$ is an arbitrary function and ς is a constant of integration. g/T is a function of z given by

$$\frac{g}{T} = \frac{F}{z} + \int \frac{F}{z^2} dz - \varsigma \tag{2.28}$$

Note that, by $(2.27)_1$, the thermal equation of state $p_{\rm E} = p(\varrho, T)$ is determined by a function of a *single* variable. We shall discuss special cases later; see Section 4.

2.3.6 The Potentials near Equilibrium (Step v.)

The potentials h' and h'_k, being defined by (2.18), are isotropic scalar and vectorial functions of $\lambda, \lambda_{ij}, \lambda_{ppi}$. We write their representations in polynomial form in the neighborhood of equilibrium by omitting higher-order terms in the variables $\lambda_{<ij>}$ and λ_{qqi}, which vanish in equilibrium.

$$\begin{aligned} h' &= h'_{\rm E} + h'_1 \lambda_{<ij>}\lambda_{<ij>} + h'_2 \lambda_{ppi}\lambda_{qqi} + h'_3 \lambda_{<in>}\lambda_{<nj>}\lambda_{<ji>} + h'_4 \lambda_{<kj>}\lambda_{ppk}\lambda_{qqj} + \mathcal{O}(4) \\ h'_k &= \varphi_1 \lambda_{ppk} + \varphi_2 \lambda_{<ke>}\lambda_{ppe} + \mathcal{O}(3). \end{aligned} \tag{2.29}$$

All coefficients may be functions of $\lambda_{\rm E} = -g/T$ and $\lambda_{ii}|_{\rm E} = \frac{3}{2}\frac{1}{T}$. From the definition $(2.18)_1$ of h' it follows that its equilibrium value is given in terms of the equilibrium pressure by

$$h'_{\rm E} = -\frac{p_{\rm E}}{T}. \tag{2.30}$$

We exploit the equation $(2.19)_1$ and obtain with a modicum of calculation

$$\begin{aligned} \varrho &= \frac{\partial h'}{\partial \lambda} &&= \varrho_{\rm E} - \frac{\partial h'_1}{\partial g/T}\lambda_{<ij>}\lambda_{<ij>} - \frac{\partial h'_2}{\partial g/T}\lambda_{ppi}\lambda_{qqi} + \mathcal{O}(3) \\ \frac{1}{3}\varrho_{pp} &= \frac{\partial h'}{\partial \lambda_{pp}} &&= p_{\rm E} + \frac{2}{3}\frac{\partial h'_1}{\partial 1/T}\lambda_{<ij>}\lambda_{<ij>} + \frac{2}{3}\frac{\partial h'_2}{\partial 1/T}\lambda_{ppi}\lambda_{qqi} + \mathcal{O}(3) \\ \varrho_{<ij>} &= \frac{\partial h'}{\partial \lambda_{<ij>}} &&= 2h'_1\lambda_{<ij>} + 3h'_3(\lambda_{<ik>}\lambda_{<kj>} - \frac{1}{3}\lambda_{<ln>}\lambda_{<ln>}\delta_{ij}) + h'_4\lambda_{qq<i}\lambda_{j>pp} + \mathcal{O}(3) \\ \varrho_{ppi} &= \frac{\partial h'}{\partial \lambda_{ppi}} &&= 2h'_2\lambda_{ppi} + 2h'_4\lambda_{<ik>}\lambda_{kpp} + \mathcal{O}(3). \end{aligned} \tag{2.31}$$

We conclude from $(2.31)_{1,2}$ that the density ϱ and the pressure $p = \frac{1}{3}\varrho_{pp}$ are equal to the equilibrium values $\varrho_{\rm E}$ and $p_{\rm E}$ to within second-order terms. In the sequel this fact enables us to drop the distinction.

Next we exploit $(2.19)_2$. We have

$$a_{ei} = \frac{1}{\varrho}\left(4h'_1\lambda_{<ei>} + 5p_{\rm E}\delta_{ei}\right) + \mathcal{O}(2) \tag{2.32}$$

and obtain

$$
\begin{aligned}
0 &= \left(\frac{\partial\varphi_1}{\partial g/T} - 5pT\right)\lambda_{qqk} + \left(\frac{\partial\varphi_2}{\partial g/T} - 10\frac{p}{\varrho}\frac{\partial h_1'}{\partial g/T} - 4h_1'T\right)\lambda_{<kl>}\lambda_{qql} + \mathcal{O}(3)\\
0 &= \varrho_{<rsk>} + \mathcal{O}(2)\\
0 &= \left(\frac{\partial\varphi_1}{\partial 1/T} - h_2' + \frac{25}{2}\frac{p^2T}{\varrho}\right)\lambda_{ppk} + \left(\frac{\partial\varphi_2}{\partial 1/T} - h_4' - 10\frac{p}{\varrho}\frac{\partial h_1'}{\partial 1/T} - 10h_1'\frac{pT}{\varrho}\right)\lambda_{ppl}\lambda_{<lk>} + \mathcal{O}(3)\\
0 &= \left(\varphi_2 - \frac{4}{5}h_2' - 10\frac{p}{\varrho}h_1'\right)\left(\lambda_{qqr}\delta_{ks} + \lambda_{qqs}\delta_{kr} - \frac{2}{3}\lambda_{qqk}\delta_{rs}\right) + \mathcal{O}(2)\\
\varrho_{ppnk} &= \varphi_1\delta_{nk} + \varphi_2\lambda_{<nk>} + \mathcal{O}(2).
\end{aligned}
\tag{2.33}
$$

Finally we exploit (2.20) and obtain from its traceless part

$$
\begin{aligned}
0 = 2(h_1' + pT)\lambda_{<tk>} \;&+\; (4h_1'T + 3h_3')\left(\lambda_{<tj>}\lambda_{<jk>} - \frac{1}{3}\lambda_{<mn>}\lambda_{<mn>}\delta_{tk}\right) +\\
&+\; \left(\frac{18}{5}h_2'T + h_4'\right)\lambda_{qq<k}\lambda_{t>pp} + \mathcal{O}(3).
\end{aligned}
\tag{2.34}
$$

From (2.34) we see that

$$h_1' = -pT \qquad \text{and} \qquad h_3' = \frac{4}{3}pT^2 \tag{2.35}$$

hold and therefore the equations $(2.33)_{1,3}$ provide equations for the derivatives of φ_1 and φ_2, viz.

$$
\begin{aligned}
\frac{\partial\varphi_1}{\partial g/T} &= 5pT & \qquad \frac{\partial\varphi_1}{\partial 1/T} &= h_2' - \frac{25}{2}\frac{p^2T}{\varrho}\\
\frac{\partial\varphi_2}{\partial g/T} &= -14pT^2 & \qquad \frac{\partial\varphi_2}{\partial 1/T} &= h_4' + 45\frac{p^2T^2}{\varrho}.
\end{aligned}
\tag{2.36}
$$

Again from (2.34) we know that $h_4' = -\frac{18}{5}h_2'T$ holds, and therefore we obtain from $(2.33)_4$

$$h_2' = \frac{5}{4}\left(\varphi_2 + 10\frac{p^2T}{\varrho}\right) \qquad \text{and} \qquad h_4' = -\frac{9}{2}\left(\varphi_2 + 10\frac{p^2T}{\varrho}\right)T. \tag{2.37}$$

Elimination of h_2' and h_4' between (2.36) and (2.37) provides the equations

$$\frac{\partial\varphi_1}{\partial g/T} = -\frac{5}{14}\frac{\partial\frac{\varphi_2}{T}}{\partial g/T} \qquad \text{and} \qquad \frac{\partial\varphi_1}{\partial 1/T} = -\frac{5}{14}\frac{\partial\frac{\varphi_2}{T}}{\partial 1/T}$$

which can readily be integrated to give

$$\varphi_1 = -\frac{5}{14}\frac{\varphi_2}{T} + C. \tag{2.38}$$

On the other hand, by (2.36) and (2.27) we have

$$\begin{aligned} \frac{\partial \varphi_1}{\partial g/T} &= 5T^{7/2}F(z) \\ \frac{\partial \varphi_2}{\partial g/T} &= -14T^{9/2}F(z) \end{aligned} \qquad \text{or, by (2.28)} \qquad \begin{aligned} \frac{\partial \varphi_1}{\partial z} &= 5T^{7/2}\frac{FF'}{z} \\ \frac{\partial \varphi_2}{\partial z} &= -14T^{9/2}\frac{FF'}{z}, \end{aligned}$$

where as before z stands for $\frac{\varrho}{T^{3/2}}$. We integrate and obtain

$$\begin{aligned} \varphi_1 &= 5T^{7/2}\left(\int \frac{FF'}{z}dz + c\right) + C \\ \varphi_2 &= -14T^{9/2}\left(\int \frac{FF'}{z}dz + c\right). \end{aligned} \qquad (2.39)$$

With this knowledge we calculate the coefficients h' from (2.35), (2.37), and (2.27)

$$\begin{aligned} h'_1 &= -T^{7/2}F(z) \\ h'_2 &= T^{9/2}\left[\frac{25}{2}\frac{F^2(z)}{z} - \frac{35}{2}\left(\int \frac{FF'}{z}dz + c\right)\right] \\ h'_3 &= \frac{4}{3}T^{9/2}F(z) \\ h'_4 &= T^{11/2}\left[-45\frac{F^2(z)}{z} + 63\left(\int \frac{FF'}{z}dz + c\right)\right]. \end{aligned} \qquad (2.40)$$

We thus conclude that all coefficients h' and φ that determine the potentials are in turn determined by the function $F(\frac{\varrho}{T^{3/2}})$, which first appeared in the thermal equation of state (2.27).

2.3.7 Residual Inequality (Step vi)

Of course as a further consequence of the entropy principle there is the residual inequality (2.16), which entails linear relations between the productions $s_{<ij>}, s_{ppi}$ and the Lagrange multipliers $\lambda_{<ij>}$ and λ_{ppi} of the form

$$s_{<ij>} = s\lambda_{<ij>} \qquad \text{and} \qquad s_{ppi} = t\lambda_{ppi}, \qquad (2.41)$$

where s and t are positive-valued functions of ϱ and T (say)

$$s \geq 0 \qquad \text{and} \qquad t \geq 0. \qquad (2.42)$$

2.3.8 Summary of Results of the Entropy Principle (Step vii)

We recall from Section 2.2 that $\varrho_{<ijk>}, \varrho_{ppik}, s_{<ij>}, s_{ppi}$ are the only constitutive quantities that occur in the balance equations. Now we have obtained (see $(2.33)_2$, $(2.33)_5$ with (2.39) and (2.41))

$$\varrho_{<rsk>} = \mathcal{O}(2)$$

$$\begin{aligned}
\varrho_{ppnk} &= \left[5T^{7/2}\left(\int \frac{FF'}{z}dz + c\right) + C\right]\delta_{nk} - 14T^{9/2}\left(\int \frac{FF'}{z}dz + c\right)\lambda_{<nk>} + \mathcal{O}(2) \\
s_{<ij>} &= s\lambda_{<ij>} + \mathcal{O}(2) \\
s_{ppi} &= t\lambda_{ppi} + \mathcal{O}(2).
\end{aligned} \tag{2.43}$$

We may use (2.31) and (2.40) to replace the Lagrange multipliers in (2.43) by the original variables $\varrho_{<ij>}$ and ϱ_{ppi}, because we have, to within terms of $\mathcal{O}(2)$,

$$\begin{aligned}
\lambda_{<ij>} &= -\frac{1}{2pT}\varrho_{<ij>} = -\frac{1}{2T^{7/2}F(z)}\varrho_{<ij>} \\
\lambda_{ppi} &= \frac{1}{2h_2'}\varrho_{ppi} = \frac{1}{5T^{9/2}\left[5\frac{F^2}{z} - 7\left(\int \frac{FF'}{z}dz + c\right)\right]}\varrho_{ppi}.
\end{aligned} \tag{2.44}$$

Therefore we may write the constitutive relations (2.43) in the final forms

$$\begin{aligned}
\varrho_{<ijk>} &= \mathcal{O}(2) \\
\varrho_{ppnk} &= \left[5T^{7/2}\left(\int \frac{FF'}{z}dz + c\right) + C\right]\delta_{ij} + 7\frac{T}{F}\left[\int \frac{FF'}{z}dz + c\right]\varrho_{\langle ij\rangle} + \mathcal{O}(2) \\
s_{\langle ij\rangle} &= -\frac{s}{2T^{7/2}F(z)}\varrho_{\langle ij\rangle} + \mathcal{O}(2) \\
s_{ppi} &= \frac{t}{5T^{9/2}\left[5\frac{F^2(z)}{z} - 7\left(\int \frac{FF'}{z}dz + c\right)\right]}\varrho_{ppi} + \mathcal{O}(2).
\end{aligned} \tag{2.45}$$

Of course, we may also be interested in the entropy density and the entropy flux. We have by (2.18)

$$h = -h' + \lambda\varrho + \lambda_{ij}\varrho_{ij} + \lambda_{ppi}\varrho_{qqi} \qquad \text{and} \qquad \varphi_k = -\varphi_k' + \lambda_{ij}\varrho_{ijk} + \lambda_{ppi}\varrho_{ppik}. \tag{2.46}$$

From these equations, the representations (2.29) with (2.27), and with the coefficients (2.39), (2.40) we obtain to within terms of order 3

$$\begin{aligned}
h &= \varrho\left[\frac{3}{2}\int\left(\frac{F'}{z} - \frac{5}{3}\frac{F}{z^2}\right)dz + \varsigma\right] - \frac{\varrho_{\langle ij\rangle}\varrho_{\langle ij\rangle}}{4T^{7/2}F(z)} - \frac{\varrho_{ppi}\varrho_{qqi}}{10T^{7/2}F(z)\left[7\frac{T}{F(z)}\left(\int \frac{FF'}{z}dz + c\right) - 5T\frac{F(z)}{z}\right]} \\
\varphi_k &= \frac{1}{T}\frac{1}{2}\varrho_{ppk} - \frac{2}{5}\frac{1}{T^{7/2}F(z)}\varrho_{\langle kj\rangle}\frac{1}{2}\varrho_{ppj}.
\end{aligned} \tag{2.47}$$

The equations (2.45), (2.47) are the final results obtained from the entropy principle. We conclude that $\varrho_{\langle ijk\rangle}$, ϱ_{ppnk}, h, and φ_k are fully explicit to within the constants c, C, and ς, if only the function $F(z)$ is known. That function determines the thermal equation of state and is thus known either from experiment or from statistical mechanics.

The productions $s_{\langle ij\rangle}$ and s_{ppi} are determined to within two positive-valued functions of ϱ and T.

These amazingly restrictive results first derived by Liu & Müller [21] in a slightly different and more laborious manner.

2.4 Exploitation of the Requirement of Convexity and Causality

In Chapter 3—within the framework of the formal theory of extended thermodynamics—we have shown that the requirement of convexity of $-h$ with respect to the densities F, F_i, F_{ij}, F_{ppi} implies that $-h$ is also a convex function of the densities $\varrho, \varrho_{pp}, \varrho_{\langle ij \rangle}$, and ϱ_{ppi}. Hence follows by inspection of $(2.47)_1$ that we must have

$$F(z) > 0 \quad \text{and} \quad 7\left(\int \frac{FF'}{z} dz + c\right) - 5\frac{F^2}{z} > 0. \tag{2.48}$$

But this is not all; the matrix of second derivatives of $-h_E$ with respect to ϱ and ϱ_{ll} must be positive definite. Hence follows

$$\begin{aligned}
&\left\| \begin{matrix} -\left(\dfrac{\partial^2 h_E}{\partial \varrho^2}\right)_{\varrho_{pp}} & -\dfrac{\partial^2 h_E}{\partial \varrho \partial \varrho_{pp}} \\ -\dfrac{\partial^2 h_E}{\partial \varrho \partial \varrho_{pp}} & -\left(\dfrac{\partial^2 h_E}{\partial \varrho_{pp}^2}\right)_{\varrho} \end{matrix} \right\| = \left\| \begin{matrix} \left(\dfrac{\partial \frac{g}{T}}{\partial \varrho}\right)_{\varrho_{pp}} & -\dfrac{1}{2}\left(\dfrac{\partial \frac{1}{T}}{\partial \varrho}\right)_{\varrho_{pp}} \\ -\dfrac{1}{2}\left(\dfrac{\partial \frac{1}{T}}{\partial \varrho}\right)_{\varrho_{pp}} & -\dfrac{1}{2}\left(\dfrac{\partial \frac{1}{T}}{\partial \varrho_{pp}}\right)_{\varrho} \end{matrix} \right\| \\
&= \left\| \begin{matrix} \dfrac{1}{T^{3/2}} \dfrac{\frac{5}{3}\frac{FF'}{z^2}}{\frac{5}{3}\frac{F}{z} - F'} & -\dfrac{1}{\varrho T} \dfrac{\frac{1}{3}F'}{\frac{5}{3}\frac{F}{z} - F'} \\ -\dfrac{1}{\varrho T} \dfrac{\frac{1}{3}F'}{\frac{5}{3}\frac{F}{z} - F'} & \dfrac{1}{\varrho T^2} \dfrac{\frac{1}{9}}{\frac{5}{3}\frac{F}{z} - F'} \end{matrix} \right\| \text{ - pos. def.}
\end{aligned} \tag{2.49}$$

The first equation follows from (2.25), the second from (2.27). The condition is satisfied if we have

$$0 < F' < \frac{5}{3}\frac{F}{z}. \tag{2.50}$$

It is clear that the above exploitation of the requirement of convexity is appropriate to the present near-equilibrium approximation. In Chapter 8 we shall prove that it is possible to determine the region of hyperbolicity around equilibrium.

3 Field Equations and the Thermodynamic Limit

3.1 Field Equations

In the equations (1.15) we replace the $\mathbf{F}^*$'s by ϱ^*'s as dictated by (1.6) and eliminate the explicit dependence of all quantities on the velocity in the manner that was described for Galilean systems in Paragraph 1.3. Thus we obtain the balance laws of extended thermodynamics in the following form:

$$
\begin{aligned}
&\dot{\varrho} + \varrho \frac{\partial v_i^*}{\partial x_i^*} = 0 \\
&\varrho \left[\dot{v}_i^* - i_i^0 - 2W_{ik} v_k^* \right] + \frac{\partial \varrho_{ij}^*}{\partial x_j^*} = 0 \\
&\varrho \left(\frac{\varrho_{pp}}{\varrho} \right)^{\bullet} + \frac{\partial \varrho_{ppi}^*}{\partial x_j^*} + 2(\varrho_{\langle ij \rangle}^* + \frac{1}{3} \varrho_{pp} \delta_{ij}) \frac{\partial v_i^*}{\partial x_j^*} = 0 \\
&\varrho \left[\left(\frac{\varrho_{\langle ik \rangle}^*}{\varrho} \right)^{\bullet} + 2 \varrho_{l\langle i}^* \frac{\partial v_{k\rangle}^*}{\partial x_l} - 4 \varrho_{l\langle i}^* W_{k\rangle l} \right] + \frac{\partial \varrho_{\langle ik \rangle j}^*}{\partial x_j^*} = s_{\langle ik \rangle}^* \\
&\varrho \left[\left(\frac{\varrho_{ppi}^*}{\varrho} \right)^{\bullet} + \frac{\varrho_{ppk}^*}{\varrho} \frac{\partial v_i^*}{\partial x_k^*} - 2 \varrho_{ppk}^* W_{ik} \right] + \frac{\partial \varrho_{ppik}^*}{\partial x_k^*} + \\
&\qquad\qquad + 2 \varrho_{ijk}^* \frac{\partial v_j^*}{\partial x_k^*} - 2 \frac{\varrho_{ij}^*}{\varrho} \frac{\partial \varrho_{jk}^*}{\partial x_k^*} - \frac{\varrho_{pp}}{\varrho} \frac{\partial \varrho_{ik}^*}{\partial x_k^*} = s_{ppi}^* .
\end{aligned}
\tag{3.1}
$$

For simplicity in notation we shall now drop the stars that characterize quantities referred to noninertial frames. The presence of the inertial terms with i_i^0 and W_{ik} must suffice as a reminder that the frame is noninertial. Also we introduce the suggestive notation (1.11), (1.12) for ϱ_{ij}, ϱ_{pp}, and ϱ_{ppi}. Furthermore we insert the explicit expressions (2.45) for $\varrho_{\langle ijk \rangle}$, ϱ_{ppij}, $s_{\langle ij \rangle}$, and s_{ppi} that are dictated by the entropy principle. In this manner we obtain from (3.1), always with $z = \frac{\varrho}{T^{3/2}}$,

$$
\begin{aligned}
&\dot{\varrho} + \varrho \frac{\partial v_i}{\partial x_i} = 0 \\
&\varrho \left[\dot{v}_i - i_i^0 - 2W_{ik} v_k \right] - \frac{\partial (t_{\langle ij \rangle} - T^{5/2} F(z) \delta_{ij})}{\partial x_j} = 0 \\
&\varrho \frac{3}{2} \left(T \frac{F(z)}{z} \right)^{\bullet} + \frac{\partial q_j}{\partial x_j} - \left(t_{\langle ij \rangle} - T^{5/2} F(z) \delta_{ij} \right) \frac{\partial v_i}{\partial x_j} = 0 \\
&\varrho \left[\left(\frac{t_{\langle ik \rangle}}{\varrho} \right)^{\bullet} + 2 \frac{t_{l\langle i}}{\varrho} \frac{\partial v_{k\rangle}}{\partial x_l} - 4 \frac{t_{l\langle i}}{\varrho} W_{k\rangle l} \right] - \frac{4}{5} \frac{\partial q_{\langle i}}{\partial x_{k\rangle}} = -\frac{1}{\tau_t} t_{\langle ik \rangle} \\
&\varrho \left[\left(\frac{q_i}{\varrho} \right)^{\bullet} + \frac{q_k}{\varrho} \frac{\partial v_i}{\partial x_k} - 2 \frac{q_k}{\varrho} W_{ik} \right] - \frac{\partial \left\{ \frac{7}{2} \left(\frac{T}{F(z)} t_{\langle ij \rangle} - \frac{5}{7} T^{7/2} \delta_{ij} \right) \left(\int \frac{FF'}{z} dz + c \right) \right\}}{\partial x_j} + \\
&\qquad + \frac{2}{5} q_i \frac{\partial v_k}{\partial x_k} + \frac{4}{5} q_k \frac{\partial v_{(i}}{\partial x_{k)}} - \frac{\left(t_{\langle ij \rangle} - T^{5/2} F(z) \delta_{ij} \right)}{\varrho} \frac{\partial \left(t_{\langle jk \rangle} - T^{5/2} F(z) \delta_{jk} \right)}{\partial x_k} + \\
&\qquad + \frac{3}{2} \frac{T^{5/2} F(z)}{\varrho} \frac{\partial \left(t_{\langle ik \rangle} - T^{5/2} F(z) \delta_{ik} \right)}{\partial x_k} = -\frac{1}{\tau_q} q_i .
\end{aligned}
\tag{3.2}
$$

We have introduced

$$
\frac{1}{\tau_t} \quad \text{for} \quad \frac{s}{2T^{7/2} F(z)} \quad \text{and} \quad \frac{1}{\tau_q} \quad \text{for} \quad \frac{t}{T^{9/2} \left[7 \left(\int \frac{FF'}{z} dz + c \right) - 5 \frac{F^2}{z} \right]} . \tag{3.3}
$$

The τ's may be considered as relaxation times for stress and heat flux respectively.

Inspection of (3.2) shows that this system of equations for the determination of the thirteen variables

$$\varrho, v_i, T, t_{\langle ij \rangle}, q_i$$

is quite explicit. What remains to be determined is

the constant c
the function $F(z)$ from the thermal equation of state
the coefficients $\tau_t(\varrho, T)$ and $\tau_q(\varrho, T)$.

The functions $\tau_t(\varrho, T)$ and $\tau_q(\varrho, T)$ are related to the viscosity and the heat conductivity of the gas. We proceed to show this in the next paragraph.

The equations $(3.2)_{2,4,5}$ all contain inertial terms and yet the equations as they stand hold in all Euclidean frames, because the expressions

$$\dot{v}_i - i_i^0 - 2W_{ik}v_k \quad \text{and} \quad \left(\frac{t_{\langle ik \rangle}}{\varrho}\right)^{\bullet} + 2\frac{t_{l\langle i}}{\varrho}\frac{\partial v_{k\rangle}}{\partial x_l} - 4\frac{t_{l\langle i}}{\varrho}W_{k\rangle l} \quad \text{and} \quad \left(\frac{q_i}{\varrho}\right)^{\bullet} + \frac{q_k}{\varrho}\frac{\partial v_i}{\partial x_k} - 2\frac{q_k}{\varrho}W_{ik}, \tag{3.4}$$

while frame-dependent, are objective vectors and tensors.

3.2 The Thermodynamic Limit

A natural question to be asked from extended thermodynamics is how it relates to ordinary thermodynamics. In ordinary thermodynamics of viscous, heat-conducting monatomic gases the stress deviator and the heat flux are given by the constitutive relations of Navier-Stokes and Fourier, viz.

$$\begin{aligned} t_{\langle ij \rangle} &= 2\mu \; \frac{\partial v_{\langle i}}{\partial x_{j\rangle}} \\ q_i &= -\kappa \; \frac{\partial T}{\partial x_i}, \end{aligned} \tag{3.5}$$

with μ and κ being viscosity and heat conductivity respectively. In extended thermodynamics we have replaced that simple set of equations for $t_{\langle ij \rangle}, q_i$ by the complicated set of differential equations $(3.2)_{4,5}$. However, there is a relation between the two:

This relation is provided by a formal iterative scheme, by which the equations (3.5) emerge from $(3.2)_{4,5}$ as first iterates. The scheme is called Maxwellian iteration, because Maxwell invented it when faced with a similar problem in the kinetic theory of gases; see Ikenberry & Truesdell [55]. The Maxwellian iteration proceeds as follows:

In the first step we calculate first iterates $t_{\langle ij \rangle}^{(1)}$ and $q_i^{(1)}$ from $(3.2)_{4,5}$ by using the "zeroth" iterates $t_{\langle ij \rangle}^{(0)} = 0$ and $q_i^{(0)} = 0$ on the left-hand sides of the equations. The zeroth iterates are the equilibrium values, of course. In the second step we calculate second iterates $t_{\langle ij \rangle}^{(2)}$ and $q_i^{(2)}$ by insertion of $t_{\langle ij \rangle}^{(1)}$

and $q_i^{(1)}$ into the left-hand sides of $(3.2)_{4,5}$, etc.[3] In this manner we obtain as first iterates after a brief calculation

$$\begin{aligned} t_{\langle ij\rangle}^{(1)} &= 2T^{5/2}F(z)\tau_t\frac{\partial v_{\langle i}}{\partial x_{j\rangle}} \\ q_i^{(1)} &= -\frac{5}{4}T^{5/2}\left\{5\frac{F^2}{z}-7\left(\int\frac{FF'}{z}dz+c\right)\right\}\tau_q\frac{\partial T}{\partial x_i}. \end{aligned} \tag{3.6}$$

Thus we conclude that the classical laws of Navier-Stokes and Fourier result from the equations of extended thermodynamics in an approximate manner as first iterates in the Maxwellian scheme. The viscosity and the heat conductivity of the gas result from the comparison of (3.5) and (3.6)

$$\mu = T^{5/2}F(z)\tau_t \qquad \kappa = \frac{5}{4}T^{5/2}\left\{5\frac{F^2}{z}-7\left(\int\frac{FF'}{z}dz+c\right)\right\}\tau_q. \tag{3.7}$$

Since μ and κ may be measured as functions of ϱ and T—and are in fact tabulated—we see that the coefficients τ_t and τ_q may be determined.

Insertion of (3.6) on the left-hand side of $(3.2)_{4,5}$ furnishes the second iterates $t_{\langle ij\rangle}^{(2)}$ and $q_i^{(2)}$, viz.

$$\begin{aligned} t_{\langle ij\rangle}^{(2)} &= 2p\tau_t\frac{\partial v_{\langle i}}{\partial x_{j\rangle}} - \varrho\tau_t\left[\left(\frac{2p\tau_t}{\varrho}\frac{\partial v_{\langle i}}{\partial x_{j\rangle}}\right)^{\bullet} + \frac{4p\tau_t}{\varrho}\frac{\partial v_{\langle l}}{\partial x_{\langle i\rangle}}\frac{\partial v_{k\rangle}}{\partial x_l} - \frac{8p\tau_t}{\varrho}\frac{\partial v_{\langle l}}{\partial x_{\langle i\rangle}}W_{k\rangle l}\right] - \frac{4}{5}\tau_t\frac{\partial\left(\pi\tau_q\frac{\partial T}{\partial x_{\langle i}}\right)}{\partial x_{j\rangle}} \\ q_i^{(2)} &= -\pi\tau_q\frac{\partial T}{\partial x_i} + \varrho\tau_q\left[\left(\frac{\pi\tau_q}{\varrho}\frac{\partial T}{\partial x_i}\right)^{\bullet} + \frac{\pi\tau_q}{\varrho}\frac{\partial T}{\partial x_k}\frac{\partial v_i}{\partial x_k} - 2\frac{\pi\tau_q}{\varrho}\frac{\partial T}{\partial x_k}W_{ik}\right] + \\ &\quad +\tau_q\frac{\partial\left\{7T^{7/2}\tau_t\frac{\partial v_{\langle i}}{\partial x_{j\rangle}}\left(\int\frac{FF'}{z}dz+c\right)\right\}}{\partial x_j} + \\ &\quad +\frac{2}{5}\pi\tau_q^2\frac{\partial T}{\partial x_i}\frac{\partial v_k}{\partial x_k} + \frac{4}{5}\pi\tau_q^2\frac{\partial T}{\partial x_k}\frac{\partial v_{\langle i}}{\partial x_{k\rangle}} + 2\frac{p}{\varrho}\tau_q\tau_t\frac{\partial v_{\langle i}}{\partial x_{j\rangle}}\frac{\partial\left(2p\tau_t\frac{\partial v_{j\rangle}}{\partial x_{k\rangle}}\right)}{\partial x_k} - \\ &\quad -2\frac{p}{\varrho}\tau_q\tau_t\frac{\partial v_{\langle i}}{\partial x_{j\rangle}}\frac{\partial p}{\partial x_j} - \frac{p}{\varrho}\tau_q\frac{\partial\left(2p\tau_t\frac{\partial v_{\langle i}}{\partial x_{k\rangle}}\right)}{\partial x_k} - \frac{3}{2}\frac{p}{\varrho}\tau_q\frac{\partial\left(2p\tau_t\frac{\partial v_{\langle i}}{\partial x_{k\rangle}}\right)}{\partial x_k}. \end{aligned} \tag{3.8}$$

Here p of course is equal to $T^{5/2}F(z)$ and π has been introduced as an abbreviation for

$$\frac{5}{4}T^{5/2}\left\{5\frac{F^2}{z}-7\left(\int\frac{FF'}{z}dz+c\right)\right\}.$$

While we shall not discuss the second iterates in any detail, we point out three features that seem worthy of a comment.

[3]This is a slight misrepresentation of the Maxwell iteration in the kinetic theory. Indeed in the kinetic theory each of the consecutive iterates pulls further moment equations into the scheme, while here we use the same equations for all iterates. For the first iterate there is no difference, and the essential terms of the second iterates—those which we are going to consider—are not affected either by our simplification.

First of all we observe that the first iterates are linear in τ_t and τ_q, while the contributions of the second iterative step are quadratic or cubic in the relaxation times. It is easy to extrapolate and conclude that the contributions of the nth iterative step are of nth order at least in τ_t and τ_q. Thus the iteration may be considered as a kind of expansion in powers of the relaxation times and, since these are expected to be small, one may thus justify the break-off of the iteration. Of course, this is far from a rigorous argument; convergence is a open problem in this field.

Secondly we see that in the second step the stress depends on the temperature gradient as well as on the velocity gradient. Similarly the heat flux depends on the velocity gradient. That observation was already made by Maxwell in the context of the kinetic theory of gases. To my knowledge such dependencies have never been observed, because they are too small.

Thirdly we conclude from (3.8) that the stress $t^{(2)}_{\langle ij\rangle}$ and the heat flux $q^{(2)}_i$ are affected by the angular velocity $\mathbf{W}$ of the frame. This fact may be the most noteworthy aspect of the second iterates and we shall investigate it in a separate paragraph. The effect was first described by Müller [26] in the kinetic theory of gases.

3.3 The Frame Dependence of the Heat Flux

The effect of the frame-dependent terms in (3.8) is hard to evaluate in general, because the equations are too complicated. In order to understand the nature and origin of the frame dependence we shall therefore concentrate on the heat flux and consider the special case of a rigid rotation in which the fluid elements have constant temperature. That case is characterized by

$$\frac{\partial v_{(i}}{\partial x_{j)}} = 0 \qquad \text{and} \qquad \dot{T} = 0. \tag{3.9}$$

It is easy to see that in such a case the equation $(3.8)_2$ reduces to the simple form[4]

$$q_i = -\pi\tau_q \left[\frac{\partial T}{\partial x_i} + 2\tau_q \left(W_{ik} - \frac{\partial v_{[k}}{\partial x_{i]}}\right)\frac{\partial T}{\partial x_k}\right]. \tag{3.10}$$

This is a remarkable formula, because it says that in general the heat flux is not parallel to the temperature gradient. Indeed, if the gas rotates with respect to an inertial frame, the heat flux contains the term $2\pi\tau_q^2 \frac{\partial v_{[k}}{\partial x_{i]}}\frac{\partial T}{\partial x_k}$, which is perpendicular to the temperature gradient. Or else, when the gas is at rest in a noninertial frame there is the contribution $-2\pi\tau_q^2 W_{ik}\frac{\partial T}{\partial x_k}$ to the heat flux. In general the term perpendicular to the temperature gradient reads

$$-2\pi\tau_q^2\left(W_{ik} - \frac{\partial v_{[k}}{\partial x_{i]}}\right)\frac{\partial T}{\partial x_k} \tag{3.11}$$

which is a frame-dependent objective vector.

In order to understand the nature of the frame-dependent term we consider a gas at rest in a cylinder with a radial temperature gradient as depicted in Figure 4.1a. We concentrate on a small

[4]Square brackets denote antisymmetric tensors.

volume element of the linear dimensions of the mean free path of the atoms. First we consider the gas at rest in an inertial frame. In that case the free paths of the atoms are straight lines and because of the temperature gradient the atoms flying from bottom to top carry a bigger energy upwards across the plane $s - s$ than is carried downwards by the atoms moving down; see Figure 4.1b. Therefore a net flux of energy accompanies the passage of a pair of atoms through $s - s$, and that flux is proportional to the temperature gradient and opposite to it just as predicted by (3.10) for $W_{ik} = 0$ and $\frac{\partial v_{[k}}{\partial x_{i]}} = 0$. Next we consider the same situation for a gas at rest in a noninertial frame. Now the paths of free flight are curved by the Coriolis force, and there is a flux of energy through the plane $H - H$ as well as through the plane $s - s$; see Figure 4.1c. Thus the flux has a component perpendicular to the gradient of temperature as predicted by (3.10) in the case $W_{ik} \neq 0, \frac{\partial v_{[k}}{\partial x_{i]}} = 0$.

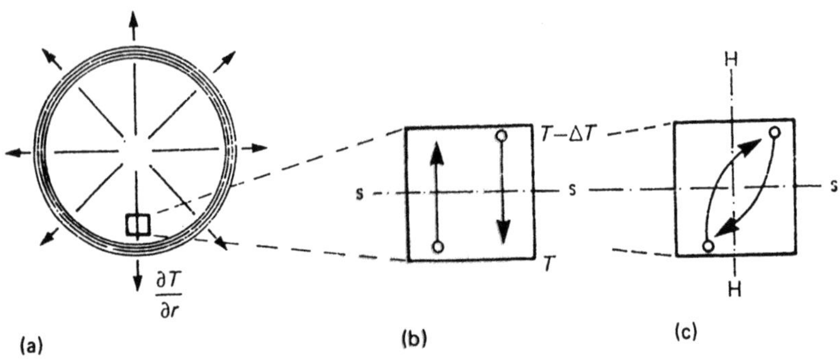

Figure 4.1 On the frame dependence of the heat flux.

It is clear from the above argument that the frame-dependent term in the heat flux is small, because it depends on the extent to which the Coriolis force can bend the free path of an atom. Also formally the smallness is apparent from equation (3.10), since the components of $\tau_q \mathbf{W}$ are quotients of the relaxation time τ_q and the period of rotation of the frame. If we consider τ_q to be of the order of a mean time of free flight of an atom, this quotient is very small indeed for practical frequencies of revolution.

The above suggestive explanation is akin to the explanation of the Hall effect in a conductor under a magnetic field. Indeed, the Lorentz force on the free-moving electrons in the conductor bends their paths in much the same manner as the Coriolis force bends the paths of the atoms. The Hall effect is greater than the Coriolis effect, because the electrons have a much smaller mass than an atom; also the magnetic flux density can easily be made big while the Coriolis force is limited since a number of revolutions of the frame in the order 10^6 Hz is impractical. Therefore the Hall effect is well documented whereas the Coriolis effect has never been observed. Müller [56] has treated both effects jointly, and Chapter 15 gives an account of his arguments.

According to $(3.8)_1$ there is also a frame-dependent term in the stress. That term can also be explained suggestively much in the same manner as the term in the heat flux, but we do not do this here.

3.4 Material Frame Indifference in Ordinary and Extended Thermodynamics

At this stage it is easy to be confused by the question whether or not material frame indifference is valid: In Paragraph 2.1 we use it and in Paragraph 3.3 we reject it. A clarifying remark is in order so as to explain the different roles of material frame indifference in extended thermodynamics and in ordinary thermodynamics.

In extended thermodynamics with its local and instantaneous constitutive functions material frame indifference under Euclidean transformations provides no more restrictive conditions than material frame indifference under Galilean transformations does. We need only the former because we wish to use the Galilei-invariant functions in noninertial, i.e. Euclidean frames.

In ordinary thermodynamics with its nonlocal and—generally—history-dependent constitutive functions material frame indifference under Euclidean transformations is *more* restrictive than Galilean invariance; e.g. it forbids the occurrence of the spin tensor $\frac{\partial v_{[i}}{\partial x_{j]}}$, and that is the point where material frame indifference is wrong—*in ordinary thermodynamics*!

Extended thermodynamics uncovers this error, because—via the Maxwellian iteration—it can *derive* the constitutive relations of ordinary thermodynamics from balance equations. Those balance equations are naturally dependent on frame and therefore extended thermodynamics finds the constitutive of ordinary thermodynamics frame dependent—except in the first iterative step.

4 Thermal Equations of State and Specific Equation for Ideal Gases

4.1 The Classical Ideal Gas

In (3.2) it remains to insert the function $F(z)$ which determines the thermal equation of state, because $p(\varrho, T) = T^{5/2}F(z)$ holds with $z = \frac{\varrho}{T^{3/2}}$. The thermal equation of state of ideal gases has been found empirically in the early history of thermodynamics, and it reads

$$p = \varrho \frac{k}{m} T, \tag{4.1}$$

where m is the atomic mass and $k = 1.38 \cdot 10^{-23} \frac{J}{K}$ is the Boltzmann constant. It follows that in a classical ideal gas we must have

$$F(z) = \frac{k}{m} z. \tag{4.2}$$

Therefore the field equations (3.2) of extended thermodynamics for a classical ideal gas read (henceforth the constant c is set equal to zero; nobody has been able to suggest a physical meaning for c and it is zero in the kinetic theory of gases).

$$\begin{aligned}
&\dot{\varrho} + \varrho \frac{\partial v_i}{\partial x_i} = 0. \\
&\varrho \left[\dot{v}_i - i_i^0 - 2W_{ik} v_k \right] - \frac{\partial (t_{\langle ij \rangle} - p\delta_{ij})}{\partial x_j} = 0 \\
&\varrho \frac{3}{2} \frac{k}{m} \dot{T} + \frac{\partial q_j}{\partial x_j} - (t_{\langle ij \rangle} - p\delta_{ij}) \frac{\partial v_i}{\partial x_j} = 0 \qquad (4.3) \\
&t_{\langle ij \rangle} = 2p\tau_t \left\{ \frac{\partial v_{\langle i}}{\partial x_{j \rangle}} - \frac{1}{2p} \left[\dot{t}_{\langle ij \rangle} + 2t_{\langle n \langle i \rangle} \frac{\partial v_{j \rangle}}{\partial n_n} - 4t_{\langle n \langle i \rangle} W_{j \rangle n} \right] + \frac{2}{5p} \frac{\partial q_{\langle i}}{\partial x_{j \rangle}} \right\} - \tau_t t_{\langle ij \rangle} \frac{\partial v_n}{\partial x_n} \\
&q_i = -\frac{5}{2} \frac{k}{m} p\tau_q \left\{ \frac{\partial T}{\partial x_i} + \frac{1}{\frac{5}{2} \frac{k}{m} p} \left[\dot{q}_i + q_k \frac{\partial v_i}{\partial x_k} - 2q_k W_{ik} \right] - \frac{\frac{k}{m} T}{\frac{5}{2} \frac{k}{m} p} \frac{\partial t_{\langle ij \rangle}}{\partial x_j} \right\} - \\
&\qquad -\tau_q \left(\frac{7}{5} q_i \frac{\partial v_k}{\partial x_k} + \frac{4}{5} q_k \frac{\partial v_i}{\partial x_k} - \frac{7}{2} \frac{k}{m} t_{\langle ij \rangle} \frac{\partial T}{\partial x_j} - \frac{t_{\langle ij \rangle}}{\varrho} \frac{\partial t_{\langle jk \rangle}}{\partial x_k} + \frac{t_{\langle ij \rangle}}{\varrho} \frac{\partial p}{\partial x_j} \right).
\end{aligned}$$

The relaxation times τ_t and τ_q are related to the viscosity μ and the heat conductivity κ by the equations (3.7), which read for a classical ideal gas

$$\mu = \varrho \frac{k}{m} T\tau_t \quad \text{and} \quad \kappa = \frac{5}{2} \varrho \left(\frac{k}{m} \right)^2 T\tau_q. \qquad (4.4)$$

4.2 Comparison with the Kinetic Theory

We compare the equations $(4.3)_{4,5}$ with the corresponding equations (2.3.34) of the kinetic theory of gases and find nearly complete agreement.

Actually the only discrepancy occurs in the equations $(4.3)_{4,5}$ and (2.3.34), and that concerns the relaxation times. While τ_t and τ_q are independent, the corresponding values in the kinetic theory are related by the factor $\frac{2}{3}$. This is due to the assumption of a Maxwellian interatomic potential in the derivation of (2.3.34). For other more realistic potentials there is no such relation, even though τ_t/τ_q is close to 2/3 for all monatomic gases.

The fact that the phenomenological theory of extended thermodynamics has led to the same results as the kinetic theory of gases in the Grad approximation is truly amazing. It shows how little of the complicated apparatus of the kinetic theory is needed for these results and that they follow from a few macroscopic phenomenological principles like Galilean and Euclidean invariance, the entropy principle and the balance type of the field equations.

Nor does the agreement between extended thermodynamics and the kinetic theory stop with the 13 field equations. The entropy and its flux agree as well. Indeed, from (2.47) we obtain with (4.2)

$$\begin{aligned}
\frac{h}{\varrho} &= \left[\frac{k}{m} ln \frac{T^{3/2}}{\varrho} + \varsigma \right] - \frac{t_{\langle ij \rangle} t_{\langle ij \rangle}}{4p\varrho T} - \frac{q_i q_i}{5p^2 T} \\
\varphi_i &= \frac{q_i}{T} + \frac{2}{5pT} t_{\langle ij \rangle} q_j. \qquad (4.5)
\end{aligned}$$

provided we set $c = 0$. These formulae agree completely with (2.3.29) and (2.3.30) from the kinetic theory.

4.3 Comparison with Extended TIP

Comparison of $(4.3)_{4,5}$ with the "phenomenological relations" (2.2.13) of extended TIP shows some agreement, but also differences. The differences are the same ones as those between the kinetic theory and extended TIP, namely

i.) λ in (2.2.13) must be set equal to zero as is appropriate for an ideal gas which does not exhibit bulk viscosity.

ii.) The time derivatives $\dot{q}_i$ and $\dot{t}_{\langle ij\rangle}$ in (2.2.13) must be replaced by

$$\dot{q}_i + q_k \frac{\partial v_i}{\partial x_k} - 2q_k W_{ik} \quad \text{and} \quad \dot{t}_{\langle ij\rangle} + 2t_{\langle n\langle i\rangle} \frac{\partial v_{j\rangle}}{\partial x_n} - 4t_{\langle n\langle i\rangle} W_{j\rangle n}.$$

iii.) Nonlinear terms of the type

$$q_i \frac{\partial v_k}{\partial x_k} \qquad \text{or} \qquad t_{\langle ij\rangle} \frac{\partial t_{\langle ik\rangle}}{\partial x_k} \qquad \text{or} \qquad t_{\langle ij\rangle} \frac{\partial p}{\partial x_i}$$

in $(4.3)_{4,5}$ must be ignored.

iv.) The coefficients a, b, and K in (2.2.13) are now specific. We have

$$a = -\frac{1}{5p^2T}, \qquad b = -\frac{1}{4p\varrho T}, \qquad K = \frac{2}{5pT}.$$

All these points together show to what extent the present systematic version of extended thermodynamics is superior to extended TIP. Indeed, in extended TIP there is no way to calculate the coefficients a, b, and K so that that theory has five unknown coefficients, viz. κ, μ, a, b, and K in the case of monatomic ideal gases, whereas the present theory has only two, namely κ and μ.

4.4 Degenerate Ideal Gases

The thermal equation of state (4.1) is an empirical relation originally, but it can also be derived from molecular considerations in statistical mechanics. That theory reveals that $p = \varrho \frac{k}{m} T$ holds only for high temperatures and small densities where quantum effects may be neglected. At temperatures and densities where such effects are important we say that the gas is degenerate and the particles must be distinguished as Bosons and Fermions. We may then calculate the thermal equation of state by means of statistical mechanics of equilibrium.

The equilibrium phase density f_{E} is then not given by the Maxwell distribution (2.3.23); rather it reads (e.g. see Müller [45])

$$f_{\mathrm{E}} = \frac{1}{e^{\alpha + \frac{m(\mathbf{c}-\mathbf{v})^2}{2kT}} \mp 1} \quad \text{for} \quad \begin{array}{l}\text{Bosons,}\\ \text{Fermions.}\end{array} \tag{4.6}$$

Thus the density $\varrho = \int mf dc$ and the pressure $p = \frac{1}{3}\int mC^2 f dc$ are given by

$$\varrho = 4\pi m \sqrt{\frac{2kT}{m}}^3 i_2^{\mp}(\alpha) \quad \text{and} \quad p = \frac{4\pi}{3} m \sqrt{\frac{2kT}{m}}^5 i_4^{\mp}(\alpha), \tag{4.7}$$

where the i's stand for

$$i_n^{\mp}(\alpha) = \int_0^\infty \frac{x^n}{e^{\alpha+x^2} \mp 1} dx. \tag{4.8}$$

It is easy to show that the i's satisfy the recurrence relation

$$\frac{di_n(\alpha)}{d\alpha} = -\frac{n-1}{2} i_{n-2}(\alpha). \tag{4.9}$$

Note that by $(4.7)_1$ α is a function of $z = \frac{\varrho}{T^{3/2}}$ so that $(4.7)_2$ is a thermal equation of state of the form

$$p = T^{5/2} F(z) \qquad \text{with} \qquad F(z) = \frac{4\pi}{3} m \sqrt{\frac{2k}{m}}^5 i_4^{\mp}(\alpha). \tag{4.10}$$

Since we are unable in general to solve $(4.7)_1$ for α in terms of $z = \frac{\varrho}{T^{3/2}}$, it is impossible to calculate the thermal equation of state $p = p(\varrho, T)$ explicitly, except in the following limiting cases

i.) $\alpha \gg 1$, nondegenerate gas.
In that case the i's assume the values

$$i_n^{\mp}(\alpha) = e^{-\alpha} \int_0^\infty x^n e^{-x^2} dx = e^{-\alpha} \frac{1}{2} \Gamma\left(\frac{n+1}{2}\right) \tag{4.11}$$

and the equilibrium phase density (4.6) tends to the Maxwell distribution

$$f_{\mathrm{E}} = \frac{\varrho}{m} \sqrt{\frac{m}{2\pi kT}}^3 e^{-\frac{m(\mathbf{c}-\mathbf{v})^2}{2kT}}. \tag{4.12}$$

$\Gamma(n)$ in (4.11) is the Γ-function. We obtain $F(z) = \frac{k}{m} z$. This case has already been discussed in Paragraphs 4.1. through 4.3.

ii.) $\alpha \ll -1$, strongly degenerate Fermi gas.
The i's may be calculated to give

$$i_n^{+}(\alpha) = \int_0^{\sqrt{-\alpha}} x^n dx = \frac{1}{n+1} (\sqrt{-\alpha})^{n+1}. \tag{4.13}$$

Thus by (4.10) we have

$$F(z) = \frac{1}{5} \sqrt{\frac{3}{4\pi m}}^3 \left(\frac{\varrho}{T^{3/2}}\right)^{5/3} \qquad \text{hence} \qquad p = \frac{1}{5} \sqrt{\frac{3}{4\pi m}}^3 \varrho^{5/3}. \tag{4.14}$$

In the limit of the strongly degenerate Fermi gas the pressure is thus independent of T; it depends only on ϱ.

iii.) $0 < \alpha \ll 1$, strongly degenerate Bose gas.
The i's assume the form

$$i_n^-(0) = \int_0^\infty \frac{x^n}{e^{x^2}-1} = \frac{1}{2}\Gamma\left(\frac{n+1}{2}\right)\zeta_{\frac{n+1}{2}},$$

where Γ denotes the Γ-function and $\zeta_n = \sum_{r=1}^{\infty}\frac{1}{r^n}$. The values of this sum for different values of n may be found in mathematical handbooks (e.g. see Abramowitz & Stegun [57]). It follows that $F(z)$ and $p = p(\varrho, T)$ have the forms

$$F(z) = \frac{1}{3}4\pi m\sqrt{\frac{2k}{m}}^5 i_4^-(0) \quad \text{and} \quad p = \frac{1}{3}4\pi m\sqrt{\frac{2k}{m}}^5 i_4^-(0)T^{5/2}. \tag{4.15}$$

In the strongly degenerate Bose gas the pressure is independent of ϱ; it depends only on T.

iv.) $\alpha = 0$, Bose-Einstein condensation.
In this case f_{E} has a singularity for $\mathbf{c} - \mathbf{v} = 0$ which corresponds to an indefinite number of atoms in the state of zero energy. We must replace the equation $(4.7)_1$ for Bosons by

$$\varrho = \varrho_0 + 4\pi m\sqrt{\frac{2kT}{m}}^3 i_2^+(0) \tag{4.16}$$

where ϱ_0 is the density of the "condensate" that is formed by the atoms in the zero-energy state. $F(z)$ and p are still given by (4.15) in this case.

For $\alpha = 0$ the function F is a constant given by $(4.15)_1$, and it follows from $(2.47)_1$ that the specific entropy has the form

$$s = \frac{5}{6}4\pi m\sqrt{\frac{2k}{m}}^5 i_4^+(0)\frac{T^{3/2}}{\varrho} + \varsigma$$

or, by (4.16)

$$s = \frac{5}{3}\frac{k}{m}\frac{i_4^+(0)}{i_2^+(0)}(1-\zeta) + \varsigma, \tag{4.17}$$

where ζ is the fraction of atoms in the state of zero energy. Note that for the fully condensed gas the entropy s is equal to the constant ς which is usually taken to be zero.

The field equations for degenerate gases result from insertion of (4.10) into the equations (3.2). We do not write out these results. The limiting cases characterized heretofore lead to several degeneracies in the field equations. It is also obvious from (2.50) that these limiting cases are singular. Indeed, for the strongly degenerate Fermi gas we have $F' = \frac{5}{3}\frac{F}{z}$, while for the strongly degenerate Bose gas we have $F' = 0$. Thus the limiting cases violate the requirement of convexity, and hence the corresponding degeneracies in the field equations are not surprising.

Chapter 5

Thermodynamics of Mixtures of Euler Fluids

The paradoxa of heat conduction and shear wave propagation described in Chapter 2 are not alone. A similar phenomenon occurs in diffusion as we shall demonstrate in this chapter. This fact calls for the formulation of an extended thermodynamic theory of mixtures.

The extension of the ordinary thermodynamic theory of mixtures is conceptually simpler than the extension of the theory of heat conduction and shear stresses that was described in Chapter 4. The reason for this simplicity is that no new – unconventional – balance laws need to be postulated. Indeed, the extension consists of considering the velocities of all constituents as basic fields and the only new balance laws are the equations of balance of the momenta of the constituents.

This chapter presents ordinary thermodynamic of mixtures of fluids and extended thermodynamics of mixtures of Euler fluids. The particular case so characterized is chosen for demonstrative purposes: Onsager relations for diffusion and thermal diffusion can be proved in this case, and the speeds of first and second sound can be calculated. Liquid Helium II turns out to a special case of mixtures of Euler fluids.

The inertial contributions that have appeared in the new balance equations of extended thermodynamics will also appear in this theory of mixtures. They represent the inertial forces on the relative motion of the constituents.

1 Ordinary Thermodynamics of Mixtures (TIP)

1.1 Constitutive Equations

We consider a homogeneous mixture of ν fluids in which all volume elements are occupied by particles of all constituents. One may say that the objective of thermodynamics of such mixtures is the determination of the $\nu + 4$ fields of

$$\begin{array}{ll} \text{mass densities} & \varrho_\alpha \ (\alpha = 1, 2, \ldots \nu) \\ \text{velocity} & v_i \\ \text{temperature} & T. \end{array} \tag{1.1}$$

In thermodynamics of irreversible processes (TIP) the necessary equations are based on the equations of balance of masses of the constituents and of momentum and energy of the mixture, viz.

$$\begin{aligned} \frac{\partial \varrho_\alpha}{\partial t} + \frac{\partial \varrho_\alpha v_j^\alpha}{\partial x_j} &= \tau_\alpha \\ \frac{\partial \varrho v_i}{\partial t} + \frac{\partial (\varrho v_i v_j - t_{ij})}{\partial x_j} &= 0 \\ \frac{\partial \varrho \varepsilon}{\partial t} + \frac{\partial (\varrho \varepsilon v_j + q_j)}{\partial x_j} &= t_{ij} \frac{\partial v_i}{\partial x_j} \end{aligned} \quad \text{or} \quad \begin{aligned} \dot{\varrho} + \varrho \frac{\partial v_i}{\partial x_i} &= 0. \\ \varrho \dot{c}_\alpha + \frac{\partial J_j^\alpha}{\partial x_j} &= \tau_\alpha \\ \varrho \dot{v}_i + \frac{\partial t_{ij}}{\partial x_j} &= 0 \\ \varrho \dot{\varepsilon} + \frac{\partial q_j}{\partial x_j} &= t_{ij} \frac{\partial v_i}{\partial x_j}. \end{aligned} \tag{1.2}$$

The dot denotes the rate of change seen by the observer who moves with the velocity $v_i = \sum_{\alpha=1}^{\nu} \frac{\varrho_\alpha}{\varrho} v_i^\alpha$ of the mixture; $c_\alpha = \frac{\varrho_\alpha}{\varrho}$ is the concentration and $J_i^\alpha = \varrho_\alpha (v_i^\alpha - v_i)$ is the diffusion flux of constituent α. τ_α is its mass production due to chemical reactions. Among the ν productions there are only n independent ones, where n is the number of independent chemical reactions. Indeed, we may express the τ_α's in terms of n reaction rate densities Λ^a by the relations

$$\tau_\alpha = \sum_{a=1}^{n} \gamma_\alpha^a m_\alpha \Lambda^a. \tag{1.3}$$

γ_α^a is the stoichiometric coefficient of constituent α in reaction a and m_α is the molecular mass of constituent α. Note that $\sum_{\alpha=1}^{\nu} \tau_\alpha = 0$ by virtue of the equation $\sum_{\alpha=1}^{\nu} \gamma_\alpha^a m_\alpha = 0$, which represents the conservation of mass in every molecular reaction.

In order to close the system (1.2) we need constitutive equations for

$$\begin{array}{ll} \text{diffusion fluxes} & J_i^\alpha \\ \text{reaction rate densities} & \Lambda^a \\ \text{stress tensor} & t_{ij} \\ \text{internal energy} & \varepsilon \\ \text{heat flux} & q_i. \end{array} \tag{1.4}$$

In TIP such relations are derived from the Gibbs equation for mixtures, viz.

$$\begin{aligned}(\varrho s)^{\bullet} &= \frac{1}{T}\left((\varrho\varepsilon)^{\bullet} - \sum_{\alpha=1}^{\nu}\mu_\alpha \dot{\varrho}_\alpha\right) \quad \text{or equivalently} \\ \dot{s} &= \frac{1}{T}\left(\dot{\varepsilon} - \frac{p}{\varrho^2}\dot{\varrho} - \sum_{\alpha=1}^{\nu-1}(\mu_\alpha - \mu_\nu)\dot{c}_\alpha\right).\end{aligned} \tag{1.5}$$

s is the specific entropy and p is the pressure of the mixture in equilibrium and μ_α are the chemical potentials. $\varepsilon, p,$ and μ_α are considered to be known functions of ϱ, c_α, T from measurements in mixtures in equilibrium.

Elimination of $\dot{\varepsilon}, \dot{\varrho}, \dot{c}_\alpha$ between (1.5) and (1.2) and some rearrangement leads to the equation

$$\begin{aligned}\varrho\dot{s} + \frac{\partial}{\partial x_j}\left(\frac{q_j - \sum_{\alpha=1}^{\nu-1}(\mu_\alpha - \mu_\nu)J_j^\alpha}{T}\right) &= \frac{1}{T}\sum_{a=1}^{n}\left(\sum_{\alpha=1}^{\nu-1}(\mu_\alpha - \mu_\nu)\gamma_\alpha^a m_\alpha\right)\Lambda^a + \\ &+ q_j\frac{\partial\frac{1}{T}}{\partial x_j} - \sum_{\alpha=1}^{\nu-1}J_j^\alpha\frac{\partial\frac{\mu_\alpha-\mu_\nu}{T}}{\partial x_j} + \frac{1}{T}t_{\langle ij\rangle}\frac{\partial v_{\langle i}}{\partial x_{j\rangle}} + \frac{1}{T}\left(\frac{1}{3}t_{ii} + p\right)\frac{\partial v_i}{\partial x_i}\end{aligned} \tag{1.6}$$

which may be interpreted as a balance of entropy if we identify

$$\begin{aligned}\varphi_i &= \frac{1}{T}\left(q_i - \sum_{\alpha=1}^{\nu-1}(\mu_\alpha - \mu_\nu)J_i^\alpha\right) \\ \Sigma &= q_j\frac{\partial\frac{1}{T}}{\partial x_j} - \sum_{\alpha=1}^{\nu-1}J_i^\alpha\frac{\partial\frac{\mu_\alpha-\mu_\nu}{T}}{\partial x_i} + \frac{1}{T}t_{\langle ij\rangle}\frac{\partial v_{\langle i}}{\partial x_{j\rangle}} + \frac{1}{T}\left(\frac{1}{3}t_{ii} + p\right)\frac{\partial v_i}{\partial x_i} + \frac{1}{T}\sum_{a=1}^{n}\left(\sum_{\alpha=1}^{\nu-1}(\mu_\alpha - \mu_\nu)\gamma_\alpha^a m_\alpha\right)\Lambda^a\end{aligned} \tag{1.7}$$

as entropy flux and entropy production respectively.

We observe that the entropy production is a sum of products of

thermodynamic fluxes	and	thermodynamic forces
heat flux q_i		temperature gradient $\frac{\partial\frac{1}{T}}{\partial x_i}$
diffusion fluxes J_i^α		chemical potential gradients $\frac{\partial\frac{\mu_\alpha-\mu_\nu}{T}}{\partial x_i}$
stress deviator $t_{\langle ij\rangle}$		deviatoric velocity gradient $\frac{\partial v_{\langle i}}{\partial x_{j\rangle}}$
dynamic pressure $\pi = -\frac{1}{3}t_{ii} - p$		divergence of velocity $\frac{\partial v_i}{\partial x_i}$
reaction rate densities Λ^a		chemical "force" $\sum_{\alpha=1}^{\nu-1}(\mu_\alpha - \mu_\nu)\gamma_\alpha^a m_\alpha$.

The entropy production must be nonnegative. Assuming linear relations between forces and fluxes

TIP ensures this inequality by constitutive equations of the type

$$\begin{aligned}
\Lambda^a &= \sum_{b=1}^{n} L^{ab}\left(\sum_{\alpha=1}^{\nu-1}(\mu_\alpha-\mu_\nu)\gamma_\alpha^b m_\alpha\right)+L^a\frac{\partial v_i}{\partial x_i}\\
\frac{1}{3}t_{ii}+p &= \sum_{b=1}^{n}\tilde{L}^b\left(\sum_{\alpha=1}^{\nu-1}(\mu_\alpha-\mu_\nu)\gamma_\alpha^b m_\alpha\right)+\lambda\,\frac{\partial v_i}{\partial x_i}
\end{aligned} \tag{1.8}$$

$$\begin{aligned}
q_i &= L\,\frac{\partial\frac{1}{T}}{\partial x_i}+\sum_{\beta=1}^{\nu-1}L_\beta\left(-\frac{\partial\frac{\mu_\beta-\mu_\nu}{T}}{\partial x_i}\right)\\
J_i^\alpha &= \tilde{L}_\alpha\frac{\partial\frac{1}{T}}{\partial x_i}+\sum_{\beta=1}^{\nu-1}L_{\alpha\beta}\left(-\frac{\partial\frac{\mu_\beta-\mu_\nu}{T}}{\partial x_i}\right)
\end{aligned} \tag{1.9}$$

$$t_{\langle ij\rangle} = 2\mu\frac{\partial v_{\langle i}}{\partial x_{j\rangle}}. \tag{1.10}$$

where the *phenomenological coefficients* L^{ab} through μ must satisfy the following inequalities.

$$\begin{bmatrix} L^{ab} & L^a \\ \tilde{L}^b & \lambda \end{bmatrix} \quad \text{— positive definite} \tag{1.11}$$

$$\begin{bmatrix} L & L_\beta \\ \tilde{L}_\alpha & L_{\alpha\beta} \end{bmatrix} \quad \text{— positive definite} \tag{1.12}$$

$$\mu \geq 0. \tag{1.13}$$

The equations (1.9) are generalizations of the original phenomenological laws of Fourier and Fick, according to which the heat flux and the diffusion flux depend on the gradients of temperature and concentrations respectively *but not on both.* TIP permits the temperature gradient to influence the diffusion fluxes and concentration gradients to influence the heat flux; both effects are indeed observed, and they are called the thermo-diffusion effect and the diffusion-thermo effect respectively.

μ and λ are shear- and bulk viscosities respectively. The coefficients $L, L_\beta, \tilde{L}_\alpha$, and $L_{\alpha\beta}$ are the transport coefficients of heat conduction and diffusion, while $\tilde{L}^b$ is called the chemical viscosity.

If the coefficients L and λ, μ are known, the balance equations (1.2) and the constitutive relations (1.8) through (1.10) combine to provide a set of field equations for the fields $\varrho_\alpha, v_i^\alpha$, and T.

In conclusion of this short description of TIP in mixtures we briefly mention Onsager relations. These are symmetry or antisymmetry relations for the matrices of phenomenological coefficients. As generic notation for fluxes and forces we write J_A and X_A so that the constitutive relations (1.8) through (1.10) assume the schematic form

$$J_A = \sum_B L_{AB}X_B. \tag{1.14}$$

Onsager has postulated that either $L_{AB} = L_{BA}$ or $L_{AB} = -L_{BA}$ holds depending on the transformation property of the forces X_A and X_B under time reversal $t^* = -t$: If

$$X_C^* = \varepsilon_C X_C \tag{1.15}$$

holds, with $\varepsilon_C = \pm 1$ for even and odd forces under time reversal, the Onsager relations read

$$L_{AB} = \varepsilon_A \varepsilon_B L_{BA}. \tag{1.16}$$

In the present case we have

$$\sum_{\alpha=1}^{\nu} \mu_\alpha \overset{a}{\gamma_\alpha} m_\alpha, \quad \frac{\partial \frac{1}{T}}{\partial x_i}, \quad -\frac{\partial \frac{\mu_\alpha - \mu_\nu}{T}}{\partial x_i}$$

as even forces and

$$\frac{\partial v_i}{\partial x_i}, \quad \frac{\partial v_{<i}}{\partial x_{j>}}$$

as odd forces, so that (1.16) implies

$$\begin{aligned} L^{ab} &= L^{ba} & \tilde{L}^b &= -L^b \\ L_{\alpha\beta} &= L_{\beta\alpha} & \tilde{L}_\beta &= +L_\beta \end{aligned} \tag{1.17}$$

There is no good proof for the Onsager postulate in general but we shall see that $(1.17)_2$ can be proved in extended thermodynamics of mixtures of Euler fluids.

1.2 Paradox of Diffusion

We investigate the special case of a nonreacting binary mixture at rest, i.e. with $v_i \equiv 0$, in which the temperature is uniform. In that case the only equations of interest among (1.2) and (1.9) are

$$\varrho \frac{\partial c}{\partial t} + \frac{\partial J_i}{\partial x_i} = 0 \qquad \text{with} \qquad J_i = -L_{11} \frac{\mu_c}{T} \frac{\partial c}{\partial x_i}, \tag{1.18}$$

where c and J_i stand for c_1 and J_i^1 and μ_c is an abbreviation for $\frac{\partial(\mu^1 - \mu^2)}{\partial c}$. Elimination of J_i provides a differential equation for c, viz.

$$\frac{\partial c}{\partial t} = L_{11} \frac{\mu_c}{\varrho T} \frac{\partial^2 c}{\partial x_i^2}. \tag{1.19}$$

This equation is called the *diffusion equation*; it is a differential equation of parabolic type analogous to the equations (2.1.9) of heat conduction and shear diffusion. Consequently it predicts the propagation of disturbances in concentration with infinite speed. This is the same kind of paradox that was discussed in Chapter 2 in the context of heat conduction and shear diffusion.

We proceed to develop extended thermodynamics of mixtures in which no such paradox arises.

2 Extended Thermodynamics of Mixtures of Euler Fluids

2.1 Balance Equations

Again we consider a homogeneous mixture of ν constituents. The equations of balance of mass, momentum, and energy of the constituents read as follows:

$$\begin{aligned} \frac{\partial \varrho_\alpha}{\partial t} &+ \frac{\partial}{\partial x_j}(\varrho_\alpha v_j^\alpha) &&= \tau_\alpha \\ \frac{\partial \varrho_\alpha v_i^\alpha}{\partial t} &+ \frac{\partial}{\partial x_j}(\varrho_\alpha v_i^\alpha v_j^\alpha - t_{ij}^\alpha) &&= m_i^\alpha \ (\alpha = 1, 2, \ldots \nu) \\ \frac{\partial \varrho_\alpha(\varepsilon_\alpha + \frac{1}{2}v_\alpha^2)}{\partial t} &+ \frac{\partial}{\partial x_j}\left(\varrho_\alpha(\varepsilon_\alpha + \frac{1}{2}v_\alpha^2)v_j^\alpha - t_{ij}^\alpha v_i^\alpha + q_j^\alpha\right) &&= e_\alpha. \end{aligned} \tag{2.1}$$

They have the same form as the corresponding equations of balance for a single body, except for the nonzero right-hand sides, which represent the production densities of mass, momentum, and energy of constituent α; these terms are due to interaction with the other constituents. Of course, since the mass, momentum, and energy of the mixture are conserved, we must have

$$\sum_{\alpha=1}^{\nu} \tau_\alpha = 0, \qquad \sum_{\alpha=1}^{\nu} m_i^\alpha = 0, \qquad \sum_{\alpha=1}^{\nu} e_\alpha = 0. \tag{2.2}$$

If we sum the equations $(2.1)_1$ over all constituents and introduce

$$\text{the density} \quad \varrho = \sum_{\alpha=1}^{\nu} \varrho_\alpha \qquad \text{and velocity} \quad v_i = \sum_{\alpha=1}^{\nu} \frac{\varrho_\alpha}{\varrho} v_i^\alpha \tag{2.3$_1$}$$

of the mixture, we obtain the mass balance of the mixture, viz.

$$\frac{\partial \varrho}{\partial t} + \frac{\partial}{\partial x_j}(\varrho v_j) \quad = 0. \tag{2.4$_1$}$$

It is convenient to introduce the diffusion velocity u_i^α and the diffusion flux J_i^α by

$$u_i^\alpha = v_i^\alpha - v_i \qquad J_i^\alpha = \varrho_\alpha(v_i^\alpha - v_i). \tag{2.5}$$

With $(2.5)_1$ and the definition of

$$\text{the stress} \qquad t_{ij} = \sum_{\alpha=1}^{\nu}(t_{ij}^\alpha - \varrho_\alpha u_i^\alpha u_j^\alpha) \tag{2.3$_2$}$$

of the mixture we obtain, by summing the equations $(2.1)_2$,

$$\frac{\partial \varrho v_i}{\partial t} + \frac{\partial}{\partial x_j}(\varrho v_i v_j - t_{ij}) = 0. \tag{2.4$_2$}$$

This is the balance equation of momentum of the mixture.

Summing over all equations $(2.1)_3$ and introducing

$$\text{the internal energy density} \qquad \varrho\varepsilon = \sum_{\alpha=1}^{\nu} \varrho_\alpha(\varepsilon_\alpha + \frac{1}{2}u_\alpha^2) \quad \text{and}$$

$$\text{the flux of internal energy} \qquad q_i = \sum_{\alpha=1}^{\nu}(q_i^\alpha + \varrho_\alpha(\varepsilon_\alpha + \frac{1}{2}u_\alpha^2)u_i^\alpha - t_{ij}^\alpha u_j^\alpha) \tag{2.3$_3$}$$

we obtain the balance of energy of the mixture

$$\frac{\partial \varrho(\varepsilon + \frac{1}{2}v^2)}{\partial t} + \frac{\partial}{\partial x_j}\left(\varrho(\varepsilon + \frac{1}{2}v^2)v_j - t_{ij}v_i + q_j\right) = 0. \tag{2.4$_3$}$$

Note that the equations $(2.4)_{1,2,3}$ for the mixture have the same form as those for a single fluid. It was in order to achieve that structure that the definitions $(2.3)_{1,2,3}$ were chosen. This choice goes back to Truesdell [58]; it seems to be the most rational procedure to follow despite the fact that Truesdell summarized these arguments as the "three *metaphysical* principles" of mixture theory.

Of course $(2.4)_{2,3}$ imply a balance of internal energy, which we may write in the form

$$\frac{\partial \varrho\varepsilon}{\partial t} + \frac{\partial}{\partial x_j}(\varrho\varepsilon v_j + q_j) = t_{ij}\frac{\partial v_i}{\partial x_j}. \tag{2.6}$$

2.2 Thermodynamic Processes

The objective of extended thermodynamics of mixtures of fluids is the determination of the $4\nu + 1$ fields of

$$\begin{array}{lll} \text{mass densities} & \varrho_\alpha & \\ \text{velocities} & v_i^\alpha & (\alpha = 1, 2, \ldots \nu) \\ \text{temperature} & T. & \end{array} \tag{2.7}$$

For the determination of these fields we need the appropriate number of field equations. They are based on the equations of balance of masses and momenta of the constituents and of the energy – or internal energy – of the mixture, viz.

$$\begin{array}{llll} \text{masses} & \dfrac{\partial \varrho_\alpha}{\partial t} & + \dfrac{\partial \varrho_\alpha v_j^\alpha}{\partial x_j} & = \tau_\alpha \\ \text{momenta} & \dfrac{\partial \varrho_\alpha v_i^\alpha}{\partial t} & + \dfrac{\partial(\varrho_\alpha v_i^\alpha v_j^\alpha - t_{ij}^\alpha)}{\partial x_j} & = m_i^\alpha \\ \text{(internal) energy} & \dfrac{\partial \varrho\varepsilon}{\partial t} & + \dfrac{\partial(\varrho\varepsilon v_j + q_j)}{\partial x_j} & = t_{ij}\dfrac{\partial v_i}{\partial x_j}. \end{array} \tag{2.8}$$

Among the mass productions τ_α there are only n independent ones as we have argued in Section 1. Indeed we have

$$\tau_\alpha = \sum_{a=1}^{n} \gamma_\alpha^a m_\alpha \Lambda^a \tag{2.9}$$

where – as before – the Λ^a's are the reaction rate densities of the n independent reactions.

The momentum productions m_i^α are not objective vectors, because the momentum of the chemically produced mass contributes to them. However, the combinations $m_i^\alpha - \tau_\alpha v_i^\alpha$ are objective vectors. We call these expressions the interaction forces and denote them by M_i^α:

$$M_i^\alpha = m_i^\alpha - \tau_\alpha v_i^\alpha. \tag{2.10}$$

By $(2.3)_3$ the specific internal energy ε and its flux q_i contain explicit expressions in terms of ϱ_α and v_i^α, viz.

$$\sum_{\alpha=1}^{\nu} \frac{1}{2}\frac{\varrho_\alpha}{\varrho} u_\alpha^2 \quad \text{and} \quad \sum_{\alpha=1}^{\nu} \frac{1}{2}\varrho_\alpha u_\alpha^2 u_i^\alpha. \tag{2.11}$$

It is convenient to split these expressions off and work with the intrinsic values ε^I and q_i^I of the internal energy and its flux, viz.

$$\varrho\varepsilon^I = \sum_{\alpha=1}^{\nu} \varrho_\alpha\varepsilon_\alpha \quad \text{and} \quad q_i^I = \sum_{\alpha=1}^{\nu} \left(q_i^\alpha + \varrho_\alpha\varepsilon_\alpha u_i^\alpha - t_{ij}^\alpha u_j^\alpha\right). \tag{2.12}$$

After these remarks we may close the system (2.8) of equations of balance by constitutive equations for $\Lambda^a, t_{ij}^\alpha, M_i^\alpha, \varepsilon^I$, and q_i^I. In extended thermodynamics the constitutive quantities depend only on the local and present values of the fields, here ρ_α, v_i^α, and T. Therefore the generic form of the constitutive relations in extended thermodynamics of mixtures reads

$$C = \hat{C}(\varrho_\alpha, v_i^\alpha, T). \tag{2.13}$$

Mixture like this are neither viscous nor heat conducting, and therefore we call them mixtures of Euler fluids. Despite their simplicity they form an interesting class of materials, as we shall see. If viscosity and heat conductivity were to be included, the gradients of v_i^α and T should be among the variables—and possibly the gradient of ϱ_α; or else the fields should contain the stress deviators and the flux of internal energy as in Chapter 4.

If the constitutive functions $\hat{C}$ were known explicitly, between (2.8) and (2.13) we should have a system of explicit field equations for the fields $\varrho_\alpha, v_i^\alpha, T$. Every solution of that system is called a *thermodynamic process*.

2.3 Constitutive Theory

As always in extended thermodynamics we adopt the principle of material frame indifference, which states that the constitutive functions $\hat{C}$ are invariant under Euclidean transformations. In the present case that principle requires that the velocities v_i^α cannot appear separately among the variables but only as relative velocities V_i^α defined by

$$V_i^\alpha = v_i^\alpha - v_i^\nu.[1] \tag{2.14}$$

If we restrict the constitutive quantities Λ^a through q_i^I to be at most linearly dependent on V_i^α, we have

$$\begin{aligned} \Lambda^a &= \Lambda^a(\varrho_\gamma, T) \\ t_{ij}^\alpha &= -p^\alpha(\varrho_\gamma, T)\delta_{ij} \end{aligned}$$

[1] It is convenient to single out one constituent—here ν—and refer the motion of the others to it.

$$
\begin{aligned}
M_i^\alpha &= \sum_{\beta=1}^{\nu-1} M_V^{\alpha\beta}(\varrho_\gamma, T) V_i^\beta \\
\varepsilon_I &= \varepsilon_I(\varrho_\gamma, T) \\
q_i^I &= \sum_{\beta=1}^{\nu-1} q_V^\beta(\varrho_\gamma, T) V_i^\beta .
\end{aligned}
\tag{2.15}
$$

Note that the intrinsic flux of internal energy is determined by the relative velocities V_i^β. Therefore the flux of internal energy in a mixture of Euler fluids is due to the diffusion of the constituents.

The entropy principle will provide restrictions on the functions $\Lambda^\alpha(\varrho_\gamma, T)$ through $q_V^\beta(\varrho_\gamma, T)$ that appear in (2.15). The entropy principle requires that the inequality

$$
\frac{\partial \varrho s}{\partial t} + \frac{\partial(\varrho s v_j + \varphi_j)}{\partial x_j} \geq 0 \tag{2.16}
$$

must hold for *all thermodynamic processes.* Here s is the specific entropy and φ_i is the (non-convective) entropy flux. Both are considered as constitutive quantities of the type (2.13), subject to the principle of material frame indifference and at most linearly dependent on V_i^α. Thus we have

$$
\begin{aligned}
s &= s(\varrho_\gamma, T) \\
\varphi_i &= \sum_{\alpha=1}^{\nu-1} \varphi_V^\alpha(\varrho_\gamma, T) V_i^\alpha .
\end{aligned}
\tag{2.17}
$$

The exploitation of the inequality (2.16) makes use of the method of Lagrange multipliers (see Liu [20]). We form the new inequality

$$
\begin{aligned}
\frac{\partial \varrho s}{\partial t} + \frac{\partial(\varrho s v_i + \varphi_i)}{\partial x_i} &- \sum_{\alpha=1}^{\nu} \Lambda^{\varrho_\alpha} \left(\frac{\partial \varrho_\alpha}{\partial t} + \frac{\partial \varrho_\alpha v_j^\alpha}{\partial x_j} - \tau_\alpha \right) - \\
&- \sum_{\alpha=1}^{\nu} \Lambda^{v_i^\alpha} \left(\frac{\partial \varrho_\alpha v_i^\alpha}{\partial t} + \frac{\partial(\varrho_\alpha v_i^\alpha v_j^\alpha - t_{ij}^\alpha)}{\partial x_j} - m_i^\alpha \right) - \\
&- \Lambda^\varepsilon \left(\frac{\partial \varrho \varepsilon}{\partial t} + \frac{\partial(\varrho \varepsilon v_j + q_j)}{\partial x_j} - t_{ij} \frac{\partial v_i}{\partial x_j} \right) \geq 0
\end{aligned}
\tag{2.18}
$$

which must hold for *all fields* $\varrho_\alpha, v_i^\alpha, T$. The Λ's are the Lagrange multipliers; they may be functions of $\varrho_\alpha, v_i^\alpha, T$.

The key to the exploitation of the inequality (2.18) is that—after insertion of the constitutive equations and after all indicated differentiations have been carried out—its left-hand side is explicitly linear in the derivatives

$$
\frac{\partial \varrho_\beta}{\partial t}, \frac{\partial v_i^\beta}{\partial t}, \frac{\partial T}{\partial t}, \frac{\partial \varrho_\beta}{\partial x_k}, \frac{\partial v_i^\beta}{\partial x_k}, \frac{\partial T}{\partial x_k}. \tag{2.19}
$$

The factors of these derivatives must then vanish lest the inequality be violated for some choice of the values (2.19). This requirement leads to the following conditions:

$$\begin{aligned}
&\frac{\partial s}{\partial \varrho_\beta} - \Lambda^\varepsilon \frac{\partial \varepsilon}{\partial \varrho_\beta} + \frac{1}{\varrho}(s - \Lambda^\varepsilon \varepsilon) - \frac{1}{\varrho}\Lambda^{\varrho_\beta} - \frac{1}{\varrho} v_k^\beta \Lambda^{v_k^\beta} = 0 \\
&\frac{\partial s}{\partial V_i^\beta} - \Lambda^\varepsilon \frac{\partial \varepsilon}{\partial V_i^\beta} - \frac{\varrho_\beta}{\varrho}\Lambda^{v_i^\beta} = 0 \qquad (\beta = 1, 2, \ldots \nu - 1) \\
&-\sum_{\beta=1}^{\nu-1}\left(\frac{\partial s}{\partial V_i^\beta} - \Lambda^\varepsilon \frac{\partial \varepsilon}{\partial V_i^\beta}\right) - \frac{\varrho_\nu}{\varrho}\Lambda^{v_i^\nu} = 0. \\
&\frac{\partial s}{\partial T} - \Lambda^\varepsilon \frac{\partial \varepsilon}{\partial T} = 0. \qquad\qquad (2.20) \\
&\frac{\partial \varphi_k}{\partial \varrho_\beta} - \Lambda^\varepsilon \frac{\partial q_k}{\partial \varrho_\beta} - \varrho\left(\frac{\partial s}{\partial \varrho_\beta} - \Lambda^\varepsilon \frac{\partial \varepsilon}{\partial \varrho_\beta}\right) u_k^\beta - \sum_{\alpha=1}^{\nu} \Lambda^{v_k^\alpha} \frac{\partial p^\alpha}{\partial \varrho_\beta} + \Lambda^\varepsilon t_{ik} \frac{1}{\varrho} u_i^\beta = 0. \\
&\left(\frac{\partial s}{\partial \varrho_\beta} - \Lambda^\varepsilon \frac{\partial \varepsilon}{\partial \varrho_\beta}\right)\delta_{ik} - \frac{1}{\varrho\varrho_\beta}\left(\frac{\partial \varphi_k}{\partial V_i^\beta} - \Lambda^\varepsilon \frac{\partial q_k}{\partial V_i^\beta}\right) + \frac{1}{\varrho}\Lambda^{v_i^\beta} u_k^\beta - \frac{1}{\varrho^2}\Lambda^\varepsilon t_{ik} = 0 \quad (\beta = 1, 2, \ldots \nu - 1) \\
&\left(\frac{\partial s}{\partial \varrho_\nu} - \Lambda^\varepsilon \frac{\partial \varepsilon}{\partial \varrho_\nu}\right)\delta_{ik} + \sum_{\beta=1}^{\nu-1} \frac{1}{\varrho\varrho_\nu}\left(\frac{\partial \varphi_k}{\partial V_i^\beta} - \Lambda^\varepsilon \frac{\partial q_k}{\partial V_i^\beta}\right) - \frac{1}{\varrho}\Lambda^{v_i^\nu} u_k^\nu - \frac{1}{\varrho^2}\Lambda^\varepsilon t_{ik} = 0 \\
&\left(\frac{\partial \varphi_k}{\partial T} - \Lambda^\varepsilon \frac{\partial q_k}{\partial T} - \sum_{\alpha=1}^{\nu} \Lambda^{v_k^\alpha} \frac{\partial p_\alpha}{\partial T}\right) = 0.
\end{aligned}$$

There remains the residual inequality

$$\sum_{\alpha=1}^{\nu} \Lambda^{v_i^\alpha} m_i^\alpha + \sum_{a=1}^{n}\left(\sum_{\alpha=1}^{\nu} \Lambda^{\varrho_\alpha} \gamma_\alpha^a m_\alpha\right)\Lambda^a \geq 0. \qquad (2.21)$$

We proceed to evaluate the conditions (2.20).

First of all we list two useful identities that will allow us to switch between the relative velocities V_i^α and the diffusion fluxes $J_i^\alpha = \varrho_\alpha u_i^\alpha$. They read

$$\varrho_\beta u_i^\beta = \sum_{\gamma=1}^{\nu-1} \underbrace{\left(\varrho_\beta \delta_{\beta\gamma} - \frac{\varrho_\beta \varrho_\gamma}{\varrho}\right)}_{\overset{-1}{F}_{\beta\gamma}} V_i^\gamma \iff V_i^\gamma = \sum_{\beta=1}^{\nu-1} \underbrace{\left(\frac{1}{\varrho_\beta}\delta_{\gamma\beta} + \frac{1}{\varrho_\nu}\right)}_{F_{\beta\gamma}} \varrho_\beta u_i^\beta. \qquad (2.22)$$

Now we recall that s is independent of V_i^γ and that ε depends on V_i^γ only through the explicit term $\sum_{\alpha=1}^{\nu} \frac{1}{2}\frac{\varrho_\alpha}{\varrho} u_\alpha^2$. We find

$$\frac{\partial \varepsilon}{\partial V_i^\beta} = -\frac{\varrho_\beta}{\varrho} u_i^\beta$$

and thus obtain from $(2.20)_{2,3}$

$$\Lambda^{v_i^\beta} = -\Lambda^\varepsilon u_i^\beta. \qquad (2.23)$$

The equations $(2.20)_{4,1}$ may be written in the forms

$$\begin{aligned}\frac{\partial s}{\partial T} &= \Lambda^\varepsilon \frac{\partial \varepsilon_I}{\partial T} \\ \frac{\partial \varrho s}{\partial \varrho_\beta} &= \Lambda^\varepsilon \frac{\partial \varrho \varepsilon_I}{\partial \varrho_\beta} + \underbrace{\Lambda^{\varrho_\beta} + \Lambda^\varepsilon \left(\frac{1}{2}u_\beta^2 - v_k^\beta u_k^\beta\right)}_{\Lambda_I^{\varrho_\beta}}\end{aligned}$$

which we may combine to read

$$d(\varrho s) = \Lambda^\varepsilon d(\varrho \varepsilon_I) + \sum_{\beta=1}^{\nu} \Lambda_I^{\varrho_\beta} d\varrho_\beta. \tag{2.24}$$

$\Lambda_I^{\varrho_\beta}$ is the velocity-independent part of the Lagrange multiplier Λ^{ϱ_β}. Compairing equation (2.24) with the Gibbs relation of thermostatics (see $(1.5)_1$) we conclude that

$$\Lambda^\varepsilon = \frac{1}{T} \quad \text{and} \quad \Lambda_I^{\varrho_\beta} = -\frac{\mu_\beta}{T} \tag{2.25}$$

holds. Thus all Lagrange multipliers have been identified. T and μ_β are the absolute temperature of the mixture and the chemical potentials of its constituents. For future reference we rewrite the Gibbs equation in two forms, viz.

$$d(\varrho s) = \frac{1}{T}\left(d(\varrho\varepsilon_I) - \sum_{\alpha=1}^{\nu} \mu_\alpha d\varrho_\alpha\right) \quad \text{or} \quad d\left[\varrho(\varepsilon_I - Ts)\right] = -s dT + \sum_{\alpha=1}^{\nu} \mu_\alpha d\varrho_\alpha. \tag{2.26}$$

From $(2.20)_6$ and $(2.20)_7$ we obtain by summation and subtraction respectively

$$\varrho(\varepsilon_I - Ts) + p = \sum_{\alpha=1}^{\nu} \varrho_\alpha \mu_\alpha \quad \text{and} \tag{2.27}$$

$$\varphi_j = \frac{q_j^I - \sum_{\alpha=1}^{\nu} \varrho_\alpha \mu_\alpha u_j^\alpha}{T}, \tag{2.28}$$

where $p = \sum_{\alpha=1}^{\nu} p_\alpha$ has been introduced. It remains to exploit the conditions $(2.20)_5$ and $(2.20)_8$. From $(2.20)_5$ we obtain after some rearrangement.

$$\frac{\partial \mu_\alpha}{\partial \varrho_\beta} = \frac{1}{\varrho_\alpha}\frac{\partial p_\alpha}{\partial \varrho_\beta}, \tag{2.29}$$

a relation that has a severely restrictive consequence. Indeed, the integrability condition implied by (2.29) reads

$$\frac{\partial p_\alpha}{\partial \varrho_\beta} = 0 \, (\alpha \neq \beta), \text{hence by (2.29)} \quad \frac{\partial \mu_\alpha}{\partial \varrho_\beta} = 0 \quad (\alpha \neq \beta). \tag{2.30}$$

It follows that a mixture of Euler fluids characterized by generic constitutive relations of the form (2.13) is simple in the sense that the partial pressure p_α and the chemical μ_α of constituent α

depend only on their own density ϱ_α—and on T, of course. Another aspect of this simplicity reveals itself when we make use of the relation

$$\mu_\alpha = \frac{\partial\left[\varrho(\varepsilon_I - Ts)\right]}{\partial \varrho_\alpha} \tag{2.31}$$

which follows from $(2.26)_2$. Insertion into $(2.30)_2$ gives

$$\frac{\partial^2 \varrho[\varepsilon_I - Ts]}{\partial \varrho_\alpha \partial \varrho_\beta} = 0 \quad \text{whence follows} \quad \varrho[\varepsilon_I - Ts] = \sum_{\alpha=1}^{\nu} \varrho_\alpha \psi_\alpha(\varrho_\alpha, T) \tag{2.32}$$

so that the free energy density $\varrho[\varepsilon_I - Ts]$ is a sum of functions of which each one depends only on one density. Obviously, by (2.26) this same summational character then holds for ϱs and hence for $\varrho\varepsilon_I$

$$\varrho s = \sum_{\alpha=1}^{\nu} \varrho_\alpha s_\alpha(\varrho_\alpha, T) \qquad \varrho\varepsilon_I = \sum_{\alpha=1}^{\nu} \varrho_\alpha \varepsilon_\alpha(\varrho_\alpha, T). \tag{2.33}$$

Clearly $s_\alpha(\varrho_\alpha, T)$ and $\varepsilon_\alpha(\varrho_\alpha, T)$ must be interpreted as the specific entropy and internal energy of the constituent α. Also we may conclude that the relation (2.27) and the Gibbs equation (2.26) may be decomposed into expressions for the single constituents, viz.

$$\varrho_\alpha(\varepsilon_\alpha - Ts_\alpha) + p_\alpha = \varrho_\alpha \mu_\alpha \tag{2.34}$$

$$ds_\alpha = \frac{1}{T}(d\varepsilon_\alpha - \frac{p_\alpha}{\varrho_\alpha^2} d\varrho_\alpha). \tag{2.35}$$

The only relation in (2.20) yet to be exploited is $(2.20)_8$. We insert whatever we have learned so far and obtain after a little calculation

$$q_k^I = \sum_{\alpha=1}^{\nu} \left(\varepsilon_\alpha + \frac{p_\alpha}{\varrho_\alpha}\right) \varrho_\alpha u_k^\alpha. \tag{2.36}$$

By comparison with $(2.3)_3$ we conclude that the intrinsic flux of internal energy represents only the convection of the partial enthalpy $\varepsilon_\alpha + \frac{p_\alpha}{\varrho_\alpha}$ with the diffusive motion; in particular, in a simple mixture with the generic constitutive relations (2.13) the partial fluxes q_i^α are all zero.

It remains to exploit the residual inequality (2.21). An easy calculation shows that this inequality may be written in the form

$$-\sum_{\alpha,\beta=1}^{\nu-1} M_V^{\alpha\beta} V_i^\alpha V_i^\beta - \sum_{a=1}^{n} \left(\sum_{\alpha=1}^{\nu-1} \gamma_\alpha^a m_\alpha (\mu_\alpha - \mu_\nu + \frac{1}{2} V_\alpha^2)\right) \Lambda^a \geq 0. \tag{2.37}$$

We conclude that in equilibrium, where V_i^α and Λ^a all vanish, we must have

$$M_V^{\alpha\beta}|_E \quad \text{— negative definite} \qquad \text{and} \qquad \sum_{\alpha=1}^{\nu} \gamma_\alpha^a m_\alpha \mu_\alpha|_E = 0. \tag{2.38}$$

The second relation is the law of mass action which at a given temperature provides n constraints on the equilibrium values of the densities ϱ_β—as many as there are independent reactions. If the law of mass action is not satisfied, chemical reactions will occur and in a linear theory we must set

$$\Lambda^a = \sum_{b=1}^{n} L_{ab} \left(\sum_{\alpha=1}^{\nu} \gamma_\alpha^b m_\alpha \mu_\alpha \right), \tag{2.39}$$

where the coefficients L_{ab} may depend on $\nu - n$ densities and T.

2.4 Summary of Results

We summarize the results of the previous two paragraphs. These results are applicable to mixtures of Euler fluids.

For the determination of the $4\nu + 1$ fields $\varrho_\beta, v_i^\beta, T$ we have the equations of balance

$$\frac{\partial \varrho_\alpha}{\partial t} + \frac{\partial \varrho_\alpha v_j^\alpha}{\partial x_j} = \tau_\alpha \quad \text{with } \tau_\alpha = \sum_{a=1}^{n} \gamma_\alpha^a m_\alpha \Lambda^a \tag{2.40}$$

$$\frac{\partial \varrho_\alpha v_i^\alpha}{\partial t} + \frac{\partial (\varrho_\alpha v_i^\alpha v_j^\alpha - t_{ij}^\alpha)}{\partial x_j} = m_i^\alpha \quad \text{with } t_{ij}^\alpha = -p_\alpha(\varrho_\alpha, T)\delta_{ij} \tag{2.41}$$

$$m_i^\alpha = \tau_\alpha v_i^\alpha + \textstyle\sum_{\beta=1}^{\nu-1} M_V^{\alpha\beta}(\varrho_\gamma, T) V_i^\beta$$

$$\frac{\partial \varrho \varepsilon}{\partial t} + \frac{\partial (\varrho \varepsilon v_j + q_j)}{\partial x_j} = t_{ij} \frac{\partial v_i}{\partial x_j} \text{ with } \varrho = \textstyle\sum_{\alpha=1}^{\nu} \varrho_\alpha,\ v_i = \sum_{\alpha=1}^{\nu} \frac{\varrho_\alpha}{\varrho} v_i^\alpha \tag{2.42}$$

$$t_{ij} = -\textstyle\sum_{\alpha=1}^{\nu} \left(p_\alpha(\varrho_\alpha T)\delta_{ij} + \varrho_\alpha u_i^\alpha u_j^\alpha \right)$$

$$\varepsilon = \textstyle\sum_{\alpha=1}^{\nu} \frac{\varrho_\alpha}{\varrho} \left(\varepsilon_\alpha(\varrho_\alpha, T) + \frac{1}{2} u_\alpha^2 \right) = \varepsilon_I + \sum_{\alpha=1}^{\nu} \frac{1}{2} \frac{\varrho_\alpha}{\varrho} u_\alpha^2$$

$$q_i = \textstyle\sum_{\alpha=1}^{\nu} \varrho_\alpha \left(\varepsilon_\alpha + \frac{p_\alpha}{\varrho_\alpha} + \frac{1}{2} u_\alpha^2 \right) u_i^\alpha.$$

The specific entropy s and the entropy flux φ_i are given in the forms

$$s = \sum_{\alpha=1}^{\nu} \frac{\varrho_\alpha}{\varrho} s_\alpha(\varrho_\alpha, T) \tag{2.43}$$

$$\varphi_i = \frac{\sum_{\alpha=1}^{\nu} \varrho_\alpha \left(T s_\alpha + \frac{1}{2} u_\alpha^2 \right) u_j^\alpha}{T}. \tag{2.44}$$

Gibbs equations hold for the mixture and for all constituents

$$d(\varrho s) = \frac{1}{T} \left(d(\varrho \varepsilon_I) - \sum_{\alpha=1}^{\nu} \mu_\alpha d\varrho_\alpha \right) \tag{2.45}$$

$$ds_\alpha = \frac{1}{T} \left(d\varepsilon_\alpha - \frac{p_\alpha}{\varrho_\alpha^2} d\varrho_\alpha \right), \quad \text{hence} \quad \frac{\partial \varepsilon_\alpha}{\partial \varrho_\alpha} = -\frac{1}{\varrho_\alpha^2} T^2 \frac{\partial \frac{p_\alpha}{T}}{\partial T} \tag{2.46}$$

and the chemical potential of constituent α may be written as

$$\mu_\alpha(\varrho_\alpha, T) = \varepsilon_\alpha - Ts_\alpha + \frac{p_\alpha}{\varrho_\alpha}, \quad \text{hence} \quad \frac{\partial \mu_\alpha}{\partial \varrho_\alpha} = \frac{1}{\varrho_\alpha}\frac{\partial p_\alpha}{\partial \varrho_\alpha}. \tag{2.47}$$

The coefficients $M_V^{\alpha\beta}$ form a negative-definite matrix and in equilibrium the densities ϱ_α and temperature T are related by the law of mass action, viz.

$$\sum_{\alpha=1}^{\nu} \gamma_\alpha^a m_\alpha \mu_\alpha|_E = 0. \tag{2.48}$$

In nonequilibrium the linear expression for the reaction rate densities reads

$$\Lambda^a = \sum_{b=1}^{n} L_{ab}\left(\sum_{\alpha=1}^{\nu} \gamma_\alpha^b m_\alpha \mu_\alpha\right). \tag{2.49}$$

2.5 Wave Propagation in a Nonreacting Binary Mixture

In this section we wish to demonstrate explicitly that all wave speeds are finite in mixtures according to extended thermodynamics. As an illustrative special case we treat only the case of a nonreacting binary mixture. In such a mixture it is appropriate for intuitive reasons to replace the variables $\varrho_1, \varrho_2, v_i^1, v_i^2, T$ by the equivalent set

$$\varrho = \varrho_1 + \varrho_2,\; v_i = \frac{\varrho_1}{\varrho}v_i^1 + \frac{\varrho_2}{\varrho}v_i^2,\; T,\; c = \frac{\varrho_1}{\varrho},\; J_i = \varrho_1(v_i^1 - v_i). \tag{2.50}$$

c is called the concentration of constituent 1 and J_i is its diffusion flux. It is then proper to replace the governing balance laws by the set.

$$\begin{aligned}
&\frac{\partial \varrho}{\partial t} + \frac{\partial \varrho v_j}{\partial x_j} = 0.\\
&\frac{\partial \varrho v_i}{\partial t} + \frac{\partial(\varrho v_i v_j - t_{ij})}{\partial x_j} = 0\\
&\frac{\partial \varrho\varepsilon}{\partial t} + \frac{\partial(\varrho\varepsilon v_j + q_j)}{\partial x_j} = t_{ij}\frac{\partial v_i}{\partial x_j}\\
&\varrho\left(\frac{\partial c}{\partial t} + v_j\frac{\partial c}{\partial x_j}\right) + \frac{\partial J_j}{\partial x_j} = 0.\\
&\frac{\partial J_i}{\partial t} + \frac{\partial}{\partial x_j}\left(J_i v_j - \frac{1}{\varrho c}J_i J_j - (t_{ij}^1 - c t_{ij})\right) = t_{ij}\frac{\partial c}{\partial x_j} - J_j\frac{\partial v_i}{\partial x_j} + m_i^1.
\end{aligned} \tag{2.51}$$

The constitutive relations are still the same ones as those summarized in Paragraph 2.4. But the new choice of variables makes some of the restrictive conditions inappropriate. In particular, the Gibbs equations (2.45) should now be written in the new form

$$ds = \frac{1}{T}(d\varepsilon_I - \frac{p}{\varrho^2}d\varrho - \mu dc), \tag{2.52}$$

where μ stands for $\mu_1 - \mu_2$. The integrability conditions implied by (2.52) read

$$\frac{\partial \varepsilon_I}{\partial \varrho} = -\frac{1}{\varrho^2} T^2 \frac{\partial \frac{p}{T}}{\partial T}, \qquad \frac{\partial \varepsilon_I}{\partial c} = -T^2 \frac{\partial \frac{\mu}{T}}{\partial T}, \qquad \frac{\partial p}{\partial c} = \varrho^2 \frac{\partial \mu}{\partial \varrho}. \tag{2.53}$$

We linearize the equations in all deviations $\bar{\varrho}, \bar{v}_i, \bar{T}, \bar{c}, \bar{J}_i$ from a homogeneous and constant rest state $\tilde{\varrho}, \tilde{v}_i = 0, \tilde{T}, \tilde{c}, \tilde{J}_i = 0$ and obtain

$$\begin{bmatrix} \frac{\partial}{\partial t} & \tilde{\varrho}\frac{\partial}{\partial x_i} & 0 & 0 & 0 \\ \tilde{p}_\varrho \frac{\partial}{\partial x_i} & \tilde{\varrho}\frac{\partial}{\partial t} & \tilde{p}_{\mathrm{T}}\frac{\partial}{\partial x_i} & \tilde{p}_c \frac{\partial}{\partial x_i} & 0 \\ -\frac{\tilde{T}}{\tilde{\varrho}^2}\tilde{p}_{\mathrm{T}}\frac{\partial}{\partial t} & 0 & \tilde{\varepsilon}_{\mathrm{T}}\frac{\partial}{\partial t} & 0 & \frac{\tilde{T}}{\tilde{\varrho}}\left(\frac{\tilde{p}^1_{\mathrm{T}}}{\tilde{\varrho}_1} - \frac{\tilde{p}^2_{\mathrm{T}}}{\tilde{\varrho}_2}\right)\frac{\partial}{\partial x_i} \\ 0 & 0 & 0 & \tilde{\varrho}\frac{\partial}{\partial t} & \frac{\partial}{\partial x_i} \\ -\frac{\tilde{\varrho}_1\tilde{\varrho}_2}{\tilde{\varrho}}\tilde{\mu}_\varrho\frac{\partial}{\partial x_i} & 0 & -\frac{\tilde{\varrho}_1\tilde{\varrho}_2}{\tilde{\varrho}}\left(\frac{\tilde{p}^1_{\mathrm{T}}}{\tilde{\varrho}_1} - \frac{\tilde{p}^2_{\mathrm{T}}}{\tilde{\varrho}_2}\right)\frac{\partial}{\partial x_i} & -\frac{\tilde{\varrho}_1\tilde{\varrho}_2}{\tilde{\varrho}}\tilde{\mu}_c\frac{\partial}{\partial x_i} & \frac{\partial}{\partial t} - M^{11}_V \frac{\tilde{\varrho}}{\tilde{\varrho}_1\tilde{\varrho}_2} \end{bmatrix} \begin{bmatrix} \bar{\varrho} \\ \bar{v}_i \\ \bar{T} \\ \bar{c} \\ \bar{J}_i \end{bmatrix} = 0. \tag{2.54}$$

The quantities $p_\varrho, \varepsilon_{\mathrm{T}}, \mu_c$ etc. denote derivatives of p, ε, and μ with respect to the indicated variables.

We eliminate v_i, T, and J_i from this linear first-order system of partial differential equations and obtain a coupled set of second-order differential equations for $\bar{\varrho}$ and $\bar{c}$, viz.

$$\begin{aligned} &\frac{\partial^2 \bar{\varrho}}{\partial t^2} - \left[p_\varrho + \frac{T p_T^2}{\varrho^2 \varepsilon_T}\right] \Delta\bar{\varrho} - \left[p_c + \frac{T}{\varepsilon_T} p_T \left(\frac{p^1_T}{\varrho_1} - \frac{p^2_T}{\varrho_2}\right)\right] \Delta\bar{c} = 0 \\ &\frac{\partial^2 \bar{c}}{\partial t^2} + M^{11}_V \frac{\varrho}{\varrho_1\varrho_2}\frac{\partial \bar{c}}{\partial t} - \frac{\varrho_1\varrho_2}{\varrho^2}\left[\mu_c + \frac{T}{\varepsilon_T}\left(\frac{p^1_T}{\varrho_1} \quad \frac{p^2_T}{\varrho_2}\right)^2\right] \Delta\bar{c} - \frac{\varrho_1\varrho_2}{\varrho^4}\left[p_c + \frac{T}{\varepsilon_T} p_T \left(\frac{p^1_T}{\varrho_1} - \frac{p^2_T}{\varrho_2}\right)\right] \Delta\bar{\varrho} = 0. \end{aligned} \tag{2.55}$$

The tilde has been dropped from the coefficients for convenience.

The coupling of the fields $\bar{\varrho}$ and $\bar{c}$ is provided by the last term in each equation with the coefficient

$$W = p_c + \frac{T}{\varepsilon_T} p_T \left(\frac{p^1_T}{\varrho_1} - \frac{p^2_T}{\varrho_2}\right). \tag{2.56}$$

Let us suppose for a moment that that term were absent. In that case the first equation is the usual equation for the propagation of sound with the speed

$$V_\varrho^2 = p_\varrho + \frac{T p_T^2}{\varrho^2 \varepsilon_T}. \tag{2.57}$$

The second equation is the telegraph equation—always assuming $W = 0$—and it describes the propagation of what has been called *second sound*, a strongly damped mode of sound propagation because of the term with M_V^{11}. According to $(2.55)_2$ the speed of the second sound is given by

$$V_c^2 = \frac{\varrho_1\varrho_2}{\varrho^2}\left(\mu_c + \frac{T}{\varepsilon_T}\left(\frac{p_T^1}{\varrho_1} - \frac{p_T^2}{\varrho_2}\right)^2\right) = \frac{\varrho_1\varrho_2}{\varrho^2}\left(\mu_c - \frac{p_c}{p_T}\left(\frac{p_T^1}{\varrho_1} - \frac{p_T^2}{\varrho_2}\right)\right). \tag{2.58}$$

where $(2.58)_2$ only holds for $W = 0$.

We continue to consider the case of no interaction and note that without the term $\frac{\partial J_i}{\partial t}$ in (2.51), which describes the acceleration of the relative motion of the two constituents, the second equation in (2.55) would read

$$-M_V^{11}\frac{\partial \bar{c}}{\partial t} + \frac{\varrho_1^2\varrho_2^2}{\varrho^3}\left[\mu_c + \frac{T}{\varepsilon_T}\left(\frac{p_T^1}{\varrho_1} - \frac{p_T^2}{\varrho_2}\right)^2\right]\Delta\bar{c} = 0. \tag{2.59}$$

This is a parabolic equation. It provides infinite speeds of propagation of disturbances in concentration. We recall the arguments of Section 1 about the paradox of diffusion and conclude that the paradox arises because we have ignored the inertia of the diffusive motion, i.e. the term with $\frac{\partial J_i}{\partial t}$.

Now, with coupling, i.e. with $W \neq 0$, we proceed to eliminate $\bar{c}$ between the equations (2.55) and obtain a single equation of fourth order in time, viz.

$$\frac{\partial^4 \bar{c}}{\partial t^4} + M_V^{11}\frac{\varrho}{\varrho_1\varrho_2}\frac{\varrho^3\bar{c}}{\partial t^3} - (V_\varrho^2 + V_c^2)\frac{\partial^2\Delta\bar{c}}{\partial t^2} - M_V^{11}\frac{\varrho}{\varrho_1\varrho_2}V_\varrho^2\frac{\partial\Delta\bar{c}}{\partial t} + \left(V_\varrho^2V_c^2 - \frac{\varrho_1\varrho_2}{\varrho^4}W^2\right)\Delta\Delta c = 0. \tag{2.60}$$

This is again a hyperbolic equation. It predicts the propagation of the first and second sound with the speeds

$$V_\pm = \sqrt{2}\frac{\sqrt{\frac{V_\varrho^2V_c^2 - \frac{\varrho_1\varrho_2}{\varrho^4}W^2}{V_\varrho^2+V_c^2}}}{\sqrt{\left\{1 \pm \sqrt{1 - 4\frac{V_\varrho^2V_c^2 - \frac{\varrho_1\varrho_2}{\varrho^4}W^2}{(V_\varrho^2+V_c^2)^2}}\right\}}}. \tag{2.61}$$

Additional insight into the nature of first and second sound is gained by the consideration of plane harmonic waves of small amplitude. We introduce

$$\bar{\varrho} = \hat{\varrho}E \quad \bar{v}_i = \hat{v}_iE \quad \bar{T} = \hat{T}E \quad \bar{c} = \hat{c}E \quad \bar{J}_i = \hat{J}_iE \tag{2.62}$$

into the linearized equations (2.54). E stands for

$$E = e^{i(\omega t - k\mathbf{n}\cdot\mathbf{x})} \tag{2.63}$$

and $\hat{\varrho}$ through $\hat{J}_i$ are complex amplitudes. We obtain with $u = \omega/k$.

$$\begin{bmatrix} u & -\varrho n_i & 0 & 0 & 0 \\ -p_\varrho n_i & \varrho u & -p_T n_i & -p_c n_i & 0 \\ -\frac{T}{\varrho^2}p_T u & 0 & \varepsilon_T u & 0 & -\frac{T}{\varrho}\left(\frac{p_T^1}{\varrho_1}-\frac{p_T^2}{\varrho_2}\right)n_i \\ 0 & 0 & 0 & \varrho u & -n_i \\ \frac{\varrho_1\varrho_2}{\varrho}\mu_\varrho n_i & 0 & \frac{\varrho_1\varrho_2}{\varrho}\left(\frac{p_T^1}{\varrho_1}-\frac{p_T^2}{\varrho_2}\right)n_i & \frac{\varrho_1\varrho_2}{\varrho}\mu_c n_i & u\left(1+iM_V^{11}\frac{\varrho}{\varrho_1\varrho_2}\frac{1}{\omega}\right) \end{bmatrix} \begin{bmatrix} \hat{\varrho} \\ \hat{v}_i \\ \hat{T} \\ \hat{c} \\ \hat{J}_i \end{bmatrix} = 0. \tag{2.64}$$

Inspection of $(2.64)_{2,5}$ shows that both v_i and J_i are longitudinal waves. Without loss of generality we may then take n_i, v_i, and J_i to point in the x_1-direction. Elimination of $\hat{\varrho}, \hat{T}$, and $\hat{c}$ provides a linear system of equations for $\hat{v}$ and $\hat{J}$, viz. with $\pi = \dfrac{p_T^1}{\varrho_1} - \dfrac{p_T^2}{\varrho_2}$

$$\begin{aligned} -\varrho\left(p_\varrho - \frac{Tp_T^2}{\varrho^2\varepsilon_T} - u^2\right) \hat{v} - \frac{1}{\varrho}\left(p_c + T\frac{p_T}{\varepsilon_T}\pi\right) \hat{J} &= 0 \\ -\varrho_1\varrho_2\left(\mu_\varrho - \frac{1}{p_T}\pi(p_\varrho - u^2)\right)\hat{v} - \left(\frac{\varrho_1\varrho_2}{\varrho^2}\left(\mu_c - \frac{p_c}{p_T}\pi\right) - u^2\left(1+iM_V^{11}\frac{\varrho}{\varrho_1\varrho_2}\frac{1}{\omega}\right)\right)\hat{J} &= 0. \end{aligned} \tag{2.65}$$

We investigate the uncoupled case in which W, defined by (2.56), vanishes and obtain

$$\begin{aligned} \left(p_\varrho + \frac{Tp_T^2}{\varrho^2\varepsilon_T} - u^2\right)\hat{v} &= 0 \\ \frac{\varrho_1\varrho_2}{p_T}\pi\left(p_\varrho + \frac{Tp_T^2}{\varrho^2\varepsilon_T} - u^2\right)\hat{v} - \left(\frac{\varrho_1\varrho_2}{\varrho^2}\left(\mu_c - \frac{p_c}{p_T}\pi\right) - u^2\left(1+iM_V^{11}\frac{\varrho}{\varrho_1\varrho_2}\frac{1}{\omega}\right)\right)\hat{J} &= 0. \end{aligned} \tag{2.66}$$

For $\omega \to \infty$ the solutions $u_{1,2}$ are the pulse speeds of disturbances. From (2.66) we conclude

i.) Either $\hat{v} = 0$ holds, in which case $\hat{J}$ may be unequal to zero and pulses propagate with the speed

$$u = V_c \quad \text{where } V_c \text{ is given by (2.58),}$$

ii.) or $\hat{J} = 0$ holds, in which case $\hat{v}$ may be unequal to zero and pulses propagate with the speed

$$u = V_\varrho \quad \text{where } V_\varrho \text{ is given by (2.57).}$$

In the first case we speak of the second sound: The two constituents oscillate with respect to each other and the mixture as a whole is at rest; we have $\hat{\varrho} = 0$, while $\hat{c} \neq 0$. The second case corresponds to the propagation of the first sound: The two constituents are at rest with respect to each other while the mixture as a whole oscillates; we have $\hat{c} = 0$ and $\hat{\varrho} \neq 0$. In both cases the temperature oscillates; indeed, from $(2.64)_2$ we conclude that

$$\hat{T} = \frac{1}{p_T}\left((u^2 - p_\varrho)\hat{\varrho} - p_c\hat{c}\right) \tag{2.67}$$

holds always for the uncoupled case.

Note that for the second sound u is complex, according to (2.66), so that the second sound is damped. In fact the damping is so strong that the second sound has never been observed in mixtures of gases.

2.6 Landau Equations. First and Second Sound in He II

The only case in which the second sound has been observed in a fluid is the case of liquid He II. This of course is not a mixture in the proper sense of the word, because it consists only of Helium atoms. We may consider He II as a *phase* mixture consisting of the superfluid component s, whose atoms are in the state of Bose-Einstein condensation, and the normal fluid n, whose atoms have nonzero energies. We have given a brief description of a gas in the state of Bose-Einstein condensation in Chapter 4, Paragraph 4.4.4, and we recall that the entropy of the fully condensed gas vanishes. Thus in the theory of mixtures between the constituents s and n we have to set

$$s_s = 0. \tag{2.68}$$

The most striking observation that can be made in He II is that there is no dissipation accompanying the diffusion and the transition between the phases s and n, i.e. there is no entropy production. We recall from (2.37) that in a binary mixture of Euler fluids there are two "mechanisms" of entropy production, namely diffusion and "chemical reaction," i.e. here: transition between the phases s and n. Since there is only one "reaction," and since the stoichiometric coefficients are 1 and -1 we may write (2.37) in the form

$$-M_V^{ss}V_s^2 - m\left(\mu_s - \mu_n + \frac{1}{2}V_s^2\right)\Lambda \geq 0. \tag{2.69}$$

The only way that the left-hand side can vanish in the presence of diffusion and phase transition is that $M_V^{ss} = 0$ and $\mu_s - \mu_n + \frac{1}{2}V_s^2 = 0$ holds. In other words, the interaction force between the diffusing phases must vanish and—ignoring the nonlinear term V_s^2—the chemical potentials of the phases must always be equal. We may therefore write

$$M_V^{ss} = 0 \quad \text{and} \quad \mu_s(\varrho_s, T) = \mu_n(\varrho_n, T). \tag{2.70}$$

We proceed to find the special forms of the equations (2.40) through (2.42) appropriate for He II. The following argument about how to reduce the equations of a mixture of Euler fluids to the case of He II is due to Dreyer [59]; see also Müller [45]. First of all we recall (2.39), or (2.49), which in the present case reads

$$\Lambda = Lm(\mu_s - \mu_n). \tag{2.71}$$

Since $\mu_s = \mu_n$ holds and Λ does not vanish, L must be infinite. Thus the equations (2.40) have an indeterminate right-hand side and the only useful consequence is given by their sum, viz. in linearized form

$$\boxed{\frac{\partial \varrho}{\partial t} + \tilde{\varrho}\frac{\partial v_i}{\partial x_i} = 0}. \tag{2.72}$$

From (2.41) we derive the balance of total momentum by summation

$$\boxed{\tilde{\varrho}\frac{\partial v_i}{\partial t} + \frac{\partial p}{\partial x_i} = 0} \tag{2.73}$$

and the balance of relative momentum by subtraction

$$\frac{\partial V_i^s}{\partial t} + \frac{1}{\varrho_s}\frac{\partial p^s}{\partial x_i} - \frac{1}{\varrho_n}\frac{\partial p^n}{\partial x_i} = 0. \tag{2.74}$$

We rewrite (2.74) by use of (2.29). Hence follows

$$\begin{aligned}
\frac{1}{\varrho_s}\frac{\partial p^s}{\partial x_i} - \frac{1}{\varrho_n}\frac{\partial p^n}{\partial x_i} &= \frac{\partial(\mu^s - \mu^n)}{\partial x_i} - \left(\frac{\partial\left(\mu^s - \frac{p_s}{\varrho_s}\right)}{\partial T} - \frac{\partial\left(\mu^n - \frac{p_n}{\varrho_n}\right)}{\partial T} \right)\frac{\partial T}{\partial x_i} \quad \text{with (2.70), (2.47):} \\
&= -\left[\left(\frac{\partial \varepsilon_s}{\partial T} - T\frac{\partial s_s}{\partial T} - s_s\right) - \left(\frac{\partial \varepsilon_n}{\partial T} - T\frac{\partial s_n}{\partial T} - s_n\right)\right]\frac{\partial T}{\partial x_i} \quad \text{with (2.68), (2.46):} \\
&= -s_n\frac{\partial T}{\partial x_i} \quad \text{with} \quad \varrho s = \varrho_n s_n, \text{ see (2.68):} \\
&= -\frac{\varrho}{\varrho_n} s \frac{\partial T}{\partial x_i}.
\end{aligned}$$

Thus (2.74) may be written in the form

$$\boxed{\frac{\tilde{\varrho}_n}{\tilde{\varrho}}\frac{\partial V_i^s}{\partial t} - \tilde{s}\frac{\partial T}{\partial x_i} = 0}. \tag{2.75}$$

There remains the energy equation (2.42) which, when linearized, assumes the form

$$\tilde{\varrho}\frac{\partial \varepsilon}{\partial t} + \frac{\tilde{\varrho}_s\tilde{\varrho}_n}{\tilde{\varrho}}\left(\tilde{\varepsilon}_s + \frac{\tilde{p}_s}{\tilde{\varrho}_s} - \tilde{\varepsilon}_n - \frac{\tilde{p}_n}{\tilde{\varrho}_n}\right)\frac{\partial V_i^s}{\partial x_i} + \tilde{p}\frac{\partial v_i}{\partial x_i} = 0. \tag{2.76}$$

We use (2.52) and $(2.51)_4$ to write

$$\begin{aligned}
\tilde{\varrho}\frac{\partial \varepsilon}{\partial t} + \tilde{p}\frac{\partial v_i}{\partial x_i} &= \tilde{\varrho}\tilde{T}\frac{\partial s}{\partial t} + \tilde{\varrho}(\tilde{\mu}^s - \tilde{\mu}^n)\frac{\partial c^s}{\partial t} \\
&= \tilde{\varrho}\tilde{T}\frac{\partial s}{\partial t} + (\tilde{\mu}^s - \tilde{\mu}^n)\frac{\tilde{\varrho}_s\tilde{\varrho}_n}{\tilde{\varrho}}\frac{\partial V_i^s}{\partial x_i}.
\end{aligned}$$

Hence follows an alternative of (2.76) with (2.47) and (2.68), viz.

$$\boxed{\frac{\partial s}{\partial t} - \frac{\tilde{\varrho}_s}{\tilde{\varrho}}\tilde{s}\frac{\partial V_i^s}{\partial x_i} = 0}. \tag{2.77}$$

Some of the preceding reformulations may seem arbitrary; they have been made so as to arrive at the framed equations, which are those that Landau postulated for He II. We conclude that the Landau bi-fluid theory is a special case of a mixture of Euler fluids. The special character is provided by the assumptions (2.68) and (2.70). The recognition of this fact has long been delayed. The issue was confused, because one of the mass balance equations is missing from the set of Landau equations. We have seen that that equation is not needed for the calculation of the fields $\varrho_s, \varrho_n, v_i^s, v_i^n$, and T because, by $(2.70)_2$, there is an algebraic equation between ϱ_s, ϱ_n and T. Once the fields $\varrho_s, \varrho_n, v_i^s, v_i^n$ and T have been calculated from $(2.70)_2$ and the Landau equations, we may use the "missing mass balance" to calculate the indeterminate reaction rate density Λ.

We conclude this section with a treatment of first and second sound in He II. We choose p, T and v_i^s, v_i^n as variables and write

$$\frac{\partial \varrho}{\partial t} = \left(\frac{\partial \varrho}{\partial p}\right)_T \frac{\partial p}{\partial t} + \left(\frac{\partial \varrho}{\partial T}\right)_p \frac{\partial T}{\partial t}, \qquad \frac{\partial s}{\partial t} = \left(\frac{\partial s}{\partial p}\right)_T \frac{\partial p}{\partial t} + \left(\frac{\partial s}{\partial T}\right)_p \frac{\partial T}{\partial t}. \tag{2.78}$$

We recall that in an arbitrary binary mixture ϱ and s should depend on three variables, e.g. on p, T and c^s. But in Helium only two of those are independent because of the constraint $\mu_s = \mu_n$, see $(2.70)_2$.

We assume that there is no thermal expansion, so that

$$\left(\frac{\partial \varrho}{\partial T}\right)_p = 0 \quad \text{and} \quad \left(\frac{\partial s}{\partial p}\right)_T = \frac{1}{\varrho^2}\left(\frac{\partial \varrho}{\partial T}\right)_p = 0,$$

and obtain the Landau equations in the form

$$\begin{bmatrix} 0 & \left(\frac{\partial \varrho}{\partial p}\right)_T \frac{\partial}{\partial t} & \varrho_s \frac{\partial}{\partial x_i} & \varrho_n \frac{\partial}{\partial x_i} \\ 0 & \frac{\partial}{\partial x_i} & \varrho_s \frac{\partial}{\partial t} & \varrho_n \frac{\partial}{\partial t} \\ -\varrho s \frac{\partial}{\partial x_i} & 0 & \varrho_n \frac{\partial}{\partial t} & -\varrho_n \frac{\partial}{\partial t} \\ \frac{1}{\varrho T} c_p \frac{\partial}{\partial t} & 0 & -\varrho_s s \frac{\partial}{\partial x_i} & \varrho_s s \frac{\partial}{\partial x_i} \end{bmatrix} \begin{bmatrix} T \\ p \\ v_i^s \\ v_i^n \end{bmatrix} = \begin{bmatrix} 0 \\ 0 \\ 0 \\ 0 \end{bmatrix}, \tag{2.79}$$

where $c_p = T\left(\frac{\partial s}{\partial T}\right)_p$ is the specific heat. We investigate plane harmonic waves and introduce

$$T = \tilde{T} + \hat{T}E \qquad p = \tilde{p} + \hat{p}E \qquad v_i^s = \hat{v}_i^s E \qquad v_i^n = \hat{v}_i^n E, \tag{2.80}$$

where $E = exp\left[i(\omega t - k\mathbf{n}\mathbf{x})\right]$. It turns out that the waves are longitudinal, i.e. v_i^s and v_i^n point in the direction of propagation $\mathbf{n}$ which we choose to be the x_1-direction. We insert (2.80) into

(2.79) and obtain with $u = \frac{\omega}{k}$ as the phase speed

$$\begin{bmatrix} 0 & \left(\frac{\partial \varrho}{\partial p}\right)_T u & -\varrho_s & -\varrho_n \\ 0 & -1 & \varrho_s u & \varrho_n u \\ \varrho s & 0 & \varrho_n u & -\varrho_n u \\ \frac{c_p}{\varrho T} u & 0 & \varrho_s s & -\varrho_s s \end{bmatrix} \begin{bmatrix} \hat{T} \\ \hat{p} \\ \hat{v}_1^s \\ \hat{v}_i^n \end{bmatrix} = \begin{bmatrix} 0 \\ 0 \\ 0 \\ 0 \end{bmatrix}. \tag{2.81}$$

Elimination of $\hat{T}$ and $\hat{p}$ between those four equations results in the equations

$$\begin{aligned} \left(\frac{\varrho_s}{\varrho_n}\frac{s^2 T}{c_p} - u^2\right)(v_1^s - v_1^n) = 0 \\ \left(\left(\frac{\partial p}{\partial \varrho}\right)_T - u^2\right)\left(\frac{\varrho_s}{\varrho} v_1^s + \frac{\varrho_n}{\varrho} v_1^n\right) = 0. \end{aligned} \tag{2.82}$$

We conclude that there are two modes of propagation:

i.) Either $V_i^s = v_1^s - v_1^n = 0$ holds, in which case $v_1 = \frac{\varrho_s}{\varrho} v_1^s + \frac{\varrho_n}{\varrho} v_1^n$ may be unequal to zero. The phase speed is

$$u = \sqrt{\left(\frac{\partial p}{\partial \varrho}\right)_T}. \tag{2.83}$$

ii.) Or $v_1 = 0$ holds, in which case V_1^s may be unequal to zero and the wave moves with the speed

$$u = \sqrt{\frac{\varrho_s}{\varrho_n}\frac{s^2 T}{c_p}}. \tag{2.84}$$

These two cases correspond to the first and second sound respectively in He II. In the first sound the two phases move jointly and $(2.81)_{1,4}$ show that the pressure oscillates while the temperature remains constant. In the second sound the two phases oscillate with respect to each other but the mixture as a whole does not move; it follows from $(2.81)_{1,4}$ that now the pressure is constant while the temperature oscillates. For that reason the second sound must be excited by temperature oscillations rather than pressure oscillations.

3 Ordinary and Extended Thermodynamics of Mixtures

3.1 The Laws of Fick and Fourier in Extended Thermodynamics

We recall the equations (1.9), which are the generalizations offered by TIP of the laws of Fourier and Fick. In extended thermodynamics we have a rather different equation for q_i—see (2.42)—while J_i^α or V_i^α are variables rather than constitutive quantities. The question arises how the two theories compare and, in particular, what the status of the laws of Fourier and Fick is in extended thermodynamics. For simplicity we answer that question for a nonreacting mixture.

It turns out that Fick's law results from a truncation of the balance equations of momenta. In order to see this we recall $(2.41)_1$ and, for generality, we refer it to a noninertial frame so that there are inertial forces. We have

$$\dot{v}_i^\alpha + u_j^\alpha \frac{\partial v_i^\alpha}{\partial x_j} - i_i^0 - 2W_{ik} v_k^\alpha - \frac{1}{\varrho_\alpha}\frac{\partial t_{ij}^\alpha}{\partial x_j} = \frac{1}{\varrho_\alpha} m_i^\alpha, \qquad (\alpha = 1, 2, \ldots \nu) \tag{3.1}$$

where i_i^0 is the velocity independent part of the inertial acceleration and $\mathbf{W}$ is the angular velocity of the frame. We subtract the νth equation from all others and obtain with $(2.41)_{2,3}$ and $(2.47)_2$

$$\left[\dot{V}_i^\alpha + u_j^\alpha \frac{\partial v_i^\alpha}{\partial x_j} - u_j^\nu \frac{\partial v_i^\nu}{\partial x_j} - 2W_{ik} V_k^\alpha\right] +$$

$$+T\frac{\partial \frac{\mu_\alpha - \mu_\nu}{T}}{\partial x_i} + T^2\left(T\frac{\partial \frac{\mu_\alpha - \mu_\nu}{T}}{\partial T} - \frac{\partial\left(\frac{p_\alpha}{\varrho_\alpha} - \frac{p_\nu}{\varrho_\nu}\right)}{\partial T}\right)\frac{\partial \frac{1}{T}}{\partial x_i} = \sum_{\beta=1}^{\nu-1} (\mathbf{F}\mathbf{M}_V\mathbf{F})_{\alpha\beta} J_i^\beta, \tag{3.2}$$

where $\mathbf{M}_V$ is the matrix consisting of the $M_V^{\alpha\beta}$'s and $\mathbf{F}$ is the matrix defined by $(2.22)_2$. The equations (3.2) represent the equations of relative motion in the mixture and they are exact within the present theory of mixtures of Euler fluids.

In order to obtain from (3.2) the constitutive equation for J_i^β, which TIP requires, we rely on the Maxwellian iteration again, which was explained in a different context in Paragraph 4.3.2: On the left-hand side of (3.2) we replace V_i^α by its equilibrium value, namely zero, and use the equation to calculate $\overset{(1)}{J_i^\beta}$, the first iterate. This, or the corresponding $\overset{(1)}{V_i^\beta}$, is then used on the left-hand side to give $\overset{(2)}{J_i^\beta}$, etc. Here we shall be content with the first iterate because that gives us Fick's law. Indeed with $V_i^\alpha \equiv 0$ on the left-hand side of (3.2) that equation may be written in the form

$$\boxed{J_i^\alpha = \sum_{\beta=1}^{\nu-1} \underbrace{\left\{-T\,[\mathbf{A}]_{\alpha\beta}\right\}}_{L_{\alpha\beta}} \frac{\partial\left(-\frac{\mu^{\beta\nu}}{T}\right)}{\partial x_i} + \underbrace{\left\{\sum_{\beta=1}^{\nu-1} T^2\,[\mathbf{A}]_{\alpha\beta}\left(T\frac{\partial \frac{\mu_{\beta\nu}}{T}}{\partial T} - \frac{\partial \pi^{\beta\nu}}{\partial T}\right)\right\}}_{\tilde{L}_\alpha} \frac{\partial \frac{1}{T}}{\partial x_i},} \tag{3.3}$$

where $\mu^{\beta\nu}$ stands for $\mu^\beta - \mu^\nu$, $\pi^{\beta\nu}$ for $\frac{p_\beta}{\varrho_\beta} - \frac{p_\nu}{\varrho_\nu}$, and $\mathbf{A}$ for $\left(\overset{-1}{\mathbf{F}}\overset{-1}{\mathbf{M}_V}\overset{-1}{\mathbf{F}}\right)$. This equation is equivalent to TIP's equation $(1.9)_2$. Comparison of the coefficients in the two equations leads to the identification of $L_{\alpha\beta}$ and $\tilde{L}_\alpha$ as indicated in (3.3).

Fourier's law, or its generalization $(1.9)_1$ can now also be derived from the arguments of extended thermodynamics. If we ignore a nonlinear term we have by (2.42)

$$q_i = \sum_{\alpha=1}^{\nu-1}\left(\varepsilon_\alpha + \frac{p_\alpha}{\varrho_\alpha} - \varepsilon_\nu - \frac{p_\nu}{\varrho_\nu}\right) J_i^\alpha \qquad \text{or by } (2.22)_1 \text{ and with} \qquad h_\alpha = \varepsilon_\alpha + \frac{p_\alpha}{\varrho_\alpha}$$

$$\boxed{q_i = \sum_{\beta=1}^{\nu-1} \underbrace{\left\{ \sum_{\alpha=1}^{\nu-1} -h_{\alpha\nu} T \left[\mathbf{A}\right]_{\alpha\beta} \right\}}_{L_\beta} \frac{\partial\left(-\frac{\mu^{\beta\nu}}{T}\right)}{\partial x_i} + \underbrace{\left\{ \sum_{\alpha,\beta=1}^{\nu-1} h_{\alpha\nu} T^2 \left[\mathbf{A}\right]_{\alpha\beta} \left(T \frac{\partial \frac{\mu^{\beta\nu}}{T}}{\partial T} - \frac{\partial \pi^{\beta\nu}}{\partial T} \right) \right\}}_{L} \frac{\partial \frac{1}{T}}{\partial x_i},} \quad (3.4)$$

where $h_{\alpha\nu}$ stands for $h_\alpha - h_\nu$. This equation is analogous to TIP's equation $(1.9)_1$ if we identify the coefficients L_β and L as indicated.

We conclude that the constitutive laws of Fick and Fourier—or TIP's generalizations (1.9) of these classical laws—arise in extended thermodynamics in an approximate manner as a result of a formal iterative scheme. Roughly speaking Fick's law for the diffusion fluxes J_i^α comes out by ignoring the acceleration of the relative motions of the constituents. The flux of internal energy in extended thermodynamics is solely due to the convection of energy with the diffusive motion. Thus Fourier's law is merely a corollary of Fick's law. This of course is a peculiarity of mixtures of Euler fluids.

3.2 Onsager Relations

It is true that we treat a rather special case of mixtures in extended thermodynamics of Euler fluids. Not only are they nonviscous, which means that the partial stresses are isotropic tensors, but also they are not heat conducting, because the flux of internal energy is due solely to the convection of energy with the diffusive motion. And yet the laws of Fick and Fourier, viz. (3.3) and (3.4), have the same general form here and in ordinary thermodynamics. Except that we now have some information about the coefficients $L_{\alpha\beta}$, $\tilde{L}_\alpha$, and L_β. We proceed to show that we may use this knowledge to prove the validity of the Onsager relations $(1.17)_2$ for diffusion and thermal diffusion.

Inspection of (3.3) shows that $L_{\alpha\beta}$ is symmetric if $M_V^{\alpha\beta}$ is. Truesdell [60] has made some eminently reasonable assumptions under which the symmetry of $M_V^{\alpha\beta}$ can be proved. We follow his argument. First of all we introduce the new coefficient $M_V^{\alpha\nu}$ by $M_V^{\alpha\nu} = -\sum_{\beta=1}^{\nu-1} M_V^{\alpha\beta}$ and rewrite the constitutive equation $(2.15)_3$ in the form

$$m_i^\alpha = -\sum_{\beta=1}^{\nu} M_V^{\alpha\beta}(v_i^\alpha - v_i^\beta). \quad (3.5)$$

Chemical reactions are ignored for this argument, so that $\sum_{\alpha=1}^{\nu} M_V^{\alpha\beta} = 0$ holds. The advantage of the form (3.5) is that the coefficients $M_V^{\alpha\beta}$ may now be interpreted suggestively as coefficients that determine how the force on constituent α is influenced by the relative motion of that constituent and constituent β. It seems therefore reasonable to assume that the following two properties hold:

i.) $M_V^{\alpha\beta}$ is independent of ϱ_γ for $\gamma \neq \alpha, \beta$

ii.) $M_V^{\alpha\beta}$ tends to zero for $\varrho_\beta \to 0$.

We say that the mixture experiences binary drags only if i.) is true, and ii.) states that there is no drag on constituent α from constituent β if the latter is absent. We proceed to prove the symmetry of $M_V^{\alpha\beta}$ under these conditions.

Because of conservation of momentum we must have

$$\sum_{\alpha,\beta=1}^{\nu} M_V^{\alpha\beta}(v_i^\alpha - v_i^\beta) = 0 \quad \text{or} \quad \sum_{\alpha,\beta=1}^{\nu} M_V^{[\alpha\beta]} v_i^\alpha = 0, \tag{3.6}$$

where the square bracket denotes the antisymmetric part of $M_V^{\alpha\beta}$. Since (3.6) must hold for all v_i^α we have

$$\sum_{\beta=1}^{\nu} M_V^{[\alpha\beta]} = 0.$$

This means that the sum over all elements in every row, and every column, must vanish. In particular in a binary mixture we have

$$M_V^{[\alpha\beta]} = \begin{bmatrix} 0 & \alpha \\ -\alpha & 0 \end{bmatrix}$$

and since all rows and columns must add up to zero, we must conclude that $\alpha = 0$ holds. In a mixture of three constituents we have

$$M_V^{[\alpha\beta]} = \begin{bmatrix} 0 & \alpha & -\alpha \\ -\alpha & 0 & \alpha \\ \alpha & -\alpha & 0 \end{bmatrix}.$$

Now, if the above conditions i.) and ii.) hold for $M_V^{\alpha\beta}$, they must also hold for its symmetric and antisymmetric parts. Thus $M_V^{[12]}$ is independent of $\varrho_3, M_V^{[31]}$ independent of ϱ_2 and $M_V^{[23]}$ is independent of ϱ_1. But all of these components are equal to α. Hence α is a constant and by condition ii.) that constant is zero, so that $M_V^{[\alpha\beta]} = 0$ holds for a ternary mixture. The corresponding proof for more than three constituents can easily be reduced to the case of three constituents by condition ii.). Thus $M_V^{[\alpha\beta]} = 0$ is generally true for our mixtures and we have proved the symmetry of $M_V^{\alpha\beta}$, whence follows the symmetry of $L_{\alpha\beta}$ by (3.3).

We continue to establish the validity of the Onsager relations $(1.17)_2$ and proceed to show that by (3.3) and (3.4) the coefficients $\tilde{L}_\alpha$ and L_α are equal. Indeed, by (2.47) we have with $\frac{\partial \varepsilon_\alpha}{\partial T} = T\frac{\partial s_\alpha}{\partial T}$

$$T\frac{\partial \frac{\mu_\beta - \mu_\nu}{T}}{\partial T} - \frac{\partial\left(\frac{p_\beta}{\varrho_\beta} - \frac{p_\nu}{\varrho_\nu}\right)}{\partial T} = -\frac{1}{T}(h_\beta - h_\nu)$$

from which we conclude that $\tilde{L}_\alpha = L_\alpha$ holds. This was first proved by Müller [61].

The important conclusion to be drawn here is that Onsager relations hold as a consequence not of microscopic considerations—as in Onsager's motivations, e.g. see de Groot & Mazur [62]—but as a consequence of the macroscopic structure of the coefficients $L_{\alpha\beta}, L_\alpha$, and $\tilde{L}_\alpha$, a structure that is revealed by extended thermodynamics of mixtures of Euler fluids. It is true though, that this structure is not maintained for mixtures of non-Eulerian fluids and properly heat-conducting mixtures; it may therefore be that in such more complicated—and more realistic—mixtures Onsager relations do not hold.

3.3 Inertial Contribution to the Laws of Diffusion

We return to the consideration of the equations (3.2), which represent the equations of relative motion of the constituents. For the present argument we restrict the attention to a binary mixture with a uniform temperature, in which case there is only one equation, namely, with $\mu = \mu^1 - \mu^2$

$$\left[\dot{V}_i^1 + V_j^1 \frac{\partial v_i}{\partial x_j} - 2W_{ji}V_j^1\right] + \underline{u_j^1 \frac{\partial u_i^1}{\partial x_j} - u_j^2 \frac{\partial u_i^2}{\partial x_j}} + \frac{\partial \mu}{\partial x_i} = M_V^{11} \frac{\varrho}{\varrho_1 \varrho_2} V_i^1. \tag{3.7}$$

We ignore the underlined terms that are nonlinear in the diffusion velocities and apply the Maxwellian iteration to (3.7). In the first step we obtain

$$\overset{(1)}{V_i^1} = \tau \frac{\partial \mu}{\partial x_i} \qquad \text{where} \qquad \tau \equiv \frac{1}{M_V^{11}} \frac{\varrho_1 \varrho_2}{\varrho} \tag{3.8}$$

and in the second step we obtain

$$\overset{(2)}{V_i^1} = \tau \frac{\partial \mu}{\partial x_i} + \tau^2 \left[\frac{1}{\tau}\left(\tau \frac{\partial \mu}{\partial x_i}\right)^{\bullet} + \frac{\partial v_i}{\partial x_j}\frac{\partial \mu}{\partial x_j} - 2W_{ji}\frac{\partial \mu}{\partial x_j}\right]. \tag{3.9}$$

Note that the terms in square brackets in (3.7) and (3.9) are objective vectors but frame-dependent ones, since they contain the matrix $\mathbf{W}$ of angular velocity of the frame.

We simplify (3.9) by looking at the special case of a rigid rotation with $\dot{\mu} = 0$. In that case we obtain

$$\overset{(2)}{V_i^1} = \tau \left\{\frac{\partial \mu}{\partial x_i} - 2\tau \left(\frac{\partial v_{[j}}{\partial x_{i]}} - W_{ji}\right) \frac{\partial \mu}{\partial x_j}\right\}. \tag{3.10}$$

It follows that the diffusion flux created by a chemical potential gradient in a mixture at rest in a noninertial frame has a component perpendicular to $\frac{\partial \mu}{\partial x_i}$. Equation (3.10) is analogous to the equation (4.3.10) for heat conduction and indeed, the interpretation of the phenomena is very similar, as we proceed to explain.

Consider a binary mixture of gases with $v_i = 0$ in a cylinder with a radial gradient of μ as depicted in Figure 5.1a. The radial gradient of μ may be created by a gradient of the concentration. We focus attention on a small volume element of the linear dimensions of the mean free paths of the atoms, and first we consider a gas at rest in an inertial frame. In that case the free paths of the atoms are straight lines and—because of the concentration gradient—there are more atoms of constituent 1 (say) flying from top to bottom than in the other direction; see Figure 5.1b. Therefore we observe a diffusion flux, or a nonzero vector V_i^1 which is proportional to grad μ as it should be according to (3.10) in an inertial frame and when the mixture is at rest. Next we consider the same situation for a mixture at rest in a noninertial frame. Now the free paths between collisions are curved by the Coriolis force as shown in Figure 5.1c. Consequently there is now not only a radial mass flux, but also a flux perpendicular to grad μ, and to the angular velocity of the frame as predicted by (3.10).

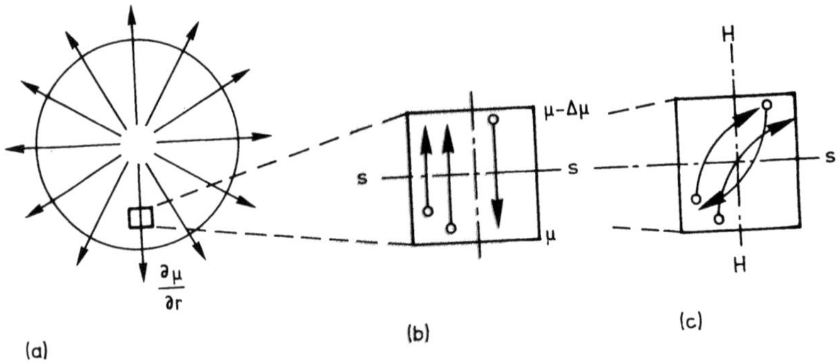

Figure 5.1 On the frame dependence of the diffusion flux

This argument about the nature of the frame dependence is entirely analogous to the corresponding argument about the heat flux in Chapter 4. It seems convincing in the sense that nobody will doubt the effect of the Coriolis force on the relative motion of two constituents of a mixture. In contrast to that, the corresponding argument about heat flux (see Paragraph 4.3.3) has drawn a lot of criticism at the time when it was presented; see Wang [63] and Truesdell [64]. That criticism was always unfounded and the present argument confirms this fact.

Chapter 6

Relativistic Extended Thermodynamics of Gases

Relativists have shown a keen interest in the development of extended thermodynamics, because the paradoxa of infinite speeds implied by the parabolic equations of ordinary thermodynamics disturbed them acutely. Indeed immediately after the formulation of the early version of extended themodynamics Müller [7] formulated that theory relativistically but this work was not properly published until Israel [29] presented a very similar theory. Those theories had all the shortcomings of extended TIP that were discussed in Chapter 2.

After the development of the systematic version of extended thermodynamics – the theory presented in Chapter 3 and 4 – relativistic thermodynamics could be revisited and properly formulated. This was done by Liu, Müller & Ruggeri [30] and the present chapter gives a detailed account of their work, see also Ruggeri [65], and Müller [66]. The basic fields are the particle flux vector and the energy-momentum tensor, 14 fields altogether. The necessary field equations are based on the conservation laws of the number of particles, of energy and momentum and on a balance of fluxes. Constitutive equations are needed for the flux tensor and the flux production. The constitutive theory makes use of the principle of relativity, the entropy principle and the requirement of convexity and causality and it derives explicit field equations in which only four functions must be determined – either experimentally or from statistical mechanics, – viz. the thermal equation of state, the heat conductivity and the shear- and bulk viscosities. Specific results are given for some special cases: The non-degenerate relativistic gas, the degenerate non-relativistic gas and the ultrarelativistic gas.

We recapitulate Chandrasekhar's treatment of white dwarfs [67] so as to demonstrate the adequacy of the equations of state.

It is not obvious how relativistic extended thermodynamics reduces to the non-relativistic theory. The easiest way to investigate that question is by reference to the kinetic theory. Therefore we give a brief account of the relativistic version of the kinetic theory of gases and investigate its non-relativistic limit. It turns out that in that limit the 14 variables of the relativistic theory reduce to 13, the dynamic pressure drops out, because it is $\mathcal{O}(\frac{1}{c^2})$.

1 Balance Equations and Constitutive Restrictions

1.1 Thermodynamic Processes

The principal objective of relativistic extended thermodynamics is the determination of the 14 fields

$$\begin{array}{ll} A^A & \text{- particle flux vector} \\ A^{AB} & \text{- energy-momentum tensor} \end{array} \tag{1.1}$$

in all events x^D. The energy-momentum tensor is assumed symmetric so that it has 10 independent components.

For the determination of the fields (1.1) we need field equations, and these are formed by the conservation laws of particle number and energy-momentum, viz.

$$A^A{}_{,A} = 0 \tag{1.2$_1$}$$

$$A^{AB}{}_{,B} = 0 \tag{1.2$_2$}$$

and by the equations of balance of fluxes

$$A^{ABC}{}_{,C} = I^{AB}. \tag{1.2$_3$}$$

A^{ABC} is a completely symmetric tensor of fluxes and I^{AB} is its production density. The equations (1.2) are 15 in number for only 14 fields. Therefore we assume

$$I^A{}_A = 0 \qquad \text{and} \qquad A^{AC}{}_C \sim A^A, \tag{1.3}$$

so that the trace of $(1.2)_3$ reduces to $(1.2)_1$. We are then left with only 14 independent equations (1.2).

The components of A^A and A^{AB} have the following interpretations:

$$\begin{aligned} A^0 &= c \cdot \text{rest mass density} \\ A^a &= \text{flux of rest mass} \\ A^{00} &= \text{energy density} \\ A^{0a} &= \frac{1}{c} \cdot \text{energy flux} \\ A^{a0} &= c \cdot \text{momentum density} \\ A^{ab} &= \text{momentum flux.} \end{aligned} \tag{1.4}$$

The motivation for the choice of the equations (1.2), and in particular for the equation $(1.2)_3$ stems from the kinetic theory of mon-atomic gases of which a brief account is given in Section 6. Indeed, A^A and A^{AB} are the first two moments in the kinetic theory and $A^A{}_{,A} = 0$ and $A^{AB}{}_{,B} = 0$ are the first two equations of transfer. Therefore it seems reasonable to take the further equations from the equation of transfer for the third moment A^{ABC} and these have the form $(1.2)_3$. In the kinetic theory the two conditions (1.3) are satisfied and the factor of proportionality in $(1.3)_2$ turns out to be c^2, see (6.11). Therefore we rewrite (1.3) in the form

$$I^A{}_A = 0 \qquad \text{and} \qquad A^{AC}{}_C = c^2 A^A. \tag{1.5}$$

The kinetic theory makes it also clear that the assumed symmetry of A^{ABC} is characteristic for mon-atomic gases. The present theory is therefore restricted to that case. This is not much of a restriction, because molecular compounds cannot exist at temperatures that are high enough for relativistic effects to matter.

Of course the set of equations (1.2) is not a set of field equations for the basic fields (1.1), because two new quantities appear in them, namely the flux tensor A^{ABC} and the flux production I^{AB}. In this situation we let ourselves be guided by the arguments of nonrelativistic continuum mechanics and thermodynamics and assume that A^{ABC} and I^{AB} are constitutive quantities. Such quantities are related to the basic fields in a materially dependent manner by constitutive functions. In particular, when the constitutive relations are of the general form

$$\begin{aligned} A^{ABC} &= \hat{A}^{ABC} \quad (A^M, A^{MN}) \\ I^{AB} &= \hat{I}^{AB} \quad (A^M, A^{MN}), \end{aligned} \tag{1.6}$$

we say that they characterize a viscous, heat-conducting gas.

If the constitutive function $\hat{A}^{ABC}$ and $\hat{I}^{AB}$ were known, we could eliminate A^{ABC} and I^{AB} between (1.2) and (1.6) and come up with an explicit set of 14 field equations for the 14 fields A^A and A^{AB}. Each solution of this set of equations is called a *thermodynamic process.*

1.2 Principles of the Constitutive Theory

In reality of course the constitutive functions $\hat{A}^{ABC}$ and $\hat{I}^{AB}$ are not known, and therefore the thermodynamicist tries to determine them or at least to restrict their generality. These efforts are guided by universal physical principles to which the constitutive functions must conform. The most important such principles are

i.) the entropy principle,

ii.) the principle of relativity, and

iii.) the requirement of convexity and causality.

The entropy principle postulates the existence of the entropy flux vector h^A, which is a constitutive quantity, so that in a viscous, heat-conducting gas we have

$$h^A = \hat{h}^A(A^M, A^{MN}). \tag{1.7}$$

Moreover, the principle requires that h^A satisfy the entropy inequality

$$h^A{}_{,A} \geq 0 \tag{1.8}$$

for all thermodynamic processes.

The principle of relativity assumes that the field equation and the entropy inequality are invariant under space-time transformations

$$\overset{*}{x}{}^A = \overset{*}{x}{}^A\,(x^B). \tag{1.9}$$

Since the balance equations (1.2) are tensor equations which are naturally invariant, the principle in effect requires that the constitutive functions be invariant. If C and C^* are generic expressions for the constitutive quantities A^{ABC}, I^{AB}, and h^A in the two frames connected by the transformation (1.9), the principle of relativity can be stated as follows:

$$C = \hat{C}(A^M, A^{MN}) \qquad \text{and} \qquad C^* = \hat{C}(A^{*M}, A^{*MN}). \tag{1.10}$$

Note that $\hat{C}$ is the *same* function in both equations.

Causality and convexity is a condition on the character of the field equations. It requires that these equations be symmetric hyperbolic. This property guarantees that Cauchy problems are well posed and that all wave speeds are finite.

2 Constitutive Theory

2.1 Scope and Structure

The exploitation of the above restrictive principles is the subject of the thermodynamic constitutive theory. The details of this theory are cumbersome and require much work. So as to maintain the clarity of the argument we proceed in simple steps as follows.

- i.) Lagrange multipliers and the vector potential,
- ii.) Linear representations,
- iii.) Suggestive variables,
- iv.) Equilibrium. Fugacity and absolute temperature,
- v.) Linear expressions for the Lagrange multipliers,
- vi.) The linear flux tensor,
- vii.) The entropy flux vector,
- viii.) Residual inequality,
- ix.) Causality and convexity,
- x.) Summary of results.

2.2 Lagrange Multipliers and the Vector Potential. Step i.

The key to the exploitation of the entropy principle is the observation that the entropy inequality (1.8) must not hold for arbitrary fields $A^M(x^D), A^{MN}(x^D)$ but only for thermodynamic processes, i.e. solutions of the field equations. One may say that the field equations form constraints on the fields that satisfy (1.8). By a lemma of Liu [20] we may take care of these constraints by use of Lagrange multipliers. Application of Liu's lemma in this case shows that the new inequality

$$h^A{}_{,A} + \Lambda A^A{}_{,A} + \Lambda_B A^{BA}{}_{,A} + \Lambda_{BC}(A^{BCA}{}_{,A} - I^{BC}) \geq 0 \tag{2.1}$$

must hold for arbitrary fields A^A, A^{BA}. The Λ's in (2.1) are Lagrange multipliers which may depend on the same variables as A^{BCA}, I^{BC}, and h^A, i.e. on A^M and A^{MN}. The multiplier Λ_{BC} may be considered traceless, because its factor in (2.1) is traceless according to (1.5). Obviously the inequality (2.1) may be written in the form

$$h'^A{}_{,A} - A^A \Lambda_{,A} - A^{AB}\Lambda_{B,A} - A^{BCA}\Lambda_{BC,A} - \Lambda_{BC} I^{BC} \geq 0, \tag{2.2}$$

where h'^A has been introduced as an abbreviation for

$$h'^A = h^A + \Lambda A^A + \Lambda_B A^{BA} + \Lambda_{BC} A^{BCA}. \tag{2.3}$$

The inequality (2.2) is most easily evaluated after a change of independent variables

$$\text{from} \quad A^A, A^{AB} \quad \text{to} \quad \Lambda, \Lambda_B, \Lambda_{BC}. \tag{2.4}$$

In Chapter 3 we have proved that this transformation is globally invertible for reasons of convexity. With this change of variables A^A and A^{AB} become constitutive quantities just as h^A, or h'^A, and A^{BCA} and I^{BC}. If the chain rule is used on $h'^A(\Lambda, \Lambda_B, \Lambda_{BC})$ in (2.2) we obtain

$$\left(\frac{\partial h'^A}{\partial \Lambda} - A^A\right)\Lambda_{,A} + \left(\frac{\partial h'^A}{\partial \Lambda_B} - A^{AB}\right)\Lambda_{B,A} + \left(\frac{\partial h'^A}{\partial \Lambda_{BC}} - A^{BCA}\right)\Lambda_{BC,A} - \Lambda_{BC} I^{BC} \geq 0. \tag{2.5}$$

This inequality must hold for all fields Λ, Λ_B, and Λ_{BC} and therefore in particular for arbitrary values of the derivatives $\Lambda_{,A}$, $\Lambda_{B,A}$, and $\Lambda_{BC,A}$ in one event. But these derivatives occur linearly on the left-hand side of (2.5); the inequality could thus be violated if the brackets were contributing to its left-hand side. Therefore we must have

$$A^A = \frac{\partial h'^A}{\partial \Lambda}, \quad A^{BA} = \frac{\partial h'^A}{\partial \Lambda_B}, \quad A^{\langle BC\rangle A} = \frac{\partial h'^A}{\partial \Lambda_{BC}} - \frac{1}{4} g^{BC} g_{MN} \frac{\partial h'^A}{\partial \Lambda_{MN}}. \tag{2.6}$$

The last condition is complicated by the fact that Λ_{BC} is traceless. g is the metric tensor with the signature (1, -1, -1, -1).

There remains the residual inequality on the entropy production $\sum$

$$\sum = -\Lambda_{BC} I^{BC} \geq 0. \tag{2.7}$$

We conclude from (2.6) that the dependence of A^A, A^{BC}, and A^{BCA} on the Lagrange multipliers $\Lambda, \Lambda_B, \Lambda_{BC}$ is determined by the derivatives of the single vector function $h'^A(\Lambda, \Lambda_B, \Lambda_{BC})$. Therefore we call this function the *vector potential*. The equations (2.6) represent the main result of the entropy principle and, rather obviously, they imply a great number of integrability conditions. Those integrability conditions lead to all the restrictive conditions which we are about to derive.

From the residual inequality we may deduce a condition concerning equilibrium. Equilibrium is defined as a process in which all productions I^{AB} vanish. Therefore inspection of (2.7) shows that the entropy production—which may be considered as a function of I^{AB}—has its minimum in equilibrium, namely zero. Of necessity then we must have

$$\frac{\partial \sum}{\partial I^{BC}}\Big|_E = -\Lambda_{BC}|_E = 0. \tag{2.8}$$

Thus the Lagrange multiplier Λ_{BC} vanishes in equilibrium.

2.3 Principle of Relativity and Linear Representations. Step ii.

We recall from (2.6) that the vector h'^A, a constitutive quantity dependent on Λ, Λ_B, and Λ_{BC}, plays the dominant role in the determination of A^A, A^{AB}, and A^{ABC}. The principle of relativity requires that we have

$$h'^A = \hat{h}'^A(\Lambda, \Lambda^M, \Lambda^{MN}) \quad \text{and} \quad \overset{*}{h'^A} = \hat{h}'^A(\Lambda^*, \Lambda^{*M}, \Lambda^{*MN}). \tag{2.9}$$

Since h'^A and $\Lambda, \Lambda_M, \Lambda_{MN}$ are tensors of rank zero, one, and two, the equations (2.9) may be combined to read

$$\frac{\partial x^{*A}}{\partial x^B}\hat{h}'^B(\Lambda, \Lambda^M, \Lambda^{MN}) = \hat{h}'^A\left(\Lambda, \frac{\partial x^{*M}}{\partial x^P}\Lambda^P, \frac{\partial x^{*M}}{\partial x^P}\frac{\partial x^{*N}}{\partial x^Q}\Lambda^{PQ}\right), \tag{2.10}$$

which means that $\hat{h}'^B$ is an isotropic vector function.

In nonrelativistic continuum mechanics there are representation theorems for isotropic functions, e.g. see Smith [54], and Pennisi [68] has shown how these carry over to the present case. Thus the most general form of h'^A that satisfies (2.10) reads

$$h'_A = \gamma_0\Lambda_A + \gamma_1\Lambda_{AB}\Lambda^B + \gamma_2\Lambda^2_{AB}\Lambda^B + \gamma_3\Lambda^3_{AB}\Lambda^B, \tag{2.11}$$

where the γ's may depend on the following 8 scalar combinations of variables.

$$\begin{array}{lcl lcl} H_0 & = & \Lambda & G_0 & = & \Lambda^A\Lambda_A \\ H_2 & = & tr(\Lambda^2_{AB}) & G_1 & = & \Lambda^A\Lambda_{AB}\Lambda^B \\ H_3 & = & tr(\Lambda^3_{AB}) & G_2 & = & \Lambda^A\Lambda^2_{AB}\Lambda^B \\ H_4 & = & tr(\Lambda^4_{AB}) & G_3 & = & \Lambda^A\Lambda^3_{AB}\Lambda^B. \end{array} \tag{2.12}$$

The indices on H and G indicate the order to which these scalars depend on the components of Λ_{AB}.

In this book we describe a *linear* theory, so that A^A, A^{BA}, and A^{BCA} are linear functions of the multiplier, Λ_{BC} which by (2.8) vanishes in equilibrium. Inspection of (2.6) shows that we need h'^A to within third-order terms in Λ_{BC}. In this approximation we may ignore γ_3 in the representation (2.11) and set

$$h'_A = [\Gamma_0 + \Gamma_{01}G_1 + \Gamma_{02}G_1^2 + \Gamma_{03}G_2 + \Gamma_{04}H_2]\,\Lambda_A + [\Gamma_1 + \Gamma_{11}G_1]\,\Lambda_{AB}\Lambda^B + \Gamma_2\Lambda^2_{AB}\Lambda^B, \tag{2.13}$$

where the Γ's may depend only on Λ and G_0.

In order to obtain linear representations for A^A, A^{BA}, and A^{BCA} we insert (2.13) into (2.6) and ignore nonlinear terms after the differentiation. In this manner we obtain expressions for A^A, A^{AB}, and A^{ABC} which do not automatically satisfy the requirement that the tensors A^{AB} and A^{BCA} be symmetric, nor is the trace condition (1.5) satisfied, of course. These conditions can be fulfilled, however, by a proper choice of the coefficients (2.13).

We must have

$$\begin{aligned} h'_A \; = \; & \left[\Gamma_0 + \frac{\partial\Gamma_1}{\partial G_0}G_1 + \frac{1}{2}\frac{\partial^2\Gamma_2}{\partial G_0^2}G_1^2 + \frac{\partial\Gamma_2}{\partial G_0}G_2 + \frac{1}{4}\Gamma_2H_2\right]\Lambda_A + \\ & + \left[\Gamma_1 + \frac{\partial\Gamma_2}{\partial G_0}G_1\right]\Lambda_{AB}\Lambda^B + [\Gamma_2]\,\Lambda^2_{AB}\Lambda^B. \end{aligned} \tag{2.14}$$

Insertion into (2.6) now provides

$$A^A = \left(\frac{\partial \Gamma_0}{\partial \Lambda} + \frac{\partial^2 \Gamma_1}{\partial \Lambda \partial G_0} G_1\right) \Lambda^A + \frac{\partial \Gamma_1}{\partial \Lambda} \Lambda^{AB} \Lambda_B, \tag{2.15}$$

$$\begin{aligned} A^{BA} = & \left(\Gamma_0 + \frac{\partial \Gamma_1}{\partial G_0} G_1\right) g^{BA} + \Gamma_1 \Lambda^{BA} + 2\left(\frac{\partial \Gamma_0}{\partial G_0} + \frac{\partial^2 \Gamma_1}{\partial G_0^2} G_1\right) \Lambda^B \Lambda^A \\ & + 2\frac{\partial \Gamma_1}{\partial G_0}\left(\Lambda^B \Lambda^{AC} \Lambda_C + \Lambda^A \Lambda^{BC} \Lambda_C\right), \end{aligned} \tag{2.16}$$

$$\begin{aligned} A^{BCA} = & \frac{1}{2}\left(\Gamma_1 + \frac{\partial \Gamma_2}{\partial G_0} G_1\right)\left(g^{BC}\Lambda^A + g^{CA}\Lambda^B + g^{AB}\Lambda^C\right) + \\ & + \frac{1}{2}\Gamma_2\left(g^{BC}\Lambda^{AD}\Lambda_D + g^{CA}\Lambda^{BD}\Lambda_D + g^{AB}\Lambda^{CD}\Lambda_D + \right. \\ & \left. + \Lambda^{BC}\Lambda^A + \Lambda^{CA}\Lambda^B + \Lambda^{AB}\Lambda^C\right) + \\ & + \left(\frac{\partial \Gamma_1}{\partial G_0} + \frac{\partial^2 \Gamma_2}{\partial G_0^2} G_1\right) \Lambda^B \Lambda^C \Lambda^A + \\ & + \frac{\partial \Gamma_2}{\partial G_0}\left(\Lambda^B \Lambda^C \Lambda^{AD}\Lambda_D + \Lambda^C \Lambda^A \Lambda^{BD}\Lambda_D + \Lambda^A \Lambda^B \Lambda^{CD}\Lambda_D\right). \end{aligned} \tag{2.17}$$

The tensors (2.16) and (2.17) are fully symmetric, and the trace condition (1.5) will be satisfied if we have

$$\frac{\partial G_0^3 \Gamma_1}{\partial G_0} = \frac{\partial c^2 G_0^2 \Gamma_0}{\partial \Lambda} \quad \text{and} \quad \frac{\partial G_0^4 \Gamma_2}{\partial G_0} = \frac{\partial c^2 G_0^3 \Gamma_1}{\partial \Lambda}. \tag{2.18}$$

A linear expression for the entropy flux vector follows from (2.3), (2.14) and from the above linear equations for A^A, A^{BA}, and A^{BCA}. We get

$$\begin{aligned} h^A = & -\left[\Lambda \frac{\partial \Gamma_0}{\partial \Lambda} + 2\frac{\partial \Gamma_0}{\partial G_0} G_0 + \left(\Lambda \frac{\partial^2 \Gamma_1}{\partial \Lambda \partial G_0} + 2G_0 \frac{\partial^2 \Gamma_1}{\partial G_0^2} + 3\frac{\partial \Gamma_1}{\partial G_0}\right) G_1\right] \Lambda^A - \\ & \left[\Gamma_1 + \Lambda \frac{\partial \Gamma_1}{\partial \Lambda} + 2G_0 \frac{\partial \Gamma_1}{\partial G_0}\right] \Lambda^{AB} \Lambda_B. \end{aligned} \tag{2.19}$$

From (2.18) we conclude that if only $\Gamma_0(\Lambda, G_0)$ is known, we may calculate Γ_1 and Γ_2 by integration to within two functions of Λ alone. Therefore the knowledge of the single scalar function $\Gamma_0(\Lambda, G_0)$ goes a long way in determining the linear constitutive quantities A^A, A^{BA}, A^{BCA}, and h^A in (2.15), (2.16), (2.17), and (2.19), albeit only in terms of the Lagrange multipliers Λ, Λ_B, and Λ_{BC}.

This last-mentioned fact is a drawback, because the Lagrange multipliers have no suggestive meaning at all—at least not at this stage. It is therefore imperative that we determine the $\mathbf{\Lambda}$'s in terms of meaningful, suggestive, and measurable variables.

2.4 Stress Deviator, Heat Flux, and Dynamic Pressure. Step iii.

By use of the four-velocity U^A, with $U^A U_A = c^2$, and of the projector

$$h_{AB} = \frac{1}{c^2} U_A U_B - g_{AB} \tag{2.20}$$

we may decompose the particle flux vector A^A, the energy-momentum tensor A^{BA}, and the entropy flux vector h^A as follows:

$$\begin{aligned} A^A &= nmU^A \\ A^{BA} &= t^{\langle BA\rangle} + (p(n,e)+\pi)h^{BA} + \frac{1}{c^2}(U^B q^A + U^A q^B) + \frac{e}{c^2}U^A U^B \\ h^A &= hU^A + \Phi^A. \end{aligned} \tag{2.21}$$

m is the atomic rest mass. These equations are nothing but identities, because they define the new quantities

$$n, U^A, t^{\langle BA\rangle}, \pi, q^A, e, h, \Phi^A \tag{2.22}$$

in terms of A^A and A^{BA} as follows:

$$\begin{aligned} n &= \frac{1}{mc^2}A^A U_A && \text{- number density} \\ t^{\langle BA\rangle} &= (h^B_M h^A_N - \frac{1}{3}h^{BA}h_{MN})A^{MN} && \text{- stress deviator} \\ p(n,e)+\pi &= \frac{1}{3}h_{MN}A^{MN} && \text{- pressure} \\ q^A &= -h^A_M U_N A^{MN} && \text{- heat flux} \\ e &= \frac{1}{c^2}U_M U_N A^{MN} && \text{- internal energy density} \\ h &= \frac{1}{c^2}h^A U_A && \text{- entropy density} \\ \Phi^A &= -h^A_M h^M && \text{- entropy flux.} \end{aligned} \tag{2.23}$$

The names which we have attached to these definitions in (2.23) are motivated by the observation that in the Lorentz frame, which is locally and momentarily at rest in the fluid, the nonvanishing components of $n, t^{\langle BA\rangle}, p(n,e)+\pi, q^A, e$ and h, Φ^A have the indicated physical meaning as number density, stress deviator, etc.

The pressure is decomposed into an equilibrium part p, which depends on n and e, and a non-equilibrium part π, which we call the dynamic pressure. Stress deviator, heat flux, and dynamic pressure are the nine quantities that vanish in equilibrium.

Compared to the original variables A^A, A^{BA}, the quantities $n, U^A, t^{\langle AB\rangle}, \pi, q^A, e$ are more suggestive of physical meaning. Therefore we shall use them in the sequel.

2.5 Fugacity and Absolute Temperature. Step iv.

We eliminate A^A and A^{BA} between the equations (2.15), (2.16), and $(2.21)_{1,2}$ to obtain

$$\begin{aligned}
nmU^A &= \left(\frac{\partial\Gamma_0}{\partial\Lambda} + \frac{\partial^2\Gamma_1}{\partial\Lambda\partial G_0}G_1\right)\Lambda^A + \frac{\partial\Gamma_1}{\partial\Lambda}\Lambda^{AB}\Lambda_B \\
t^{\langle BA\rangle} &+ (p+\pi)h^{BA} + \frac{1}{c^2}(U^B q^A + U^A q^B) + \frac{1}{c^2}eU^B U^A = \\
&= \left(\Gamma_0 + \frac{\partial\Gamma_1}{\partial G_0}G_1\right) g^{BA} + \Gamma_1\Lambda^{BA} + 2\left(\frac{\partial\Gamma_0}{\partial G_0} + \frac{\partial^2\Gamma_1}{\partial G_0^2}G_1\right)\Lambda^B\Lambda^A \\
&\quad + 2\frac{\partial\Gamma_1}{\partial G_0}\left(\Lambda^B\Lambda^{AC}\Lambda_C + \Lambda^A\Lambda^{BC}\Lambda_C\right).
\end{aligned} \tag{2.24}$$

This is a system of 14 equations from which, in principle, we may calculate the Lagrange multipliers Λ, Λ_B, and Λ_{BC} in terms of $n, U^A, t^{\langle AB\rangle}, \pi, q^A$, and e. To do so explicitly is, however, quite impossible in general, because the system of equations is strongly nonlinear. We can successfully perform that calculation only in a *linear* theory, i.e. in equilibrium and close to it.

For that purpose we recall that $\Lambda^{AB}|_E = 0$ holds and that $t^{\langle BC\rangle}, \pi$, and q^A all vanish in equilibrium. Thus we may hope to determine the equilibrium values of Λ and Λ_A in terms of n, U^A, and e from the equilibrium parts of (2.24), viz.

$$\begin{aligned}
nmU^A &= \left(\frac{\partial\Gamma_0}{\partial\Lambda}\Lambda^A\right)|_E \\
ph^{AB} + \frac{1}{c^2}eU^A U^B &= \left(\Gamma_0 g^{AB} + 2\frac{\partial\Gamma_0}{\partial G_0}\Lambda^A\Lambda^B\right)|_E.
\end{aligned} \tag{2.25}$$

Even this is impossible, however, unless $\Gamma_0 = \Gamma_0(\Lambda|_E, G_0|_E)$ is specified as a function of $\Lambda|_E$ and $G_0|_E$.

The best way to proceed is to identify $\Lambda|_E$ and $G_0|_E$ with the fugacity α and the absolute temperature T—two suggestive and measurable quantities—and to determine $\Gamma_0(\alpha, T)$ in terms of the thermal equation of state $p = p(n, e)$. To achieve this we first write the equilibrium version of (2.19) with $(2.21)_3$. We obtain

$$h|_E U^A = -\left(\Lambda\frac{\partial\Gamma_0}{\partial\Lambda} + 2\frac{\partial\Gamma_0}{\partial G_0}G_0\right)\Lambda^A|_E. \tag{2.26}$$

The equations (2.25), (2.26) imply

$$nmc = \sqrt{G_0}\frac{\partial\Gamma_0}{\partial\Lambda}|_E; \quad p = -\Gamma_0|_E; \quad p + e = 2G_0\frac{\partial\Gamma_0}{\partial G_0}|_E; \quad \frac{h|_E}{nm} = -\left(\Lambda + 2\frac{G_0\frac{\partial\Gamma_0}{\partial G_0}}{\frac{\partial\Gamma_0}{\partial\Lambda}}\right)|_E \tag{2.27}$$

and from these we derive easily

$$d\left(\frac{h|_E}{nm}\right) = -\frac{\sqrt{G_0|_E}}{c}\left(d\left(\frac{e}{nm}\right) + pd\left(\frac{1}{nm}\right)\right) \quad \text{and} \tag{2.28}$$

$$\Lambda|_E = -\frac{\sqrt{G_0|_E}}{c}\left(\frac{e}{nm} + \frac{c}{\sqrt{G_0|_E}}\frac{h|_E}{nm} + \frac{p}{nm}\right). \tag{2.29}$$

By comparison of (2.28) with the Gibbs equation of thermostatics we thus identify the absolute temperature T by the equation

$$\frac{1}{T} = -\frac{\sqrt{G_0|_E}}{c}. \tag{2.30}$$

Knowing this we see that the right-hand side of (2.29) is the fugacity α with

$$\alpha = -\frac{1}{T}\frac{e - Th|_E + p}{nm}. \tag{2.31}$$

We summarize the foregoing arguments about equilibrium and write

$$\Lambda|_E = -\alpha \qquad \Lambda_A|_E = -\frac{U_A}{T}, \qquad \Lambda_{AB}|_E = 0 \tag{2.32}$$

so that the equilibrium values of the Lagrange multipliers are now known as simple combinations of α, T, and U_A.

In many ways the fugacity and the absolute temperature T are more useful variables than n and e, because they are easier to measure by virtue of their continuity at diathermal walls and permeable surfaces such as phase boundaries. Moreover, α and T are the natural variables of statistical thermodynamics.[1] If $\Gamma_0|_E$, or by $(2.27)_2$, if p is known as a function of α and T, one may use $(2.27)_{1,3}$ to calculate

$$nm = -\frac{1}{T}\left(\frac{\partial p(\alpha,T)}{\partial \alpha}\right)_T \quad \text{and} \quad e = T^2\left(\frac{\partial}{\partial T}\left(\frac{p(\alpha,T)}{T}\right)\right)_\alpha. \tag{2.33}$$

These two relations may be used to switch back and forth between the sets of variables (n, e) and (α, T) provided, of course, that the thermal equation of state $p = p(\alpha, T)$ is known.

2.6 Linear Relations Between Lagrange Multipliers and $\mathbf{n}, \mathbf{U^A}, \mathbf{t^{\langle AB \rangle}}, \pi, \mathbf{q^A}, \mathbf{e}$. Step v.

We shall now use the equations (2.24) to calculate the Lagrange multipliers as linear functions of $t^{\langle AB \rangle}, q^A$, and π. In order to linearize the system (2.24) we introduce

$$\begin{aligned} \Lambda \quad &= -\alpha + \lambda \\ \Lambda_A \quad &= -\frac{U_A}{T} + \tau_A + \tau U_A \\ \Lambda_{AB} &= \sigma_{\langle AB \rangle} + \sigma h_{AB} + \frac{1}{c^2}(U_A\sigma_B + U_B\sigma_A) + \frac{3}{c^2}\sigma U_A U_B \end{aligned} \tag{2.34}$$

on the right-hand side of (2.24) and ignore all products of the quantities λ, τ, and σ. In (2.34) the terms with $\lambda, \tau, \sigma, \tau_A, \sigma_A$, and $\sigma_{\langle AB \rangle}$ represent the nonequilibrium parts of the Lagrange multipliers, which have been decomposed into space-like and time-like components. Due account has been taken of the constraint that Λ_{AB} is traceless.

---

[1] It should be said that the present nomenclature deviates from the common one in thermodynamics in the definition (2.31) of the fugacity. Our α is equal to $-1/T$ times the chemical potential, as indicated by (2.31).

Insertion of (2.34) into (2.24) provides a system of linear equations for $\lambda, \tau, \sigma, \tau^A, \sigma^A$, and $\sigma^{\langle AB\rangle}$. We set $\Gamma_0|_E = -p$, according to $(2.27)_2$, and define the derivatives

$$R' = \frac{\partial R}{\partial \ln T} \qquad \text{and} \qquad \dot{R} = \frac{\partial R}{\partial \alpha}. \tag{2.35}$$

Thus we obtain

$$\begin{bmatrix} -\ddot{p} & \dot{p}-\dot{p}' & 3\left(-\dot{\Gamma}_1+\frac{1}{2}\dot{\Gamma}_1'\right) \\ -\dot{p} & -p' & \left(\frac{3}{2}\Gamma_1'+\Gamma_1\right) \\ \dot{p}-\dot{p}' & p'-p'' & 3\left(\Gamma_1-\frac{3}{2}\Gamma_1'+\frac{1}{2}\Gamma_1''\right) \end{bmatrix} \begin{bmatrix} \lambda \\ (-T\tau) \\ \sigma \end{bmatrix} = \begin{bmatrix} 0 \\ \pi \\ 0 \end{bmatrix} \tag{2.36}$$

$$\begin{bmatrix} \dot{p} & -\dot{\Gamma}_1 \\ p' & (\Gamma_1-\Gamma_1') \end{bmatrix} \begin{bmatrix} (-T\tau^A) \\ \sigma^A \end{bmatrix} = \begin{bmatrix} 0 \\ q^A \end{bmatrix} \tag{2.37}$$

$$\Gamma_1\sigma^{\langle AB\rangle} = t^{\langle AB\rangle}. \tag{2.38}$$

The index E on Γ_1 has been dropped for simplicity.

The solutions of these equations read

$$\begin{aligned}
\lambda &= -\frac{\begin{bmatrix} \dot{p}-\dot{p}' & 3(-\dot{\Gamma}_1+\frac{1}{2}\dot{\Gamma}_1') \\ p'-p'' & 3(\Gamma_1-\frac{3}{2}\Gamma_1'+\frac{1}{2}\Gamma_1'') \end{bmatrix}}{D_1^\pi}\pi = \lambda_\pi\pi \\
\tau &= -\frac{1}{T}\frac{\begin{bmatrix} -\ddot{p} & 3(-\dot{\Gamma}_1+\frac{1}{2}\dot{\Gamma}_1') \\ \dot{p}-\dot{p}' & 3(\Gamma_1-\frac{3}{2}\Gamma_1'+\frac{1}{2}\Gamma_1'') \end{bmatrix}}{D_1^\pi}\pi = \tau_\pi\pi \\
\sigma &= -\frac{\begin{bmatrix} -\ddot{p} & \dot{p}-\dot{p}' \\ \dot{p}-\dot{p}' & p'-p'' \end{bmatrix}}{D_1^\pi}\pi = \sigma_\pi\pi \\
\tau^A &= -\frac{1}{T}\frac{\dot{\Gamma}_1}{D_3}q^A = \tau_q q^A \\
\sigma^A &= +\ \frac{\dot{p}}{D_3}q^A = \sigma_q q^A \\
\sigma^{\langle BA\rangle} &= \frac{1}{\Gamma_1}t^{\langle BA\rangle}.
\end{aligned} \tag{2.39}$$

D_1^π and D_3 are abbreviations for the determinants of the linear system (2.36) and (2.37) respectively. The coefficients $\lambda_\pi, \tau_\pi, \sigma_\pi, \tau_q, \sigma_q$, whose definitions may be read off from (2.39), will serve as useful abbreviations in subsequent formulae.

Combining (2.34) and (2.39), we obtain linear expressions for the Lagrange multipliers Λ, Λ_A, and Λ_{AB} in terms of π, q_A, and $t_{\langle AB\rangle}$, viz.

$$\Lambda = -\alpha + \lambda_\pi\pi$$

$$\begin{aligned}\Lambda_A &= (-\frac{1}{T} + \tau_\pi \pi) U_A + \tau_q q_A \\ \Lambda_{AB} &= \frac{1}{\Gamma_1} t_{\langle AB \rangle} + \sigma_\pi \pi h_{AB} + \frac{1}{c^2} \sigma_q (U_A q_B + U_B q_A) + \frac{3}{c^2} \sigma_\pi \pi U_A U_B.\end{aligned} \tag{2.40}$$

2.7 The Linear Flux Tensor. Step vi.

In order to obtain a linear expression for the flux tensor A^{ABC}, we insert (2.34) into the representation (2.17) and ignore all products of the quantities λ, τ, and σ. The resulting expression reads

$$\begin{aligned}&A^{BCA} = \\ &= -\frac{1}{T}\left\{\frac{1}{c^2}\left[-\frac{1}{2}\Gamma_1' + \frac{1}{2}\dot{\Gamma}_1'\lambda + \left(\frac{1}{2}\Gamma_1'' - \frac{1}{2}\Gamma_1'\right)(-T\tau) + 3\left(\frac{1}{4}\Gamma_2'' - \Gamma_2' + 2\Gamma_2\right)\sigma\right] U^A U^B U^C + \right.\\ &+ \left[\frac{1}{2}\Gamma_1 - \frac{1}{2}\dot{\Gamma}_1\lambda + \left(\frac{1}{2}\Gamma_1 - \frac{1}{2}\Gamma_1'\right)(-T\tau) + \left(-\frac{3}{4}\Gamma_2' + \Gamma_2\right)\sigma\right](g^{AB}U^C + g^{BC}U^A + g^{CA}U^B) + \\ &+ \frac{1}{2c^2}\Gamma_1' T\left[U^A U^B \tau^C + U^B U^C \tau^A + U^C U^A \tau^B\right] \\ &\qquad + \frac{1}{c^2}(\Gamma_2 - \frac{1}{2}\Gamma_2')\left[U^A U^B \sigma^C + U^B U^C \sigma^A + U^C U^A \sigma^B\right] + \\ &- \frac{1}{2}\Gamma_1 T\left[g^{AB}\tau^C + g^{BC}\tau^A + g^{CA}\tau^B\right] + \frac{1}{2}\Gamma_2\left[g^{AB}\sigma^C + g^{BC}\sigma^A + g^{CA}\sigma^B\right] + \\ &\left. + \frac{1}{2}\Gamma_2\left[\sigma^{\langle AB\rangle}U^C + \sigma^{\langle BC\rangle}U^A + \sigma^{\langle CA\rangle}U^B\right]\right\}.\end{aligned} \tag{2.41}$$

Hence we calculate A^{ABC} as a linear function of π, q^A, and $t^{\langle AB\rangle}$ by introducing (2.39). After a little calculation and by use of the conditions (2.18) we obtain

$$\begin{aligned}A^{ABC} &= (C_1^0 + C_1^\pi \pi) U^A U^B U^C + \frac{c^2}{6}(nm - C_1^0 - C_1^\pi \pi)(g^{AB}U^C + g^{BC}U^A + g^{CA}U^B) + \\ &+ C_3(g^{AB}q^C + g^{BC}q^A + g^{CA}q^B) - \frac{6}{c^2}C_3(U^A U^B q^C + U^B U^C q^A + U^C U^A q^B) + \\ &+ C_5(t^{\langle AB\rangle}U^C + t^{\langle BC\rangle}U^A + t^{\langle CA\rangle}U^B),\end{aligned} \tag{2.42}$$

where the coefficients C have been introduced as convenient abbreviations defined as follows:

$$\begin{aligned}C_1^0 &= \frac{1}{2c^2}\frac{\Gamma_1'}{T} \\ C_1^\pi &= -\frac{3}{c^2}\frac{1}{T}\frac{\begin{bmatrix} -\dot{\Gamma}_1 & \Gamma_1 - \Gamma_1' & (-\frac{3}{2}\Gamma_2' + 2\Gamma_2) \\ -\ddot{p} & \dot{p} - \dot{p}' & 3(-\dot{\Gamma}_1 + \frac{1}{2}\dot{\Gamma}_1') \\ \dot{p} - \dot{p}' & +p' - p'' & 3(\Gamma_1 - \frac{3}{2}\Gamma_1' + \frac{1}{2}\Gamma_1'') \end{bmatrix}}{D_1^\pi} \\ C_3 &= \frac{1}{2}\frac{1}{T}\frac{\begin{bmatrix} \Gamma_1 & \Gamma_2 \\ \dot{p} & -\dot{\Gamma}_1 \end{bmatrix}}{D_3} \\ C_5 &= -\frac{1}{T}\frac{1}{2}\frac{\Gamma_2}{\Gamma_1}.\end{aligned} \tag{2.43}$$

We conclude from (2.43) that the linear expression for the flux tensor A^{ABC} is determined by the three functions p, Γ_1, and Γ_2, which all depend on the fugacity α and the absolute temperature T. Among these three functions, Γ_1 and Γ_2 are closely related to p by the trace conditions (2.18), which we shall exploit later.

2.8 The Entropy Flux Vector. Step vii.

From (2.3) and (2.5) we conclude that the entropy flux vector satisfies the condition

$$dh^A = -\Lambda dA^A - \Lambda_B dA^{BA} - \Lambda_{BC} dA^{BCA}. \tag{2.44}$$

We insert (2.40) to eliminate the Lagrange multipliers and $(2.21)_{1,2}$, (2.42) to eliminate A^A, A^{BA}, and A^{BCA}. Thus on the right-hand side of (2.44) there appear terms with $d\pi, dq^E$, and $dt^{\langle EF\rangle}$, whose coefficients must be interpreted as derivatives of h^A with respect to π, q^E, and $t^{\langle EF\rangle}$. We get

$$\begin{aligned}
\frac{\partial h^A}{\partial \pi} &= -2c^2 C_1^\pi \sigma_\pi \pi U^A + (\tau_q + \frac{c^2}{3} C_1^\pi \sigma_q) q^A \\
\frac{\partial h^A}{\partial q^E} &= -\left(\frac{1}{T} + (\tau_\pi - 20C_3\sigma_\pi)\pi\right) h_E^A - (\tau_q - 10C_3\sigma_q)\frac{1}{c^2} U^A q_E - 2C_3 \frac{1}{\Gamma_1} t_{\langle E}^{A\rangle} \\
\frac{\partial h^A}{\partial t^{\langle EF\rangle}} &= \frac{1}{2}(\tau_q - 2C_5\sigma_q)\left(q_E h_F^A + q_F h_E^A - \frac{2}{3} q^A h_{EF}\right) - C_5 \frac{1}{\Gamma_1} t_{\langle EF\rangle} U^A.
\end{aligned} \tag{2.45}$$

Integration gives after a short calculation

$$h^A = (h|_E + A_1^{\pi^2}\pi^2 + A_1^q q_E q^E + A_1^t t^{\langle EF\rangle} t_{\langle EF\rangle}) U^A + (A_2^0 + A_2^\pi \pi) q^A + A_3^0 t^{\langle AE\rangle} q_E, \tag{2.46}$$

where the coefficients A have been introduced as abbreviations for the following expressions.

$$\begin{aligned}
A_1^{\pi^2} &= -c^2 C_1^\pi \sigma_\pi \\
A_1^q &= -\frac{1}{2c^2}(\tau_q - 10C_3\sigma_q) \\
A_1^t &= -\frac{1}{2} C_5 \frac{1}{\Gamma_1} \\
A_2^0 &= \frac{1}{T} \\
A_2^\pi &= -(\tau_\pi - 20C_3\sigma_\pi) = \tau_q + \frac{c^2}{3} C_1^\pi \sigma_q, \\
A_3^0 &= -(\tau_q - 2C_5\sigma_q) = -C_3 \frac{2}{\Gamma_1}.
\end{aligned} \tag{2.47}$$

The definition of $\tau_\pi, \sigma_\pi, \tau_q, \sigma_q$ in terms of the functions p, Γ_1 and the definition of C_1^π, C_3, C_5 in terms of the functions p, Γ_1, Γ_2 may be read off from (2.39) and (2.43). We conclude that the entropy flux vector is determined by the functions p, Γ_1, Γ_2 to within third-order terms in π, q^E, and $t^{\langle EF\rangle}$.

The second equations in $(2.47)_{5,6}$ result from the integrability conditions implied by (2.45). These equations are, of course, identically satisfied by the explicit forms (2.39) and (2.43) for $\tau_\pi, \tau_q, \sigma_\pi, \sigma_q$ and C_1^π, C_3, C_5.

2.9 Residual Inequality. Step viii.

We recall from (2.7) the residual inequality for the entropy production

$$\sum = -\Lambda_{BC} I^{BC} \geq 0 \tag{2.48}$$

and remember that the flux production I^{BC} is a constitutive quantity subject to the principle of relativity, so that we must have

$$\frac{\partial x^{*A}}{\partial x^C}\frac{\partial x^{*B}}{\partial x^D}\hat{I}^{CD}(n, U^P, t^{\langle PQ\rangle}, \pi, q^P, e) = \hat{I}^{AB}\left(n, \frac{\partial x^{*P}}{\partial x^R}U^R, \frac{\partial x^{*P}}{\partial x^R}\frac{\partial x^{*Q}}{\partial x^S}t^{\langle RS\rangle}, \pi, \frac{\partial x^{*P}}{\partial x^R}q^R, e\right). \tag{2.49}$$

Thus $\hat{I}^{AB}$ is an isotropic function. From the representation theorems for isotropic functions (see Smith [54], Pennisi [68]) we obtain a fairly explicit form for I^{BC} if we restrict the attention to linear terms in π, q^C, and $t^{\langle CD\rangle}$.

$$I^{BC} = B_1^\pi \pi g^{BC} - \frac{4}{c^2}B_1^\pi \pi U^B U^C + B_3 t^{\langle BC\rangle} + \frac{1}{c^2}\hat{B}_4(q^B U^C + q^C U^B). \tag{2.50}$$

In writing (2.50) we have also used the fact that I^{BC} must vanish in equilibrium and that, by (1.5), I^{BC} must be traceless. The coefficients B may depend on n and e.

We insert Λ_{BC} from (2.40) and I^{BC} from (2.50) into the residual inequality and calculate its left-hand side excluding third order terms, because these would not be reliable in the present linear theory. We obtain after a little calculation

$$\sum = 12B_1^\pi \sigma_\pi \pi^2 - 2\frac{1}{c^2}\hat{B}_4\sigma_q q^C q_C - B_3\frac{1}{\Gamma_1}t^{\langle BC\rangle}t_{\langle BC\rangle} \geq 0 \tag{2.51}$$

and it follows that B_1^π, $\hat{B}_4$, and B_3 must satisfy the inequalities

$$B_1^\pi \sigma_\pi \geq 0 \qquad \hat{B}_4\sigma_q \geq 0 \qquad B_3\frac{1}{\Gamma_1} \leq 0. \tag{2.52}$$

2.10 Causality and Convexity. Step ix.

For a compact statement of the requirement of causality and convexity it is appropriate to introduce a synthetic form of the balance equations (1.2). We let

$$\left.\begin{matrix} A^A_\alpha \\ I_\alpha \\ \Lambda_\alpha \end{matrix}\right] \quad \text{stand for} \quad \left[\begin{matrix} \{A^A, A^A_B, A^A_{\langle BC\rangle}\} \\ \{0,\ 0,\ I_{BC}\} \\ \{\Lambda,\ \Lambda_B, \Lambda_{BC}\} \end{matrix}\right. \tag{2.53}$$

and rewrite the field equations (1.2) and the restrictions (2.6) from the entropy principle in the form

$$A^A_{\alpha,A} = I_\alpha \qquad \text{and} \qquad A^A_\alpha = \frac{\partial h'^A}{\partial \Lambda_\alpha}. \tag{2.54}$$

The Greek indices range over the values 1, 2, ...14. Combination of the two equations (2.54) gives

$$\frac{\partial^2 h'^A}{\partial \Lambda_\alpha \partial \Lambda_\beta} \Lambda_{\beta,A} = I_\alpha. \tag{2.55}$$

This is a symmetric system of equations, because the matrix $\frac{\partial^2 h'^A}{\partial \Lambda_\alpha \partial \Lambda_\beta}$ is symmetric in α and β. The system is said to be symmetric *hyperbolic* if the temporal part of this matrix is positive definite; see Chapter 3. The temporal part is obtained by multiplication of $\frac{\partial^2 h'^A}{\partial \Lambda_\alpha \partial \Lambda_\beta}$ by the four-velocity U_A of the gas. Thus the requirement of causality reads

$$\frac{\partial^2 h'^A}{\partial \Lambda_\alpha \partial \Lambda_\beta} U_A \delta\Lambda_\alpha \delta\Lambda_\beta > 0 \quad \text{or} \quad \frac{\partial A^A_\alpha}{\partial \Lambda_\beta} U_A \delta\Lambda_\alpha \delta\Lambda_\beta > 0, \tag{2.56}$$

and it must hold for arbitrary choices of $\delta\Lambda_\gamma$.

In [69] Ruggeri has proved that for a nondegenerate gas this symmetrization condition is equivalent to the convexity of $-h^A U_A$—the entropy density in the rest frame—with respect to the densities $A^A U_A, A^{AB} U_A, A^{ABC} U_A$.

With $(2.54)_2$ we may rewrite (2.56) in the form

$$U_A \delta A^A_\alpha \delta\Lambda_\alpha > 0 \qquad \text{or explicitly with (2.53)}$$

$$U_A \delta A^A \delta\Lambda + U_A \delta A^{AB} \delta\Lambda_B + U_A \delta A^{ABC} \delta\Lambda_{BC} > 0. \tag{2.57}$$

This requirement can best be exploited by insertion of the Lagrange multipliers (2.40) and of the representations $(2.21)_{1,2}$ for A^A, A^{BA} and (2.42) for A^{ABC}. There results a quadratic form in $\delta\alpha, \delta T, \delta\pi, \delta U^C, \delta q^C$, and $\delta t^{\langle CD \rangle}$ which must be positive. The matrix of this quadratic form contains nonequilibrium terms, which must be ignored since they are not reliable in the present theory where the $\mathbf{\Lambda}$'s and A^{ABC} are only known linearly. The matrix must be positive definite and a little calculation shows that this requirement implies

$$C_5 \frac{1}{\Gamma_1} > 0$$

$$\begin{bmatrix} \frac{1}{T} p' & -\frac{1}{2}\left[-\frac{1}{T} - \frac{1}{2}\frac{1}{T}\left(-\frac{3}{2}\Gamma_1' - m^2c^2\dot{p}\right) + p'\tau_q\right] \\ \odot & -(\tau_q - 10 C_3 \sigma_q) \end{bmatrix} \text{- pos. def.}$$

$$\begin{bmatrix} \frac{1}{T}\ddot{p} & -\frac{1}{2T}(\dot{p} - \dot{p}') & \frac{1}{2}\left[\tau_\pi(\dot{p} - \dot{p}') - \frac{1}{T}\sigma_\pi(\dot{\Gamma}_1' + m^2c^2\ddot{p})\right] \\ \odot & -\frac{1}{2T}\left(-\frac{1}{2}p'' + \frac{1}{2}p'\right) & \frac{1}{2}\left[-\frac{1}{T}\sigma_\pi\left(-\frac{1}{2}\Gamma_1' + \frac{1}{2}\Gamma_1'' - \frac{1}{2}m^2c^2(\dot{p} - \dot{p}')\right)\right] \\ \odot & \odot & 2c^2\sigma_\pi C_1^\pi \end{bmatrix} \text{- pos. def.} \tag{2.58}$$

Only the upper part of the matrices is written here; the symbols $\odot$ are supposed to indicate that the matrices are symmetric. In particular it follows from (2.58) and (2.47) that we have

$$A_1^{\pi^2} < 0 \qquad A_1^q > 0 \qquad A_1^t < 0 \tag{2.59}$$

so that the equilibrium entropy $\frac{1}{c^2} h^A U_A$ is bigger than the nonequilibrium entropy.

2.11 Summary of Results. Step x.

It has proved convenient to introduce the four-velocity U^A with $U_A U^A = c^2$ and the projector $h_{AB} = \frac{1}{c^2} U_A U_B - g_{AB}$. By use of these the particle flux vector, the energy-momentum tensor, and the entropy flux vector may be decomposed into time-like and space-like components. We have

$$\begin{aligned} A^A &= nmU^A \\ A^{AB} &= t^{\langle AB\rangle} + (p(n,e)+\pi)h^{AB} + \frac{1}{c^2}(U^A q^B + U^B q^A) + \frac{e}{c^2} U^A U^B, \\ h^A &= hU^A + \Phi^A, \end{aligned} \tag{2.60}$$

and the components have suggestive meaning as follows:

$$\begin{array}{ll} n & \text{- number density} \\ t^{\langle AB\rangle} & \text{- stress deviator} \\ p+\pi & \text{- pressure} \\ q^A & \text{- heat flux} \\ e & \text{- energy density} \\ h & \text{- entropy density} \\ \Phi^A & \text{- (non-convective) entropy flux} \end{array} \tag{2.61}$$

The stress deviator, the heat flux, and the dynamic pressure π vanish in equilibrium.

The constitutive theory has provided expressions for the flux tensor A^{ABC} and the flux production I^{AB} which are linear in π, q^A, and $t^{\langle AB\rangle}$ and have the forms

$$\begin{aligned} A^{ABC} &= (C_1^0 + C_1^\pi \pi)U^A U^B U^C + \frac{c^2}{6}(nm - C_1^0 - C_1^\pi \pi)(g^{AB}U^C + g^{BC}U^A + g^{CA}U^B) + \\ &\quad + C_3(g^{AB}q^C + g^{BC}q^A + g^{CA}q^B) - \frac{6}{c^2}C_3(U^A U^B q^C + U^B U^C q^A + U^C U^A q^B) + \\ &\quad + C_5(t^{\langle AB\rangle}U^C + t^{\langle BC\rangle}U^A + t^{\langle CA\rangle}U^B), \\ I^{AB} &= B_1^\pi \pi g^{AB} - \frac{4}{c^2} B_1^\pi \pi U^A U^B + B_3 t^{\langle AB\rangle} + \frac{1}{c^2}\hat{B}_4(q^A U^B + q^B U^A). \end{aligned} \tag{2.62}$$

Also the entropy flux vector h^A has emerged to within third-order terms in π, q^A, and $t^{\langle AB\rangle}$, viz.

$$h^A = (h|_E + A_1^{\pi^2}\pi^2 + A_1^q q_E q^E + A_1^t t_{\langle EF\rangle} t^{\langle EF\rangle})U^A + (A_2^0 + A_2^\pi \pi)q^A + A_3^0 t^{\langle AE\rangle} q_E. \tag{2.63}$$

$h|_E$ is the equilibrium entropy density, which satisfies the Gibbs equation

$$d\left(\frac{h|_E}{nm}\right) = \frac{1}{T}\left(d\left(\frac{e}{nm}\right) + pd\left(\frac{1}{nm}\right)\right) \tag{2.64}$$

where T is the absolute temperature.

The coefficients C in the flux tensor (2.62) and the coefficients A in the entropy flux vector (2.63) are related to the three functions p, Γ_1, and Γ_2 of the two variables

$$\text{fugacity} \quad \alpha = -\frac{1}{T}\frac{e - Th_E + p}{nm} \qquad \text{and} \qquad \text{absolute temperature} \quad T$$

in the following manner:

$$C_1^0 = \frac{1}{2c^2}\frac{\Gamma_1'}{T} \tag{2.65$_1$}$$

$$C_1^\pi = -\frac{3}{c^2}\frac{1}{T}\frac{\begin{bmatrix} -\ddot{p} & \dot{p}-\dot{p}' & \dot{\Gamma}_1 \\ \dot{p}-\dot{p}' & p'-p'' & \Gamma_1'-\Gamma_1 \\ \dot{\Gamma}_1 & \Gamma_1'-\Gamma_1 & \frac{5}{3}\Gamma_2 \end{bmatrix}}{\begin{bmatrix} -\ddot{p} & \dot{p}-\dot{p}' & \dot{\Gamma}_1 \\ \dot{p}-\dot{p}' & p'-p'' & \Gamma_1'-\Gamma_1 \\ -\dot{p} & -p' & \frac{5}{3}\Gamma_1 \end{bmatrix}} \tag{2.65$_2$}$$

$$C_3 = -\frac{1}{2}\frac{1}{T}\frac{\begin{bmatrix} \dot{p} & -\dot{\Gamma}_1 \\ \Gamma_1 & \Gamma_2 \end{bmatrix}}{\begin{bmatrix} \dot{p} & -\dot{\Gamma}_1 \\ p' & \Gamma_1-\Gamma_1' \end{bmatrix}} \tag{2.65$_3$}$$

$$C_5 = -\frac{1}{2}\frac{1}{T}\frac{\Gamma_2}{\Gamma_1} \tag{2.65$_4$}$$

$$A_1^{\pi^2} = -\frac{c^2}{6}C_1^\pi\frac{\begin{bmatrix} -\ddot{p} & \dot{p}-\dot{p}' \\ \dot{p}-\dot{p}' & p'-p'' \end{bmatrix}}{\begin{bmatrix} -\ddot{p} & \dot{p}-\dot{p}' & \dot{\Gamma}_1 \\ \dot{p}-\dot{p}' & p'-p'' & \Gamma_1'-\Gamma_1 \\ -\dot{p} & -p' & \frac{5}{3}\Gamma_1 \end{bmatrix}} \tag{2.66$_1$}$$

$$A_1^q = -\frac{1}{2c^2}\frac{-\frac{1}{T}\dot{\Gamma}_1 - 10C_3\dot{p}}{\begin{bmatrix} \dot{p} & -\dot{\Gamma}_1 \\ p' & \Gamma_1-\Gamma_1' \end{bmatrix}} \tag{2.66$_2$}$$

$$A_1^t = -\frac{1}{2}C_5\frac{1}{\Gamma_1} \tag{2.66$_3$}$$

$$A_2^0 = \frac{1}{T} \tag{2.66$_4$}$$

$$A_2^\pi = -\frac{\frac{1}{T}\begin{bmatrix} -\ddot{p} & \dot{\Gamma}_1 \\ \dot{p}-\dot{p}' & \Gamma_1'-\Gamma_1 \end{bmatrix} - \frac{10}{3}C_3\begin{bmatrix} -\ddot{p} & \dot{p}-\dot{p}' \\ \dot{p}-\dot{p}' & p'-p'' \end{bmatrix}}{\begin{bmatrix} -\ddot{p} & \dot{p}-\dot{p}' & \dot{\Gamma}_1 \\ \dot{p}-\dot{p}' & p'-p'' & \Gamma_1'-\Gamma_1 \\ -\dot{p} & -p' & \frac{5}{3}\Gamma_1 \end{bmatrix}} = -\frac{\frac{1}{T}\dot{\Gamma}_1 - \frac{c^2}{3}C_1^\pi\dot{p}}{\begin{bmatrix} \dot{p} & -\dot{\Gamma}_1 \\ p' & \Gamma_1-\Gamma_1' \end{bmatrix}} \tag{2.66$_5$}$$

$$A_3^0 = \frac{\frac{1}{T}\dot{\Gamma}_1 - 2C_5\dot{p}}{\begin{bmatrix} \dot{p} & -\dot{\Gamma}_1 \\ p' & \Gamma_1-\Gamma_1' \end{bmatrix}} = -C_3\frac{2}{\Gamma_1}. \tag{2.66$_6$}$$

Dots and primes denote the derivatives with respect to α and $\ln T$ respectively.

Fugacity and absolute temperature are the natural variables of statistical thermodynamics, as we shall see later. Also they are easier to measure—at least in principle—than the original variables n and e. However, if we wish to switch between the two sets of variables (α, T) and (n, e), that may be by use of the formulae

$$nm = -\frac{1}{T}\dot{p} \qquad \text{and} \qquad e = p' - p \tag{2.67}$$

provided that the thermal equation of state $p = p(\alpha, T)$ is known.

The results (2.65) and (2.66) seem to indicate that the three functions $p(\alpha, T), \Gamma_1(\alpha, T)$, and $\Gamma_2(\alpha, T)$ are needed for the determination of the flux tensor A^{ABC} and I^{AB}. This is not so, however, because the functions Γ_1 and Γ_2 are dependent on $p = p(\alpha, T)$. Indeed we have (see (2.18))

$$-\frac{1}{2}\Gamma_1' + 3\Gamma_1 = c^2\dot{p} \qquad \text{and} \qquad -\frac{1}{2}\Gamma_2' + 4\Gamma_2 = -c^2\dot{\Gamma}_1, \tag{2.68}$$

so that Γ_1 and Γ_2 can be related to p by integration. We obtain

$$\begin{aligned} \Gamma_1 &= T^6 \left[-2c^2 \int \frac{\dot{p}}{T^7} dT + A_1(\alpha)\right] \\ \Gamma_2 &= T^8 \left\{2c^2 \int \frac{1}{T^3}\left[-2c^2 \int \frac{\ddot{p}}{T^7} dT + \dot{A}_1(\alpha)\right] dT + A_2(\alpha)\right\}. \end{aligned} \tag{2.69}$$

and we conclude that Γ_1 and Γ_2 may be determined from $p(\alpha, T)$, each to within a function of the single variable α, the fugacity.

It follows from (2.65), (2.66) that all coefficients of the flux tensor A^{ABC} and of the entropy flux vector h^A can be determined to within two functions of α *if only* the thermal equation of state $p = p(\alpha, T)$ is known.

The entropy production $\sum$ in (2.51) contains three terms, one each due to the dynamic pressure, the heat flux, and the stress deviator. Thus there are three dissipative "mechanisms" in the gas. The conditions for a nonnegative entropy production are listed in (2.52); more explicitly they read

$$B_1^{\pi} \frac{\begin{bmatrix} -\ddot{p} & \dot{p} - \dot{p}' \\ \dot{p} - \dot{p}' & p' - p'' \end{bmatrix}}{\begin{bmatrix} -\ddot{p} & \dot{p} - \dot{p}' & \dot{\Gamma}_1 \\ \dot{p} - \dot{p}' & p' - p'' & \Gamma_1' - \Gamma_1 \\ -\dot{p} & -p' & \frac{5}{3}\Gamma_1 \end{bmatrix}} \geq 0 \qquad \hat{B}_4 \frac{\dot{p}}{\begin{bmatrix} \dot{p} & -\Gamma_1' \\ p' & \Gamma_1 - \Gamma_1' \end{bmatrix}} \geq 0 \qquad B_3 \frac{1}{\Gamma_1} \leq 0. \tag{2.70}$$

These inequalities provide restrictions on the coefficients B in the flux production.

In the next section we shall show that the coefficients B may be related to the transport coefficients of ordinary thermodynamics, namely the heat conductivity, the viscosity, and the bulk viscosity. These are measurable quantities, at least in principle.

The requirement of causality and convexity furnishes restrictions on the coefficients C and A in the flux tensor and in the entropy flux vector; see (2.58), (2.59). In effect these conditions restrict the form of the thermal equation of state $p = p(\alpha, T)$ and of the functions $A_1(\alpha), A_2(\alpha)$. Most of the restrictions are difficult to interpret and we only list a few that can be interpreted, viz.

$$\begin{bmatrix} \frac{1}{T}\ddot{p} & -\frac{1}{2T}(\dot{p} - \dot{p}') \\ -\frac{1}{2T}(\dot{p} - \dot{p}') & \frac{1}{4T}(p'' - p') \end{bmatrix} \text{ - pos. def.,} \quad A_1^{\pi^2} < 0, A_1^q > 0, A_1^t < 0. \tag{2.71}$$

Inspection of (2.63) shows that the inequalities $(2.71)_{2,3,4}$ are necessary for the entropy density $h = \frac{1}{c^2}h^A U_A$ to have a maximum in equilibrium. The positive-definitness of the matrix $(2.71)_1$ ensures that the compressibility $\frac{1}{n}\left(\frac{\partial n}{\partial p}\right)_T$ and the specific heat $\left(\frac{\partial e}{\partial T}\right)_n$ are positive. These are two well-known *thermodynamic stability conditions*, i.e. requirements on the constitutive functions which guarantee that the entropy of a body in homogeneous equilibrium is bigger than in a nonhomogeneous equilibrium.

3 Identification of Viscosities and Heat-Conductivity

3.1 Extended Thermodynamics and Ordinary Thermodynamics

The purpose of this section is to identify the coefficients $B_1^\pi, B_3, \hat{B}_4$ in the flux production by relating them to the bulk viscosity, the shear viscosity, and the heat conductivity. These are the transport coefficients of ordinary thermodynamics, which can in principle be measured. In order to find the relation we make a brief detour and show how extended thermodynamics relates to ordinary thermodynamics.

First of all we recall the objective of *extended* thermodynamics, which is the determination of the 14 fields

$$A^A, A^{BA} \quad \text{or} \quad n, U^A, e, \pi, q^A, t^{\langle AB\rangle}. \tag{3.1}$$

The field equations are based upon the equations of balance (1.2), which by $(2.60)_{1,2}$ and (2.62) may be written as

$$\begin{aligned}
&(nmU^A)_{,A} = 0 \\
&(t^{\langle BA\rangle} + (p+\pi)h^{BA} + \frac{1}{c^2}(q^B U^A + q^A U^B) + \frac{e}{c^2}U^B U^A)_{,A} = 0 \\
&A^{BCA}{}_{,A} = B_1^\pi \pi g^{BC} - \frac{4}{c^2}B_1^\pi \pi U^B U^C + B_3 t^{\langle BC\rangle} + \frac{1}{c^2}\hat{B}_4(q^B U^C + q^C U^B)
\end{aligned} \tag{3.2}$$

where A^{BCA} must be inserted from (2.62).

In contrast to this, the objective of *ordinary* thermodynamics is the determination of the five fields

$$n, U^A, e. \tag{3.3}$$

The necessary field equations are based upon the balance equations

$$(nmU^A)_{,A} = 0 \tag{3.4$_1$}$$

$$\left(t^{\langle BA\rangle} + (p+\pi)h^{BA} + \frac{1}{c^2}(q^B U^A + q^A U^B) + \frac{e}{c^2}U^B U^A\right)_{,A} = 0 \tag{3.4$_2$}$$

and constitutive relations must be found for $\pi, t^{\langle BA\rangle}$, and q^A which relate these quantities to the fields (3.3). For viscous, heat-conducting fluids the constitutive quantities $\pi, t^{\langle BA\rangle}$, and q^A in one event x^D are related to the values of the fields n, U^A, e in that event and to the values of n, U^A, e in the immediate neighborhood of x^D.

3.2 Transition from Extended to Ordinary Thermodynamics

We recall the balance laws (1.2) of extended thermodynamics and use these to create the constitutive relations of ordinary thermodynamic in a manner akin to the Maxwell iteration, which was described in Paragraph 4.3.2.

The iterates $\overset{(n)}{I}{}^{AB}$ result from

$$\begin{array}{lll} A^A{}_{,A} &= 0 & \\ \overset{(n-1)}{A}{}^{AB}{}_{,B} &= 0 & \\ \overset{(n-1)}{A}{}^{ABC}{}_{,C} &= \overset{(n)}{I}{}^{AB} \end{array} \quad \text{with the initiation agreement} \quad \begin{array}{ll} A^A{}_{,A} &= 0 \\ \overset{E}{A}{}^{AB}{}_{,B} &= 0 \\ \overset{E}{A}{}^{ABC}{}_{,C} &= \overset{(1)}{I}{}^{AB} \end{array} \tag{3.5}$$

$\overset{E}{A}{}^{AB}$ and $\overset{E}{A}{}^{ABC}$ are equilibrium values, i.e.

$$\begin{aligned} \overset{E}{A}{}^{AB} &= p h^{AB} + \frac{e}{c^2} U^A U^B \\ &= p h^{AB} - (p - p') U^A U^B, \\ \overset{E}{A}{}^{ABC} &= C_1^0 U^A U^B U^C + \frac{c^2}{6}\left(nm - C_1^0\right)\left(g^{AB}U^C + g^{BC}U^A + g^{CA}U^B\right) \\ &= \frac{1}{2c^2}\frac{\Gamma_1'}{T} U^A U^B U^C - \frac{1}{2}\frac{\Gamma_1}{T}\left(g^{AB}U^C + g^{BC}U^A + g^{CA}U^B\right). \end{aligned}$$

Insertion into $(3.5)_2$—and decomposition into time-like and space-like parts of the equations—provides the set

$$\begin{array}{llllllll} 0 & = -\ddot{p} & \frac{d\alpha}{d\tau} & -(\dot{p}' - \dot{p}) & \frac{d\ln T}{d\tau} & -\dot{p} & U^A{}_{,A} \\ 0 & = (\dot{p}' - \dot{p}) & \frac{d\alpha}{d\tau} & +(p'' - p') & \frac{d\ln T}{d\tau} & +p' & U^A{}_{,A} \\ 0 & = \dot{p} & h_M^A \alpha_{,A} & +p' & h_M^A (\ln T)_{,A} & +\frac{p'}{c^2} & \frac{dU_M}{d\tau} \\ -3B_1^\pi \pi & = \frac{\dot{\Gamma}_1' - 3\dot{\Gamma}_1}{2T} & \frac{d\alpha}{d\tau} & +\frac{\Gamma_1'' - 4\Gamma_1' + 3\Gamma_1}{2T} & \frac{d\ln T}{d\tau} & +\frac{\Gamma_1' - \Gamma_1}{2T} & U^A{}_{,A} \\ -\frac{1}{c^2}\hat{B}_4 q_M & = -\frac{\dot{\Gamma}_1}{2T} & h_M^A \alpha_{,A} & -\frac{\Gamma_1' - \Gamma_1}{2T} & h_M^A (\ln T)_{,A} & -\frac{1}{c^2}\frac{\Gamma_1' - \Gamma_1}{2T} & \frac{dU_M}{d\tau} \\ B_3 t_{\langle MN\rangle} & = -\frac{\Gamma_1}{T} h_M^B h_N^C U_{\langle B,C\rangle}. & & & & & \end{array} \tag{3.6}$$

The equations (2.67) have been used here to replace $\dot{n}, \dot{e}, n'$, and e' by derivatives of p. The symbol $\frac{dR}{d\tau}$ stands for the material time derivative $U^A R_{,A}$.

Elimination of the derivatives $\frac{d\alpha}{dT}, \frac{d\ln T}{d\tau}$ between $(3.6)_{1,2,4}$ and of the derivatives $h^A_M \alpha_{,A}$ between $(3.6)_{3,5}$ allows us to determine π, q_M, and $t_{\langle MN\rangle}$ as first iterates from $(3.6)_{4,5,6}$ as follows.

$$\overset{(1)}{\pi} = -\frac{1}{2T}\frac{1}{B_1^\pi}\frac{\begin{bmatrix} -\ddot{p} & \dot{p}-\dot{p}' & \dot{\Gamma}_1 \\ \dot{p}-\dot{p}' & p'-p'' & \Gamma_1'-\Gamma_1 \\ -\dot{p} & -p' & \frac{5}{3}\Gamma_1 \end{bmatrix}}{\begin{bmatrix} -\ddot{p} & \dot{p}-\dot{p}' \\ \dot{p}-\dot{p}' & p'-p'' \end{bmatrix}}[U^A{}_{,A}] \tag{3.7$_1$}$$

$$\overset{(1)}{q}_M = -\frac{1}{2T}\frac{c^2}{\hat{B}_4}\begin{bmatrix} \dot{p} & -\dot{\Gamma}_1 \\ +p' & \Gamma_1-\Gamma_1' \end{bmatrix}\left[h^A_M\left((\ln T)_{,A}-\frac{1}{c^2}\frac{dU_A}{d\tau}\right)\right]\frac{1}{\dot{p}} \tag{3.7$_2$}$$

$$\overset{(1)}{t}_{\langle MN\rangle} = -\frac{1}{T}\frac{1}{B_3}\Gamma_1\left[h^B_M h^C_N U_{\langle B,C\rangle}\right]. \tag{3.7$_3$}$$

The equations $(3.7)_{2,3}$ are the relativistic analogues to the phenomenological equations of Fourier and Navier-Stokes. Equation $(3.7)_1$ gives a linear relation between the dynamic pressure and the divergence of velocity. In analogy to the nomenclature of ordinary irreversible thermodynamics, we may call the quantities

$$\overset{(1)}{\pi}, \overset{(1)}{q}_M, \overset{(1)}{t}_{\langle MN\rangle} \quad \text{and} \quad [U^A{}_{,A}], \left[h^A_M T_{,A}+\frac{T}{c^2}\frac{dU_M}{d\tau}\right], [h^B_M h^C_N U_{B,C}] \tag{3.8}$$

thermodynamic forces and fluxes respectively. The factors of proportionality between these forces and fluxes are called transport coefficients. There are three of them, viz. the bulk viscosity λ, the heat conductivity κ, and the shear viscosity μ. We have

$$\begin{aligned} \lambda &= +\frac{1}{2T}\frac{1}{B_1^\pi}\frac{\begin{bmatrix} -\ddot{p} & \dot{p}-\dot{p}' & \dot{\Gamma}_1 \\ \dot{p}-\dot{p}' & p'-p'' & \Gamma_1'-\Gamma_1 \\ -\dot{p} & -p' & \frac{5}{3}\Gamma_1 \end{bmatrix}}{\begin{bmatrix} -\ddot{p} & \dot{p}-\dot{p}' \\ \dot{p}-\dot{p}' & p'-p'' \end{bmatrix}} \geq 0 \\ \kappa &= +\frac{1}{2T^2}\frac{c^2}{\hat{B}_4}\frac{\begin{bmatrix} \dot{p} & -\dot{\Gamma}_1 \\ p' & \Gamma_1-\Gamma_1' \end{bmatrix}}{\dot{p}} \geq 0 \\ \mu &= -\frac{1}{2T}\frac{1}{B_3}\Gamma_1 \geq 0. \end{aligned} \tag{3.9}$$

By comparison of (3.9) with the inequalities (2.70) we conclude that the requirement of a non-negative entropy production determines the sign of the transport coefficients λ, κ, and μ as indicated in (3.9).

Since in principle viscosity, heat conductivity, and bulk viscosity can be measured, we may use (3.9) to identify the coefficients B_1^π, B_3, and $\hat{B}_4$.

The first iterates (3.7) may be used to calculate $\overset{(1)}{A}{}^{AB}, \overset{(1)}{A}{}^{ABC}$ from $(2.60)_2$, $(2.62)_1$, and if those are inserted into $(3.5)_1$, one obtains equations for the second iterates $\overset{(2)}{I}{}^{AB}$ —or $\overset{(2)}{\pi}$, $\overset{(2)}{q}_M$, $\overset{(2)}{t}_{\langle MN\rangle}$. In Paragraph 7.4 we shall use this scheme to calculate the second iterate $\overset{(2)}{\pi}$ of the dynamic pressure.

Such a scheme is akin to the Maxwellian iteration in the kinetic theory of gases that has been explained before in this book; see Paragraph 4.3.2. Purely formal as this iterative scheme may be, it achieves some plausibility by producing Fourier's law and the Navier-Stokes equations as first iterates. Also the corresponding scheme in the kinetic theory provides progressively higher powers of the mean free path in higher iterates.

Note that, by $(3.7)_2$, the thermodynamic force driving the heat flux is not just equal to the temperature gradient. Rather there is an additional acceleration term which was first derived by Eckart [37]. This term has the interesting consequence that a gas in equilibrium within a gravitational field has a non-uniform temperature. A discussion of this effect and related ones may be found in Müller [45], p. 427 ff., or elsewhere in the literature.

4 Specific Results for Relativistic and Degenerate Gases

4.1 Equilibrium Distribution Function

We recall that the coefficients C and A in A^{ABC} and h^A have been related to the thermal equation of state $p = p(\alpha, T)$; see $(2,62)_1$, (2.65) and (2.63), (2.66). Specific expressions for those coefficients require the knowledge of a specific expression for the thermal equation of state. In principle that equation may be determined experimentally, but in practice it is difficult to come up with an analytic expression in this manner, particularly for a relativistic gas. There is an alternative, however. Indeed, we may calculate the function $p(\alpha, T)$ from statistical thermodynamics of equilibrium.

Statistical thermodynamics defines the particle flux vector A^A, the energy-momentum tensor A^{AB}, and the flux tensor A^{ABC} as moments of the distribution function $F(x^A, p^q)$.

$$\begin{aligned} A^A &= mc \int p^A \, F dP \\ A^{AB} &= c \int p^A p^B \, F dP \\ A^{ABC} &= \tfrac{1}{m} c \int p^A p^B p^C \, F dP. \end{aligned} \tag{4.1}$$

p^A is the four-momentum of a particle and $dP = \sqrt{-g}\frac{1}{p_0} dp^1 dp^2 dp^3$ is the invariant element of momentum space. For more detail see Section 6, where we give a brief account of the relativistic kinetic theory of gases.

In particular, in equilibrium the equations (4.1) read with $(2.60_{1,2}, (2.62)_1$ and (2.67), (2.68)

$$\begin{aligned} nmU^A &= mc \int p^A \, F_E dP \\ ph^{AB} + \frac{e}{c^2} U^A U^B &= c \int p^A p^B \, F_E dP \\ \frac{1}{2c^2}\frac{\Gamma_1'}{T} U^A U^B U^C - \frac{1}{2}\frac{\Gamma_1}{T}(g^{AB}U^C + g^{BC}U^A + g^{CA}U^B) &= \frac{1}{m} c \int p^A p^B p^C F_E dP, \end{aligned} \tag{4.2}$$

where F_E is the equilibrium distribution. From (4.2) we obtain

$$\begin{aligned} n &= \frac{1}{c}U_A && \int p^A && F_E dP \\ e &= \frac{1}{c}U_A U_B && \int p^A p^B && F_E dP \\ p &= \frac{c}{3}h_{AB} && \int p^A p^B && F_E dP \\ \frac{\Gamma_1}{T} &= \frac{2}{3mc}h_{AB}U_C && \int p^A p^B p^C && F_E dP. \end{aligned} \tag{4.3}$$

The general form of F_E is the Jüttner distribution; see Jüttner [70], [71].

$$F_E = \frac{y}{e^{\frac{m}{k}\alpha + \frac{U_A p^A}{kT}} \mp 1}. \tag{4.4}$$

k is the Boltzmann constant and y is equal to $\frac{w}{h^3}$, where h is Planck's constant and w is equal to $2s+1$ for particles with spin $s\frac{h}{2\pi}$. Thus $1/y$ is the smallest element of phase space that can acommodate a particle. The $\mp$ signs refer to Bosons and Fermions.

The equilibrium distribution function (4.4) assumes different forms for different degrees of degeneracy of the gas and for different degrees to which relativistic effects become important. An appreciation of the whole spectrum of gas properties is obtained by an inspection of Table 6.1 on the next page, which exhibits the equilibrium distribution functions—in a rest Lorentz frame—appropriate to various limits. Jüttner's formula is the master formula; it stands in the frame in the center of the table. In the upper left corner we have the classical Maxwell distribution and in the lower right corner we have the Planck distribution appropriate for a gas of massless Bosons. γ in the table is defined as $\frac{mc^2}{kT}$ and we shall refer to it as the *relativistic coldness*. If γ is much larger than 1, we have a nonrelativistic gas; if it is much smaller than 1, we have an ultrarelativistic gas.

We proceed to give specific expressions for the constitutive coefficients in particular cases.

4.2 The Degenerate Relativistic Gas

First of all we consider the generic case which results from insertion of the Jüttner distribution into (4.3). We obtain

$$\begin{aligned} n &= 4\pi y m^3 c^3 \frac{1}{4} \quad I_{3,1} \\ e &= 4\pi y m^4 c^5 \frac{1}{8} \quad I_{4,0} \\ p &= 4\pi y m^4 c^5 \frac{1}{24}(I_{4,2} - 3I_{2,0}) \\ \frac{\Gamma_1}{T} &= 4\pi y m^4 c^5 \frac{1}{24}(I_{5,3} - 2I_{3,1}). \end{aligned} \tag{4.5}$$

	Non-relativistic	relativistic	ultrarelativistic
non-degenerate	$y \exp\left\{-\frac{m}{k}\alpha+\gamma\right\} \exp\left\{-\frac{p^2}{2mkT}\right\}$	$y \exp\left\{-\frac{m}{k}\alpha+\gamma\right\} \left\{-\gamma\sqrt{1+\frac{p^2}{m^2c^2}}\right\}$	$y \exp\left\{-\frac{m}{k}\alpha\right\} \exp\left\{-\frac{cp}{kT}\right\}$
degenerate	$\frac{y}{\exp\left\{\left(\frac{m}{k}\alpha+\gamma\right)\right\} \exp\left\{-\frac{p^2}{2mkT}\right\} \mp 1}$	$\frac{y}{\exp\left\{\frac{m}{k}\alpha+\gamma\sqrt{1+\frac{p^2}{m^2c^2}}\right\} \mp 1}$	$\frac{y}{\exp\left\{\frac{m}{k}\alpha+\frac{cp}{kT}\right\} \mp 1}$
strongly degenerate FERMI	$\begin{bmatrix} y & 0 \leq p \leq \sqrt{-2mkT\left(\frac{m}{k}\alpha+\gamma\right)} \\ 0 & p > \sqrt{-2mkT\left(\frac{m}{k}\alpha+\gamma\right)} \end{bmatrix}$	$\begin{bmatrix} y & 0 \leq p \leq mc\sqrt{-2mkT\left(\frac{m}{k}\alpha\frac{1}{\gamma}\right)^2-1} \\ 0 & p > mc\sqrt{-2mkT\left(\frac{m}{k}\alpha\frac{1}{\gamma}\right)^2-1} \end{bmatrix}$	$\begin{bmatrix} y & 0 \leq p \leq -\alpha\frac{T}{c} \\ 0 & p > -\alpha\frac{T}{c} \end{bmatrix}$
strongly degenerate BOSE	$\frac{y}{\exp\left\{\frac{p^2}{2mkT}\right\}-1} \quad p \neq 0$	$\frac{y}{\exp\left\{\gamma\left(\sqrt{1+\frac{p^2}{m^2c^2}}-1\right)\right\}-1} \quad p \neq 0$	$\frac{y}{\exp\left\{\frac{cp}{kT}\right\}-1} \quad p \neq 0$

Table 6.1: Characterization of limiting cases. Equilibrium distribution function in rest Lorentz frame.

The right-hand sides of (4.5) have been calculated in the rest Lorentz frame, where U_A equals $(c,0,0,0)$ and the identity $p_A p^A = m^2c^2$ has been used. The expressions $I_{m,n}$ are defined by

$$I_{m,n} = I_m - I_n \quad \text{with} \quad I_m(\alpha,\gamma) = \int_0^\infty \frac{\cosh m\,\varrho}{exp\left\{\frac{m}{k}\alpha + \gamma\cosh\varrho\right\} \mp 1} d\varrho, \tag{4.6}$$

where the integrand results from the one in (4.3) by the substitution $\sinh\varrho = \frac{p}{mc}$, and p is the magnitude of the momentum. The $I_{m,n}$ are functions

$$\text{of the } \textit{fugacity } \alpha \qquad \text{and of} \qquad \text{the } \textit{relativistic coldness } \gamma = \frac{mc^2}{kT}. \tag{4.7}$$

Note that it is impossible to invert the equations (4.5) analytically so as to obtain $\alpha = \alpha(n,e)$ and $T = T(n,e)$. This can only be done numerically.

For subsequent calculations with the functions I_m it is often useful to know two recurrence relations which can easily be derived from (4.6). We have

$$I'_{n,n-2} = nI_n + (n-2)I_{n-2} \quad \text{and} \quad \dot{I}_{n,n-2} = -\frac{m}{k}\frac{2}{\gamma}(n-1)I_{n-1} \tag{4.8}$$

and by use of these relations we can prove the following particular ones

$$\gamma^5 I_{3,1} = \frac{1}{12}\left[\gamma^5(I_{5,3} - 2I_{3,1})\right]' \quad \text{and} \quad \gamma^6 I_{4,2} = \frac{1}{40}\left[\gamma^6(3I_{6,4} - 5I_{4,2})\right]'. \tag{4.9}$$

From all we have said before, it is clear that we only need $p = p(\alpha,T)$, i.e. equation $(4.5)_3$, from statistical mechanics, because $n(\alpha,T), e(\alpha,T), \Gamma_1(\alpha,T)$, and $\Gamma_2(\alpha,T)$ follow from $p(\alpha,T)$ by thermodynamic arguments; see (2.67) and (2.69). But, of course, the statistical equations $(4.5)_{1,2,4}$ must be compatible with the thermodynamic equations (2.67) and (2.69). This is the case for (2.67) without any further requirement, while insertion of $p(\alpha,T)$ from $(4.5)_3$ into $(2.69)_1$ gives

$$\frac{\Gamma_1}{T} = 4\pi y m^4 c^5 \frac{1}{24}(I_{5,3} - 2I_{3,1}) + T^5 A_1(\alpha) \tag{4.10}$$

when $(4.9)_1$ is used. Comparison with $(4.5)_4$ shows that we must have

$$A_1(\alpha) = 0 \tag{4.11}$$

for compatibility of thermodynamics and statistical mechanics.

Apart from $p(\alpha,T)$ and $\Gamma_1(\alpha,T)$, we need Γ_2 if we wish to determine the coefficients C and A in (2.65), (2.66). Γ_2 may be calculated from $(2.69)_2$ by insertion of $p(\alpha,T)$. The integration makes use of the identity $(4.9)_2$ and we obtain

$$\frac{\Gamma_2}{T} = -\frac{1}{60}4\pi y m^5 c^7 \frac{1}{k\gamma}\left(3I_{6,4} - 5I_{4,2} + \frac{1}{\gamma^6}A(\alpha)\right)$$

where $A(\alpha)$ is an arbitrary function of α closely related to the function $A_2(\alpha)$ in $(2.69)_2$.

We introduce $p(\alpha,T), \Gamma_1(\alpha,T)$ from $(4.5)_{3,4}$ and $\Gamma_2(\alpha,T)$ into (2.65), (2.66) and get

$$C_1^0 = 4\pi y m^4 c^3 \frac{1}{8} I_{5,3} \tag{4.12$_1$}$$

$$C_1^\pi = -\frac{3}{2}\frac{1}{c^2}\frac{\begin{bmatrix} I_2 & 3I_3+I_1 & 2I_4 \\ 3I_3+I_1 & 8(I_4+I_2) & 5I_5+3I_3 \\ 2I_4 & 5I_5+3I_3 & 3I_6+I_2+\frac{A}{\gamma^6} \end{bmatrix}}{\begin{bmatrix} I_1 & 2I_2 & 3I_3-I_1 \\ 2I_2 & 3I_3+I_1 & 4I_4 \\ 3I_3-I_1 & 4I_4 & 5I_5-3I_3+2I_1 \end{bmatrix}} \tag{4.12$_2$}$$

$$C_3 = -\frac{4}{5}\frac{\begin{bmatrix} 3I_{3,1} & 2I_{4,2} \\ 5I_{5,3}+2I_{3,1} & 3I_{6,2}+\frac{A}{\gamma^6} \end{bmatrix}}{\begin{bmatrix} 12I_{3,1} & 8I_{4,2} \\ 8I_{4,2} & 5I_{5,3}+2I_{3,1} \end{bmatrix}} \tag{4.12$_3$}$$

$$C_5 = \frac{1}{5}\frac{3I_{6,4}-5I_{4,2}+\frac{A}{\gamma^6}}{I_{5,3}-2I_{3,1}} \tag{4.12$_4$}$$

$$A_1^{\pi^2} = 3\frac{1}{T}\frac{1}{4\pi y}\frac{1}{m^4c^3}C_1^\pi\frac{\begin{bmatrix} 4I_2 & 3I_3+I_1 \\ 3I_3+I_1 & 2(I_4+I_2) \end{bmatrix}}{\begin{bmatrix} I_1 & 2I_2 & 3I_3-I_1 \\ 2I_2 & 3I_3+I_1 & 4I_4 \\ 3I_3-I_1 & 4I_2 & 5I_5-3I_3+2I_1 \end{bmatrix}} \tag{4.13$_1$}$$

$$A_1^q = -192\frac{1}{T}\frac{1}{4\pi y}\frac{1}{m^4c^7}\frac{I_{4,2}+\frac{15}{2}C_3I_{3,1}}{\begin{bmatrix} 12I_{3,1} & 8I_{4,2} \\ 8I_{4,2} & 5I_{5,3}+2I_{3,1} \end{bmatrix}} \tag{4.13$_2$}$$

$$A_1^t = -\frac{12}{5}\frac{1}{T}\frac{1}{4\pi y}\frac{1}{m^4c^5}\frac{3I_{6,4}-5I_{4,2}+\frac{A}{\gamma^6}}{(I_{5,3}-2I_{3,1})^2} \tag{4.13$_3$}$$

$$A_2^0 = \frac{1}{T} \tag{4.13$_4$}$$

$$\begin{aligned} A_2^\pi &= -3\frac{1}{T}\frac{1}{4\pi y}\frac{1}{m^4c^5}\frac{\begin{bmatrix} 4I_2 & 3I_3+I_1 \\ 8I_4 & 5I_5+3I_3 \end{bmatrix}+20C_3\begin{bmatrix} 4I_2 & 3I_3+I_1 \\ 3I_3+I_1 & 2(I_4+I_2) \end{bmatrix}}{\begin{bmatrix} I_1 & 2I_2 & 3I_3-I_1 \\ 2I_2 & 3I_3+I_1 & 4I_2 \\ 3I_3-I_1 & 4I_4 & 5I_5-3I_3+2I_1 \end{bmatrix}} = \\ &= 128\frac{1}{T}\frac{1}{4\pi y}\frac{1}{m^4c^5}\frac{3I_{4,2}-\frac{3}{4}c^2C_1^\pi I_{3,1}}{\begin{bmatrix} 12I_{3,1} & 8I_{4,2} \\ 8I_{4,2} & 5I_{5,3}+2I_{3,1} \end{bmatrix}} \end{aligned} \tag{4.13$_5$}$$

$$A_3^0 = -192\frac{1}{T}\frac{1}{4\pi y}\frac{1}{m^4c^5}\frac{3I_{4,2}-\frac{9}{2}C_5I_{3,1}}{\begin{bmatrix} 12I_{3,1} & 8I_{4,2} \\ 8I_{4,1} & 5I_{5,3}+2I_{3,1} \end{bmatrix}} = -48\frac{1}{T}\frac{1}{4\pi y}\frac{1}{m^4c^5}C_3\frac{1}{I_{5,3}-2I_{3,1}}. \tag{4.13$_6$}$$

Insertion of $p(\alpha,T)$ and $\Gamma_1(\alpha,T)$ into the inequalities (2.70) gives

$$B_1^\pi \frac{\begin{bmatrix} 4I_2 & 3I_3+I_1 \\ 3I_3+I_1 & 2(I_4+I_2) \end{bmatrix}}{\begin{bmatrix} I_1 & 2I_2 & 3I_3-I_1 \\ 2I_2 & 3I_3+I_1 & 4I_4 \\ 3I_3-I_1 & 4I_4 & 5I_5-3I_3+2I_1 \end{bmatrix}} \leq 0,\ \hat{B}_4 \frac{I_{3,1}}{\begin{bmatrix} 12I_{3,1} & 8I_{4,2} \\ 8I_{4,2} & 5I_{5,3}+2I_{3,1} \end{bmatrix}} \leq 0,\ B_3 \frac{1}{I_{5,3}-2I_{3,1}} \leq 0. \tag{4.14}$$

It can easily be shown that $I_{5,3} - 2I_{3,1}$ is positive valued, so that $(4.14)_3$ implies $B_3 \leq 0$. In the inequalities $(4.14)_{1,2}$ the complexity of the factors of B_1^π and $\hat{B}_4$ makes it difficult to evaluate the sign of those coefficients.

With the knowledge $p(\alpha,T)$ and $\Gamma_1(\alpha,T)$ by $(4.5)_{3,4}$ we can also make the expressions for the transport coefficients (3.9) more explicit. We obtain

$$\lambda = -\frac{1}{36}\frac{1}{B_1^\pi}(4\pi y)m^4c^5 \frac{\begin{bmatrix} I_1 & 2I_2 & 3I_3-I_1 \\ 2I_2 & 3I_3+I_1 & 4I_4 \\ 3I_3-I_1 & 4I_4 & 5I_5-3I_3+2I_1 \end{bmatrix}}{\begin{bmatrix} 4I_2 & 3I_3+I_1 \\ 3I_3+I_1 & 2(I_4+I_2) \end{bmatrix}}, \tag{4.15$_1$}$$

$$\kappa = -\frac{1}{9\cdot 64}\frac{1}{T}\frac{c^2}{\hat{B}_4}(4\pi y)m^4c^5 \frac{\begin{bmatrix} 12I_{3,1} & 8I_{4,2} \\ 8I_{4,2} & 5I_{5,3}+2I_{3,1} \end{bmatrix}}{I_{3,1}}, \tag{4.15$_2$}$$

$$\mu = -\frac{1}{48}\frac{1}{B_3}(4\pi y)m^4c^5(I_5 - 3I_3 + 2I_1). \tag{4.15$_3$}$$

Recall that by (4.14) all three coefficients are positive.

The relations (4.12) through (4.15) clearly exhibit the main result of the theory so far: All coefficients of the flux tensor A^{ABC} and of the entropy flux vector h^A are determined to within a single function $A(\alpha)$. The coefficients of the flux production I^{AB} are related to the bulk viscosity λ, the heat conductivity κ, and the shear viscosity μ.

We shall now proceed to investigate limiting cases, in which the state functions and the constitutive coefficients assume a more specific form.

4.3 Nondegenerate Relativistic Gas

The nondegenerate gas is characterized by large values of the fugacity

$$\frac{m}{k}\alpha \gg 1 \tag{4.16}$$

so that the term ± 1 in the denominator of (4.6) may be neglected. In the rest Lorentz frame we may write

$$I_n(\alpha,\gamma) \approx e^{-\frac{m}{k}\alpha} \int_0^\infty \cosh(n\varrho) e^{-\gamma \cosh \varrho} d\varrho. \tag{4.17}$$

The integrals in (4.17) define Bessel functions of the second kind, which we shall denote by K_n.

$$K_n(\gamma) = \int_0^\infty \cosh(n\varrho) e^{-\gamma \cosh \varrho} d\varrho. \tag{4.18}$$

The $K's$ are the natural special functions of nondegenerate relativistic gases. They satisfy the recurrence relations

$$K_{n+1}(\gamma) - K_{n-1}(\gamma) = \frac{2n}{\gamma} K_n(\gamma), \tag{4.19}$$

$$\frac{d}{d\gamma}[\gamma^n K_n(\gamma)] = -\gamma^n K_{n-1}(\gamma), \qquad \frac{d}{d\gamma}[\gamma^{-n} K_n(\gamma)] = -\gamma^{-n} K_{n+1}(\gamma).$$

With this knowledge it is a trivial though cumbersome task to write the formulae (4.5) and (4.12) through (4.15) in a form appropriate to a nondegenerate relativistic gas. We obtain from (4.17), (4.18), and (4.19)

$$\begin{aligned} n &= e^{-\frac{m}{k}\alpha} 4\pi y m^3 c^3 \frac{K_2}{\gamma} \\ e &= nmc^2 \left(G - \frac{1}{\gamma} \right) \\ p &= nmc^2 \frac{1}{\gamma} = nkT \\ \frac{\Gamma_1}{T} &= nmc^2 \frac{2}{\gamma} G \end{aligned} \tag{4.20}$$

where G stands for $\frac{K_3}{K_2}$.

Note that the pressure is related to the particle density and the temperature in the same manner as in a classical ideal gas. This coincidence was first remarked on by Jüttner [70].

The coefficients C and A in the flux tensor A^{ABC} and the entropy flux vector h^A read[2]

$$C_1^0 = nm\left(1 + \frac{6}{\gamma} G\right) \tag{4.21$_1$}$$

$$C_1^\pi = -\frac{6}{c^2} \frac{\left(2 - \frac{5}{\gamma^2}\right) + \left(\frac{19}{\gamma} - \frac{30}{\gamma^3}\right) G - \left(2 - \frac{45}{\gamma^2}\right) G^2 - \frac{9}{\gamma} G^3 + \left(1 - \frac{1}{\gamma^2} + \frac{5}{\gamma} G - G^2\right) \frac{1}{48} m^3 c^3 \frac{A(\alpha)}{n\gamma^5}}{\frac{3}{\gamma} - \left(2 - \frac{20}{\gamma^2}\right) G - \frac{13}{\gamma} G^2 + 2G^3} \tag{4.21$_2$}$$

$$C_3 = -\frac{1}{\gamma} \frac{1 + \frac{6}{\gamma} G - G^2 + \frac{1}{240} m^3 c^3 \frac{A(\alpha)}{n\gamma^5}}{1 + \frac{5}{\gamma} G - G^2} \tag{4.21$_3$}$$

$$C_5 = \left(\frac{6}{\gamma} + \frac{1}{G}\right) + \frac{1}{G} \frac{1}{240} m^3 c^3 \frac{A(\alpha)}{n\gamma^5} \tag{4.21$_4$}$$

$$A_1^{\pi^2} = \frac{1}{4} \frac{k}{m^3 c^2} \frac{\gamma^2}{n} \frac{1 - \frac{1}{\gamma^2} + \frac{5}{\gamma} G - G^2}{\frac{3}{\gamma} - \left(2 - \frac{20}{\gamma}\right) G - \frac{13}{\gamma} G^2 + 2G^3} \cdot C_1^\pi \tag{4.22$_1$}$$

[2]We are indebted to S. Pennisi and G. M. Kremer, who have corrected several errors in these formulae.

$$
\begin{aligned}
A_1^q &= \frac{1}{2}\frac{k}{m^2c^6}\frac{\gamma}{n}\frac{\frac{5}{\gamma}+\left(\frac{30}{\gamma^2}-1\right)G-\frac{10}{\gamma}G^2+G^3+\frac{1}{48}m^3c^3\frac{A}{n\gamma^6}}{\left(1+\frac{5}{\gamma}G-G^2\right)^2} && (4.22)_2\\
A_1^t &= -\frac{1}{4}\frac{k}{m^2c^4}\frac{\gamma^2}{n}\frac{1}{G^2}\left(1+\frac{6}{\gamma}G+\frac{1}{240}m^3c^3\frac{A}{n\gamma^5}\right) && (4.22)_3\\
A_2^0 &= \frac{1}{T} && (4.22)_4\\
A_2^\pi &= \frac{k}{m^2c^4}\frac{\gamma}{n}\cdot && (4.22)_5\\
&\cdot\frac{-2+\frac{5}{\gamma^2}-\left(\frac{22}{\gamma}-\frac{30}{\gamma^3}\right)G+\left(4-\frac{65}{\gamma^2}\right)G^2+\frac{22}{\gamma}G^3-2G^4+\left(1-\frac{1}{\gamma^2}+\frac{5}{\gamma}G-G^2\right)\frac{1}{48}m^3c^3\frac{A}{n\gamma^5}}{\left(1+\frac{5}{\gamma}G-G^2\right)\left(-\frac{3}{\gamma}+(2-\frac{20}{\gamma^2})G+\frac{13}{\gamma}G^2-2G^3\right)}\\
A_3^0 &= \frac{k}{m^2c^4}\frac{\gamma}{n}\frac{1+\frac{6}{\gamma}G-G^2+\frac{1}{240}m^3c^3\frac{A}{n\gamma^5}}{G\left(1+\frac{5}{\gamma}G-G^2\right)} && (4.22)_6
\end{aligned}
$$

In the sequel we shall drop the function $A(\alpha)$, since we suspect that it is zero. That suspicion is based on the kinetic theory of gases. The transport coefficients (4.15) for a nondegenerate gas assume the forms

$$
\begin{aligned}
\lambda &= \frac{1}{3}\frac{nmc^2}{\gamma}\frac{1}{B_1^\pi}\frac{-\frac{3}{\gamma}+(2-\frac{20}{\gamma^2})G+\frac{13}{\gamma}G^2-2G^3}{1-\frac{1}{\gamma^2}+\frac{5}{\gamma}G-G^2}\\
\kappa &= -nkT\ \frac{c^2}{\hat{B}_4}(\gamma+5G-\gamma G^2) \qquad\qquad (4.23)\\
\mu &= nkT\ \frac{1}{B_3}G.
\end{aligned}
$$

Unlike the general case, the nondegenerate one has the particle density n depending explicitly on α, so that we were able to eliminate the fugacity from the equations $(4.20)_{2,3,4}$ and (4.21), (4.22), and (4.23).

4.4 Degenerate Nonrelativistic Gas

The nonrelativistic gas is characterized by large values of the relativistic coldness

$$\gamma \gg 1. \qquad (4.24)$$

In that case the integrals (4.6) may be approximated by polynomials in $1/\gamma$ so that we have

$$\begin{aligned}
I_0 &= \frac{\sqrt{2mkT}}{mc}\left(i_0 - \frac{1}{4}\frac{1}{\gamma}i_2 + \frac{3}{32}\frac{1}{\gamma^2}i_4 + \ldots\right)\\
I_1 &= \frac{\sqrt{2mkT}}{mc}\left(i_0 + \frac{3}{4}\frac{1}{\gamma}i_2 - \frac{5}{32}\frac{1}{\gamma^2}i_4 + \ldots\right)\\
I_2 &= \frac{\sqrt{2mkT}}{mc}\left(i_0 + \frac{15}{4}\frac{1}{\gamma}i_2 + \frac{35}{32}\frac{1}{\gamma^2}i_4 + \ldots\right)\\
I_3 &= \frac{\sqrt{2mkT}}{mc}\left(i_0 + \frac{35}{4}\frac{1}{\gamma}i_2 + \frac{315}{32}\frac{1}{\gamma^2}i_4 + \ldots\right)\\
I_4 &= \frac{\sqrt{2mkT}}{mc}\left(i_0 + \frac{63}{4}\frac{1}{\gamma}i_2 + \frac{1155}{32}\frac{1}{\gamma^2}i_4 + \ldots\right)\\
I_5 &= \frac{\sqrt{2mkT}}{mc}\left(i_0 + \frac{99}{4}\frac{1}{\gamma}i_2 + \frac{3003}{32}\frac{1}{\gamma^2}i_4 + \ldots\right)\\
I_6 &= \frac{\sqrt{2mkT}}{mc}\left(i_0 + \frac{143}{4}\frac{1}{\gamma}i_2 + \frac{6435}{32}\frac{1}{\gamma^2}i_4 + \ldots\right)
\end{aligned} \tag{4.25}$$

where the i_n are defined by

$$i_n\left(\frac{m}{k}\alpha + \gamma\right) = \int_0^\infty \frac{x^n}{exp\left\{\frac{m}{k}\alpha + \gamma + x^2\right\} \mp 1}dx. \tag{4.26}$$

The functions i_n are the special functions familiar from nonrelativistic statistical mechanics of degenerate gases; see Paragraph 7.4.4. They satisfy the recurrence relation

$$\frac{di_n(a)}{da} = -\frac{n-1}{2}i_{n-2}(a). \tag{4.27}$$

With the knowledge of the limiting forms (4.25) of the functions $I_n(\alpha, \gamma)$, it is easy to calculate from (4.5) the equations of state appropriate to a degenerate nonrelativistic gas, viz.

$$\begin{aligned}
n &= y8\pi\sqrt{2}(kT)^{3/2}m^{3/2} \left(i_2\left(\frac{m}{k}\alpha + \gamma\right) + \frac{5}{4}i_4\left(\frac{m}{k}\alpha + \gamma\right)\frac{1}{\gamma} + \frac{7}{32}i_6\left(\frac{m}{k}\alpha + \gamma\right)\frac{1}{\gamma^2}\right)\\
e &= nmc^2 + y8\pi\sqrt{2}(kT)^{5/2}m^{3/2}\left(i_4\left(\frac{m}{k}\alpha + \gamma\right) + \frac{5}{4}i_6\left(\frac{m}{k}\alpha + \gamma\right)\frac{1}{\gamma}\right)\\
p &= \frac{2}{3}y8\pi\sqrt{2}(kT)^{5/2}m^{3/2}\left(i_4\left(\frac{m}{k}\alpha + \gamma\right) + \frac{3}{4}i_6\left(\frac{m}{k}\alpha + \gamma\right)\frac{1}{\gamma}\right)\\
\frac{\Gamma_1}{T} &= \frac{4}{3}y8\pi\sqrt{2}(kT)^{5/2}m^{3/2}\left(i_4\left(\frac{m}{k}\alpha + \gamma\right) + \frac{7}{4}i_6\left(\frac{m}{k}\alpha + \gamma\right)\frac{1}{\gamma}\right).
\end{aligned} \tag{4.28}$$

Higher-order terms in $1/\gamma$ are neglected.

The coefficients C and A in the flux tensor A^{ABC} and the entropy flux vector h^A read in the nonrelativistic case

$$C_1^0 = nm + 4my8\pi\sqrt{2}(kT)^{3/2}m^{3/2}\left(i_4\left(\frac{m}{k}\alpha + \gamma\right)\frac{1}{\gamma} + \frac{7}{4}i_6\left(\frac{m}{k}\alpha + \gamma\right)\frac{1}{\gamma^2}\right) + \mathcal{O}\left(\frac{1}{\gamma^3}\right) \quad (4.29)_1$$

$$C_1^\pi = -6\frac{m}{kT}\frac{1}{\gamma} + \mathcal{O}\left(\frac{1}{\gamma^2}\right) \tag{4.29$_2$}$$

$$C_3 = -\frac{2}{5} - \frac{27i_2i_8 - 35i_4i_6}{105i_2i_6 - 125i_4^2}\frac{1}{\gamma} + \mathcal{O}\left(\frac{1}{\gamma^2}\right) \tag{4.29$_3$}$$

$$C_5 = 1 + \frac{7}{5}\frac{i_6}{i_4}\frac{1}{\gamma} + \mathcal{O}\left(\frac{1}{\gamma^2}\right) \tag{4.29$_4$}$$

$$A_1^{\pi^2} = -\frac{9}{nkT^2}\gamma^2\frac{5i_0i_2i_4 - 8i_2^3}{5i_0i_4i_8 - 49i_0i_6^2 - 81i_2^2i_8 + 210i_2i_4i_6 - 125i_4^3} + \mathcal{O}(\gamma) \tag{4.30$_1$}$$

$$A_1^q = \frac{9}{2}\frac{m}{nk^2T^3}\frac{i_2^2}{21i_2i_6 - 25i_4^2} + \mathcal{O}\left(\frac{1}{\gamma}\right) \tag{4.30$_2$}$$

$$A_1^t = -\frac{3}{8}\frac{1}{nkT^2}\frac{i_2}{i_4} + \mathcal{O}\left(\frac{1}{\gamma}\right) \tag{4.30$_3$}$$

$$A_2^0 = \frac{1}{T} \tag{4.30$_4$}$$

$$A_2^\pi = \frac{18}{nkT^2}\gamma\frac{i_2^2}{21i_2i_6 - 25i_4^2} + \mathcal{O}(\gamma^0) \tag{4.30$_5$}$$

$$A_3^0 = \frac{3}{5}\frac{1}{nkT^2}\frac{i_2}{i_4} + \mathcal{O}\left(\frac{1}{\gamma}\right) \tag{4.30$_6$}$$

The transport coefficients λ, κ, and μ assume the forms

$$\lambda = -\frac{1}{9}\frac{1}{B_1^\pi c^4}\frac{nk^3T^3}{m^2}\frac{45i_0i_4i_8 - 49i_0i_6^2 - 81i_2^2i_8 + 210i_2i_4i_6 - 125i_4^3}{5i_0i_2i_4 - 9i_2^3}, \tag{4.31$_1$}$$

$$\kappa = -\frac{1}{9}\frac{1}{\hat{B}_4}\frac{nk^2T}{m}\frac{21i_2i_6 - 25i_4^2}{i_2^2} \tag{4.31$_2$}$$

$$\mu = -\frac{2}{3}\frac{1}{B_3}nkT\frac{i_4}{i_2}. \tag{4.31$_3$}$$

4.5 Nondegenerate Nonrelativistic Gas

It is an easy matter to calculate the constitutive coefficients in the purely classical case, i.e. for the nondegenerate, nonrelativistic gas. This can be achieved in two ways.

i.) Either we start from the equations (4.20) through (4.23) for the nondegenerate relativistic case and let the relativistic coldness γ go to infinity, thus getting the nonrelativistic limit. In this case we make use of the asymptotic expansion

$$K_n(\gamma)\underset{\gamma\to\infty}{\longrightarrow}\sqrt{\frac{\pi}{2\gamma}}e^{-\gamma}\left(1 + \frac{4n^2-1}{1!8\gamma} + \frac{(4n^2-1)(4n^2-9)}{2!(8\gamma)^2} + \ldots\right) \tag{4.32}$$

for the Bessel functions (4.18).

ii.) Or we start from the equations (4.28) through (4.31) and let $\frac{m}{k}\alpha + \gamma$ go to infinity as is appropriate for the nondegenerate limit. In this case we make use of the asymptotic formula

$$i_{2n}(a)\underset{a\to\infty}{\longrightarrow} e^{-a}\int_0^\infty x^{2n}e^{-x^2}dx = e^{-a}\frac{\sqrt{\pi}}{2}\frac{(2n-1)!!}{2^n} \tag{4.33}$$

for the functions (4.26).

In either way we obtain for the equations of state

$$\begin{aligned} n &= (4\pi y)\sqrt{2mkT}^3 e^{-(\frac{m}{k}\alpha+\gamma)}\frac{3}{8}\sqrt{\pi} \\ e &= nmc^2 + \frac{3}{2}nkT \\ p &= nkT \\ \frac{\Gamma_1}{T} &= 2nkT. \end{aligned} \tag{4.34}$$

The equations $(4.34)_{2,3}$ represent the classical caloric and thermal equations of state for a monatomic ideal gas. The only difference to those classical constitutive equations is the term nmc^2 in e. This term renders specific the constant of integration that remains unknown in classical thermodynamics.

The coefficients C and A in the flux tensor A^{ABC} and in the entropy flux vector h^A read

$$C_1^0 = nm(1+\frac{6}{\gamma}+\frac{15}{\gamma^2}) \qquad (4.35)_1$$

$$C_1^\pi = -\frac{1}{c^2}\left(6+\frac{63}{\gamma}+\frac{693}{4}\frac{1}{\gamma^2}\right) \qquad (4.35)_2$$

$$C_3 = -\left(\frac{2}{5}+\frac{7}{5}\frac{1}{\gamma}+\frac{21}{20}\frac{1}{\gamma^2}\right) \qquad (4.35)_3$$

$$C_5 = 1+\frac{7}{2}\frac{1}{\gamma}+\frac{35}{8}\frac{1}{\gamma^2} \qquad (4.35)_4$$

$$A_1^{\pi^2} = -\frac{3}{5}\frac{m^2}{nk^3T^4}c^4 \qquad (4.36)_1$$

$$A_1^q = \frac{1}{5}\frac{m}{nk^2T^3} \qquad (4.36)_2$$

$$A_1^t = -\frac{1}{4}\frac{1}{nkT^2} \qquad (4.36)_3$$

$$A_2^0 = \frac{1}{T} \qquad (4.36)_4$$

$$A_2^\pi = \frac{4}{5}\frac{m}{nk^2T^3}c^2 \qquad (4.36)_5$$

$$A_3^0 = \frac{2}{5}\frac{1}{nkT^2} \qquad (4.36)_6$$

The transport coefficients λ, κ, and μ are given by

$$\lambda = -\frac{5}{6}\frac{1}{B_1^\pi c^4}\frac{nk^3T^3}{m^2} \qquad (4.37)_1$$

$$\kappa = -\frac{5}{2}\frac{1}{\hat{B}_4}\frac{nk^2T^2}{m} \tag{4.37$_2$}$$

$$\mu = -\frac{1}{B_3}\, nkT. \tag{4.37$_3$}$$

4.6 Strongly Degenerate Relativistic Fermi Gas

For Fermi particles strong degeneration means

$$\frac{m}{k}\alpha + \gamma \ll -1 \tag{4.38}$$

so that $\frac{g_c}{kT} = -\left(\frac{m}{k}\alpha + \gamma\right)$—with g_c being the free enthalpy—tends to infinity as the degeneration increases. In that case the distribution function (4.4) has a nearly constant value, equal to y for

$$0 \leq p < mc\sqrt{(1 + \frac{g_c}{mc^2})^2 - 1},$$

and it practically vanishes for

$$p > mc\sqrt{(1 + \frac{g_c}{mc^2})^2 - 1}.$$

Therefore the function $F(\varrho) = \left(exp\{\frac{\alpha}{k} + \gamma\cosh\varrho\} + 1\right)^{-1}$ in the integrand of $I_m(\alpha, \gamma)$ in (4.6) assumes the form shown in Figure 6.1. The sudden decrease occurs at ϱ_F, where the argument of the exponential vanishes.

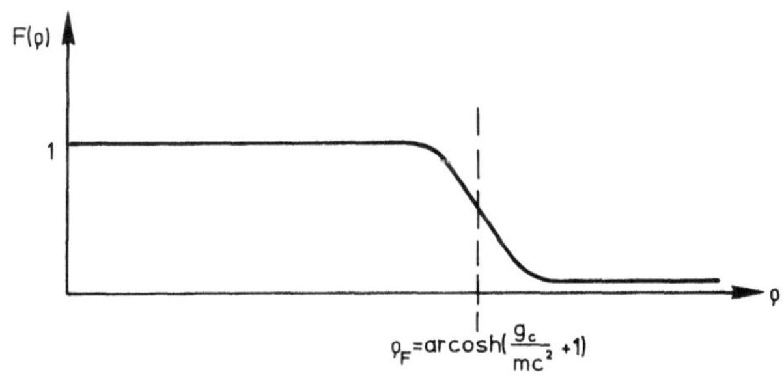

Figure 6.1 Strong degeneration in Fermions

The slope of the function at ϱ_F is equal to $-\sqrt{\left(1 + \frac{g_c}{mc^2}\right)^2 - 1}$ so that, by (4.38), the curve is very steep there. This fact enables us to calculate approximate expressions for the integrals $I_n(\alpha, \gamma)$ as follows. By partial integration we may write these integrals in the form

$$I_n(\alpha, \gamma) = -\frac{1}{n}\int_0^\infty \sinh\varrho \left[-\gamma\sinh\varrho \frac{e^{(\frac{m}{k}\alpha + \gamma\cosh\varrho)}}{(e^{\frac{m}{k}\alpha + \gamma\cosh\varrho} + 1)^2}\right] d\varrho, \tag{4.39}$$

where the square bracket encloses the derivative of $F(\varrho)$. Since that derivative has appreciable values only in the neighborhood of ϱ_F, we obtain approximate values for $I_n(\alpha,\gamma)$ by expanding $\sinh n\varrho$ in terms of $\xi = \gamma(\cosh\varrho - \cosh\varrho_F)$ up to second order:

$$\sinh n\varrho = \sinh n\varrho_F\left(1+\frac{n}{\gamma}\frac{\cosh n\varrho_F}{\sinh n\varrho_F\sinh\varrho_F}\xi+\frac{1}{2}\frac{n}{\gamma^2}\frac{n\sinh n\varrho_F\sinh\varrho_F-\cosh n\varrho_F\cosh\varrho_F}{\sinh n\varrho_F\sinh^3\varrho_F}\xi^2\right) \tag{4.40}$$

Insertion into (4.39) provides integrals of the type

$$\int_{\frac{\alpha}{k}+\gamma}^{\infty}\xi^n\frac{e^\xi}{(e^\xi+1)^2}d\xi \approx \int_{-\infty}^{\infty}\xi^n\frac{e^\xi}{(e^\xi+1)^2}d\xi = \begin{cases} 0 & \text{n odd} \\ 2n(n-1)!(1-2^{1-n})\zeta_n & \text{n even,}\end{cases} \tag{4.41}$$

where the first equation is approximately valid by (4.38) and the second one is taken from a mathematical handbook. We obtain

$$I_n(\alpha,\gamma) = \frac{\sinh n\varrho_F}{n}\left(1+\frac{\pi^2}{6}\frac{n}{\gamma^2}\frac{n\sinh n\varrho_F\sinh\varrho_F-\cosh n\varrho_F\cosh\varrho_F}{\sinh n\varrho_F\sinh^3\varrho_F}\right). \tag{4.42}$$

ζ_n in (4.41) stands for $\sum_{r=1}^{\infty}\frac{1}{r^n}$. Note that with increasing degeneration g_c and hence ϱ_F becomes bigger and the second term in (4.42) becomes negligible.

Inspection of the results (4.5) and (4.12) through (4.15) shows that we need the function I_n for $n = 0$ through $n = 6$. For those few values of n it is easy to write $I_n(\alpha,\gamma)$ as functions of the variables

$$\gamma \quad \text{and} \quad x = \sinh\varrho_F = \sqrt{(1+\frac{g_c}{mc^2})^2-1}. \tag{4.43}$$

With a little calculation we obtain

$$\begin{aligned}
I_0(\alpha,\gamma) &= ar\sinh x - \frac{\pi^2}{6}\frac{1}{\gamma^2}\frac{\sqrt{1+x^2}}{x^3} \\
I_1(\alpha,\gamma) &= x - \frac{\pi^2}{6\gamma^2}\frac{1}{x^3} \\
I_2(\alpha,\gamma) &= \sqrt{1+x^2}\left(x-\frac{\pi^2}{6\gamma^2}\left(\frac{1}{x^3}-\frac{2}{x}\right)\right) \\
I_3(\alpha,\gamma) &= \left(x+\frac{4}{5}x^3\right)-\frac{\pi^2}{6\gamma^2}\left(\frac{1}{x^3}-\frac{4}{x}-8x\right) \\
I_4(\alpha,\gamma) &= \sqrt{1+x^2}\left(x+2x^3-\frac{\pi^2}{6\gamma^2}\left(\frac{1}{x^3}-\frac{8}{x}-24x\right)\right) \\
I_5(\alpha,\gamma) &= \frac{16}{5}x^5+4x^3+x-\frac{\pi^2}{6\gamma^2}\left(\frac{1}{x^3}-\frac{12}{x}-72x-64x^3\right) \\
I_6(\alpha,\gamma) &= \sqrt{1+x^2}\left(\frac{16}{3}x^5+\frac{16}{3}x^3+x-\frac{\pi^2}{6\gamma^2}\left(\frac{1}{x^3}-\frac{18}{x}-144x-160x^3\right)\right).
\end{aligned} \tag{4.44}$$

This knowledge enables us to write the state functions (4.5) in a form appropriate to a strongly degenerate relativistic Fermi gas.

$$
\begin{aligned}
n &= 4\pi y m^3 \; c^3 \; \frac{1}{3} \; x^3 \; \left[1+\frac{\pi^2}{6\gamma^2}\frac{6x^2+3}{x^4}\right] \\
e &= 4\pi y m^4 \; c^5 \; \frac{1}{24} \; g(x) \; \left[1+\frac{\pi^2}{6\gamma^2}\frac{24(3x^2+1)\sqrt{1+x^2}}{g(x)x}\right] \\
p &= \frac{1}{3} \; 4\pi y m^4 \; c^5 \; \frac{1}{8} \; f(x) \; \left[1+\frac{\pi^2}{6\gamma^2}\frac{24x\sqrt{1+x^2}}{f(x)}\right] \\
\frac{\Gamma_1}{T} &= \frac{2}{3} \; 4\pi y m^5 \; c^5 \; \tfrac{1}{5} \; x^5 \; \left[1+\frac{\pi^2}{6\gamma^2}\frac{4x^6+3x^4}{\frac{1}{5}x^8}\right].
\end{aligned}
\tag{4.45}
$$

Following Chandrasekhar [67], we have introduced $g(x)$ and $f(x)$ as abbreviations defined by

$$
g(x) = 3x(2x^2+1)\sqrt{1+x^2} - 3\mathrm{arsinh}x \quad \text{and} \quad f(x) = x(2x^2-3)\sqrt{1+x^2} + 3\mathrm{arsinh}x. \tag{4.46}
$$

Note that in the case of complete degeneration the second term in the square brackets of (4.45) may be neglected,

$$
n \to 4\pi y m^3 c^3 \frac{1}{3}x^3, e \to 4\pi y m^4 c^5 \frac{1}{24} g(x), p \to 4\pi y m^4 c^5 \frac{1}{8} f(x), \frac{\Gamma_1}{T} \to 4\pi y m^5 c^5 \frac{2}{15} x^5. \tag{4.47}
$$

The temperature is then no longer a variable, because it is effectively equal to zero at full degeneration.

With the functions (4.44) known it is an easy though cumbersome task to calculate the relations (4.12) through (4.15) in a form appropriate to a strongly degenerate relativistic Fermi gas. We leave that to the reader.

4.7 A Remark on the Strongly Degenerate Relativistic Bose Gas

For Bose particles strong degeneration means

$$
0 < \frac{m}{k}\alpha + \gamma \ll 1 \tag{4.48}
$$

In that case the integrals $I_m(\alpha,\gamma)$ in (4.6) may be approximately represented in terms of the Bessel functions (4.32) as follows:

$$
\begin{aligned}
I_m(\alpha,\gamma) &= e^{-\frac{m}{k}\alpha}\int_0^\infty \cosh m\varrho e^{-\gamma\cosh\varrho}\frac{1}{1-e^{-(\frac{m}{k}\alpha+\gamma\cosh\varrho)}}d\varrho \\
I_m(\alpha,\gamma) &\approx e^{-\frac{m}{k}\alpha}\int_0^\infty \cosh m\varrho e^{-\gamma\cosh\varrho}\sum_{r=0}^{\infty}(-1)^r e^{-r(\frac{m}{k}\alpha+\gamma\cosh\varrho)}d\varrho \\
I_m(\alpha,\gamma) &\approx \sum_{r=0}^{\infty}(-1)^r e^{-(1+r)\frac{m}{k}\alpha}K_m\left((1+r)\gamma\right).
\end{aligned}
\tag{4.49}
$$

However, the sum must contain many terms if the approximation is to be good. Therefore no simple formulae can be had for the constitutive coefficients in this case.

4.8 Equilibrium Properties of an Ultrarelativistic Gas

The ultrarelativistic gas is characterized by a very small relativistic coldness

$$\gamma \ll 1 \tag{4.50}$$

This condition is realized in a gas at very high temperatures. The distribution function (4.4) in the local rest frame assumes the form

$$F_{\mathrm{E}} = \frac{y}{e^{\frac{m}{k}\alpha + \frac{cp}{kT}} \mp 1} \tag{4.51}$$

and instead of (4.5) we obtain from (4.3)

$$\begin{aligned} n &= 4\pi y \left(\frac{kT}{c}\right)^3 j_2(\alpha) \\ e &= 4\pi y \, c\left(\frac{kT}{c}\right)^4 j_3(\alpha) \\ p &= \frac{1}{3} 4\pi y \, c\left(\frac{kT}{c}\right)^4 j_3(\alpha) \\ \frac{\Gamma_1}{T} &= \frac{2}{3} 4\pi y \left(\frac{kT}{c}\right)^5 j_4(\alpha). \end{aligned} \tag{4.52}$$

The functions $j_n(\alpha)$ are defined by

$$j_n(\alpha) = \int_0^\infty \frac{x^n}{e^{\frac{m}{k}\alpha + x} \mp 1} dx \tag{4.53}$$

and they satisfy the recurrence formula

$$\frac{dj_n(\alpha)}{d\alpha} = -\frac{m}{k} n j_{n-1}(\alpha). \tag{4.54}$$

5 An Application: The Mass Limit of a White Dwarf

We assume that a star as a white dwarf is fully ionized and that the free electrons are the only component that is thermodynamically active. Due to the high density the electrons form a fully degenerate Fermi gas, and the particle number, i.e. electron number and the pressure are

$$\begin{aligned} n &= 4\pi y m^3 c^3 \frac{1}{3} x^3 = A x^3 \\ p &= \frac{1}{24} 4\pi y m^4 c^5 f(x) = B f(x) \end{aligned} \tag{5.1}$$

according to (4.45). $f(x)$ is given by $(4.46)_2$ and for a later purpose we note that

$$\frac{df}{dx} = \frac{8x^4}{\sqrt{1+x^2}} \tag{5.2}$$

holds. x, or n, is the only relevant variable here, since the star is considered to be at rest. The temperature plays no role in a fully degenerate gas and stress deviator and heat flux are ignored. Thus this consideration is not really an example for relativistic extended thermodynamics but rather for ordinary relativistic thermostatics. However, it does serve to show the relevance of relativistic effects in thermodynamics and that is why it is included. The arguments presented here are due to Chandrasekhar; see [67].

While the ions of hydrogen, helium, and traces of heavier elements are unimportant thermodynamically, they do of course create the gravitational field, and therefore we must find the relation between n, the number density of electrons, and ϱ, the mass density of the star. If X, Y denote the mass fractions of hydrogen and helium, and if Z is the mass fraction of the heavier elements with a mean molecular weight of M_z (say) and order number $\frac{1}{2}M_z$, we have

$$n = \varrho\frac{X}{m_H} + 2\varrho\frac{Y}{4m_H} + \frac{1}{2}M_z\varrho\frac{Z}{M_z m_H}.$$

m_H is the mass of the hydrogen nucleus. We ignore X in comparison with Y and Z, since the star will have very little hydrogen left, and obtain with $Y + Z = 1$ and $(5.1)_1$

$$\varrho = 2m_H n = 2m_H A x^3. \tag{5.3}$$

The radial component of the momentum balance is the only relevant equation for the single variable x, or n, or p. It reads

$$\begin{aligned} \frac{dp(r)}{dr} &= -\varrho(r)G\frac{M(r)}{r^2} \\ \frac{dp(r)}{dr} &= -\frac{\varrho(r)}{r^2}4\pi G\int_0^r \varrho(r')r'^2 dr', \end{aligned} \tag{5.4}$$

where G is the gravitation constant. By use of (5.3) and of the state equations we convert this equation into an equation that determines x as a function of r. Several substitutions are made so as to give the equation its canonical form. First we differentiate, introduce $(5.1)_1$ and (5.3), and use (5.2). Thus we obtain

$$\begin{aligned} \frac{1}{r^2}\frac{d}{dr}\left(\frac{r^2}{\varrho}\frac{dp}{dr}\right) &= -4\pi G\varrho \\ \frac{1}{r^2}\frac{d}{dr}\left(\frac{r^2}{x^3}\frac{df(x)}{dr}\right) &= -4\pi G(2m_H)^2\frac{A^2}{B}x^3 \\ \frac{1}{r^2}\frac{d}{dr}\left(r^2\frac{d\sqrt{1+x^2}}{dr}\right) &= -\frac{2\pi G m_H^2 A^2}{B}x^3. \end{aligned}$$

We eliminate the square root on the left-hand side by introducing the new dependent variable $y = \sqrt{1+x^2}$

$$\frac{1}{r^2}\frac{d}{dr}\left(r^2\frac{dy}{dr}\right) = -\underbrace{\frac{2\pi G m_H^2 A^2}{B}}_{1/l^2}(y^2-1)^{3/2}. \tag{5.5}$$

We define the length l as indicated in (5.5) and use it to introduce a dimensionless radius. It is also useful for the solution of the differential equation to normalize y by its (unknown) central value y_0 and introduce

$$\Phi = \frac{y}{y_0} \quad \text{and} \quad \eta = y_0 \frac{r}{l}. \tag{5.6}$$

In this manner the momentum balance reads

$$\frac{1}{\eta^2} \frac{d}{d\eta} \left(\eta^2 \frac{d\Phi}{d\eta} \right) = - \left(\Phi^2 - \frac{1}{y_0^2} \right)^{3/2}. \tag{5.7}$$

It has to be solved for the boundary conditions.

$$\Phi(0) = 1 \quad \text{and} \quad \frac{\partial \Phi}{\partial \eta}(0) = 0. \tag{5.8}$$

The latter condition is required by smoothness in the center.

Numerically found solutions for different values of y_0, or of the central density n_0, are shown in Figure 6.2a. For the outer radius R of the star we must have $n_R = 0$, hence $x_R = 0$ and $y_R = 1$. Therefore $\Phi(\eta_R) = \Phi(y_0 \frac{R}{l}) = 1/y_0$ and hence the central value y_0 determines the radius R or vice versa. We obtain

$$y_0 = y_0(R) \tag{5.9}$$

albeit only numerically. The function is shown in Figure 6.2b. But one observation is clear from Figure 6.2: If the central number density, and hence y_0, go to infinity, $\Phi = 0$ occurs at a value $\eta \approx 6.9$ and therefore R must be zero. This then represents the ultimate white dwarf, and we shall now proceed to investigate its mass.

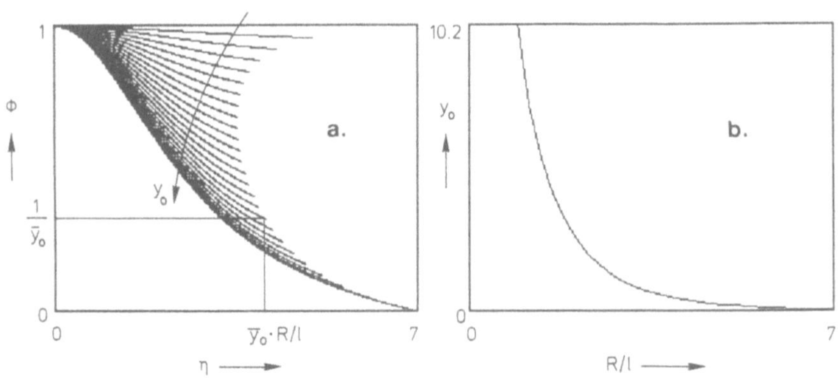

Figure 6.2 a. $\Phi(\eta, y_0)$ for different values of y_0 with an indication of how to find $y_0 = y_0(R)$. b. y_0 as a function of R.

The mass of the star is also a function of R. Indeed we have

$$
\begin{aligned}
M(r) &= 4\pi \int_0^r \varrho(r')r'^2 dr' \quad \text{with} \quad \varrho &=& 2m_H A x^3 \\
&&=& 2m_H A \sqrt{y^2-1}^3 \\
&&=& 2m_H A y_0^3 \left(\Phi^2 - \tfrac{1}{y_0^2}\right)^{3/2} : \\
M(\eta) &= 8\pi m_H A l^3 \int_0^\eta (\Phi^2 - \frac{1}{y_0^2})^{3/2} \eta'^2 d\eta' \quad \text{with (5.7):} \\
M(\eta) &= -2\sqrt{\frac{2}{\pi}} \frac{B^{3/2}}{m_H^2 G^{3/2} A^2} \int_0^\eta \frac{d}{d\eta'}\left(\eta'^2 \frac{d\Phi}{d\eta'}\right) d\eta' \\
M(\eta) &= -2\sqrt{\frac{2}{\pi}} \frac{B^{3/2}}{m_H^2 G^{3/2} A^2} \eta^2 \frac{d\Phi}{d\eta}.
\end{aligned}
$$

The total mass M is therefore given by

$$
M = -2\sqrt{\frac{2}{\pi}} \frac{B^{3/2}}{m_H^2 G^{3/2} A^2} \left(\eta^2 \frac{d\Phi}{d\eta}\right) |_{\eta = y_0 \frac{R}{l}}. \tag{5.10}
$$

Since y_0 is known as a function of R, and $\frac{d\Phi}{d\eta}$ is known for $\eta = y_0 \frac{R}{l}$ (see Figure 6.2) we may calculate

$$
M = M(R). \tag{5.11}
$$

This function has been drawn in Figure 6.3. We see that M is a decreasing function of R and that for $R = 0$—the *ultimate* dwarf—it assumes the value

$$
M_{min} \approx 2\sqrt{\frac{2}{\pi}} \frac{B^{3/2}}{m_H^2 G^{3/2} A^2} 2 \approx 1.4 M_\odot, \tag{5.12}
$$

where $M_\odot = 1.97\, 10^{30} kg$ is the mass of the sun.

It follows that the mass of a white dwarf cannot be bigger than roughly one and a half times the mass of the sun. If any star's mass *is* bigger, its final stage cannot be the white dwarf, i.e. the electrons cannot withstand the gravitational contraction.

The discussion of this example would be incomplete if no mention were made of the nonrelativistic limit. Of course, the value of the relativistic coldness cannot serve to distinguish the relativistic case from the nonrelativistic one in the case of full degeneration, because γ has dropped out of the formulae. But there is the value of x. By (4.43) we have

$$
x = \sqrt{\left(1 + \frac{g_c}{mc^2}\right)^2 - 1},
$$

where g_c is the free enthalpy. Obviously for a nonrelativistic gas we have $g_c \ll mc^2$, and x is small compared to 1. In this case $f(x) \underset{x\to\infty}{\longrightarrow} \frac{8}{5}x^5$ and we obtain from (4.45)

$$
p \sim n^{5/3}
$$

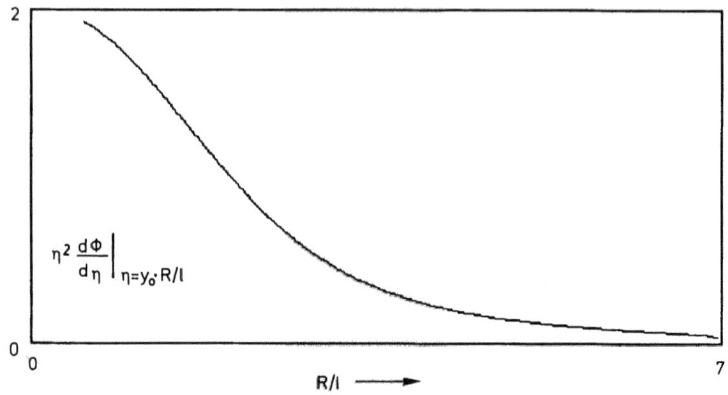

Figure 6.3 The mass of a white dwarf as a function of its radius.

which indeed conforms to the pressure-density relation (4.4.14) appropriate to the nonrelativistic strongly degenerate Fermi gas.

Instead of (5.7) we now arrive at the differential equation

$$\frac{1}{\xi^2}\frac{d}{d\xi}\left(\xi^2\frac{d\Psi}{d\xi}\right) = -\frac{2}{x_0}\Psi^{3/2} \quad \text{with} \quad \begin{aligned}\Psi &= \left(\frac{x}{x_0}\right)^2 \\ \xi &= x_0\frac{r}{l}\end{aligned} \tag{5.13}$$

which must be solved for the boundary conditions

$$\Psi(0) = 1 \qquad \frac{d\Psi}{d\xi}(0) = 0. \tag{5.14}$$

Figure 6.4 shows plots of the solution for different values of x_0. R can be calculated from the point ξ where Ψ becomes zero, and we obtain

$$x_0 = x_0(R). \tag{5.15}$$

The mass of the star is calculated as before, and we obtain

$$M = -4\pi m_H A l^3 x_0 \left(\xi^2\frac{d\Psi}{d\xi}\right)|_{\xi=x_0\frac{R}{l}} \tag{5.16}$$

The slope of Ψ is known at $\xi = x_0\frac{R}{l}$ from Figure 6.4, and we obtain

$$M = M(R)$$

as shown in Figure 6.5. We conclude that M grows to infinity as R tends to zero, so that there is no restriction on the mass of the white dwarf if the nonrelativistic equation of state is assumed to be valid.

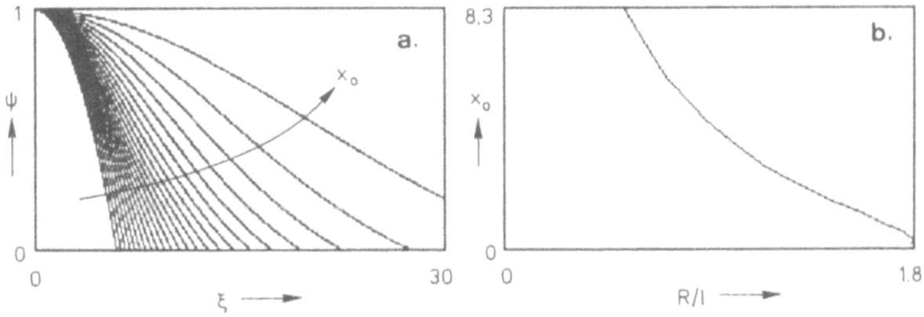

Figure 6.4 a. $\Psi(\xi; x_0)$ for different values of x_0
b. x_0 as a function of R.

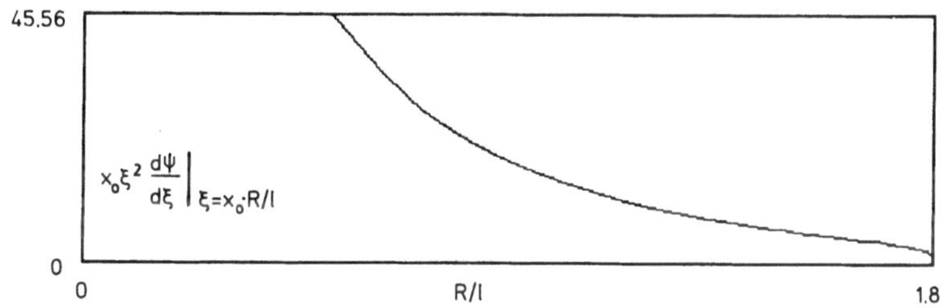

Figure 6.5 The mass as a function of R for a nonrelativistic state function.

6 The Relativistic Kinetic Theory for Nondegenerate Gases

6.1 Boltzmann-Chernikov Equation

In previous chapters we have on several occasions pointed to the close analogy between the results of extended thermodynamics and of the kinetic theory of gases in nonrelativistic gases. The same analogy can be exhibited for relativistic gases. The relativistic formulation of the Boltzmann equation has been given by Chernikov in a series of remarkable papers; see [72], [73], [74]. Here we describe only the bare bones of that theory.

In the relativistic kinetic theory a certain complexity is introduced by the fact that

$$d\mathbf{P} = \frac{\sqrt{-g}}{p_0} dp^1 dp^2 dp^3 \tag{6.1}$$

is the scalar element of momentum space. Thus

$$F(x^A, p^a) \frac{\sqrt{-g}}{p_0} dp^1 dp^2 dp^3 \tag{6.2}$$

may be called the number density of the atoms with momentum between p^a and $p^a + dp^a$.[3] The phase density F must obey the *Boltzmann-Chernikov equation*

$$p^A \frac{\partial F}{\partial x^A} - \Gamma^d_{AB} p^A p^B \frac{\partial F}{\partial p^d} = \int (F(p'_C)F(q'_C) - F(p_C)F(q_C)h\langle pq\rangle ded\mathbf{Q}. \tag{6.3}$$

The left-hand side may simply be written as the derivative of $F(x^A, p^a)$ with respect to the proper time of the atoms with four-momentum p^A. The Christoffel symbols $\mathbf{\Gamma}$ enter by use of the equation of motion of such atoms. The right-hand side is the collision integral, and it must be compared with the corresponding integral in the classical Boltzmann equation (2.3.2) in Chapter 2. p_C, q_C and p'_C, q'_C are the four-momenta of two colliding atoms before and after the collision. $\langle pq\rangle$ is the scalar relative speed of the atoms, and h is the scattering factor for scattering into the solid angle element de. Conservation of energy and momentum requires

$$p_C + q_C = p'_C + q'_C. \tag{6.4}$$

6.2 Equations of Transfer

Multiplication of the Boltzmann-Chernikov equation by a tensor $\psi^{A_1A_2\ldots A_n}$ and integration over $d\mathbf{P}$ leads to the generic equation of transfer

$$\begin{aligned}\left(\int p^A \psi^{A_1\ldots A_n} F d\mathbf{P}\right)_{,A} &- \int p^A \psi^{A_1\ldots A_n}{}_{,A} F dP + \int \Gamma^d_{AB} p^A p^B \frac{\partial \psi^{A_1\ldots A_n}}{\partial p^d} F d\mathbf{P} = \\ &= -\frac{1}{2}\int \left[\psi^{A_1\ldots A_n}; 1\right] \left(F(p'_C)F(q'_C) - F(p_C)F(q_C)\right) h\langle pq\rangle ded\mathbf{Q}d\mathbf{P}\end{aligned} \tag{6.5}$$

where [ ] stands for

$$[\psi^{A_1\ldots A_n}; 1] = \frac{1}{2}(\psi^{A_1\ldots A_n}(x^D, p^{d'}) + \psi^{A_1\ldots A_n}(x^D, q^{d'}) - \psi^{A_1\ldots A_n}(x^D, p^d) - \psi^{A_1\ldots A_n}(x^D, q^d)). \tag{6.6}$$

The nth moment of the phase density is defined by

$$A^{A_1\ldots A_n}(x^D) = m^{2-n}c\int p^{A_1}\cdots p^{A_n} F dP. \tag{6.7}$$

If $\psi^{A_1\ldots A_n}$ in (6.5) is chosen as $p^{A_1}\ldots p^{A_n}$, it turns out that the general equation of transfer assumes the form

$$A^{AA_1\ldots A_n}{}_{,A} = -\frac{1}{2}\int m^{2-n}c\left[p^{A_1}\ldots p^{A_n}; 1\right] \left(F(p'_C) + F(q'_C) - F(p_C) - F(q_C)\right) h\langle pq\rangle ded\mathbf{Q}d\mathbf{P} \tag{6.8}$$

which is an equation of balance for moments of the phase density.

[3] Actually the number of atoms between x^a, p^a and $x^a + dx^a, p^a + dp^a$ is obtained by multiplying (6.2) by $\sqrt{-g}dx^1dx^2dx^3$, but these fine points can best be studied in Chernikov's papers.

6.3 Equations of Balance for Particle Number, Energy-Momentum, Fluxes, and Entropy

For the special choices $n = 1, 2$, and 3 we obtain from (6.7)

$$\begin{aligned} A^A &= mc \int p^A \, F d\mathbf{P} \\ A^{AB} &= \int p^A p^B \, F d\mathbf{P} \\ A^{ABC} &= \frac{1}{m} c \int p^A p^B p^C \, F d\mathbf{P} \end{aligned} \tag{6.9}$$

which are the first moments of the phase density. Because of (6.4) the corresponding balance equations read

$$\begin{aligned} A^A{}_{,A} &= 0 \\ A^{BA}{}_{,A} &= 0 \\ A^{BCA}{}_{,A} &= I^{BC}. \end{aligned} \tag{6.10}$$

They represent the conservation laws of rest mass, energy, and momentum and the balance of fluxes, i.e. the 15 equations which we have chosen as the basis of relativistic extended thermodynamics; see Section 1. Since $p^A p_A = c^2$, we have from (6.7) and (6.8)

$$A^B{}_B{}^A = c^2 A^A \qquad \text{and} \qquad I^B_B = 0, \tag{6.11}$$

two relations that we had to postulate in the phenomenological theory in order to reduce the number of independent equations in (6.10) to fourteen.

If we insert the function

$$\psi = -kc \ln F \tag{6.12}$$

into the generic equation of transfer (6.5) we obtain

$$\left(-kc \int p^A \ln F F d\mathbf{P}\right)_{,A} = \frac{kc}{4} \int \ln \frac{F(p'_C)F(q'_C)}{F(p_C)F(q_C)} \left(F(p'_C)F(q'_C) - F(p_C)F(q_C)\right) h\langle pq\rangle d e d\mathbf{Q} d\mathbf{P}. \tag{6.13}$$

The right-hand side is nonnegative and therefore we come to the conclusion that

$$h^A = -kc \int p^A \ln F F d\mathbf{P} \tag{6.14}$$

represents the entropy flux vector in the kinetic theory. The right-hand side of (6.13) is the entropy production.

6.4 Maxwell-Jüttner Distribution, Equilibrium Properties

Equilibrium is characterized by a vanishing entropy production. It follows by inspection of (6.13) that $\ln F_E$ must be a summational invariant in the collisions that occur in equilibrium. Since there are only five independent summational invariants, viz. the rest mass and the components of the four-momentum p^A, $\ln F_E$ must be a linear combination of those and we obtain

$$F_E = A e^{-\lambda_B p^B}. \tag{6.15}$$

The coefficients A and λ_B may be calculated in terms of n, T, and U^A as follows: We insert $F_{\rm E}$ into (6.9) and make use of the decompositions $(2.60)_{1,2}$ to obtain

$$\begin{aligned} nmU^A &= mc \int p^A \quad Ae^{-\lambda_B p^B} d\mathbf{P} \\ p(n,e)h^{AB} + \frac{e}{c^2}U^AU^B &= c \int p^A p^B \quad Ae^{-\lambda_B p^B} d\mathbf{P}. \end{aligned} \tag{6.16}$$

We introduce $\gamma = mc\sqrt{\lambda_A \lambda^A}$ as an abbreviation and obtain

$$\lambda^A = \frac{\gamma}{mc^2}U^A, \quad A = \frac{n}{4\pi m^3 c^3}\frac{\gamma}{K_2(\gamma)}, \quad \gamma = \frac{nmc^2}{p(n,e)}, \tag{6.17}$$

where $K_2(\gamma)$ is one of the Bessel functions introduced in (4.18).

This set of equations identifies λ^A and A in terms of n, U^A, and $p(n,e)$. If we wish to relate λ^A and A to n, T, and U^A we have to calculate e from (6.16) and the entropy density

$$h|_{\rm E} = \frac{1}{c^2}h^A|_{\rm E}U_A = -\frac{k}{c}U_A \int p^A \ln F_{\rm E} F_{\rm E} d\mathbf{P}. \tag{6.18}$$

We obtain with $G = \frac{K_3}{K_2}$

$$e = nmc^2\left(G - \frac{1}{\gamma}\right), h|_{\rm E} = -kn\ln A + \frac{k\gamma}{mc^2}e. \tag{6.19}$$

From $(6.19)_2$ we derive easily

$$d\left(\frac{h|_{\rm E}}{n}\right) = \frac{k\gamma}{mc^2}\left[d\left(\frac{e}{n}\right) - \frac{1}{n^2}dp(n,e)\right], \tag{6.20}$$

whence we conclude by reference to the Gibbs equation that γ is the relativistic coldness

$$\gamma = \frac{mc^2}{kT}. \tag{6.21}$$

Thus the equilibrium phase density is given by

$$F_{\rm E} = \frac{n}{4\pi m^3c^3}\frac{\gamma}{K_2(\gamma)}e^{-\frac{U_A p^A}{kT}}. \tag{6.22}$$

This is indeed the Maxwell-Jüttner distribution for a nondegenerate gas; see (4.4) and $(4.20)_1$. Previously—in Paragraph 4.2—we have made use of that phase density to derive the equations of state.

6.5 Possible Thermodynamic Fields in Equilibrium

Inspection of the Boltzmann-Chernikov equation (6.3) shows that its right-hand side vanishes for the Maxwell-Jüttner distribution (6.15) irrespective of how A and λ_B might depend on x^B. If the equation is to be satisfied, its left-hand side must also vanish, and this requirement imposes restrictions on the possible fields $A(x^B)$ and $\lambda_A(x^B)$. In fact, insertion of (6.15) into (6.3) leads to the condition

$$p^A \frac{\partial A}{\partial x^A} - A\lambda_{B,A} p^A p^B = 0. \tag{6.23}$$

Since this equation must hold for all values of p^a, we conclude that it implies

$$\frac{\partial A}{\partial x^A} = 0 \quad \text{and} \quad \lambda_{B,A} + \lambda_{A,B} = 0. \tag{6.24}$$

It follows that A must be independent of x^B and that $\lambda_{B,A}$ must be antisymmetric in equilibrium. We recall that $\lambda_A = \frac{1}{kT} U_A$ and $A \sim n\frac{\gamma}{K_2(\gamma)}$ and conclude from (6.24) that

$$U^A \frac{\partial T}{\partial x^A} = 0 \quad \text{and} \quad \frac{\partial \ln T}{\partial x^A} = \frac{1}{c^2} U^B U_{A,B}. \tag{6.25}$$

This means that the temperature of a volume element moving with the gas must be constant in time, while the temperature gradient need not vanish. Both results differ from the nonrelativistic case, where according to Paragraph 2.3.3 the temperature in equilibrium could depend on time, but had to be independent of $\mathbf{x}$. The result $(6.25)_2$ is nicely consistent with equation $(3.7)_2$ that predicts a heat flux, unless $\frac{\partial T}{\partial x^A}$ is balanced by $T/c^2 U^B U_{A,B}$.

The equations (6.24) permit a wide variety of motions which we do not discuss in detail, e.g. see Chernikov [73]. In a gas at rest in a stationary gravitational field (6.25) implies

$$T(x^a) = \frac{c}{\sqrt{g_{00}(x^a)}} \tag{6.26}$$

so that an isothermal atmosphere does not represent a possible equilibrium condition. This observation is also confirmed by $(6.24)_1$, because, since A must be independent of x^a, T must be non-uniform to offset the barometric distribution of the particle number density n.

7 The Nonrelativistic Limit of Relativistic Thermodynamics

7.1 The Problem

There is more to the problem of finding the nonrelativistic limit of relativistic extended thermodynamics than the calculations of Paragraph 4.5 of the limiting values (4.34) and (4.35) through (4.37) of the state functions and of the coefficients C and A. Indeed, that calculation still leaves us with 14 field equations, while in the nonrelativistic theory the natural number of field equations seemed to be 13.

Also in the kinetic theory the nonrelativistic theory required moments of the distribution function of up to fourth order, while in the relativistic theory we used only first- and second-order moments.

Superficially viewed, these discrepancies seem to pose serious problems, but in fact they can be smoothly resolved. The easiest way to do so is to make use of the kinetic theory and compare the formulae of Section 6 of this chapter to those of Section 2.3 of Chapter 2. This topic was investigated by Dreyer & Weiss [75], whose analysis is condensed here into the following short paragraph.

7.2 Variables and Constitutive Quantities

We recall that in the relativistic theory as well as in the nonrelativistic one the variables and constitutive quantities were moments of the phase density. But moments were differently defined in the two theories. We recall their forms

relativistic	nonrelativistic	
$A^{A_1 A_2 \ldots A_n} = m^{2-n} c \int p^{A_1} p^{A_2} \ldots p^{A_n} F d\mathbf{P}$	$F^{a_1 a_2 \ldots a_n} = m \int c^{a_1} c^{a_2} \ldots c^{a_n} f d\mathbf{c}$	(7.1)

where c^a are velocity components and p^A are components of the four-momentum. By (6.1) we have $d\mathbf{P} = \frac{1}{p_0} dp^1 dp^2 dp^3$ in a Lorentz frame, while $d\mathbf{c}$ is equal to $dc^1 dc^2 dc^3$. The four-momentum p^A satisfies the constraint $p_A p^A = m^2 c^2$, so that we have

$$p_0 = p^0 = mc\sqrt{1 + \frac{p^2}{m^2c^2}} = mc\left(1 + \frac{1}{2}\frac{p^2}{m^2c^2} - \frac{1}{8}\frac{p^4}{m^4c^4} + \mathcal{O}\left(\frac{1}{c^6}\right)\right). \tag{7.2}$$

Momentum and velocity and the two phase densities are related by

$$p^a = mc^a + \mathcal{O}\left(\frac{1}{c^2}\right) \quad \text{and} \quad F = \frac{1}{m^3} f + \mathcal{O}\left(\frac{1}{c^2}\right). \tag{7.3}$$

It is then easy to show that the relativistic and the nonrelativistic moments are related as follows.

$$
\begin{aligned}
A^{0} &= cF + \mathcal{O}\left(\frac{1}{c}\right) & A^{c} &= F^{c} + \mathcal{O}\left(\frac{1}{c^2}\right) \\
A^{a0} &= cF^{a} + \mathcal{O}\left(\frac{1}{c}\right) & A^{ac} &= F^{ac} + \mathcal{O}\left(\frac{1}{c^2}\right) \\
A^{00} &= \underline{cA^{0}} + \frac{1}{2}F^{aa} + \mathcal{O}\left(\frac{1}{c^2}\right) & A^{0c} &= \underline{cA^{c}} + \frac{1}{2c}F^{cbb} + \mathcal{O}\left(\frac{1}{c^3}\right) \\
A^{ab0} &= cF^{ab} + \mathcal{O}\left(\frac{1}{c}\right) & A^{abc} &= F^{abc} + \mathcal{O}\left(\frac{1}{c^2}\right) \\
A^{a00} &= \underline{cA^{a0}} + \frac{1}{2}F^{abb} + \mathcal{O}\left(\frac{1}{c^2}\right) & A^{aoc} &= \underline{cA^{ac}} + \frac{1}{2c}F^{abbc} + \mathcal{O}\left(\frac{1}{c^3}\right) \\
A^{000} &= \underline{-c^2A^{0} + 2cA^{00}} + \frac{1}{4c}F^{aabb} + \mathcal{O}\left(\frac{1}{c^3}\right) & A^{00c} &= \underline{-c^2A^{c} + 2cA^{0c}} + \frac{1}{4c^2}F^{aabbc} + \mathcal{O}\left(\frac{1}{c^4}\right)
\end{aligned}
\tag{7.4}
$$

Insertion of the expansions (7.4) into the balance equations (1.2) leads to some cancellation, since the underlined terms drop out because of conservation of particle number, momentum, and energy. What remains can be written as follows up to terms of $\mathcal{O}\left(\frac{1}{c^2}\right)$:

$$
\begin{array}{lll}
\frac{1}{c}\frac{\partial A^{0}}{\partial t} + \frac{\partial A^{c}}{\partial x^{c}} = 0 & \Rightarrow \text{conservation of mass:} & \frac{\partial F}{\partial t} + \frac{\partial F^{c}}{\partial x^{c}} = 0 \\
\frac{1}{c}\frac{\partial A^{a0}}{\partial t} + \frac{\partial A^{ac}}{\partial x^{c}} = 0 & \Rightarrow \text{conservation of momentum:} & \frac{\partial F^{a}}{\partial t} + \frac{\partial F^{ac}}{\partial x^{c}} = 0 \\
\frac{1}{c}\frac{\partial A^{00}}{\partial t} + \frac{\partial A^{0c}}{\partial x^{c}} = 0 & \Rightarrow \text{conservation of energy:} & \frac{\partial \frac{1}{2}F^{aa}}{\partial t} + \frac{\partial \frac{1}{2}F^{aac}}{\partial x^{c}} = 0 \\
\frac{1}{c}\frac{\partial A^{\langle ab\rangle 0}}{\partial t} + \frac{\partial A^{\langle ab\rangle c}}{\partial x^{c}} = I^{\langle ab\rangle} & \Rightarrow \text{balance of momentum flux:} & \frac{\partial F^{\langle ab\rangle}}{\partial t} + \frac{\partial F^{\langle abc\rangle}}{\partial x^{c}} = I^{\langle ab\rangle} \\
\frac{1}{c}\frac{\partial A^{a00}}{\partial t} + \frac{\partial A^{a0c}}{\partial x^{c}} = I^{a0} & \Rightarrow \text{balance of energy flux:} & \frac{\partial \frac{1}{2}F^{abb}}{\partial t} + \frac{\partial \frac{1}{2}F^{abbc}}{\partial x^{c}} = cI^{a0} \\
\frac{1}{c}\frac{\partial A^{000}}{\partial t} + \frac{\partial A^{00c}}{\partial x^{c}} = I^{00} & \Rightarrow \text{balance of 4th moment:} & \frac{\partial F^{aabb}}{\partial t} + \frac{\partial F^{aabbc}}{\partial x^{c}} = 4c^2 I^{00}
\end{array}
\tag{7.5}
$$

The equations on the right-hand side of (7.5) are the 13 balance laws of nonrelativistic extended thermodynamics *plus* a 14th equation for the fourth moment F^{aabb}. We note that these 14 nonrelativistic equations contain moments of up to fifth order, while the relativistic equations have only first- and second-order moments. This phenomenon is due to the occurrence of p_0 in the denominator of the element $d\mathbf{P}$ of momentum space and to the expansion (7.2).

From these arguments we conclude that the relativistic theory seems to dictate a 14-moment theory for nonrelativistic extended thermodynamics as the most natural choice. Kremer [76] has been motivated by this consideration to formulate—within the nonrelativistic regime—an extended thermodynamics of 14 moments.

In order to summarize, we compare variables and constitutive quantities in the relativistic and nonrelativistic case and write

relativistic variables		nonrelativistic variables	
A^A, A^{AB} or	$n, U^A, e, t^{\langle AB \rangle}, q^A, \pi$	$F, F^a, F^{aa}, F^{\langle ab \rangle} F^{abb}, F^{aabb}$	(7.6)
relativistic constitutive quantities		nonrelativistic constitutive quantities	
A^{ABC}, I^{AB}		$F^{\langle abc \rangle}, F^{abbc}, F^{aabbc}, I^{\langle ab \rangle}, cI^{a0}, 4c^2 I^{00}$	(7.7)

7.3 The Dynamic Pressure

The physical significance of most of the nonrelativistic variables is clear from their definition (7.1). Thus F and F^a are mass- and momentum density, $\frac{1}{2}F^{aa}$ is the energy density, $F^{\langle ab \rangle}$ is the deviator of the momentum flux, and $\frac{1}{2}F^{abb}$ is the energy flux; these quantities all have a suggestive physical meaning and a clear conceptual connection to the corresponding relativistic variables n through q^A. There is no suggestive meaning, however, for the fourth moment F^{aabb}, nor is there a clear connection of this variable to the dynamic pressure π. This connection will have to be found, and we proceed to do this.

The easiest way is to write down A_a^{a0} and A^{000} in the rest frame in two ways: First by use of the representation (2.62) and then—approximately—by use of $(7.4)_{10,11}$. Equating those formulae we obtain

$$\begin{aligned} -\frac{c^3}{2}(nm - C_1^0 - C_1^\pi \pi) &= 3c(p(n,e)+\pi) + \frac{1}{2c}\varrho^{aacc} + \mathcal{O}\left(\frac{1}{c^3}\right) \\ \frac{c^3}{2}(nm + C_1^0 + C_1^\pi \pi) &= -c^3 nm + 2ce + \frac{1}{4c}\varrho^{aacc} + \mathcal{O}\left(\frac{1}{c^3}\right) \end{aligned} \tag{7.8}$$

where ϱ^{aacc} is the nonrelativistic *internal* fourth moment of the phase density. We substract the two equations (7.8) and obtain

$$p(n,e) + \pi = \frac{2}{3}(e - nmc^2) - \frac{1}{12c^2}\varrho^{aacc} + \mathcal{O}\left(\frac{1}{c^4}\right). \tag{7.9}$$

The nonequilibrium part of this equation reads

$$\pi = -\frac{1}{12c^2}\varrho^{aacc}|_{\mathrm{NE}} + \mathcal{O}\left(\frac{1}{c^4}\right). \tag{7.10}$$

We conclude that the occurrence of the dynamic pressure π in an ideal gas is a relativistic phenomenon. Therefore we were justified to ignore it in Chapters 2 and 4. The *new* result in (7.10) is that π does not appear as a pressure at all, but rather as a fourth moment of the phase density.

7.4 Order of Magnitude of the Dynamic Pressure

Some years ago there was a flurry of excitement about the dynamic pressure and the bulk viscosity. Indeed some people, knowledgeable in the fields of relativity and cosmology, thought that the dynamic pressure might affect the evolution of the early universe, e.g. see de Groot, van Leeuwen & van Weert [77], and Weinberg [78]. Yet at the same time the dynamic pressure was considered relativistically small, of order $1/c^4$.

And indeed, extended thermodyamics provides the result—for $mc^2 >> kT$—

$$\pi = \frac{1}{B_1^\pi}\frac{5}{6}nkT\left(\frac{kT}{mc^2}\right)^2 U^A{}_{,A} \tag{7.11}$$

as we have seen in $(4.37)_1$. Inspection shows that this expression is $\mathcal{O}(\frac{1}{c^4})$, since B_1^π—being the relaxation time for π—is expected to be $\mathcal{O}(1)$. [Recall that we always consider the relaxation time as of the order of magnitude of the mean time of free flight of the atoms.]

Thus, here is a dilemma: By (7.11), π is of $\mathcal{O}(\frac{1}{c^4})$, while in the previous paragraph we have concluded that π is of $\mathcal{O}(\frac{1}{c^2})$; see (7.10).

The dilemma is resolved when we realize that (7.11) is merely the first iterate of the Maxwell iteration which we have explained in Paragraph 3.1. We shall now proceed to show that the second iterate provides a term which is indeed of $\mathcal{O}(\frac{1}{c^2})$. This was recently discovered by Kremer & Müller [79], and we proceed to describe their procedure.

We recall the iterative scheme indicated in (3.5) and use it to calculate $\overset{(2)}{\pi}$. For that purpose we need only the time-like components of $(3.5)_1$ for $n = 2$, viz.

$$\begin{aligned} &A^A{}_{,A} = 0 \\ &\overset{(1)}{A}{}^{AB}{}_{,B}\, U_A = 0 \\ &\overset{(1)}{A}{}^{ABC}{}_{,C}\, U_A U_B = -3B_1^\pi c^2 \overset{(2)}{\pi} \end{aligned} \tag{7.12}$$

or else, by $(2.60)_{1,2}$ (2.62), ignoring all products of derivatives of α, T, and U^A

$$\begin{aligned} 0 &= \ddot{p}\frac{d\alpha}{d\tau} + (\dot{p}' - \dot{p})\frac{d\ln T}{d\tau} + \dot{p}U^A_{,A} \\ -\overset{(1)}{q}{}^C_{,C} &= (\dot{p}'-\dot{p})\frac{d\alpha}{d\tau} + (p'' - p')\frac{d\ln T}{d\tau} + p'U^A_{,A} \\ -2TB_1^\pi \overset{(2)}{\pi} + \frac{c^2T}{3}C_1^\pi \frac{d\overset{(1)}{\pi}}{d\tau} + \frac{10T}{3}C_3 \overset{(1)}{q}{}^C_{,C} &= \dot{\Gamma}_1\frac{d\alpha}{d\tau} + (\Gamma_1' - \Gamma_1)\frac{d\ln T}{d\tau} + \frac{5}{3}\Gamma_1 U^C_{,C} \end{aligned} \tag{7.13}$$

Hence follows by elimination of $\frac{d\alpha}{d\tau}$ and $\frac{d\ln T}{d\tau}$ and insertion $\overset{(1)}{\pi}$ and $\overset{(1)}{q}{}^C$ from $(3.7)_{1,2}$

$$
\begin{aligned}
\overset{(2)}{\pi} &= \frac{1}{2T}\frac{1}{B_1^\pi}\frac{\begin{bmatrix} -\ddot{p} & \dot{p}-\dot{p}' & \dot{\Gamma}_1 \\ \dot{p}-\dot{p}' & p'-p'' & \Gamma_1'-\Gamma_1 \\ -\dot{p} & -p' & \frac{5}{3}\Gamma_1 \end{bmatrix}}{\begin{bmatrix} -\ddot{p} & \dot{p}-\dot{p}' \\ \dot{p}-\dot{p}' & p'-p'' \end{bmatrix}} U^C{}_{,C} + \\
&+ \left(\frac{\begin{bmatrix} \ddot{p} & \dot{p}'-\dot{p} \\ \dot{\Gamma}_1 & \Gamma_1'-\Gamma_1 \end{bmatrix}}{\begin{bmatrix} \ddot{p} & \dot{p}'-\dot{p} \\ \dot{p}'-\dot{p} & p''-p' \end{bmatrix}} - \frac{5}{3}\frac{\begin{bmatrix} \dot{p} & -\dot{\Gamma}_1 \\ \Gamma_1 & \Gamma_2 \end{bmatrix}}{\begin{bmatrix} \dot{p} & -\dot{\Gamma}_1 \\ p' & \Gamma_1-\Gamma_1' \end{bmatrix}} \right) \cdot \\
&\cdot \frac{c^2}{4T^2 B_1^\pi \hat{B}_4}\begin{bmatrix} \dot{p} & -\dot{\Gamma}_1 \\ p' & \Gamma_1-\Gamma_1' \end{bmatrix}\frac{1}{\dot{p}}h^{AB}\left((\ln T)_{,A} - \frac{1}{c^2}\frac{dU_A}{d\tau}\right)_{,B} + \\
&+ \frac{1}{4T^2}\frac{1}{(B_1^\pi)^2}\frac{\begin{bmatrix} -\ddot{p} & \dot{p}-\dot{p}' & \dot{\Gamma}_1 \\ \dot{p}-\dot{p}' & p'-p'' & \Gamma_1'-\Gamma_1 \\ \dot{\Gamma}_1 & \Gamma_1'-\Gamma_1 & \frac{5}{3}\Gamma_2 \end{bmatrix}}{\begin{bmatrix} -\ddot{p} & \dot{p}-\dot{p}' \\ \dot{p}-\dot{p}' & p'-p'' \end{bmatrix}}\frac{dU^A{}_{,A}}{d\tau}.
\end{aligned}
\tag{7.14}
$$

We restrict the attention to the nondegenerate gas in the weakly relativistic limit ($mc^2 >> kT$) and obtain with (4.34)

$$
\begin{aligned}
\overset{(2)}{\pi} = \frac{1}{B_1^\pi}\frac{5}{6}\frac{nk^3T^3}{m^2}\frac{1}{c^4}U^A{}_{,A} &- \frac{1}{(B_1^\pi)^2}\frac{5}{2}\frac{nk^3T^3}{m^2}\frac{1}{c^4}\frac{dU^A{}_{,A}}{d\tau} - \\
&- \frac{1}{B_1^\pi}\frac{1}{\hat{B}_4}\frac{5}{3}\frac{nk^3T^3}{m^2}\frac{1}{c^2}\left(1+\frac{kT}{mc^2}\right)h^{AB}\left((\ln T)_{,A} - \frac{1}{c^2}\frac{dU_A}{d\tau}\right)_{,B}.
\end{aligned}
\tag{7.15}
$$

We conclude that the old result (7.11) is reproduced as the leading term here, while the two additional terms are quadratic in $1/B$ as befits second iterates. Inspection shows, however, that the leading term in powers of $\frac{1}{c^2}$ is given by—in the Lorentz frame at rest—

$$
-\frac{1}{B_1^\pi}\frac{1}{\hat{B}_4}\frac{5}{3}\frac{nk^3T^3}{m^2}\frac{1}{c^2}\Delta T \quad \text{or} \quad -\frac{1}{B_1^\pi}\frac{2}{3}\left(\frac{kT}{mc^2}\right)q^{C(1)}_{,C}. \tag{7.16}
$$

The second version of (7.16) results by elimination of the temperature gradient by use of Fourier's law $q_i = -\kappa\frac{\partial T}{\partial x_i}$ with κ given by (4.37).

We conclude from (7.15) or (7.16) that, while the dynamic pressure is still a relativistic effect, it is not as small as it is usually considered to be. Thus the dilemma is resolved: The dynamic pressure is of $\mathcal{O}(\frac{1}{c^2})$ rather than of $\mathcal{O}(\frac{1}{c^4})$. And—maybe—it did indeed contribute its bit to the evolution of the early universe.

To be sure, the leading term (7.16) in $\overset{(2)}{\pi}$ is not proportional to the expansion $U^A{}_{,A}$ but rather to ΔT, or $q^A{}_{,A}$, the heating or cooling.

Chapter 7

Extended Thermodynamics of Reacting Mixtures

A relativistic theory of reacting mixtures incorporates the heat of reaction in a more explicit manner than a non-relativistic theory does. This is so, because the heat of reaction is related to the mass-defect, which determines the production of rest mass in the reaction. Consequently the heat of reaction plays an important role in the constitutive function for the flux tensor. Once extended thermodynamics is reduced to ordinary thermodynamics of irreversible processes , and once the non-relativistic limit is taken, it turns out that the bulk viscosity and the thermal conductivity depend on the heat of reaction in an explicit manner.

1 Motivation, Results, and Discussion

1.1 Motivation

The most important phenomena that occur in mixtures and which do not occur in single fluids are chemical reactions and diffusion. Among these, chemical reactions are of particular interest in relativistic thermodynamics, because in a reacting mixture the rest mass is not conserved. Indeed, given the stoichiometric coefficients γ_α there is the *mass-defect*

$$M = \sum_\alpha \gamma_\alpha m_\alpha .$$

The mass-defect determines the heat of reaction Mc^2, a sizeable and measurable quantity, tabulated in the handbooks of chemistry.

We expect interesting phenomena, even in the nonrelativistic limit, because there are now two smallness parameters in this limit, viz. $\frac{1}{c^2}$ and M. In order to make our point we refer to (6.7.5)—for a single fluid—and rewrite the balance equation for the flux A^{000}, viz.

$$\frac{1}{c}\frac{\partial A^{000}}{\partial t} + \frac{\partial A^{00c}}{\partial x^c} = I^{00}. \tag{1.1}$$

Insertion of (6.7.4) provides the approximate equation (cf. Section 6.7)

$$-c^2\left(\frac{1}{c}\frac{\partial A^0}{\partial t} + \frac{\partial A^c}{\partial x^c}\right) + c^2\left(\frac{1}{c}\frac{\partial A^{00}}{\partial t} + \frac{\partial A^{0c}}{\partial x^c}\right) +$$

$$+\frac{1}{4}\frac{1}{c^2}\left(\frac{\partial F^{aabb}}{\partial t} + \frac{\partial F^{aabbc}}{\partial x^c}\right) = I^{00}. \tag{1.2}$$

and, of course, the two expressions in the first line vanish, because of the conservation of the number of particles—or the conservation of rest mass—and the conservation of energy. Thus the nonrelativistic equation that corresponds to the flux balance for A^{000} reads

$$\frac{\partial F^{aabb}}{\partial t} + \frac{\partial F^{aabbc}}{\partial x^c} = c^2 4\, I^{00}. \tag{1.3}$$

This is the nonrelativistic balance for the trace of the fourth moment, and since $I^{00} = -3B_1^\pi \pi$ holds—at least in the rest frame, cf. (6.2.62)—we have concluded from it that π is of $O(\frac{1}{c^2})$.

So far so good. But now think of a case when the rest mass is *not conserved*, namely a reacting mixture! In that case $\frac{1}{c}\frac{\partial A^0}{\partial t} + \frac{\partial A^c}{\partial x^c}$ equals $M\ell$, the product of the (tiny) mass-defect of the reaction and the reaction rate density. Energy is still conserved and instead of (1.3) we therefore obtain

$$\frac{\partial F^{aabb}}{\partial t} + \frac{\partial F^{aabbc}}{\partial x^c} = c^2\left(4\, I^{00} + Mc^2\ell\right). \tag{1.4}$$

So, arguing as before, we may now expect the dynamic pressure to emerge as a quantity of $O(1)$—rather than of $O(\frac{1}{c^2})$—and related to the reaction rate. [Actually I^{00} for the mixture is different from I^{00} for the single fluid, but this does not invalidate the argument; in fact it will add yet another term $Mc^2\ell$ to the dynamic pressure.]

So much for the motivation of why Kremer & Müller [31] have started the work on reacting relativistic mixtures: They were in search of a large bulk viscosity. The motivation is not compelling, perhaps, only plausible; but it has proved to have heuristic value, because it did work!

1.2 Results

The constitutive theory of a reacting mixture—albeit a nondiffusive one—is quite complex as we shall see. Therefore we summarize the results first, so that the reader may appreciate them before deciding to become immersed in the formalism—or not.

The results concern the three transport coefficients viscosity μ, bulk viscosity λ, and thermal conductivity κ. These three quantities have the form (cf. (5.9), (5.10), (5.14) below):

$$\begin{aligned}
\mu &= -\frac{1}{B_3}(n_1^E + n_2^E)kT, \\
\lambda &= \frac{-\frac{2}{3B_1^\pi}\left(\frac{Mc^2}{kT}\right)^2 (n_1^E + n_2^E)kT}{\left(\frac{Mc^2}{kT}\right)^2 + 3(\gamma_1+\gamma_2)\left(\frac{Mc^2}{kT}\right) - \frac{3}{4}\gamma_1\gamma_2\left(2\frac{m_1^2(n_1^E)^2 + m_2^2(n_2^E)^2}{m_1 m_2 n_1^E n_2^E} + 5\frac{m_1^2+m_2^2}{m_1 m_2} - 6\right)}, \qquad (1.5)\\
\kappa &= \frac{1}{\hat{B}_4}\left\{\frac{5}{8}\frac{n_1^E n_2^E}{\frac{n_1^E}{\gamma_1} - \frac{n_2^E}{\gamma_2}}\frac{(kT)^2}{m_1 m_2 \gamma_1 \gamma_2}\left[\left(2\frac{n_1^{E2} + n_2^{E2}}{n_1^E n_2^E} + 7\frac{m_1^2 + m_2^2}{m_1 m_2} - 10\right) \times \right.\right. \\
&\qquad \left.\left.(m_1\gamma_1 - m_2\gamma_2) + 4(m_2 - m_1)\left(\frac{Mc^2}{kT}\right)\right]\right\}.
\end{aligned}$$

The B's—or rather $1/B$—are relaxation times, and we consider them as closely related to the mean times of free flight of the atoms and molecules, and certainly they are of $O(1)$. $\gamma_\alpha (\alpha = 1, 2)$ are stoichiometric coefficients and n_α^E are particle number densities in chemical equilibrium.

Equation (1.5) shows the leading terms of the coefficients, i.e. it ignores terms of $O(\frac{1}{c^2})$. We conclude that indeed—as foreseen in Paragraph 1.1—the bulk viscosity, and hence the dynamic pressure, is of $O(1)$. Another interesting point occurs in the thermal conductivity, which includes a term depending on the heat of reaction Mc^2.

In order to demonstrate the size of the Mc^2 terms we have prepared Table 7.1, which refers to the simple dissociation-recombination reaction of iodine.[1]

$$I_2 \rightarrow 2I \qquad (1.6)$$

[1]It is true that I_2 is diatomic, while the state functions which we shall use are those of monatomic gases. This cannot be helped; a better example for this theory might be a fusion reaction when two deuterons combine to form an α-particle. In that case both constituents are monatomic.

which has a heat of reaction $Mc^2 = -148.7 \cdot 10^{35}\, J/mol$ according to the handbook of d'Ans & Lax [80]. The entries α^E and K_p in the table represent the equilibrium value of the extent of reaction and the chemical constant respectively; these two quantities are related by

$$\alpha^E = \sqrt{\frac{K_p}{4p + K_p}} \tag{1.7}$$

and the table refers to $p = 1atm$. We see that the reaction has barely started at $T = 800K$ and is nearly complete at $2200K$. The important entries in the table are in the last two columns. We have defined

$$V = \frac{\lambda B_1^{\pi}}{\mu B_3} \quad \text{and} \quad R = \frac{4(m_2 - m_1)\frac{Mc^2}{kT}}{\left(2\frac{n_1^{E2}+n_2^{E2}}{n_1^E n_2^E} + 7\frac{m_1^2+m_2^2}{m_1 m_2} - 10\right)(m_1\gamma_1 - m_2\gamma_2)} \tag{1.8}$$

so that V essentially represents the ratio the the bulk viscosity to the shear viscosity and R represents the influence of the heat of reaction on the thermal conductivity.

Assuming that the B's are equal, we conclude that λ reaches as much as 50% of the size of μ. Also the effect of the heat of reaction can double the size of the coefficient of thermal conductivity.

Inspection of (1.5) shows that in the limit of a single gas, i.e. when $n_1^E \to 0$ or $n_2^E \to 0$, the transport coefficients μ and κ tend to the values $(6.4.37)_{2,3}$ appropriate for a single gas, and λ tends to zero.

For a nuclear reaction, e.g. the fusion of two deuterons to an α-particle, the "heat of reaction" is more than ten million times as large as the one for the recombination of two iodine atoms. We can therefore expect a very strong influence of the new term—i.e. the Mc^2 term—on the thermal conductivity.

$T(K)$	$K_p(atm)$	$n_1^E(m^{-3})$	$n_{12}^E(m^{-3})$	a^E	V	R
800	3.16 x 10^{-5}	5.14 x 10^{22}	9.13 x 10^{24}	0.003	0.204	0.062
1000	3.16 x 10^{-3}	4.01 x 10^{23}	6.94 x 10^{24}	0.028	0.438	0.424
1200	7.08 x 10^{-2}	1.43 x 10^{24}	4.69 x 10^{24}	0.132	0.500	1.015
1400	6.17 x 10^{-1}	2.81 x 10^{24}	2.44 x 10^{24}	0.366	0.498	1.108
1500	1.51	3.36 x10^{24}	1.53 x 10^{24}	0.523	0.490	0.932
1600	3.31	3.69 x10^{24}	8.97 x 10^{23}	0.673	0.474	0.690
1800	11.2	3.77 x 10^{24}	3.11 x 10^{23}	0.858	0.420	0.311
2000	33.1	3.57 x 10^{24}	1.05 x 10^{23}	0.945	0.318	0.118
2200	75.9	3.29 x 10^{24}	4.28 x 10^{22}	0.975	0.204	0.050

Table 7.1 Effect of heat of reaction on bulk viscosity and thermal conductivity.

1.3 Discussion

How does extended thermodynamics discover such things—those which we have just discussed—when ordinary thermodynamics of irreversible processes gives no hint of them? The best answer to that question is given by pointing to the equations (6.3.7), which relate the "thermodynamic forces and fluxes." But *in contrast* to thermodynamics of irreversible processes, the relation is not just a proportionality in extended thermodynamics—with no knowledge about the factor of proportionality other than that it must be positive. Rather in extended thermodynamics the transport coefficients are explicitly related to the relaxation times, about which we have considerable ulterior knowledge. Thus, by (6.3.9) the coefficients λ, κ, μ carry a lot of information in their explicit dependence on $\dot{p}, p'$, etc; and it is this information which lets us predict properties about transport coefficients.

In the final analysis, of course, this information is due to the fact that in extended thermodynamics we are "deriving" the phenomenological equations of irreversible thermodynamics as mutilated forms of balance equations about whose fluxes we have learned a whole lot from the entropy principle.

Nowhere is this better illustrated than in the relations $(1.5)_{2,3}$; but we have seen this specificity in the transport coefficients before: e.g. in Chapter 4 (see (4.3.7)) we have identified the shear viscosity μ as proportional to the relaxation time τ_t with the pressure as the factor of proportionality. Another example: In Chapter 5—specifically in (5.3.3) and (5.3.4)—we have identified the transport coefficients of diffusion, thermal diffusion, and thermal conductivity and reduced them to the matrix of relaxation times $\overset{-1}{\mathbf{M}}_{\mathrm{V}}$ in a specific manner. The information thus gained allowed us to *derive* the Onsager relations in Paragraph 5.3.2 which are *postulated* in thermodynamics of irreversible processes.

Of course, the kinetic theory of gases does much the same thing as extended thermodynamics in making the transport coefficients more transparent. In fact, it has been argued that extended thermodynamics can only do what the kinetic theory can do better. With the present extension to chemically reacting mixtures this observation is no longer true; or at least we are unaware of any kinetic theory that could provide results as specific as the results (1.5) of extended thermodynamics.

And now we proceed to *derive* these results. Although we are most interested in the nonrelativistic results (1.5), we shall use the four-dimensional formalism of relativity, because it is formally simpler. Therefore this chapter will recall the formalism of Chapter 6; but there are cumbersome—although important—differences, which prevent us from going through all the details. For those we have to refer the reader to the paper by Kremer & Müller [31].

2 Fields

2.1 A Conventional Choice

We develop a simple theory of extended thermodynamics of a reacting binary mixture by forbidding diffusion.[2] Thus we shall have 15 fields—only one velocity—and these are

$$\begin{array}{l} \text{number densities } n_\alpha \\ \text{velocity } U^A, \\ \text{stress deviator } t^{<AB>}, \\ \text{pressure } P, \\ \text{energy density } e, \\ \text{heat flux } q^A. \end{array} \tag{2.1}$$

The general scope of the chapter is similar to Chapter 6 and, as far as possible, the notation is the same. Thus the variables (2.1) may be combined in the particle flux vectors A^A_α and the energy-momentum tensor A^{AB}, viz.

$$\begin{aligned} A^A_\alpha &= m_\alpha n_\alpha U^A, \\ A^{AB} &= t^{\langle AB\rangle} + Ph^{AB} + \frac{1}{c^2}(U^A q^B + U^B q^A) + \frac{1}{c^2} e U^A U^B. \end{aligned} \tag{2.2}$$

2.2 Absolute Temperature, Fugacities, and Chemical Affinity

While n_α and e are perfectly legitimate variables, it is useful and customary to replace them by the fugacities a_α and the absolute temperature T, because these are measurable—in principle. Also a_α,T represent the canonical choice of variables in statistical thermodynamics. *But*: In a reacting mixture the fugacities are not independent in equilibrium. Therefore it is advantageous to replace the two fugacities a_α by a *mixture fugacity* a and a *chemical affinity* Δ. The latter vanishes in equilibrium. We proceed to explain:

We recall the thermal equation of state $p(a,T)$ of a nondegenerate relativistic gas from $(6.4.20)_{1,3}$.[3] Now in the mixture we have two such relations, viz.

$$p_\alpha(a_\alpha, T) = \exp\left(-\frac{m_\alpha}{k} a_\alpha\right) 4\pi y m_\alpha^2 c (kT)^2 K_2\left(\frac{m_\alpha c^2}{kT}\right). \tag{2.3}$$

and $n_\alpha(a,T)$ and $e(a_1, a_2, T)$ may be derived from it as well as the entropy density $h(a_1, a_2 T)$, cf. (6.2.31), (6.2.33)

[2]A relativistic thermodynamics of reacting and diffusing mixtures has been formulated—within the context of ordinary thermodynamics—by Kluitenberg [81]. See also the paper by Hutter & Müller [82], where each constituent is allowed to have its own temperature.

[3]The former letter α for the fugacity is now replaced by a, so as not to confuse it with the index α that characterizes the constituent.

$$
\begin{aligned}
n_\alpha(a_\alpha, T) &= -\frac{1}{m_\alpha T}\left(\frac{\partial p_\alpha}{\partial a_\alpha}\right)_T, \\
e(a_1, a_2, T) &= \sum_\alpha T^2\left(\frac{\partial}{\partial T}\left(\frac{p_\alpha}{T}\right)\right)_{a_\alpha}, \\
h(a_1, a_2, T) &= \sum_\alpha \left[\left(\frac{\partial p_\alpha}{\partial T}\right)_{a_\alpha} - \frac{a_\alpha}{T}\left(\frac{\partial p_\alpha}{\partial a_\alpha}\right)_T\right].
\end{aligned}
\tag{2.4}
$$

In equilibrium, the fugacities a_α are not independent, because the law of mass action requires

$$\sum_\alpha \gamma_\alpha m_\alpha a_\alpha^E = 0, \tag{2.5}$$

where γ_α are the stoichiometric coefficients of the chemical reaction. Therefore in equilibrium we may choose

$$\text{the mixture fugacity:} \quad a = \gamma_1 m_1 a_1^E = -\gamma_2 m_2 a_2^E, \tag{2.6}$$

as common variable for both constituents and obtain

$$p^E_{\substack{1\\2}}(a, T) = \exp\left(\mp\frac{1}{k\gamma_{\substack{1\\2}}}a\right) 4\pi y m^2_{\substack{1\\2}} c(kT)^2 K_2\left(\frac{m_{\substack{1\\2}} c^2}{kT}\right). \tag{2.7}$$

Close to equilibrium we may thus write by virtue of a Taylor expansion

$$
\begin{aligned}
&p_{\substack{1\\2}}(a_{\substack{1\\2}}, T) = p^E_{\substack{1\\2}} \pm \gamma_{\substack{1\\2}} m_{\substack{1\\2}} \dot{p}^E_{\substack{1\\2}}(a_{\substack{1\\2}} - a^E_{\substack{1\\2}}), \\
&n_{\substack{1\\2}}(a_{\substack{1\\2}}, T) = \mp\frac{1}{T}\dot{p}^E_{\substack{1\\2}}\gamma_{\substack{1\\2}} - \frac{1}{T}\gamma^2_{\substack{1\\2}} m_{\substack{1\\2}} \ddot{p}^E_{\substack{1\\2}}(a_{\substack{1\\2}} - a^E_{\substack{1\\2}}), \\
&e(a_1, a_2, T) = \sum_\alpha (p'^E_\alpha - p^E_\alpha) + [\![\gamma_\alpha m_\alpha(\dot{p}'^E_\alpha - \dot{p}^E_\alpha)(a_\alpha - a^E_\alpha)]\!], \\
&h(a_1, a_2, T) = \frac{1}{T}\sum_\alpha (p'^E_\alpha - a\dot{p}^E_\alpha) + \frac{1}{T}[\![\gamma_\alpha m_\alpha(\dot{p}'^E_\alpha - (a\dot{p}^E_\alpha)^\bullet)(a_\alpha - a^E_\alpha)]\!].
\end{aligned}
\tag{2.8}
$$

where the derivatives $\bullet$ and $'$ in (2.8) denote differentiation with respect to a and $\ln T$, respectively. $[\![f_\alpha]\!]$ stands for $f_1 - f_2$.

In order to characterize the distance from equilibrium we introduce

$$\text{the chemical affinity:} \quad \Delta = \textstyle\sum_\alpha \gamma_\alpha m_\alpha a_\alpha, \tag{2.9}$$

Δ represents a "driving force" toward equilibrium, and by the law of mass-action we have $\Delta = 0$ in equilibrium.

In order to obtain $a_\alpha (\alpha = 1, 2)$ as functions of a and Δ we need an additional equation—additional to (2.9)—and this equation will be furnished by the observation that the entropy $(2.8)_4$ cannot have a linear term in $(a_\alpha - a^E_\alpha)$; otherwise it could not be maximal in equilibrium. Hence follows

$$\gamma_1 m_1(\dot{p}'^E_1 - (a\dot{p}^E_1)^\bullet)(a_1 - a^E_1) - \gamma_2 m_2(\dot{p}'^E_2 - (a\dot{p}^E_2)^\bullet)(a_2 - a^E_2) = 0. \tag{2.10}$$

With (2.9) and (2.10) we have two equations that permit us to switch between the two sets of variables, viz. (a_1, a_2) and (a, Δ). The equations read

$$a_1 = \frac{1}{\gamma_1 m_1}\left(a + \underbrace{\frac{(\dot{p}_2'^E - (a\dot{p}_2^E)^\bullet)}{(\dot{p}_1'^E - (a\dot{p}_1^E)^\bullet) + (\dot{p}_2'^E - (a\dot{p}_2^E)^\bullet)}}_{A_1}\Delta\right),$$

$$a_2 = \frac{1}{\gamma_2 m_2}\left(-a + \underbrace{\frac{(\dot{p}_1'^E - (a\dot{p}_1^E)^\bullet)}{(\dot{p}_1'^E - (a\dot{p}_1^E)^\bullet) + (\dot{p}_2'^E - (a\dot{p}_2^E)^\bullet)}}_{-A_2}\Delta\right). \tag{2.11}$$

Thus by $(2.8)_{2,3}$ we obtain p_α, n_α, and e as functions of a and T and as *linear* functions of Δ, viz.

$$\begin{aligned} p_\alpha(a,T,\Delta) &= p_\alpha^E + \dot{p}_\alpha^E A_\alpha \Delta, \\ n_{\frac{1}{2}}(a,T,\Delta) &= \mp\frac{1}{T}\,\dot{p}_{\frac{1}{2}}^E\gamma_{\frac{1}{2}} \mp \frac{1}{T}\,\ddot{p}_{\frac{1}{2}}^E\gamma_{\frac{1}{2}} A_\alpha\Delta, \\ e(a,T,\Delta) &= \sum_\alpha \left(p_\alpha'^E - p_\alpha^E\right) + \sum_\alpha \left(\dot{p}_\alpha'^E - \dot{p}_\alpha^E\right) A_\alpha \Delta. \end{aligned} \tag{2.12}$$

The advantage of the new variables a, Δ over the old ones a_1, a_2 is that equilibrium is now characterized by the vanishing of one variable—viz. Δ—rather than a relation between two variables.

For a similar reason it is useful and customary to decompose the pressure P in (2.2) into an equilibrium part $\sum\limits_\alpha p_\alpha^E$ and a *dynamic pressure* π such that

$$P = \sum_\alpha p_\alpha^E + \pi. \tag{2.13}$$

2.3 Summary of Fields

In summary we may now rephrase the objective of extended thermodynamics by saying that we wish to determine the 15 fields

$$a, \Delta, T, U^A, t^{<AB>}, \pi, q^A \tag{2.14}$$

which may be combined into the particle flux vector A_α^A and the energy-momentum tensor A^{AB}

$$\begin{aligned} A_\alpha^A &= m_\alpha n_\alpha(a,T,\Delta)U^A \\ A^{AB} &= t^{<AB>} + \left(\sum_\alpha p_\alpha^E(a,T) + \pi\right) h^{AB} + \frac{1}{c^2}\left(U^A q^B + U^B q^A\right) + \frac{1}{c^2} e(a,T,\Delta) U^A U^B. \end{aligned} \tag{2.15}$$

The functions $n_\alpha(a,T,\Delta)$ and $e(a,T,\Delta)$ are given by (2.12).

3 Field Equation

3.1 Balance Laws

The necessary field equations for the determination of the fields (2.13) are based upon the equations of balance of masses, and energy-momentum and the flux balance, viz.

$$\begin{aligned} A^A_{\alpha,A} &= \gamma_\alpha m_\alpha \ell, \qquad (\alpha = 1,2), \\ A^{AB}{}_{,B} &= 0 \\ A^{ABC}{}_{,C} &= I^{AB}, \qquad (I^A_A = \sum_\alpha \gamma_\alpha m_\alpha c^2 \ell, \qquad A^{AB}{}_B = \sum_\alpha c^2 A^A_\alpha). \end{aligned} \tag{3.1}$$

ℓ is the reaction rate density and I^{AB} is the flux production. A^{ABC} is the flux tensor; it is fully symmetric and its trace is proportional to $\sum\limits_\alpha A^A_\alpha$, so that $(3.1)_3$ represents only 9 equations. The constraints on the traces in $(3.1)_3$ are consistent with the kinetic theory of gases; this is not essentially different from the argument of Paragraph 6.1.1, except that now the chemical reactions have to be accounted for.

A useful alternative form of the particle conservation laws $(3.1)_1$ reads

$$\begin{aligned} \left(\sum_\alpha m_\alpha n_\alpha U^A\right)_{,A} &= \sum_\alpha \gamma_\alpha m_\alpha \ell, \\ \left(\left(\frac{n_1}{\gamma_1} - \frac{n_2}{\gamma_2}\right) U^A\right)_{,A} &= 0. \end{aligned} \tag{3.2}$$

In this set only the first equation has a production which is governed by the *mass defect* of the reaction $M = \sum\limits_\alpha \gamma_\alpha m_\alpha$; the second equation is a conservation law representing the conservation of the number of atoms in the chemical reaction, or the conservation of nucleons in a nuclear reaction.

3.2 Constitutive Theory

We close the system by the formulation of constitutive relations for

- ℓ - the reaction rate density,
- A^{ABC} - the flux tensor,
- I^{AB} - the flux production.

The constitutive equations have the generic forms

$$\begin{aligned} \ell &= \hat{\ell}\ (n_\alpha, U^M, A^{MN}) &&= \hat{\ell}\ (a, T, \Delta, U^M, t^{<MN>}, q^M, \pi) \\ A^{ABC} &= \hat{A}^{ABC}(n_\alpha, U^M, A^{MN}) &&= \hat{A}^{ABC}(a, T, \Delta, U^M, t^{<MN>}, q^M, \pi) \\ I^{AB} &= \hat{I}^{AB}\ (n_\alpha, U^M, A^{MN}) &&= \hat{I}^{AB}\ (a, T, \Delta, U^M, t^{<MN>}, q^M, \pi), \end{aligned} \tag{3.3}$$

where $\hat{\ell}, \hat{A}^{ABC}$, and $\hat{I}^{AB}$ are the constitutive functions. If these were known explicitly, we should have 15 independent equations between (3.1) and (3.3) for the 15 variables (2.14), and every solution of this set is a *thermodynamic process.* As always the form of these constitutive functions is restricted by the thermodynamic principles, namely

- the principle of relativity,
- the entropy inequality,
- the requirement of convexity and causality.

3.3 Principle of Relativity

By the principle of relativity the constitutive functions must be isotropic functions with respect to all space-time transformations. This implies linear representations of the form

$$\begin{aligned}
\ell &= \ell_\pi \pi + \ell_\Delta \Delta, \\
A^{ABC} &= \left(C_1^0 + C_1^\pi \pi + C_1^\Delta \Delta\right) U^A U^B U^C + \\
&+ \frac{c^2}{6}\left(\sum_\alpha m_\alpha n_\alpha - C_1^0 - C_1^\pi \pi - C_1^\Delta \Delta\right)\left(g^{AB}U^C + g^{BC}U^A + g^{CA}U^B\right) + \\
&+ C_3\left(g^{AB}q^C + g^{BC}q^A + g^{CA}q^B\right) - \frac{6}{c^2}C_3(U^AU^Bq^C + U^BU^Cq^A + U^CU^Aq^B) + \\
&+ C_5\left(t^{\langle AB\rangle}U^C + t^{\langle BC\rangle}U^A + t^{\langle CA\rangle}U^B\right), \qquad (3.4)\\
I^{\langle AB\rangle} &= (B_1^\pi \pi + B_1^\Delta \Delta) g^{AB} - \frac{4}{c^2}(B_1^\pi \pi + B_1^\Delta \Delta) U^A U^B + \\
&+ B_3 t^{\langle AB\rangle} + \frac{1}{c^2}\hat{B}_4(q^AU^B + q^BU^A) + M\ell U^AU^B.
\end{aligned}$$

These expressions take into account the trace conditions implied by $(3.1)_3$. Of course, ℓ in $(3.4)_3$ may be replaced by $l_\pi \pi + l_\Delta \Delta$ (cf. $(3.4)_1$) and then incorporated into the second term, but for the moment we prefer to keep it separate.

The coefficients C and B as well as l_π, l_Δ will be restricted by the entropy principle.

4 Entropy Inequality

4.1 Lagrange Multipliers

The entropy principle requires that the divergence of the entropy-entropy flux vector h^A—itself a constitutive quantity—be nonnegative for all thermodynamic processes

$$h^A_{,A} \geq 0 \quad \text{with} \quad h^A = h^A\left(a, T, \Delta, U^A, t^{<AB>}, \pi, q^A\right). \qquad (4.1)$$

Equivalent to this statement is the new inequality

$$h^A_{,A} + \lambda\left(\left(\sum_\alpha m_\alpha n_\alpha U^A\right)_{,A} - \sum_\alpha \gamma_\alpha m_\alpha \ell\right) + \Lambda\left(\left(\left(\frac{n_1}{\gamma_1} - \frac{n_2}{\gamma_2}\right)U^A\right)_{,A} + \right.$$

$$+\Lambda_A A^{AB}{}_{,B} + \Lambda_{AB}(A^{ABC}{}_{,C} - I^{AB}) \geq 0, \tag{4.2}$$

which has to hold for all fields n_α, U^A, A^{AB}. The lambdas are Lagrange multipliers and Λ_{AB} must be taken traceless so as to ensure that the trace of equation $(3.1)_3$ is not counted among the field equations.

4.2 Exploitation

The exploitation of the entropy inequality proceeds much like in Chapter 6 following the inequality (6.2.1). The addition of another mass balance—for the additional constituent—might seem innocuous, but it is not. Indeed, the analysis becomes quite cumbersome and we refer the interested reader to the paper by Kremer & Müller [31].

Necessary ingredients are, once again, the linearization of the theory in terms of the variables that vanish in equilibrium, viz. $\Delta, t^{<AB>}, \pi$, and q^A and the elimination of the Lagrange multipliers.

What results is a system of field equations which is quite explicit except for the occurrence of the coefficients B_1^π, B_1^Δ, B_3,$\hat{B}_4$ and l_π, l_Δ. In particular all the coefficients C in the flux tensor can be related to the equation of state $p^E_{\substack{1\\2}}(a,T)$ given in (2.7). In this respect the theory is similar to the one of Chapter 6.

We shall be interested in the leading terms of the field equations, i.e. those that are relevant to a nonrelativistic theory. We proceed to write these equations.

5 Nonrelativistic Limit

5.1 Discussion

There are three groups of field equations:

- four scalar equations representing the balance of total mass, the conservation laws of the number of particles, and of energy, and the flux balance for the dynamic pressure,
- two vectorial equations representing the conservation law of momentum and the heat flux balance,
- one tensorial equation representing the balance of the stress deviator.

$$\begin{bmatrix} m\dot{n} & mn' & mn & 0 & 0 & 0 \\ \ddot{p} & \dot{p}'-\dot{p} & \dot{p} & 0 & 0 & 0 \\ \dot{p}'-\dot{p} & p''-p' & p' & 0 & 0 & 1 \\ \frac{\dot{\Gamma}_1'-3\dot{\Gamma}_1}{2T} & \frac{\Gamma_1''-4\Gamma_1'+3\Gamma_1}{2T} & \frac{\Gamma_1'-\Gamma_1}{2T} & -3 & 0 & -5C_3 \end{bmatrix} \begin{bmatrix} \frac{da}{d\tau} \\ \frac{d\ln T}{d\tau} \\ U^A_{\prime A} \\ \frac{d\pi}{d\tau} \\ \frac{d\Delta}{d\tau} \\ q^A_{\prime A} \end{bmatrix} = \begin{bmatrix} M\ell \\ 0 \\ 0 \\ \frac{1}{c^2} I^{AB} U_A U_B \end{bmatrix} \tag{5.1}$$

$$\begin{bmatrix} \dot{p} & p' & 1 & \sum_\alpha \dot{p}^E_\alpha A_\alpha & 1 & p' & 1 \\ -\frac{\dot{\Gamma}_1}{T} & -\frac{\Gamma_1'-\Gamma_1}{2T} & -1 & 0 & -1 & -\frac{\Gamma_1'-\Gamma_1}{2T} & 5C_3 \end{bmatrix} \begin{bmatrix} h^{AB} a_{\prime B} \\ h^{AB} (\ln T)_{\prime B} \\ h^{AB} \pi_{\prime B} \\ h^{AB} \Delta_{\prime B} \\ t^{<AB>}{}_{\prime B} \\ \frac{1}{c^2}\frac{dU^A}{d\tau} \\ \frac{1}{c^2}\frac{dq^A}{d\tau} \end{bmatrix} = \begin{bmatrix} 0 \\ -\frac{\hat{B}_4}{c^2} q^A \end{bmatrix} \tag{5.2}$$

$$-\frac{\dot{\Gamma}_1}{T} h^{E<A} h^{B>D} U_{<D,E>} + 2C_3 h^{E<A} h^{B>D} q_{<D,E>} + \frac{dt^{<AB>}}{d\tau} = B_3 t^{<AB>}. \tag{5.3}$$

The entries in the matrices are given by

$$\begin{aligned} p &= \sum_\alpha n^E_\alpha kT \\ mn &= \sum_\alpha m_\alpha n^E_\alpha \quad \text{with} \quad n^E_{\substack{1\\2}}(a,T) = \exp\left(\mp\frac{1}{k\gamma_{\substack{1\\2}}} a\right) 4\pi y m^2_{\substack{1\\2}} ckT K_2\left(\frac{m_{\substack{1\\2}} c^2}{kT}\right) \\ \Gamma_1 &= 2c^2 \cdot T^6 \cdot \int \frac{mn}{T^6}\, dT \\ C_3 &= \frac{\frac{2}{5}\frac{(n_1^E)^2+(n_2^E)^2}{n_1^E n_2^E} + \frac{7}{5}\frac{m_1^2+m_2^2}{m_1 m_2} - 2}{\frac{\gamma_1+\gamma_2}{\gamma_1\gamma_2}\frac{Mc^2}{kT} - \left(\frac{(n_1^E)^2+(n_2^E)^2}{n_1^E n_2^E} + \frac{7}{2}\frac{m_1^2+m_2^2}{m_1 m_2} - 5\right)} + \mathcal{O}\left(\frac{1}{c^2}\right). \end{aligned} \tag{5.4}$$

There are several subtle points to be mentioned before we exploit the equations (5.1) through (5.3), because the non-relativistic limit is not altogether standard. Let us consider:

In the calculation of quantities like p', p'' and Γ_1'' , which figure in the matrices of (5.1), (5.2) the Bessel function K_2 must be differentiated and this creates again Bessel functions, cf. (6.4.19); also differentiation provides factors mc^2. In the non-relativistic limit all Bessel functions are replaced by the expansion (6.4.32) and thus for instance $p' - p''$ assumes the form

$$p' - p'' = \tfrac{c^4}{kT}\left[m_1^2 n_1^E\left(1 + 3\tfrac{kT}{m_1c^2} + \tfrac{15}{2}\left(\tfrac{kT}{m_1c^2}\right)^2\right) + \right. \\ \left. + m_2^2 n_2^E\left(1 + 3\tfrac{kT}{m_2c^2} + \tfrac{15}{2}\left(\tfrac{kT}{m_2c^2}\right)^2\right)\right] + 0\left(\tfrac{1}{c^2}\right). \tag{5.5}$$

whose leading term is $O(c^4)$! This does not mean, however, that this term – and others like it – dominate the equations in which they occur. Indeed, one of two things may happen to terms of $O(c^4)$ and $O(c^2)$:

- they may cancel out of the equations, or
- they may combine with M to form the quantity Mc^2, the heat of reaction which is $O(0)$.

Both of these phenomena do in fact occur. In this regard we should mention that the non-relativistic limit is a singular limit for the system of equations (3.2) , because in non-relativistic mechanics the rest mass is conserved. I.e. we have $\gamma_1 m_1 = -\gamma_2 m_2$ and therefore the two equations (3.2) reduce to one equation.

5.2 Dynamic Pressure and Bulk Viscosity

We use the conservation laws among (5.1) to eliminate $\frac{da}{d\tau}$ and $\frac{d\ln T}{d\tau}$ and obtain equations for the reaction rate density ℓ and the dynamic pressure, viz.

$$\frac{1}{D}\left(\frac{Mc^2}{kT}\right)\left[(n_1^E + n_2^E)U^A_{,A} - \frac{1}{kT}q^A_{,A}\right] = \ell. \tag{5.6}$$

$$-\frac{kT}{D}\left(\frac{Mc^2}{kT}\right)^2 (n_1^E + n_2^E)U^A_{,A} - \left\{\frac{1}{D}\left(\frac{Mc^2}{kT}\right)^2 + \frac{2\frac{(n_1^E)^2+(n_2^E)^2}{n_1^E n_2^E} + 7\frac{m_1^2+m_2^2}{m_1 m_2} - 10}{\frac{\gamma_1+\gamma_2}{\gamma_1\gamma_2}\frac{Mc^2}{kT} - \left(\frac{(n_1^E)^2+(n_2^E)^2}{n_1^E n_2^E} + \frac{7}{2}\frac{m_1^2+m_2^2}{m_1 m_2} - 5\right)}\right\}q^A_{,A} + \\ -3\frac{d\pi}{d\tau} = -3(B_1^\pi \pi + B_1^\Delta \Delta) + M\ell c^2. \tag{5.7}$$

D is abbreviation: it stands for

$$D = \left(\frac{Mc^2}{kT}\right)^2 + 3(\gamma_1 + \gamma_2)\left(\frac{Mc^2}{kT}\right) - \frac{3}{4}\gamma_1\gamma_2\left(2\frac{m_1^2(n_1^E)^2 + m_2^2(n_2^E)^2}{m_1 m_2 n_1^E n_2^E} + 5\frac{m_1^2 + m_2^2}{m_1 m_2} - 6\right). \tag{5.8}$$

We conclude from (5.6) that for a reaction to proceed we must have expansion $U^A_{,A}$, or heating $q^A_{,A}$, or both. And of course, no reaction can occur without a mass-defect.

Equation (5.7) is an equation of extended thermodynamics where π is a variable. We obtain a *constitutive equation* for π—as ordinary irreversible thermodynamics would have it—by the Maxwell iteration again; i.e. we set nonequilibrium quantities on the left-hand side of (5.7) equal to zero and obtain a first iterate for π, viz. $(5.9)_1$.

$$\pi = \underbrace{-\frac{B_1^\Delta}{B_1^\pi}}_{\tilde{L}}\Delta + \underbrace{\frac{1}{B_1^\pi}\frac{2}{3D}\left(\frac{Mc^2}{kT}\right)^2 (n_1^E + n_2^E)\,kT\,U^A_{,A}}_{-\lambda}$$

$$\ell = \underbrace{(l_\Delta - l_\pi)\frac{B_1^\Delta}{B_1^\pi}}_{L^{11}}\Delta + \underbrace{\frac{l_\pi}{B_1^\pi}\frac{2}{3D}\left(\frac{Mc^2}{kT}\right)^2 (n_1^E + n_2^E)}_{L} U^A_{,A}. \tag{5.9}$$

The equation for ℓ results from introducing $(5.9)_1$ into $\ell = l_\pi\pi + l_\Delta\Delta$.

The equations (5.9) correspond to the phenomenological equations (5.1.8) of TIP for mixtures. The braces in (5.9) indicate the phenomenological coefficients that were previously introduced—in Chapter 5—as factors of proportionality between forces and fluxes. In particular we identify the bulk viscosity λ: In its unabbreviated form λ reads

$$\lambda = \frac{\frac{-2}{3B_1^\pi}\left(\frac{Mc^2}{kT}\right)^2 (n_1^E + n_2^E)kT}{\left(\frac{Mc^2}{kT}\right)^2 + 3(\gamma_1+\gamma_2)\left(\frac{Mc^2}{kT}\right) - \frac{3}{4}\gamma_1\gamma_2\left(2\frac{m_1^2(n_1^E)^2 + m_2^2(n_2^E)^2}{m_1m_2n_1^En_2^E} + 5\frac{m_1^2+m_2^2}{m_1m_2} - 6\right)}, \tag{5.10}$$

The bulk viscosity depends explicitly on the heat of reaction and we have discussed it quantitatively in Section 1. The only unknown quantity in (5.10) is the relaxation time $1/B_1^\pi$, which we take to have the properties and order of magnitude of the mean time of free flight of the atoms and molecules.

5.3 Thermal Conductivity and Viscosity

Complex equations of the type (5.2) and (5.3) for q^A and $t^{<AB>}$ are known in extended thermodynamics and in the kinetic theory of gases. They are usually "boiled down" to the phenomenological equations of Fourier and Navier-Stokes by the first step in the Maxwellian iterative scheme. In that step all nonequilibrium quantities on the left-hand side of (5.2), (5.3) are neglected so that equations for q^A and $t^{<AB>}$ read, without terms of $O\left(\frac{1}{c^2}\right)$:

$$\begin{bmatrix} \dot{p}^E & p'^E \\ -\frac{\dot\Gamma_1^E}{2T} & -\frac{\Gamma_1'^E - \Gamma_1^E}{2T}\end{bmatrix}\begin{bmatrix} h^{AB}a_{,B} \\ h^{AB}(\ln T)_{,B}\end{bmatrix} = \begin{bmatrix} 0 \\ -\frac{\hat{B}_4}{c^2}q^A\end{bmatrix}, \tag{5.11}$$

$$-\frac{\Gamma_1^E}{T}h^{E\langle A}h^{B\rangle D}U_{\langle D,E\rangle} = B_3 t^{\langle AB\rangle}. \tag{5.11}$$

We use the conservation law in $(5.11)_1$ to eliminate $h^{AB}a_{,B}$ from the second equation and obtain

$$\left[\frac{\dot\Gamma_1^E}{2T}\frac{p'^E}{\dot{p}^E} - \frac{\Gamma_1'^E - \Gamma_1^E}{2T}\right]h^{AB}(\ln T)_{,B} = -\frac{\hat{B}_4}{c^2}q^A. \tag{5.12}$$

The calculation of Γ_1,$\dot\Gamma_1$,Γ_1',$\dot p$, and $\dot p$ in (5.12), $(5.11)_2$ belongs again to the things for which we refer the reader to the paper by Kremer & Müller [31], where a long list of such quantities is given. In the end we obtain

$$q_a = -\frac{1}{\hat{B}_4}\left\{\frac{5}{8}\frac{n_1^En_2^E}{\frac{n_1^E}{\gamma_1} - \frac{n_2^E}{\gamma_2}}\frac{(kT)^2}{m_1m_2\gamma_1\gamma_2}\left[\left(2\frac{n_1^{E2}+n_2^{E2}}{n_1^En_2^E} + 7\frac{m_1^2+m_2^2}{m_1m_2} - 10\right)\times\right.\right.$$

$$(m_1\gamma_1 - m_2\gamma_2) + 4(m_2 - m_1)\left(\frac{Mc^2}{kT}\right)\Bigg]\Bigg\}\frac{\partial T}{\partial x_c}, \quad (5.13)$$

$$t_{ab} = \underbrace{-\frac{1}{B_3}\left(n_1^E + n_2^E\right) kT}_{2\mu}\frac{\partial v_{<a}}{\partial x_{b>}}. \quad (5.14)$$

The factor of proportionality between the heat flux and the temperature gradient in (5.13) is the thermal conductivity κ, and the brace in (5.14) identifies the shear viscosity μ. There is little to say about μ; it is proportional to the pressure, just like in a single gas. The thermal conductivity κ, however, is interesting, since it depends on the heat of reaction Mc^2.We have discussed all three coefficients λ, μ, and κ in Section 1.2 and 1.3 and there is nothing to add.

Chapter 8

Waves in Extended Thermodynamics

Waves represent an important probe into the constitutive properties of bodies, because their speeds and attenuations may be measured as well as – in the case of shock waves – the jumps in density, temperature etc. Such observations permit the calculation of constitutive coefficients or alternatively – when the coefficients are known – they permit the evaluation of the range of validity of the theory.

The equations of extended thermodynamics are of balance type, see Paragraph 3.1.1

$$\mathbf{F}^A{}_{,A} = \mathbf{\Pi} \quad \text{with} \quad \mathbf{F}^A = \mathbf{F}^A(\mathbf{u}) \quad \text{and} \quad \mathbf{\Pi} = \mathbf{\Pi}(\mathbf{u}).$$

It follows that the field equations form a first order quasi-linear system of the generic form

$$\mathbf{A}^A(\mathbf{u})\mathbf{u}_{,A} = \mathbf{\Pi}(\mathbf{u}) \qquad \text{with} \qquad \mathbf{A}^A = \frac{\partial \mathbf{F}^A}{\partial \mathbf{u}}.$$

The purpose of this chapter is the study of the waves which this system allows to propagate. We shall consider linear waves, acceleration waves and shock waves. In each case we illustrate the general results for the special system of extended thermodynamics of 13 moments. We also determine the limits of hyperbolicity for that system.

1 Hyperbolicity and Symmetric Hyperbolic Systems

1.1 Hyperbolicity in the t-direction

The system

$$\mathbf{A}^A(\mathbf{u})\mathbf{u}_{,A} = \mathbf{\Pi}(\mathbf{u}) \tag{1.1}$$

permits the propagation of waves, in particular of weak discontinuity waves—or acceleration waves—on which the fields $\mathbf{u}$ are continuous but their derivatives are not. The possible

$$\text{speeds } V, \text{ and normal jumps } \delta\mathbf{u} = \left[\left[\frac{\partial \mathbf{u}}{\partial x^a}\right]\right] n_a$$

may be calculated from the eigenvalues λ and right eigenvectors $\mathbf{d}$ of the linear homogeneous system of algebraic equations (see Paragraph 4.1 below)

$$\left(\mathbf{A}^i n_i - \lambda \mathbf{A}^o\right)\mathbf{d} = 0. \tag{1.2}$$

n_i is the unit vector orthogonal to the wave front. The λ's are called the *characteristic speeds* and the polynomial

$$\det\left(\mathbf{A}^i n_i - \lambda \mathbf{A}^o\right) = 0 \tag{1.3}$$

is called the *characteristic polynomial.*

The system (1.1) is called *hyperbolic in the t-direction* if it has the following two properties:

- $\det\mathbf{A}^o \neq 0$,
- for all unit vectors $\mathbf{n}$ the eigenvalue problem (1.3) admits only real eigenvalues λ and a set of linearly independent right eigenvectors $\mathbf{d}$.

It is well known from linear algebra that the left eigenvector $\mathbf{l}$, to be calculated from

$$\mathbf{l}\left(\mathbf{A}^i n_i - \lambda \mathbf{A}^o\right) = 0, \tag{1.4}$$

may be chosen such that

$$\mathbf{l}^{(i)} \cdot \mathbf{d}^{(j)} = \delta^{ij}, \qquad \text{for all} \qquad i,j = 1,2,\dots n. \tag{1.5}$$

1.2 Symmetric Hyperbolic Systems

The system (1.1) is said to be *symmetric hyperbolic* (in the t- direction), or briefly symmetric—by a definition of Friedrichs—if

- the matrices $\mathbf{A}^A$ are symmetric;
- the matrix $\mathbf{A}^o$ is positive definite.

By linear algebra every symmetric system is hyperbolic, but the reverse statement is not true.

Symmetric systems play an important role in extended thermodynamics, because that theory uses hyperbolic equations and the entropy principle together with the concavity of the entropy density ensures that they form a *symmetric* hyperbolic system. Indeed, as we have shown in Chapter 3, the compatibility of the balance laws

$$\mathbf{F}^A{}_{,A} = \mathbf{\Pi} \qquad \text{with} \qquad \mathbf{F}^A = \mathbf{F}^A(\mathbf{u}) \quad \text{and} \quad \mathbf{\Pi} = \mathbf{\Pi}(\mathbf{u}) \tag{1.6}$$

with the entropy principle

$$h^A{}_{,A} \geq 0 \qquad \text{and} \qquad h^0(\mathbf{u})\text{ - concave} \tag{1.7}$$

guarantees that the system (1.6) may be written in a symmetric form. We recall this from Paragraph 3.2.3.

The advantage of having symmetric hyperbolic systems lies in the fact that there exist theorems guaranteeing the well-posedness for the Cauchy problem (e.g. see Fisher & Marsden[3]):

Theorem 1:
Each symmetric system, whose initial data are of class $H^s(R^N)$ with $s \geq 4$, has a unique solution $\in H^s(R^N)$ in the neighborhood of the initial manifold, even if the system is not strictly hyperbolic.[1]

Since time is distinguished from the spatial variables in the definition of hyperbolicity and symmetric hyperbolicity, one might expect difficulties in the relativistic case. Indeed, such difficulties occur and Ruggeri [69] has discussed them. The results of that discussion are thought-provoking but not quite conclusive. Therefore we do not introduce these arguments there.

2 Linear Waves

2.1 Plane Harmonic Waves, the Dispersion Relation

In extended thermodynamics the system (1.1) is symmetric hyperbolic, so that without loss of generality it may be written in the form

$$\frac{\partial \mathbf{u}}{\partial t} + \mathbf{A}^a(\mathbf{u})\frac{\partial \mathbf{u}}{\partial x^a} = \mathbf{\Pi}(\mathbf{u}). \tag{2.1}$$

If we consider the fields to depend on one space dimension only, we may write

$$\frac{\partial \mathbf{u}}{\partial t} + \mathbf{A}(\mathbf{u})\frac{\partial \mathbf{u}}{\partial x} = \mathbf{\Pi}(\mathbf{u}). \tag{2.2}$$

We linearize by setting

$$\mathbf{u} = \tilde{\mathbf{u}} + \bar{\mathbf{u}}, \tag{2.3}$$

[1]Strict hyperbolicity means that all characteristic speeds are different.

where $\bar{\mathbf{u}}$ is a small perturbation of an equilibrium state $\tilde{\mathbf{u}}$ for which we assume $\mathbf{\Pi}(\tilde{\mathbf{u}}) = 0$. Thus we obtain the linearized equations

$$\frac{\partial \bar{\mathbf{u}}}{\partial t} + \tilde{\mathbf{A}} \frac{\partial \bar{\mathbf{u}}}{\partial x} = \tilde{\mathbf{B}} \bar{\mathbf{u}}, \qquad \text{where} \qquad \tilde{\mathbf{B}} = \left(\frac{\partial \mathbf{\Pi}}{\partial \mathbf{u}} \right)_{\tilde{\mathbf{u}}}, \tag{2.4}$$

and we look for solutions of the form

$$\bar{\mathbf{u}} = \mathbf{w} e^{i(\omega t - kx)} \tag{2.5}$$

representing a plane harmonic wave of real frequency ω and complex wave number $k = k_r + ik_i$ with the complex amplitude $\mathbf{w}$. Equation (2.5) represents a wave travelling in the x-direction.

Substitution of (2.5) into (2.4) provides a homogeneous algebraic linear system of the form

$$\left(\mathbf{1} - z\tilde{\mathbf{A}} + \frac{i}{\omega} \tilde{\mathbf{B}} \right) \mathbf{w} = 0 \tag{2.6}$$

where z stands for k/ω.

For nontrivial solutions the *dispersion relation* must be satisfied

$$\det \left(\mathbf{1} - z\tilde{\mathbf{A}} + \frac{i}{\omega} \tilde{\mathbf{B}} \right) = 0. \tag{2.7}$$

The dispersion relation permits the calculation of the phase velocity v_{ph} and of the attenuation α in terms of the frequency ω

$$v_{ph} = \frac{\omega}{\mathcal{R}e(k)} = \frac{1}{\mathcal{R}e(z)}, \qquad \alpha = -\mathcal{I}m(k) = -\omega \mathcal{I}m(z). \tag{2.8}$$

For linear stability, $\alpha(\omega)$ must be positive for waves traveling to the right; it is negative for waves traveling to the left and, if we assume the waves to start at time $t = 0$ at $x = 0$, we have $\alpha(\omega)x > 0$. Since the path of the wave is given by $x = \tilde{\lambda} t$ under these circumstances, we may write the condition of linear stability as

$$\alpha(\omega)\tilde{\lambda} > 0. \tag{2.9}$$

2.2 The High-Frequency Limit

Our objective in this paragraph is to estimate v_{ph} and α in the limit of high frequency, i.e. for $\omega \to \infty$. For this purpose we consider the formal expansion of z and $\mathbf{w}$ in terms of powers of $\frac{1}{\omega}$, viz.

$$z = \sum_{\alpha \geq 0} \frac{z_\alpha}{\omega^\alpha}, \qquad \mathbf{w} = \sum_{\beta \geq 0} \frac{\mathbf{w}_\beta}{\omega^\beta}. \tag{2.10}$$

Insertion of (2.10) into (2.4) provides a recurrence formula, first derived by Muracchini, Ruggeri & Seccia [83] [2]

$$(\mathbf{1} - z_o \mathbf{A}) \mathbf{w}_\alpha - (z_1 \mathbf{A} - i\mathbf{B}) \mathbf{w}_{\alpha - 1} - \sum_{\beta = 2}^{\alpha} z_\beta \mathbf{A} \mathbf{w}_{\alpha - \beta} = 0 \qquad (\alpha = 0, 1, 2, ...) \tag{2.11}$$

[2]For simplicity we omit the tilde that characterizes the constant state.

where $\mathbf{w}_\gamma = 0$ for $\gamma < 0$.

For $\alpha = 0$ one obtains from (2.11)

$$(\mathbf{1} - z_o\mathbf{A})\mathbf{w}_o = 0, \quad \text{hence} \quad z_o = \frac{1}{\lambda} \text{ and } \mathbf{w}_o = \mathbf{d}, \tag{2.12}$$

where λ and $\mathbf{d}$ are an eigenvalue and the corresponding right eigenvector of $\mathbf{A}$ evaluated at the constant state. We observe that z_o and $\mathbf{w}_o$ are real.

For $\alpha = 1$ we have, recalling (2.12),

$$\left(\mathbf{1} - \frac{\mathbf{A}}{\lambda}\right)\mathbf{w}_1 - (z_1\mathbf{A} - i\mathbf{B})\mathbf{d} = 0. \tag{2.13}$$

Multiplication by the left eigenvector $\mathbf{l}$ gives with $\mathbf{lA} = \lambda\mathbf{l}$

$$z_1 = \frac{i}{\lambda}(\mathbf{l}\cdot\mathbf{B}\cdot\mathbf{d}) \tag{2.14}$$

so that z_1 is purely imaginary.

Summarizing, we have so far

$$z = \frac{k}{\omega} \approx z_o + \frac{z_1}{\omega} = \frac{1}{\lambda}\left(1 + \frac{i}{\omega}(\mathbf{l}\cdot\mathbf{B}\cdot\mathbf{d})\right) \qquad \text{for} \quad \omega\text{-large} \tag{2.15}$$

and hence, by (2.8) and (2.12),

$$\lim_{\omega\to\infty} v_{ph}(\omega) = \lambda \qquad \text{and} \qquad \lim_{\omega\to\infty} \alpha(\omega)\lambda = -\mathbf{l}\cdot\mathbf{B}\cdot\mathbf{d}. \tag{2.16}$$

The first equation in (2.16) provides the result that the phase velocity coincides—in the limit of high frequency—with the characteristic speeds. We recall that in Paragraph 2.2.3 we have anticipated that result when speaking of the maximum speed of propagation in extended TIP. The second result in (2.16) furnishes a condition of linear stability, because—by (2.9)—for stability we must have

$$\mathbf{l}\cdot\mathbf{B}\cdot\mathbf{d} < 0. \tag{2.17}$$

Later we shall see that (2.17) guarantees nonlinear stability as well; see Paragraph 4.2.

2.3 Higher-Order Terms

The exploitation of the recurrence formula (2.11) may be carried further and this has been done by Muracchini, Ruggeri & Seccia [83], but we shall not present these calculations, which become increasingly difficult. Suffice it to list two of the general results in the form of a theorem.

Theorem 2:
The phase velocity and the attenuation depend on the frequency ω through ω^2 and therefore they are even functions of ω.

The phase velocity and the attenuation corresponding to an eigenvalue λ are independent of frequency only if the left eigenvector of $\mathbf{A}$ is also a left eigenvector of $\mathbf{B}$; in that case we have for all frequencies ω

$$v_{ph} = \lambda, \qquad \alpha = -\frac{\mathbf{l} \cdot \mathbf{B} \cdot \mathbf{d}}{\lambda}. \tag{2.18}$$

2.4 Linear Waves in Extended Thermodynamics

For an example of the foregoing calculations we refer to Chapter 4, specifically to the field equation (4.4.3) for the nondegenerate gas. In a noninertial frame we assume that $\mathbf{v}$ and $\mathbf{q}$ have only x_1-components and that $t_{<ij>}$ has only diagonal components with $t_{<11>} = \sigma$. All fields are assumed to depend only on x_1. Thus we obtain with $c = \sqrt{\frac{5}{3}\frac{k}{m}T}$ and

$$x = x_1, \quad v = v_1, \quad q = q_1, \quad \sigma = t_{<11>} \tag{2.19}$$

$$\begin{bmatrix} \dot\rho \\ \dot v \\ \dot T \\ \dot\sigma \\ \dot q \end{bmatrix} + \begin{bmatrix} 0 & \rho & 0 & 0 & 0 \\ \frac{3c^2}{5\rho} & 0 & \frac{3c^2}{5T} & -\frac{1}{\rho} & 0 \\ 0 & \frac{2}{9}T(3-5\hat\sigma) & 0 & 0 & \frac{10}{9}\frac{T}{c^2\rho} \\ 0 & c^2\rho\left(\frac{7}{3}\hat\sigma - \frac{4}{5}\right) & 0 & 0 & -\frac{8}{15} \\ \frac{3}{5}c^4\hat\sigma & \frac{16}{5}c^3\rho\hat q & -\frac{3}{10}\frac{c^4\rho}{T}(5\hat\sigma - 3) & -\frac{c^2}{5}(5\hat\sigma + 3) & 0 \end{bmatrix} \cdot \begin{bmatrix} \frac{\partial\rho}{\partial x} \\ \frac{\partial v}{\partial x} \\ \frac{\partial T}{\partial x} \\ \frac{\partial\sigma}{\partial x} \\ \frac{\partial q}{\partial x} \end{bmatrix} = \begin{bmatrix} 0 \\ 0 \\ 0 \\ -\frac{3}{2\tau_q}\sigma \\ -\frac{1}{\tau_q}q \end{bmatrix}. \tag{2.20}$$

We have assumed here that $\tau_q/\tau_t = 3/2$ holds which is exactly true only for Maxwellian molecules, but it is very nearly true for all monatomic gases.

Note that the system (2.20) is not yet linearized. This fact is put in evidence by the occurrence of $\hat\sigma$ and $\hat q$ in the matrix. These quantities represent stress and heat flux in dimensionless form. We have

$$\hat\sigma = \frac{\sigma}{c^2\rho} \quad \text{and} \quad \hat q = \frac{q}{c^3\rho}. \tag{2.21}$$

The field vector $\mathbf{u}$ of the general theory must be identified as

$$\mathbf{u} = (\rho, v, T, \sigma, q) \tag{2.22}$$

and we linearize in $\bar{\mathbf{u}} = \mathbf{u} - \tilde{\mathbf{u}}$ about

$$\tilde{\mathbf{u}} = (\tilde\rho, 0, \tilde T, 0, 0). \tag{2.23}$$

The linearized system thus reads

$$\begin{bmatrix} \frac{\partial \bar\rho}{\partial t} \\ \frac{\partial \bar v}{\partial t} \\ \frac{\partial \bar T}{\partial t} \\ \frac{\partial \sigma}{\partial t} \\ \frac{\partial q}{\partial t} \end{bmatrix} + \begin{bmatrix} 0 & \tilde\rho & 0 & 0 & 0 \\ \frac{3\tilde c^2}{5\tilde\rho} & 0 & \frac{3\tilde c^2}{5\tilde T} & -\frac{1}{\tilde\rho} & 0 \\ 0 & \frac{2}{3}\tilde T & 0 & 0 & \frac{10}{9}\frac{\tilde T}{\tilde c^2\tilde\rho} \\ 0 & -\frac{4}{5}\tilde c^2\tilde\rho & 0 & 0 & -\frac{8}{15} \\ 0 & 0 & \frac{9}{10}\frac{\tilde c^4\tilde\rho}{\tilde T} & -\frac{3\tilde c^2}{5} & 0 \end{bmatrix} \cdot \begin{bmatrix} \frac{\partial \bar\rho}{\partial x} \\ \frac{\partial \bar v}{\partial x} \\ \frac{\partial \bar T}{\partial x} \\ \frac{\partial \sigma}{\partial x} \\ \frac{\partial q}{\partial x} \end{bmatrix} = \begin{bmatrix} 0 \\ 0 \\ 0 \\ -\frac{3}{2\tau_q}\sigma \\ -\frac{1}{\tau_q}q \end{bmatrix}. \tag{2.24}$$

Comparison with (2.4) identifies the matrices $\tilde{\mathbf{A}}$ and $\tilde{\mathbf{B}}$, and we may thus calculate the eigenvalues of $\tilde{\mathbf{A}}$, i.e. the characteristic speeds in equilibrium. They result as

$$\lambda_1 \simeq -1.6503\,\tilde c, \quad \lambda_2 \simeq -0.6297\,\tilde c, \quad \lambda_3 = 0, \quad \lambda_4 \simeq 0.6297\,\tilde c, \quad \lambda_5 \simeq 1.6503\,\tilde c. \tag{2.25}$$

The dispersion relation (2.7) has an explicit form in this case. We obtain with $\Omega = \tilde\tau_q\omega$ as a dimensionless frequency

$$9\Omega(z\tilde c)^4(5i - 6\Omega) + 3(z\tilde c)^2(52\Omega^2 - 80i\Omega - 25) - 50\Omega^2 + 125i\Omega + 75 = 0. \tag{2.26}$$

From this equation we may calculate $z = k/\omega$ explicitly as a function of ω and hence v_{ph} and α. The Figures 8.1 and 8.2 show the results for the "mode" with $\lim_{\omega\to\infty} z = 1/\lambda_5$, i.e. the mode of the fastest wave; see the continuous lines in the figures.

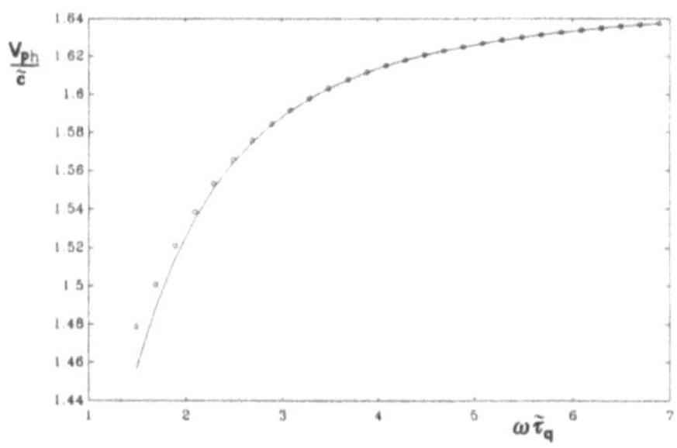

Figure 8.1 Phase speed as a function of ω.

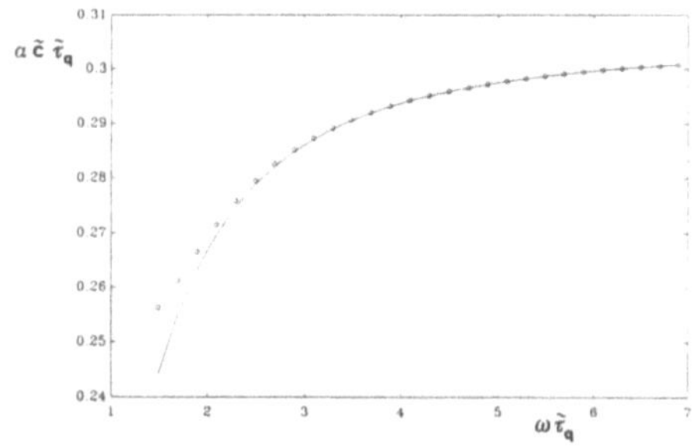

Figure 8.2 Attenuation as a function of ω.

Rather than solve (2.26) explicitly, we may use the expansion (2.10). We have calculated the first four terms for the fastest mode and obtained

$$z\tilde{c} \sim 0.6060 - \frac{0.3046i}{\Omega} + \frac{0.2295}{\Omega^2} + \frac{0.1822i}{\Omega^3} \tag{2.27}$$

whence follows by (2.8)

$$v_{ph} \simeq \left(1.6503 - \frac{0.6251}{\Omega^2}\right)\tilde{c} \quad \text{and} \quad \alpha \simeq \left(0.3046 - \frac{0.1822}{\Omega^2}\right)\frac{1}{\tilde{\tau}_q \tilde{c}}. \tag{2.28}$$

From (2.16) and (2.28) we obtain

$$\lim_{\omega \to \infty} \alpha(\omega)\lambda \simeq \frac{0.5027}{\tilde{\tau}_q}, \tag{2.29}$$

so that extended thermodynamics satisfies the condition of linear stability.

The Figures 8.1 and 8.2 show as dotted lines the phase speed and attenuation calculated from an expansion like (2.28), but with one more term. Thus we are able to get an idea about the rapid convergence of the series expansion.

3 Hyperbolicity and Nonlinear Waves in Extended Thermodynamics

3.1 The Characteristic Polynomial

We refer to the system (2.20), which is the nonlinear system of field equations of extended thermodynamics. This system is not quite in the form (2.2) because of the occurrence of the dotted quantities, like $\dot{\rho} = \frac{\partial \rho}{\partial t} + v\frac{\partial \rho}{\partial x}$. Therefore the matrix $\mathbf{A}$ in (2.2) and the matrix $\overline{\mathbf{A}}$ in (2.20) are related by the equation

$$\overline{\mathbf{A}} = \mathbf{A} - v\mathbf{1}. \tag{3.1}$$

This matrix has the same eigenvectors as $\mathbf{A}$, and the eigenvalues are related by

$$\overline{\lambda} = \lambda - v, \tag{3.2}$$

where λ are the characteristic speeds.

The characteristic polynomial from which the speeds may be calculated has the form

$$\hat{\lambda}\left\{\hat{\lambda}^4 + \hat{\lambda}^2\left(\frac{62}{15}\hat{\sigma} - \frac{78}{25}\right) - \frac{96}{25}\hat{q}\hat{\lambda} + \frac{21}{5}\hat{\sigma}^2 - \frac{18}{5}\hat{\sigma} + \frac{27}{25}\right\} = 0, \tag{3.3}$$

where $\hat{\lambda}$ is defined as $\overline{\lambda}/\tilde{c}$ with $\tilde{c} = \sqrt{\frac{5}{3}\frac{k}{m}\tilde{T}}$.

From (3.3) we conclude that—apart from the trivial solution $\hat{\lambda} = 0$—we have four waves with speeds $\hat{\lambda}$ that are roots of the fourth-order algebraic equation

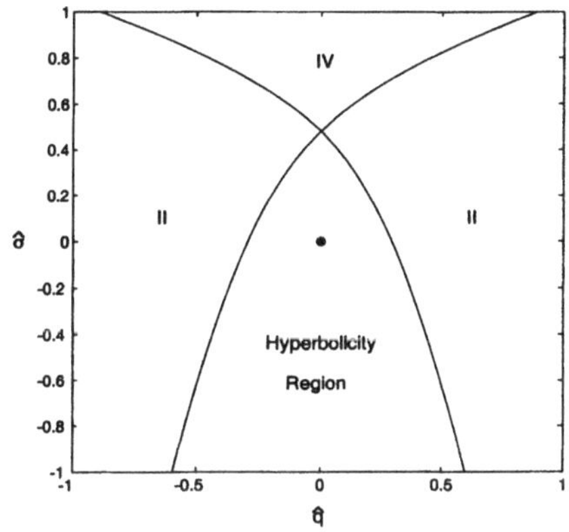

Figure 8.3: The region of hyperbolicity.

$$\hat{\lambda}^4 + \hat{\lambda}^2 \left(\frac{62}{15}\hat{\sigma} - \frac{78}{25} \right) - \frac{96}{25}\hat{q}\hat{\lambda} + \frac{21}{5}\hat{\sigma}^2 - \frac{18}{5}\hat{\sigma} + \frac{27}{25} = 0. \tag{3.4}$$

We see that the speeds depend on the nonequilibrium state characterized by the values $\hat{\sigma}$ and $\hat{q}$. In equilibrium the equation (3.4) degenerates into a biquadratic equation, viz.

$$\hat{\lambda}^4 - \frac{78}{25}\hat{\lambda}^2 + \frac{27}{25} = 0 \tag{3.5}$$

and, according to the results of Paragraph 2.2, the corresponding speeds $\hat{\lambda}$ coincide with the phase speeds in the high-frequency limit, viz., by (2.25)

$$\hat{\lambda} \simeq \pm 1.6503 \quad \text{and} \quad \hat{\lambda} \simeq \pm 0.6297. \tag{3.6}$$

3.2 Region of Hyperbolicity

We see from (3.6) that in equilibrium the characteristic speeds of extended thermodynamics are real and therefore the system of equations is hyperbolic. Actually we should like extended thermodynamics to be hyperbolic irrespective of how far from equilibrium the state is and, in order to achieve this, we have required "convexity"; e.g. see Paragraph 3.1.2. However, we were forced to make approximations in order to obtain specific results; thus the equations (4.4.3), hence (2.20), are derived from *linear* constitutive relations. It may therefore be suspected that the system (2.20) is hyperbolic only in the neighbourhood of equilibrium, i.e. in the range where the constitutive equations are valid. We proceed to investigate that suspicion. In fact, we shall find it confirmed, and we shall calculate the *region of hyperbolicity* in the $(\hat{\sigma}, \hat{q})$-plane.

For $\hat{\sigma} = 0, \hat{q} = 0$ the system is hyperbolic as we have seen, since the four roots of the characteristic equation (3.4) are real; see (3.6). This means that the function on the left-hand side of (3.4) has four points of intersection with the $\hat{\lambda}$-axis for $\hat{\sigma} = 0, \hat{q} = 0$. That function depends on the values of $\hat{\sigma}$ and $\hat{q}$, and hyperbolicity is lost if two of the four points of intersection merge into one. The mathematical condition for this to happen is that the characteristic equation has a double root μ_1 and two separate roots μ_2, μ_3, i.e. that the equation has the form

$$(\hat{\lambda} - \mu_1)^2(\hat{\lambda} - \mu_2)(\hat{\lambda} - \mu_3) = 0. \tag{3.7}$$

Comparison of this expression with (3.4) gives four conditions by equating the coefficients of $\hat{\lambda}$, viz.

$$\mu_1 - \frac{1}{2}(\mu_2 + \mu_3) = 0 \qquad (3.8)_1$$

$$\frac{62}{15}\hat{\sigma} - \frac{78}{25} + \frac{1}{2}(\mu_2 + \mu_3)^2 + \frac{1}{4}(\mu_2 - \mu_3)^2 = 0 \qquad (3.8)_2$$

$$\frac{21}{5}\hat{\sigma}^2 - \frac{18}{5}\hat{\sigma} + \frac{27}{25} - \frac{1}{4}(\mu_2 + \mu_3)^2\mu_2\mu_3 = 0 \qquad (3.8)_3$$

$$\frac{96}{25}\hat{q} = \pm\frac{1}{4}(\mu_2 + \mu_3)(\mu_2 - \mu_3)^2. \qquad (3.8)_4$$

From $(3.8)_{1,2,3}$ we obtain with some algebra

$$\begin{aligned} \mu_1 &= -\frac{1}{15}\sqrt{117 - 155\hat{\sigma} + \sqrt{2}\Delta} \\ \mu_{2,3} &= \frac{1}{15}\sqrt{117 - 155\hat{\sigma} + \sqrt{2}\Delta} \pm \frac{2}{15}\sqrt{117 - 155\hat{\sigma} - \frac{1}{\sqrt{2}}\Delta} \end{aligned} \tag{3.9}$$

where Δ stands for

$$\Delta = \sqrt{47450\hat{\sigma}^2 - 48510\hat{\sigma} + 15957}. \tag{3.10}$$

If (3.9), (3.10) are inserted into $(3.8)_4$ we obtain $\hat{q}$ as a function of $\hat{\sigma}$. That function—or rather $\hat{\sigma}$ as a function of $\hat{q}$—is shown in Figure 8.3. The area around $\hat{\sigma} = 0, \hat{q} = 0$ which is characterized as the hyperbolicity region, contains the pairs $(\hat{\sigma}, \hat{q})$ for which the characteristic speeds are real and, in fact, distinct.

There are two more regions that can be identified in Figure 8.3. The regions denoted by II correspond to the case in which μ_2 and μ_3 are real, while in region IV all eigenvalues are complex. Needless to say, the system is nonhyperbolic in regions II and IV.

On the boundary of the hyperbolicity region two roots are real and distinct while two others are real and coincident. Thus on that boundary the system, while hyperbolic, is not strictly hyperbolic.

We have constructed the equation of the maximum circle centered in the origin of the $(\hat{\sigma}, \hat{q})$-plane and contained in the region of hyperbolicity; see Figure 8.4. The radius of that circle is given by

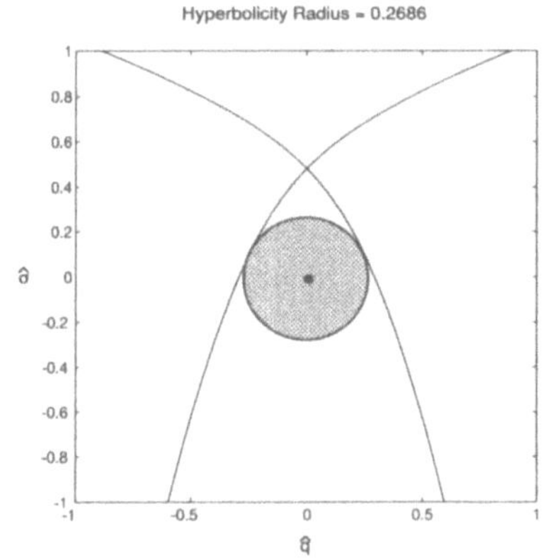

Figure 8.4: On the validity of extended thermodynamics for nonequilibrium processes.

$$r = \sqrt{\hat{q}^2 + \hat{\sigma}^2} = \frac{1}{c^2 \rho}\sqrt{\frac{q^2}{c^2} + \sigma^2} \simeq 0.2686, \tag{3.11}$$

it gives an estimate of the range of validity of extended thermodynamics concerning hyperbolicity of the linearized theory.

4 Acceleration Waves

4.1 Amplitude of Discontinuity Waves

We continue with the system (1.1) or (2.1), but we do not linearize now. Nor do we consider harmonic waves. Rather we consider a moving surface $\varphi(\mathbf{x}, t) = 0$—the wave front—with unit normal $\mathbf{n}$ which separates the unperturbed field $\tilde{\mathbf{u}}$ in front from the perturbed field behind. The field itself is continuous, but its first derivatives have jumps. We may thus write

$$[\![\mathbf{u}]\!] = 0, \qquad \left[\!\left[\frac{\partial \mathbf{u}}{\partial x^a}\right]\!\right] n_a = \mathbf{\Pi} \neq 0, \tag{4.1}$$

where $\mathbf{\Pi}$ is called the *jump vector*. For the normal speed of the wave front we have

$$V = -\frac{\frac{\partial \varphi}{\partial t}}{|\operatorname{grad}\varphi|}. \tag{4.2}$$

For the investigation of the motion and the evolution of the wave front we assume that the unperturbed field is an equilibrium state with $\mathbf{\Pi}(\mathbf{u}) = 0$. The more general case can be treated and has been treated by Boillat and Ruggeri; see [84], [85] and [86].

The calculation proceeds in two steps: identification of speed and amplitude and determination of the temporal behaviour of the amplitude.

First of all we subtract the equations (2.1) on the two sides of the front and obtain with

$$\left[\left[\frac{\partial \mathbf{u}}{\partial t}\right]\right] = -V\,\mathbf{\Pi}, \quad \left[\left[\frac{\partial \mathbf{u}}{\partial x^a}\right]\right] n_a = \mathbf{\Pi}: \quad (\mathbf{A}^a n_a - V\mathbf{1})\,\mathbf{\Pi} = 0. \tag{4.3}$$

Thus the possible speeds of the wave are given by the eigenvalues of $\mathbf{A}^a n_a$ and the jump vector is proportional to the right eigenvector of that matrix. We have

$$V = \tilde{\lambda} \equiv \lambda(\tilde{\mathbf{u}}) \qquad \text{and} \qquad \mathbf{\Pi} = \Pi\tilde{\mathbf{d}} \equiv \Pi\mathbf{d}(\tilde{\mathbf{u}}). \tag{4.4}$$

Π is called the ***amplitude of the wave.***

Secondly we calculate the evolution of the amplitude in time and, in doing this, we restrict the attention to the one-dimensional case where the wave moves in the x-direction.

We apply a left eigenvector $\mathbf{l}$ to the system (2.1) and obtain with $\mathbf{u}_t = \frac{\partial \mathbf{u}}{\partial t}$, $\mathbf{u}_x = \frac{\partial \mathbf{u}}{\partial x}$

$$\mathbf{l}\cdot(\mathbf{u}_t + \lambda \mathbf{u}_x) = \mathbf{l}\cdot\mathbf{\Pi}.$$

This equation is differentiated with respect to x to give

$$\frac{\partial \mathbf{l}}{\partial \mathbf{u}}\cdot \mathbf{u}_x\,(\mathbf{u}_t + \lambda \mathbf{u}_x) + \mathbf{l}\cdot\left\{\mathbf{u}_{tx} + \lambda \mathbf{u}_{xx} + \mathbf{u}_x \frac{\partial \lambda}{\partial \mathbf{u}}\cdot \mathbf{u}_x\right\} - \frac{\partial(\mathbf{l}\cdot\mathbf{\Pi})}{\partial \mathbf{u}}\cdot \mathbf{u}_x = 0.$$

We form the difference of this expression on the two sides of the front. The first term drops out, because, by $(4.3)_{1,2}$, we have

$$[\![\mathbf{u}_t + \lambda \mathbf{u}_x]\!] = 0, \quad \text{and} \quad \mathbf{u}_t = 0,\ \mathbf{u}_x = 0 \quad \text{before the front.}$$

What remains may be written as $\left(\text{with } \frac{d}{dt} = \frac{\partial}{\partial t} + \lambda\frac{\partial}{\partial x}\right)$

$$\mathbf{l}\cdot\frac{d\,[\![\mathbf{u}_x]\!]}{dt} + \mathbf{l}\cdot[\![\mathbf{u}_x]\!]\,\frac{\partial \lambda}{\partial \mathbf{u}}\cdot[\![\mathbf{u}_x]\!] - \frac{\partial(\mathbf{l}\cdot\mathbf{\Pi})}{\partial \mathbf{u}}\cdot[\![\mathbf{u}_x]\!] = 0$$

or, by $(4.3)_1$ and $(4.4)_2$ and with $\mathbf{f}(\mathbf{u}) = 0$ before the front

$$\frac{d\Pi}{dt} + \frac{\partial \lambda}{\partial \mathbf{u}}\cdot \mathbf{d}\,\Pi^2 - \mathbf{l}\cdot\frac{\partial \mathbf{\Pi}}{\partial \mathbf{u}}\cdot \mathbf{d}\,\Pi = 0. \tag{4.5}$$

Equation (4.5) is a Bernoulli equation for the amplitude Π and we write it as

$$\frac{d\Pi}{dt} + \alpha\Pi^2 + \beta\Pi = 0 \qquad \text{with} \qquad \alpha = \frac{\partial \lambda}{\partial \mathbf{u}}\cdot \mathbf{d}, \quad \beta = -\mathbf{l}\cdot\frac{\partial \mathbf{\Pi}}{\partial \mathbf{u}}\cdot \mathbf{d}. \tag{4.6}$$

The coefficients α and β are constants in the present simple case where α and β must be evaluated for the constant state $\tilde{\mathbf{u}}$ in front of the wave. Therefore the Bernoulli equation can be solved to give

$$\Pi(t) = \frac{\Pi(0)e^{-\beta t}}{1 - \Pi(0)\frac{\alpha}{\beta}\left(e^{-\beta t} - 1\right)}. \tag{4.7}$$

In the general case that the wave propagates into a region where the unperturbed field is not constant, α and β depend on time; see Ruggeri [86]. In that case the Bernoulli equation cannot be solved so easily. An instructive example is given by Ruggeri [86] and Muracchini & Seccia [87] for wave propagation in the gravitational atmosphere. Applications to continuum mechanics are given by Chen [88].

4.2 Growth and Decay

It is obvious from (4.6) and (4.7) that for a semi-linear system—where $\mathbf{A}^A$ and hence $\lambda, \mathbf{l}$, and $\mathbf{d}$ are constants—we have $\alpha = 0$ and therefore the amplitude will either grow or decay exponentially, depending on the sign of β.

Even for a nonlinear system α may vanish, namely if $\frac{\partial \lambda}{\partial \mathbf{u}} \cdot \mathbf{d} = 0$ holds. If this is not the case, i.e. if

$$\frac{\partial \lambda}{\partial \mathbf{u}} \cdot \mathbf{d} \neq 0 \tag{4.8}$$

for all eigenvalues we say that the system is *genuinely nonlinear*. We shall proceed on the assumption that (4.8) holds. Thus α is not zero and—without loss of generality—we assume $\alpha > 0$.

The equation (4.7) lends itself for a characterization of nonlinear stability. Stability demands that $\Pi(t)$ is bounded for all times, and inspection of (4.7) shows that this requires

$$1 + \Pi(0)\frac{\alpha}{\beta} > 0 \quad \text{and} \quad \beta > 0, \quad \text{hence} \quad \Pi(0) > -\frac{\beta}{\alpha}. \tag{4.9}$$

By $(4.6)_2$ and since $\mathbf{\Pi}(\mathbf{u}) = 0$ in front of the wave, the condition $(4.9)_2$ provides a restriction on the material. It may be written in the form

$$\beta = -\mathbf{l} \cdot \frac{\partial \mathbf{\Pi}}{\partial \mathbf{u}} \cdot \mathbf{d} > 0 \qquad \text{or, by (2.4)} \qquad \beta = -\mathbf{l} \cdot \mathbf{B} \cdot \mathbf{d} > 0 \tag{4.10}$$

which confirms the earlier remark—made in Paragraph 2.2—that the linear stability condition guarantees nonlinear stability as well. At least this is true, if the discontinuity wave travels into a region which is in equilibrium.

The condition $(4.9)_3$ provides a condition on the initial amplitude of the discontinuity. We may say that if the amplitude $\mid \Pi(0) \mid$ is too big, the attenuation does not succeed to dissipate the discontinuity and therefore stability is lost. We proceed to investigate that case.

The denominator of (4.7) may become zero at some critical time t_{cr} between 0 and ∞, viz.

$$t_{cr} = \frac{1}{\beta} \ln \frac{1}{1 + \frac{\beta}{\alpha} \frac{1}{\Pi(0))}}. \tag{4.11}$$

Since we must have $0 < t_{cr} < \infty$ this can only happen for

$$\Pi(0) < \frac{\beta}{\alpha}. \tag{4.12}$$

The physical significance of this instability is clear: if the initial discontinuity in the derivatives is too strong, it cannot be damped; instead it grows to infinity and thus the acceleration wave develops into a shock wave.

4.3 Evolution of Amplitude in Extended Thermodynamics

We wish to calculate the amplitude $\Pi(t)$ from (4.7) for extended thermodynamics of 13 fields, the theory whose system of equations is given by (2.20), and we restrict attention to one space dimension. We do not linearize now, but—as before—we consider the gas before the wave to be in equilibrium and we shall assume that $\tilde{\mathbf{v}} = 0$ holds. We must calculate the coefficients α and β defined in (4.6), i.e. we need to calculate the eigenvalues λ and the right and left eigenvectors $\mathbf{d}$ and $\mathbf{l}$ of the matrix $\mathbf{A}$ in (2.24). Fortunately the eigenvectors $\mathbf{d}$ and $\mathbf{l}$ may be evaluated in the equilibrium state, i.e. they are eigenvalues of the matrix of the linearized system (2.24). An easy though cumbersome calculation provides

$$\mathbf{d} = \begin{bmatrix} 1 \\ c\frac{\hat\lambda}{\rho} \\ \frac{T}{27\rho}\left(25\hat\lambda^2 - 27\right) \\ -\frac{4c^2\hat\lambda^2}{9} \\ \frac{c^3\hat\lambda}{6}\left(5\hat\lambda^2 - 9\right) \end{bmatrix}, \quad \mathbf{l} = a\begin{bmatrix} \frac{1}{25}c^3\left(25\hat\lambda^2 - 33\right) \\ \frac{1}{15}c^2\hat\lambda\rho\left(25\hat\lambda^2 - 33\right) \\ \frac{1}{25T}c^3\rho\left(25\hat\lambda^2 + 12\right) \\ -\frac{1}{3}c\left(5\hat\lambda^2 - 3\right) \\ 2\hat\lambda \end{bmatrix}^T \quad \text{where} \quad a = \frac{5}{4c^3\left(13\hat\lambda^2 - 9\right)}. \tag{4.13}$$

a was calculated from the requirement that $\mathbf{l}\cdot\mathbf{d} = 1$ must hold. We have given the value 1 to the first component of $\mathbf{d}$. Therefore the amplitude of the discontinuity wave is

$$\Pi = \delta\rho \quad \text{and hence} \quad \delta\mathbf{u} = \delta\rho\,\mathbf{d} \tag{4.14}$$

where δ stands for $\left[\left[\frac{\partial}{\partial x}\right]\right]$, i.e. the jump of the normal derivative.

We calculate the coefficient α from $(4.6)_2$

$$\alpha = \frac{\partial\lambda}{\partial\mathbf{u}}\cdot\mathbf{d} = \frac{1}{\delta\rho}\frac{\partial\lambda}{\partial\mathbf{u}}\cdot\delta\mathbf{u} = \frac{\delta\lambda}{\delta\rho}, \tag{4.15}$$

and this must be evaluated in equilibrium. $\delta\lambda$ follows from the characteristic equation (3.4) by differentiation. We must take into account that

$$\lambda = c\hat\lambda + v \tag{4.16}$$

and obtain

$$\delta\lambda = \frac{c}{2T}\hat\lambda\delta T + c\delta\hat\lambda + \delta v \quad \text{or, with (3.4)} \tag{4.17}$$

$$\delta\lambda = \delta v + \frac{c}{2T}\hat\lambda\delta T + \frac{1}{c\rho}\frac{18 - \frac{62}{5}\hat\lambda^2}{20\hat\lambda^3 - \frac{156}{5}\hat\lambda}\delta\sigma + \frac{1}{c^2\rho}\frac{96}{100\hat\lambda^2 - 156}\delta q. \tag{4.18}$$

From (4.14) and $(4.13)_1$ we conclude

$$\delta v = c\frac{\hat\lambda}{\rho}\delta\rho, \; \delta T = \frac{T}{\rho}\left(\frac{25}{27}\hat\lambda^2 - 1\right)\delta\rho, \; \text{etc.} \tag{4.19}$$

and thus after a little calculation

$$\alpha = \frac{c}{\rho}\hat{\lambda}\frac{625\hat{\lambda}^4 + 1400\hat{\lambda}^2 - 3537}{54\left(25\hat{\lambda}^2 - 39\right)} \simeq 5.16061\frac{c}{\rho}, \tag{4.20}$$

where $(4.20)_2$ refers to the "fast mode" with $\hat{\lambda} = 1.6503$; see (3.6).

Next we calculate β from $(4.6)_3$. Inspection of (2.20) shows that we have in equilibrium

$$\frac{\partial \mathbf{\Pi}}{\partial \mathbf{u}} = \begin{bmatrix} 0 & 0 & 0 & 0 & 0 \\ 0 & 0 & 0 & 0 & 0 \\ 0 & 0 & 0 & 0 & 0 \\ 0 & 0 & 0 & -\frac{3}{2}\frac{1}{\tau_q} & 0 \\ 0 & 0 & 0 & 0 & -\frac{1}{\tau_q} \end{bmatrix}. \tag{4.21}$$

Multiplying this from left and right by $\mathbf{l}$ and $\mathbf{d}$ respectively we obtain from (4.13)

$$\beta = \frac{1}{\tau_q}\frac{5}{4}\frac{5\hat{\lambda}^3 - 3}{13\hat{\lambda}^2 - 9} \simeq 0.5027\frac{1}{\tau_q}. \tag{4.22}$$

Note that $\beta\tau_q = 0.5027$ agrees with the high frequency limit of the attenuation of the "fast" harmonic wave that was calculated in (2.29). This fact reflects the general result about the equivalence of the criteria for linear and nonlinear stability that we have commented on in Paragraph 4.2; see (4.10) and (2.17).

If α and β from (4.20) and (4.22) are introduced into the evolution equation (4.7) of the jump—here, by (4.14), the jump $\delta\rho$ of the density gradient—we obtain

$$\delta\rho(t) = \frac{\delta\rho(0)e^{-0.5027\tau}}{1 - \frac{\delta\rho(0)}{\rho}c\tau_q 10.2658\left(e^{-0.5027\tau} - 1\right)} \quad \text{where} \quad \tau = \frac{t}{\tau_q}. \tag{4.23}$$

We conclude that the discontinuity wave grows into a shock wave under the condition

$$\delta\rho(0) < -0.0974\frac{\rho}{c\tau_q}. \tag{4.24}$$

Introducing the jump in acceleration

$$\left[\!\left[\frac{\partial v}{\partial t}\right]\!\right] = -\lambda\delta v = -\lambda c\frac{\hat{\lambda}}{\rho}\delta\rho,$$

we obtain from (4.24)

$$\left[\!\left[\frac{\partial v}{\partial t}\right]\!\right]_{t=0} > 0.265\frac{c}{\tau_q}.$$

Under normal conditions this is an enormously big acceleration which we may be sure will never be reached. If it *were* reached, and surpassed by a factor 2 (say)—so that $\delta\rho(0)$ were equal to $-0.1948\frac{\rho}{c\tau_q}$—the acceleration wave would develop into a shock wave in the time

$$t_{cr} = 1.315\tau_q$$

which is a very short time indeed under normal conditions.

4.4 Acceleration Waves in Relativistic Extended Thermodynamics

Weak discontinuity waves or acceleration waves have been considered in a relativistic fluid by Kranys [89], Stewart [90], Israel & Stewart [91] and Boillat [92].The four-dimensional notation is particularly elegant in that case. Thus the wave front is given by $\varphi(x^A) = 0$ and the two relations

$$\frac{\partial \varphi}{\partial x^a} = |\text{grad}\varphi| n_a \qquad \text{and} \qquad \frac{\partial \varphi}{\partial t} = -|\text{grad}\varphi| V, \tag{4.25}$$

valid in the rest frame of the fluid, may be combined into the single equation

$$\frac{\varphi_{,A}}{\sqrt{h^{AB}\varphi_{,A}\varphi_{,B}}} = \left(n_A - \frac{1}{c^2} V U_A\right). \tag{4.26}$$

For the definition of U^A and h^{AB} see Chapter 6. In the rest frame of the fluid we have

$$h^{AB}\varphi_{,A}\varphi_{,B} = |\text{grad}\varphi|^2, \qquad n_A = (1,0,0,0), \qquad U_A = (0,0,0,1) \tag{4.27}$$

so that (4.26) represents a decomposition of $\varphi_{,A}$ into space-like and time-like components. Obviously, by (4.27), we have $n^A n_A = -1$, $n^A U_A = 0$ and $U^A U_A = c^2$. Therefore the speed V of the wave is given by

$$\frac{V^2}{c^2} = 1 + \frac{g^{AB}\varphi_{,A}\varphi_{,B}}{h^{AB}\varphi_{,A}\varphi_{,B}}, \tag{4.28}$$

and it is less than c, if we have

$$g^{AB}\varphi_{,A}\varphi_{,B} < 0. \tag{4.29}$$

For relativistic extended thermodynamics—the 14-moment theory of Chapter 6 with the field equations (6.1.2)—the amplitudes and speeds of acceleration waves follow from the homogeneous system of algebraic equations

$$\varphi_{,A}\,\delta A^A = 0, \qquad \varphi_{,A}\,\delta A^{AB} = 0, \qquad \varphi_{,A}\,\delta A^{ABC} = 0, \tag{4.30}$$

where $\delta\psi$ for a generic ψ is defined by $\left[\frac{\partial \psi}{\partial x^a}\right] n_a$, it represents the jump of the normal component of the gradient of ψ.

The evaluation of the system (4.30) is cumbersome and long; the system was fully investigated by Seccia & Strumia [93]. These authors found a transversal wave and two longitudinal ones and they evaluated their speeds for the nondegenerate gas. We refer the reader to the Seccia & Strumia paper for details and cite only the wave speeds in the nonrelativistic limit and the ultrarelativistic limit:

$$\begin{array}{llll} \frac{mc^2}{kT} \gg 1: & V_{trans} = \sqrt{\frac{7}{5}\frac{k}{m}T}, & V^1_{long} = \sqrt{1.35\frac{k}{m}T}, & V^2_{long} = \sqrt{5.18\frac{k}{m}T} \\ \frac{mc^2}{kT} \ll 1: & V_{trans} = \sqrt{\frac{1}{5}}c, & V^1_{long} = \sqrt{\frac{1}{3}}c, & V^2_{long} = \sqrt{\frac{3}{5}}c. \end{array} \tag{4.31}$$

5 Weak Solutions and Shock Waves

5.1 Weak Solutions

If the field $\mathbf{u}$ experiences a jump across the wave front we have a ***shock wave***. Shock waves are possible only for systems of balance laws, and they are a particular class of ***weak solutions***. We recall the definition of weak solutions for systems of balance laws of the type

$$\mathbf{F}^A{}_{,A} = \mathbf{\Pi} \quad \text{with} \quad \mathbf{F}^A = \mathbf{F}^A(\mathbf{u}), \quad \mathbf{\Pi} = \mathbf{\Pi}(\mathbf{u}). \tag{5.1}$$

This involves the test function $\Phi(x^A)$ having support in a four-dimensional domain C of space-time. We multiply (5.1) by Φ and integrate over C

$$\int_C \Phi\left(\mathbf{F}^A{}_{,A} - \mathbf{\Pi}\right) dC = 0 \quad \text{or} \quad \int_C \left(\Phi \mathbf{F}^A\right)_{,A} dC - \int_C \left(\mathbf{F}^A \Phi_{,A} + \mathbf{\Pi}\Phi\right) dC = 0.$$

If Σ is the three-dimensional surface of C with the normal n_A we obtain

$$\int_\Sigma n_A \Phi \mathbf{F}^A \, d\Sigma - \int_C \left(\mathbf{F}^A \Phi_{,A} + \mathbf{\Pi}\Phi\right) dC = 0. \tag{5.2}$$

The first integral vanishes since $\Phi = 0$ on Σ, and we have

$$\int_C \left(\mathbf{F}^A \Phi_{,A} + \mathbf{\Pi}\Phi\right) dC = 0. \tag{5.3}$$

A field $\mathbf{u}(x^A)$ is said to be a *weak solution* of (5.1) if it satisfies (5.3) for every test function Φ.

The advantage of weak solutions lies in the fact that—in contrast to classical solutions—no differentiability conditions are required. Obviously all classical solutions are also weak solutions, while the reverse is not true.

5.2 Rankine-Hugoniot Equations

We consider a shock front with the normal ν_A as separating two regular solutions $\mathbf{u}_\pm(x^B)$ on the right and left respectively, and we proceed to prove that the pair $(\mathbf{u}_+(x^B), \mathbf{u}_-(x^B))$ is a weak solution of (5.1) if and only if across the shock front it satisfies the *Rankine-Hugoniot equations*, viz.

$$[[\mathbf{F}^A(\mathbf{u})]]\, \nu_A = 0. \tag{5.4}$$

We refer to Figure 8.5 and write (5.2) for both subdomains with $\mathbf{F}^A_\pm = \mathbf{F}^A(\mathbf{u}_\pm)$

$$\int_{\Sigma^\pm \cup \sigma} \nu^\pm_A \Phi \mathbf{F}^A_\pm \, d\Sigma - \int_{C^\pm} \left(\mathbf{F}^A \Phi_{,A} + \mathbf{\Pi}\Phi\right) dC = 0.$$

As Φ vanishes on Σ, we may write

$$\int_\sigma \nu^\pm_A \Phi \mathbf{F}^A_\pm \, d\Sigma - \int_{C^\pm} \left(\mathbf{F}^A \Phi_{,A} + \mathbf{\Pi}\Phi\right) dC = 0. \tag{5.5}$$

We add those two equations to obtain with $\nu_A = \nu_A^+ = -\nu_A^-$

$$\int_\sigma \nu_A \left(\mathbf{F}_+^A - \mathbf{F}_-^A\right) \Phi \, d\sigma - \int_C \left(\mathbf{F}^A \Phi_{,A} + \mathbf{\Pi}\Phi\right) dC = 0. \tag{5.6}$$

The second integral vanishes, since $\mathbf{u}$ is a weak solution in C, and we conclude the proof of (5.4), by writing

$$\nu_A \left(\mathbf{F}_+^A - \mathbf{F}_-^A\right) = 0,$$

which follows from (5.6) since that relation holds for all domains C.

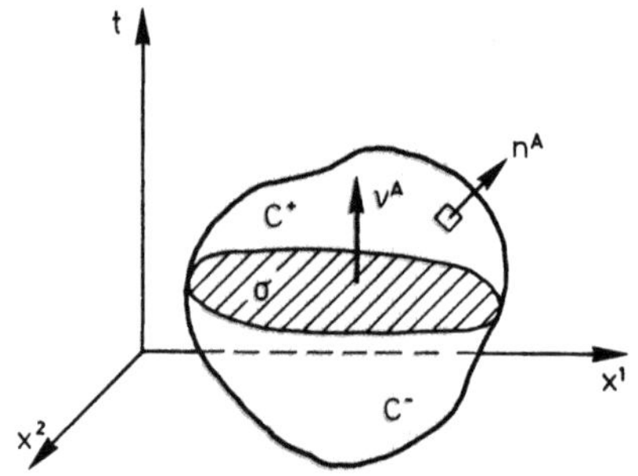

Figure 8.5 On the derivation of the Rankine-Hugoniot equations.

Therefore the Rankine-Hugoniot equations constitute the compatibility conditions for the existence of shocks in systems of balance laws of the type (5.1).

We now introduce more familiar notation. Let $\varphi(x^A)$ be the shock front, s its normal velocity, and $\mathbf{n}$ its unity normal vector in space, so that we have

$$\nu^0 = -s \mid \text{grad}\varphi \mid, \quad \nu^a = n^a \mid \text{grad}\varphi \mid. \tag{5.7}$$

Thus the Rankine-Hugoniot equations read

$$-s\,[\![\mathbf{F}^o]\!] + [\![\mathbf{F}^a]\!]\, n_a = 0. \tag{5.8}$$

If the unperturbed field $\mathbf{u}_+$ in front of the shock is known and we consider plane shocks with $\mathbf{n} = \text{const.}$, the Rankine-Hugoniot equations furnish a system of n equations for the $n+1$ unknowns $\mathbf{u}_-$ and s. Thus *one* among the $(n+1)$-tuple $(\mathbf{u}_-, s)$ may be chosen as the *shock parameter*—i.e. the quantity that characterizes the *strength* of the shock—and that one quantity may be the speed s, or any component of $\mathbf{u}_-$, or a combination of these; generically we denote the shock parameter by μ.

In a schematic way we may represent the values $\mathbf{u}_-$ for given values $\mathbf{u}_+$ as functions of s, as in Figure 8.6. That figure also indicates that *shock solutions may be seen as bifurcating from the trivial solution* $\mathbf{u}_+ = \mathbf{u}_-$ of (5.8) at the points where s is equal to one of the characteristic speeds λ. We proceed to prove this observation.

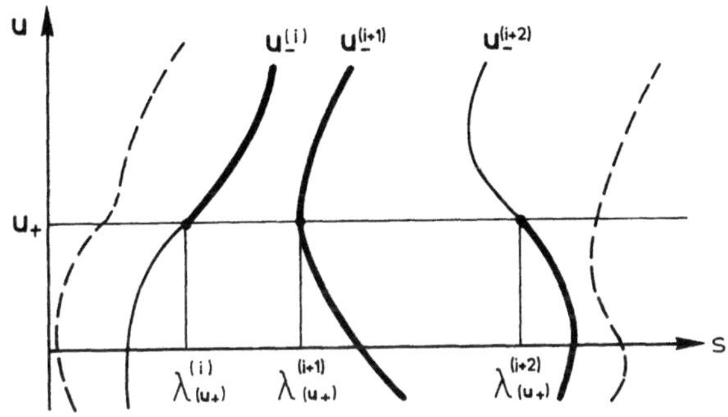

Figure 8.6 Shocks as bifurcating branches of the trivial solution with s as shock parameter.

We introduce the mapping

$$\mathbf{\Psi}_s(\mathbf{u}) = -s\mathbf{F}^o(\mathbf{u}) + \mathbf{F}^i(\mathbf{u})n_i\,,$$

and observe that, by the Rankine-Hugoniot equations, $\Psi = \Psi(\mathbf{u})$ must be locally noninvertible. In fact, by (5.8) we have $\mathbf{\Psi}_s(\mathbf{u}_-) = \mathbf{\Psi}_s(\mathbf{u}_+)$ and therefore $\det\left(\frac{\partial \Psi_s}{\partial \mathbf{u}}\right) = 0$ must hold at $\mathbf{u}_- = \mathbf{u}_+$. On the other hand, the Jacobian of the transformation $\Psi_s = \Psi_s(\mathbf{u})$, viz. $\left(\text{with } \mathbf{A}^A = \frac{\partial \mathbf{F}^A}{\partial \mathbf{u}}\right)$

$$\det\left(\frac{\partial \mathbf{\Psi}_s}{\partial \mathbf{u}}\right) = \det\left(\mathbf{A}^i n_i - s\mathbf{A}^o\right)$$

becomes singular for $s = \lambda(\mathbf{u})$; see (1.3). Therefore the local noninvertibility occurs when s is equal to the characteristic speeds.

5.3 Shocks in Extended Thermodynamics

The equations of extended thermodynamics are of balance type as put in evidence by the system (4.1.2). Therefore we may write Rankine-Hugoniot equations. We do this after using (4.1.6), (4.2.45), and $F(z) = \frac{k}{m}z$ in the system (4.1.2). Also we simplify that system by investigating the one-dimensional case with

$$x = x_1, \quad v = v_1, \quad q = q_1, \quad \sigma = t_{<11>}$$

and by introducing the relative shock speed

$$u = v - s. \tag{5.9}$$

A little calculation provides the following explicit form of the Rankine-Hugoniot equations

$$
\begin{aligned}
&[\![\rho u]\!] = 0 \\
&[\![\rho u^2 + p - \sigma]\!] = 0 \\
&[\![2q + 5pu - 2\sigma u + \rho u^3]\!] = 0 \\
&\left[\!\!\left[\frac{8}{15}q + \frac{4}{3}pu - \frac{7}{3}\sigma u + \frac{2}{3}\rho u^3\right]\!\!\right] = 0 \\
&\left[\!\!\left[\rho u^4 + 5\frac{p^2}{\rho} - 7\frac{p\sigma}{\rho} + \frac{32}{5}uq + u^2(8p - 5\sigma)\right]\!\!\right] = 0.
\end{aligned}
\tag{5.10}
$$

Before the shock we assume an equilibrium state; the shock disturbs this equilibrium and we have

$$
\begin{aligned}
\mathbf{u}_+ &= (\rho_+, 0, T_+, 0, 0), \qquad \mathbf{u}_- = (\rho_-, v_-, T_-, \sigma_-, q_-), \quad \text{or} \\
\mathbf{u}_+ &= (\rho_+, 0, p_+, 0, 0), \qquad \mathbf{u}_- = (\rho_-, v_-, p_-, \sigma_-, q_-).
\end{aligned}
\tag{5.11}
$$

For a simplified notation we introduce dimensionless fields behind the shock, viz.

$$
r = \frac{\rho_-}{\rho_+}, \quad w = \frac{u_-}{u_+}, \quad \pi = \frac{p_-}{p_+}, \quad \tau = \frac{\sigma_-}{p_+}, \quad \chi = \frac{q_-}{p_+ c_+}
\tag{5.12}
$$

where $c_+ = \sqrt{\frac{5}{3}\frac{k}{m}T_+} = \sqrt{\frac{5}{3}\frac{p_+}{\rho_+}}$. Also we define a dimensionless speed of the shock, the Mach number

$$
M_+ = \frac{s}{c_+}.
\tag{5.13}
$$

We choose M_+, i.e. s as the shock parameter and obtain—after an easy though cumbersome calculation (see Ruggeri [94])

$$
\begin{aligned}
r &= \frac{1}{w} \\
\pi &= 1 - \frac{5}{27}M_+^2\frac{(w-1)(7w-2)}{w} \\
\tau &= \frac{10}{27}M_+^2\frac{w^2-1}{w} \\
\chi &= -\frac{5}{18}M_+(w-1)\left(10M_+^2 w - 5M_+^2 - 9\right)
\end{aligned}
\tag{5.14}
$$

where $w = w(M_+)$ has to be inserted. This function results from solving the cubic equation

$$
3470M_+^4 w^3 - \left(3550M_+^2 + 4212\right)M_+^2 w^2 + \left(715M_+^4 + 2106M_+^2 + 729\right)w + 40M_+^4 = 0.
\tag{5.15}
$$

By (5.9) and $p = \frac{k}{m}\rho T$ it is then also possible to calculate

$$
V = \frac{v_-}{c_+} = M_+(1-w) \qquad \text{and} \qquad \Theta = \frac{T_-}{T_+} = \pi w.
\tag{5.16}
$$

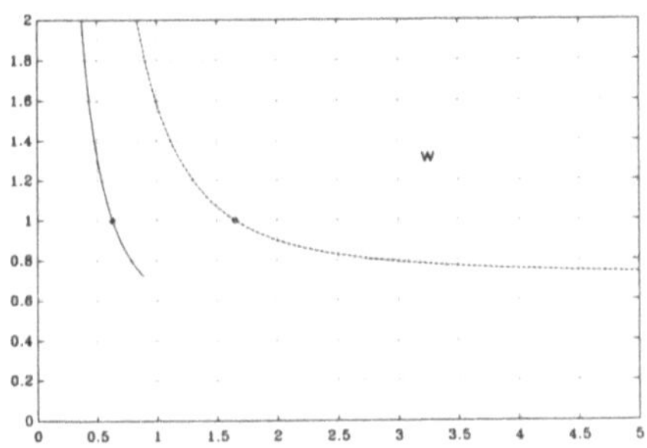

Figure 8.7 The ratio of specific volumes as a function of M_+.

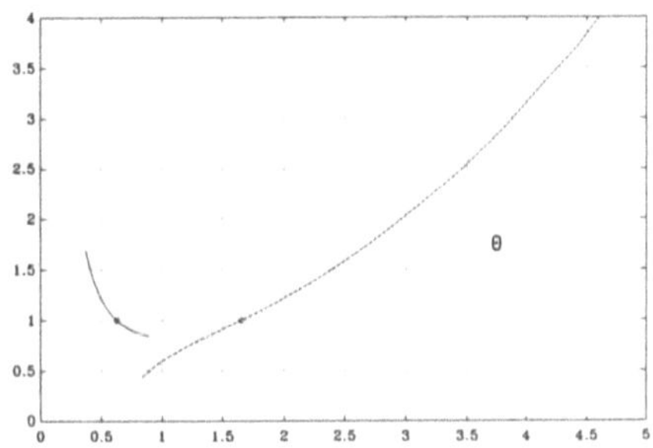

Figure 8.8 The ratio of temperatures as a function of M_+.

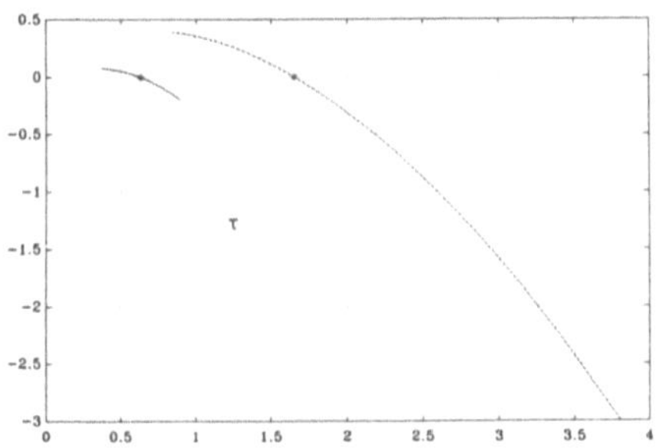

Figure 8.9 The stress behind the shock as a function of M_+.

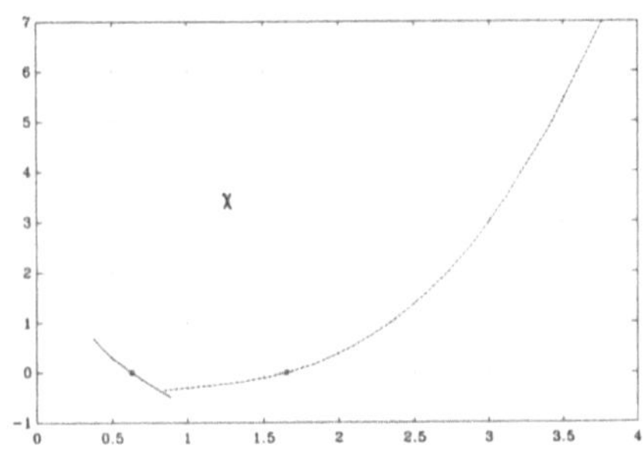

Figure 8.10 The heat flux behind the shock as a function of M_+.

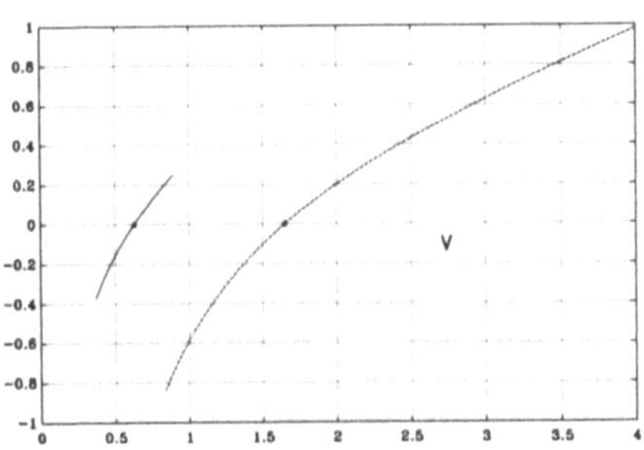

Figure 8.11 The velocity behind the shock as a function of M_+.

We proceed to discuss some properties of the equations (5.14) through (5.16) which may be read off by inspection, such as that $w = 1$ corresponds to the continuous solution $\mathbf{u}_- = \mathbf{u}_+$, the null-shock.

In accordance with the general theory presented in Paragraph 5.2 the shocks are bifurcated branches of the continuous solution occurring when s approaches the characteristic speed of the system. And indeed, for $w = 1$ (5.15) reduces to the equation

$$M_+^4 - \frac{78}{25}M_+^2 + \frac{27}{25} = 0,$$

which is identical to (3.5). From that equation we have calculated the characteristic speeds (3.6), and now we conclude that we have two families of shocks propagating to the right; they bifurcate from the null-shock when M_+ approaches, cf. (3.6)

$$M_+^{(S)} = 0.6297 \qquad \text{and} \qquad M_+^{(F)} = 1.6503. \tag{5.17}$$

These are called slow shocks and fast shocks respectively.

Another feature of the shocks—easily appreciated from $(5.14)_{3,4}$—is that behind the shock we always have a nonequilibrium state, because τ and χ are unequal to zero except in the null-shock, where $w = 1$. This result must be contrasted with the classical one of the Navier-Stokes-Fourier theory. In that theory we have *shock structure* solutions in which two states—the ones before and behind the shock—are always in equilibrium and they are connected by a continuous but very steep nonequilibrium process.[3]

There exists an upper bound for the ratio r of densities $\frac{\rho_-}{\rho_+}$. This can be read off from (5.15) which may be written as a biquadratic equation for M_+, viz.

$$5M_+^4\left(694w^3 - 710w^2 + 143w + 8\right) - 2106M_+^2 w(2w-1) + 729w = 0. \tag{5.18}$$

According to this equation M_+ is a monotonically decreasing function of w i.e., by $(5.14)_2$, an increasing function of r. Thus the biggest r corresponds to $M_+ \to \infty$ and then equation (5.18) can only be satisfied if the coefficient of M_+^4, viz. $694w^3 - 710w^2 + 143w + 8$ is zero. This happens when $w = 0.7099$, i.e. for

$$r_{\max} = \left.\frac{\rho_-}{\rho_+}\right|_{\max} = 1.4086. \tag{5.19}$$

We note that the corresponding result is $r_{\max} = 4$ if the monatomic gas is considered as an Eulerian fluid; e.g. see Landau & Lifshits [95]. So here extended thermodynamics is in a marked contrast to the classical theory.

We calculate the two positive solutions $M_+ = M_+(w)$ of (5.18) corresponding to the slow and fast shocks (see (5.17)). These solutions are introduced into the equations (5.14), (5.16) and thus provide

$$w = \frac{\rho_+}{\rho_-}, \quad \Theta = \frac{T_-}{T_+}, \quad \tau = \frac{\sigma_-}{p_+}, \quad \chi = \frac{q_-}{p_+c_+}, \quad V = \frac{v_-}{c_+} \tag{5.20}$$

as functions of s, or M_+. These functions have been numerically calculated and they are represented in Figure 8.7 through Figure 8.11. The dashed lines correspond to the fast shock, while

[3]In fact experiments show that proper shocks do not exist: There is always a shock structure, and in Chapter 12 we shall discuss what extended thermodynamics has to say about this.

the solid ones correspond to the slow shock. All curves for the slow shock have been broken off at $M_+ = 0.908$; this is the value where w reaches the value 0.7099, i.e. where the fast shock creates the maximum jump in density.

Not all parts of these plots are physically relevant, because we have selections rules which imply that some parts are unphysical, although they are perfectly valid solutions of the Rankine-Hugoniot equations. We proceed to discuss the selection criteria, of which there are two essentially equivalent ones. Afterwards we return to the curves of Figures 8.7 through 8.11 and discuss their admissible parts.

5.4 Selection Rules for Physical Shocks. The Entropy Growth Condition.

In analogy to the derivation of the Rankine-Hugoniot conditions (5.8) from the equations of balance, we may derive a jump condition for entropy. In this argument the entropy flux h^A replaces $\mathbf{F}^A$ and the entropy production Σ replaces $\mathbf{\Pi}$. But there is one important difference: where $\mathbf{\Pi}$ was supposed nonsingular in the domain $\mathcal{C}$, including the shock front σ, we cannot make that assumption for the entropy production. In fact, we must allow Σ to be singular on σ. Therefore the passage from (5.5) to (5.6) is impossible and we obtain—instead of (5.4)—

$$[\![h^A(\mathbf{u})]\!]\,\nu_A = \Sigma_S, \tag{5.21}$$

where Σ_S is the entropy production density on the shock front. Once the constitutive relation $h^A(\mathbf{u})$ is known, we may identify the shock parameter μ and introduce $\mathbf{u}_-(\mu)$ and $s(\mu)$ into (5.21). Thus we calculate $\Sigma_S(\mu)$. This quantity will generally be nonnegative for some range of the shock parameter and negative elsewhere. The *entropy growth condition* requires (see Lax [96] and Boillat [97])

$$\Sigma_S(\mu) \geq 0, \tag{5.22}$$

meaning that shocks are not admissible for those values of μ that violate (5.22).

5.5 Selection Rules for Physical Shocks. The Lax Conditions.

Another selection rule for physical shocks is the Lax condition; see Lax [98]. This conditon states that the speed of the shock lies between the characteristic speeds in front of the shock and behind it

$$\lambda(\mathbf{u}_+) < s < \lambda(\mathbf{u}_-). \tag{5.23}$$

Therefore in Figure 8.6 the admissible shocks are those lying on the branches that are emphasized by a thick line. Mostly this means that we can only have

$$\text{either} \qquad \mu > \mu_o \qquad \text{or} \qquad \mu < \mu_o \tag{5.24}$$

where μ_o is the value of the shock parameter for which $\mathbf{u}_+(\mu_o) = \mathbf{u}_-$. In exceptional cases—when the line $\mathbf{u}_-(s)$ intersects the values $\mathbf{u}_+$ vertically—no relation like (5.24) is dictated by the Lax condition.

It can be shown that for weak shocks the Lax condition and the entropy growth condition are equivalent; see Lax [96] for genuinely nonlinear systems where (4.8) holds for all eigenvalues.

We have written the Lax conditions (5.23) for a generic eigenvalue λ. Taking into account all n eigenvalues—which we suppose distinct—there are n Lax conditions, viz.

$$\lambda_+^{(1)} < \lambda_+^{(2)} < \ldots < \lambda_+^{(k)} < s < \lambda_+{}^{(k+1)} < \ldots < \lambda_+^{(n)}$$
$$\lambda_-^{(1)} < \lambda_-^{(2)} < \ldots < \lambda_-{}^{(k-1)} < s < \lambda_-{}^{(k)} < \ldots < \lambda_-^{(n)}. \tag{5.25}$$

For a fixed value k the shock that satisfies (5.25) is called a *k-shock* and, in particular, (5.23) applies to the k-shock in the form

$$\lambda_+^{(k)} < s < \lambda_-^{(k)}. \tag{5.26}$$

5.6 Lax Condition in Extended Thermodynamics

We return to Paragraph 5.3 and discuss the physical admissibility of the solutions of the Rankine-Hugoniot equations of extended thermodynamics.

We have four shocks in this case, which bifurcate from the four solutions of the characteristic equation (3.3). In front of the shock that solution is given by (see (3.6))

$$\hat{\lambda}_+^{(1)} = -1.6503, \quad \hat{\lambda}_+^{(2)} = -0.6297, \quad \hat{\lambda}_+^{(3)} = 0.6297, \quad \hat{\lambda}_+^{(4)} = 1.6503,$$

while behind the shock the four solutions are function of σ and q, which in turn—by (5.12), (5.14), and (5.15)—are functions of M_+, different ones for the fast and the slow shocks, because $w(M_+)$ is different. Thus $\hat{\lambda}_-^{(i)}$ $(i = 1, 2, 3, 4)$ are functions of M_+ which have been calculated and are represented in Figure 8.12

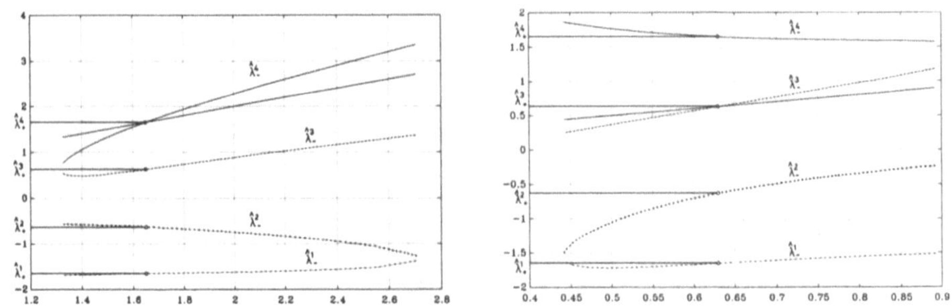

Figure 8.12 The characteristic speeds as functions of M_+. *a.* fast shock, *b.* slow shock.

The Lax condition (5.25) for the fastest wave $k = 4$ reads explicitely

$$\lambda_+^{(4)} < s \qquad \text{and} \qquad \lambda_-^{(3)} < s < \lambda_-^{(4)} \tag{5.27}$$

and of course, s/c_+ is the bisector of the $(\hat{\lambda}, M_+)$ plane shown in Figure 8.12a. Inspection shows that the Lax conditions (5.27) are satisfied provided $M_+ > 1.6503$ holds. Therefore we conclude that the dotted curves of Figures 8.7 through 8.11 are physically significant only in this range of the Mach number. In particular

- the density jumps up,
- the temperature jumps up,
- the stress jumps down,
- the heat flux jumps up, and
- the velocity jumps up.

We shall now investigate the consequences of the Lax condition (5.25) for the slow wave $k = 3$. That condition (5.25) reads explicitly

$$\lambda_+^{(3)} < s < \lambda_+^{(4)} \qquad \text{and} \qquad \lambda_-^{(2)} < s < \lambda_-^{(3)} \tag{5.28}$$

and inspection of Figure 8.12b shows that the slow shock can only propagate in the range of Mach numbers $M_+ > 0.6297$. The drawn-out curves of the Figures 8.7 through 8.11 are physically relevant only in that range and we see that

- the density jumps up,
- the temperature jumps down,
- the stress jumps down,
- the heat flux jumps down, and
- the velocity jumps up.

From the Figures 8.12 we also conclude that the system loses hyperbolicity after the passage of a fast shock with $M_+ \approx 2.7$, because two eigenvalues become equal in the disturbed region. We have discussed this break-down of hyperbolicity in Paragraph 3.2. We note that the loss of hyperbolicity at the Mach number 2.7 is not related to the critical Mach numbers found in Grad [99] or Anile & Majorana [100] for shock structure solutions.

Anyway in the small range of Mach numbers where the slow shock can propagate, it is a *cold shock*, meaning that the temperature behind the shock is smaller than in front of it. This is unusual, because we tend to think—motivated by the experience in gas dynamics—that a shock will always heat the gas up. This is not necessarily so and for more details about this situation and cases in which the entropy growth condition and the Lax conditions do not coincide the reader may turn to some recent papers by Ruggeri, Muracchini & Seccia [101], [102] concerning a simple continuum model for second sound in solids at low temperature.

In the foregoing treatment of shock waves these waves are considered as singular surfaces, i.e. as jumps of the thermodynamic fields. This is the common procedure in gas dynamics. However experiments show that the thermodynamic fields are smooth in a shock, although quite steep. Thus the shocks have a "thickness" and therefore the most appropriate treatment of shock waves is the so-called "shock structure analysis." Chapter 12 below is entirely devoted to the shock structure problem.

Chapter 9

Extended Thermodynamics of Moments

This chapter starts a new episode in this book: The extension of the number of fields does not stop with 13, 14, or 15 fields but proceeds to include many fields; – hundreds or thousands of them.

Such drastic extensions have their price and so we must restrict the attention to moments of the phase density of monatomic gases, if we wish to obtain useful results. Thus extended thermodynamics moves closer to the kinetic theory of gases. But there remain differences: Above all, closure of the system of equations is still effected by the entropy principle which will be shown to be equivalent to the method of maximization of entropy which, in turn, is an extrapolation of a well known procedure of statistical thermodynamics.

Drastic extensions to many fields raise the question of interpretability of the fields and of their controllability at the boundary, – even when they are moments. However, there are problems where the nature of the fields is unimportant, if only equilibrium is well defined. Foremost among such problems is the calculation of the characteristic speeds and of the pulse speed, – the largest characteristic speed. Other phenomena which can be treated without specifying the nature of the fields include light-scattering, sound-dispersion and shock-wave-structures. These phenomena will be considered in subsequent chapters. Invariably it will turn out in such cases that extensions to very many variables are needed in order to improve on the results of ordinary thermodynamics.

The pulse speed deserves special interest: It goes to infinity as the number of moments increases to infinity. Thus extended thermodynamics reaches an anticlimax. Indeed, it will be remembered that the theory started from the desire to have finite pulse speeds.

In the relativistic case it is possible to prove that the velocities are all smaller than the speed of light independent of the number of moments.

1 Field Equations for Moments

Sofar we have occasionally referred to the kinetic theory of gases so as to prove consistency of rational extended thermodynamics with the kinetic theory. We have also used the kinetic theory to motivate some assumptions of rational extended thermodynamics: E.g. the basic assumption of Chapter 4 that the fluxes of one equation are densities in the next equation.

But we have never made explicit use – sofar – of the moment character of densities and fluxes, nor have we used the explicit form of the kinetic entropy in nonequilibrium.

This will now change: The fields will be recognized as moments of the phase density and the entropy as a particular functional of the phase density. The entropy principle will then identify a specific form of the phase density which may be used to derive closed linear systems of field equations for the moments.

1.1 Densities, Fluxes, and Productions as Moments of the Phase Density

We consider the case when the fields $\mathbf{u} = \mathbf{F}^0$, the fluxes $\mathbf{F}_i$ and productions $\mathbf{\Pi}$ are blocks of symmetric tensors, viz.

$$\mathbf{u} = \mathbf{F}^0 = \begin{bmatrix} F \\ F_{i_1} \\ F_{i_1 i_2} \\ \vdots \\ F_{i_1 i_2 \dots i_N} \end{bmatrix}, \mathbf{F}^i = \begin{bmatrix} F_i \\ F_{i i_1} \\ F_{i i_1 i_2} \\ \vdots \\ F_{i i_1 i_2 \dots i_N} \end{bmatrix}, \mathbf{\Pi} = \begin{bmatrix} \Pi \\ \Pi_{i_1} \\ \Pi_{i_1 i_2} \\ \vdots \\ \Pi_{i_1 i_2 \dots i_N} \end{bmatrix}. \tag{1.1}$$

A similar case was discussed in Paragraph 3.4.8., except that now the fluxes and the densities are identical: The flux in one equation equals the density in the next equation. This is a property peculiar to the moment equations of the kinetic theory. And indeed we shall assume that the $\mathbf{F}$'s and $\mathbf{\Pi}$'s in (1.1) are moments of the phase density $f(\mathbf{x}, \mathbf{c}, t)$ and of the collision operator $S(\mathbf{x}, \mathbf{c}, t)$ respectively, so that we have

$$\begin{aligned} u_{i_1 \dots i_l} = \quad F_{i_1 \dots i_l} &= m \int \quad c_{i_1} \dots c_{i_l} \, f d\mathbf{c}, \\ F_{i i_1 \dots i_l} &= m \int c_i c_{i_1} \dots c_{i_l} \, f d\mathbf{c}, \\ \Pi_{i_1 \dots i_l} &= m \int \quad c_{i_1} \dots c_{i_l} \, S d\mathbf{c}. \end{aligned} \tag{1.2}$$

m is the atomic mass.

In order to simplify the notation we introduce the multi-index $\alpha = 0, \dots, N$ and write

$$u_\alpha = F_\alpha = m \int c_\alpha f d\mathbf{c}, \quad F_{i\alpha} = m \int c_i c_\alpha f d\mathbf{c}, \quad \Pi_\alpha = m \int c_\alpha S(f) d\mathbf{c}. \tag{1.3}$$

c_α stands for $c_{i_1} c_{i_2} \dots c_{i_\alpha}$ and $F_\alpha = F_{i_1 i_2 \dots i_\alpha}$. The number of independent densities is severely reduced by symmetry: There are

$$n = \frac{1}{2} \sum_{\alpha=0}^{N} (\alpha+1)(\alpha+2) = \frac{1}{6}(N+1)(N+2)(N+3) \tag{1.4}$$

independent densities. For $\alpha = 0$ we set $c_\alpha = 1$ and $F_\alpha = F$. f is the phase density and $S(f)$ is the collision operator. For a classical gas we have (cf. Section 2.3)

$$\Pi_\alpha = m \int c_\alpha (f' f^{1\prime} - f f^1)\; sg \sin\theta d\theta\; d\varepsilon\; d\mathbf{c}^1 d\mathbf{c}. \tag{1.5}$$

The field equations read

$$\frac{\partial F_\alpha}{\partial t} + \frac{\partial F_{i\alpha}}{\partial x^i} = \Pi_\alpha. \quad (\alpha = 0, \ldots, N) \tag{1.6}$$

and they are not closed on two counts:

- While the fluxes $F_{i\alpha}$ in one equation equal the density $u_{\alpha+1}$ in the next equation, the "last" flux, i.e. $F_{i\alpha}$ for $\alpha = N$, is *a priori* unrelated to the densities which are our variables.
- All nonvanishing productions are *a priori* independent of the densities.

The system will be closed by the exploitation of the entropy inequality, because this principle will provide a specific form of the phase density.

1.2 Extended Thermodynamics of Moments

We recall the entropy principle of extended thermodynamics from Chapter 3: We have the entropy inequality

$$\frac{\partial h^0}{\partial t} + \frac{\partial h^i}{\partial x^i} = \Sigma \geq 0 \tag{1.7}$$

with constitutive relations for h^0 and h^i, viz.

$$h^0 \equiv h^0(u_\alpha) \quad \text{and} \quad h^i \equiv h^i(u_\alpha) \tag{1.8}$$

and with a concave entropy density

$$\frac{\partial^2 h^0}{\partial u_\alpha \partial u_\beta} \quad - \quad \text{negative definite.} \tag{1.9}$$

From Section 3.2 we know that there exists a main field Λ^α ($\Lambda^\alpha = \Lambda$ when $\alpha = 0$, $\Lambda^\alpha = \Lambda_{i_1 i_2 \ldots i_\alpha}$ for $1 \leq \alpha \leq N$)—the field of Lagrange multipliers—and a vector potential h'^0, h'^i such that

$$u_\alpha = \frac{\partial h'^0}{\partial \Lambda^\alpha}, \quad F_{i\alpha} = \frac{\partial h'^i}{\partial \Lambda^\alpha}, \quad h^0 = u_\alpha \Lambda^\alpha - h'^0, \quad h^i = F_{i\alpha}\Lambda^\alpha - h'^i, \quad \Sigma = \Lambda^\alpha \Pi_\alpha. \tag{1.10}$$

Knowing that the u_α's and the $F_{i\alpha}$'s are moments we obtain from (1.10) with $dF(\chi) = f d\chi$

$$dh'^0 = u_\alpha\, d\Lambda^\alpha = \int d\left(\Lambda^\alpha m\, c_\alpha\right) f d\mathbf{c} = \int dF\left(\Lambda^\alpha m\, c_\alpha\right) d\mathbf{c} = d\left[\int F\left(\Lambda^\alpha m\, c_\alpha\right) d\mathbf{c}\right]$$

$$dh'^i = F_{i\alpha} d\Lambda^\alpha = \int c_i d\left(\Lambda^\alpha m\, c_\alpha\right) f d\mathbf{c} = \int c_i dF\left(\Lambda^\alpha m\, c_\alpha\right) d\mathbf{c} = d\left[\int c^i F\left(\Lambda^\alpha m\, c_\alpha\right) d\mathbf{c}\right].$$

We introduce $\chi \equiv \Lambda^\alpha m\, c_\alpha$ and draw the following conclusions from the strings of equations (1.11) and from (1.10). All these conclusions form part of the paper [25] by Boillat & Ruggeri.

- The phase density $f(\mathbf{x}, \mathbf{c}, t)$ is a function of the single variable χ

$$f(\mathbf{x}, \mathbf{c}, t) = f(\chi) \quad \text{where} \quad \chi = \Lambda^{\alpha}(\mathbf{x}, t) m c_{\alpha}. \tag{1.12}$$

which is a polynomial in $\mathbf{c}$. This means in particular that the $\mathbf{x}$- and t-dependence of f occurs only through the $\mathbf{x}$- and t-dependence of the Lagrange multipliers Λ^{α}, or—since Λ^{α} is a function of $\mathbf{u}$—through the $\mathbf{x}$- and t-dependence of the moments u^{α}.

In the kinetic theory of gases a phase density f with such a dependence on $\mathbf{x}, t$ is called a *normal solution* of the Boltzmann equation. Usually this type of solution is *assumed*, but here it is seen to be a *consequence* of the entropy law.

- The vector potential h'^0, h'^i has the form

$$\begin{aligned} h'^0 &= \int F(\chi) d\mathbf{c} \\ h'^i &= \int c_i F(\chi) d\mathbf{c} \end{aligned} \qquad \text{where} \quad \frac{dF}{d\chi} = f(\chi) \quad \text{holds,} \tag{1.13}$$

i.e. $F(\chi)$ is the generator function of $f(\chi)$.

- The entropy vector h^0, h^i has the form, by $(1.10)_{3,4}$

$$\begin{aligned} h^0 &= \int (\chi F' - F) d\mathbf{c} \\ h^i &= \int c_i (\chi F' - F) d\mathbf{c} \end{aligned} \qquad \text{where} \quad F' \equiv \frac{dF}{d\chi} = f. \tag{1.14}$$

Thus, if—like here—the variables are moments, the entropy vector h^A must have the structure shown in (1.14)—with a generic function $F(\chi)$, or $f(\chi)$.

As in Paragraph 3.2.2 we may combine $(1.10)_{1,2}$ and (1.6) to write the field equation in the form

$$\frac{\partial^2 h'^0}{\partial \Lambda^{\alpha} \partial \Lambda^{\beta}} \frac{\partial \Lambda^{\beta}}{\partial t} + \frac{\partial^2 h'^i}{\partial \Lambda^{\alpha} \partial \Lambda^{\beta}} \frac{\partial \Lambda^{\beta}}{\partial x_i} = \Pi^{\alpha}$$

or, by use of (1.13)

$$\left(m^2 \int F'' c_{\alpha} c_{\beta} d\mathbf{c} \right) \frac{\partial \Lambda^{\beta}}{\partial t} + \left(m^2 \int F'' c_i c_{\alpha} c_{\beta} d\mathbf{c} \right) \frac{\partial \Lambda^{\beta}}{\partial x_i} = \Pi^{\alpha}. \tag{1.15}$$

By the requirement of convexity the matrix $\int F'' c_{\alpha} c_{\beta} d\mathbf{c}$ must be negative definite, i.e. for all X_{α} we must have

$$\int F'' c_{\alpha} c_{\beta} d\mathbf{c} X_{\alpha} X_{\beta} = \int F'' (c_{\alpha} X_{\alpha})^2 d\mathbf{c} < 0.$$

It follows that the function $F(\chi)$ must be concave

$$F''(\chi) < 0. \tag{1.16}$$

The equation (1.15) is an important intermediate result. Note that the system is closed; it has been closed by the exploitation of the entropy inequality. However, it is not quite good enough to know that F—hence f—is a function of $\chi = \Lambda_{\alpha} m c_{\alpha}$. We need to know the form of that function. And for that purpose we need another input from statistical thermodynamics; viz.: We need to know the entropy density h^0 as a functional of the phase density.

1.3 Specific Phase Densities

For a *classical* gas, we have (cf. (2.3.18))

$$h^0 = -k \int \left(\ln \frac{f}{y} - 1 \right) f d\mathbf{c} \quad \text{hence} \quad \chi F' - F = k \left(\ln \frac{f}{y} - 1 \right) f \tag{1.17}$$

or with $f = F'$ by differentiation

$$f = y e^{-\frac{1}{k}\chi}, \quad \chi = \Lambda^\alpha m c_\alpha. \tag{1.18}$$

With (1.18) the density and flux of entropy and the vector potential come out as

$$\begin{aligned} h^0 &= \tfrac{k}{m}\, u_0 + \Lambda^\alpha u_\alpha & \qquad h'^0 &= -\tfrac{k}{m}\, u_0 \\ & & \text{and} & \\ h^i &= \tfrac{k}{m}\, F_{i0} + \Lambda^\alpha F_{i\alpha} & \qquad h'^i &= -\tfrac{k}{m}\, F_{i0} \end{aligned}.$$

Thus in particular the components of the vector potential are proportional to the density and flux of mass.

For the *degenerate* gas, we have (cf. Sommerfeld [103])

$$h^0 = -k \int \left(\ln \frac{f}{y} \pm \frac{y}{f} \left(1 \pm \frac{f}{y}\right) \ln \left(1 \pm \frac{f}{y}\right) \right) f d\mathbf{c} \quad \text{hence}$$

$$\chi F' - F = k \left(\ln \frac{f}{y} \pm \frac{y}{f} \left(1 \pm \frac{f}{y}\right) \ln \left(1 \pm \frac{f}{y}\right) \right) f \tag{1.19}$$

and with $f = F'$ we obtain, again by differentiation

$$f = \frac{y}{e^{\chi/k} \pm 1}, \qquad \chi = \Lambda^\alpha m c_\alpha. \tag{1.20}$$

The upper sign refers to Fermions, the lower one to Bosons. The entropy four vector and the vector potential have the forms

$$\begin{aligned} h^0 &= \pm k \int \; y \ln \left(1 \pm e^{-\frac{\chi}{k}}\right) + \Lambda^\alpha u_\alpha & \qquad h'^0 &= \mp k \int \; y \ln \left(1 \pm e^{-\frac{\chi}{k}}\right) \\ h^i &= \pm k \int c_i y \ln \left(1 \pm e^{-\frac{\chi}{k}}\right) + \Lambda^\alpha F_{i\alpha} & \qquad h'^i &= \mp k \int c_i y \ln \left(1 \pm e^{-\frac{\chi}{k}}\right) \end{aligned}. \tag{1.21}$$

1.4 Field Equations for Λ^α and Equations for u^α near Equilibrium

Thus we have determined the phase density, in particular the one for a classical gas, viz. $f = y e^{-\chi/k}$. Therefore $F'' = f'$ is equal to $-\frac{1}{k} f$ and we may write the system (1.15) in the form

$$\left(\int c_\alpha c_\beta e^{-\chi/k} d\mathbf{c} \right) \frac{\partial \Lambda^\beta}{\partial t} + \left(\int c_i c_\alpha c_\beta e^{-\chi/k} d\mathbf{c} \right) \frac{\partial \Lambda^\beta}{\partial x_i} = -\frac{k}{m} \frac{1}{y} \int c_\alpha S \left(y e^{-\chi/k} \right) d\mathbf{c}. \tag{1.22}$$

Since $\chi = \Lambda^\alpha m c_\alpha$ and since S is a known operator, (see (1.3), (1.5)), this system is closed and specific—*in the Lagrange multipliers* $\mathbf{\Lambda}$, which are auxiliary fields of uncertain physical interpretation.

We may feel that we know more about the moments u_α than about the Lagrange multipliers and therefore we may prefer to return to the moments.[1] This can be done, of course, because by (1.2) and (1.18) we have

$$u_\alpha = m\int c_\alpha y e^{-\frac{m}{k}\Lambda^\beta c_\beta} d\mathbf{c} \quad \text{hence} \quad \Lambda^\beta = \Lambda^\beta(u_\alpha). \tag{1.23}$$

Unfortunately it is impossible to calculate $(1.23)_2$ from $(1.23)_1$ except approximately, i.e. close to equilibrium. We proceed to show how this is done.

In equilibrium the phase density f is given by the Maxwell distribution

$$f_E = \frac{\rho}{m}\sqrt{\frac{m}{2\pi kT}}^3 e^{-\frac{m(c-v)^2}{2kT}}. \tag{1.24}$$

Thus we have, by comparison with (1.18),

$$\Lambda^E = \frac{v^2}{2T} - \frac{k}{m}\ln\left(\frac{\rho}{m}\frac{1}{y}\sqrt{\frac{m}{2\pi kT}}^3\right), \quad \Lambda_i^E = -\frac{v_i}{T}, \quad \Lambda_{ii}^E = \frac{1}{2T}, \tag{1.25}$$

while all other $\mathbf{\Lambda}$'s must vanish in equilibrium. We set

$$\Lambda_\alpha = \Lambda_\alpha^E + \lambda_\alpha \quad \text{with} \quad \lambda_\alpha^E = 0 \tag{1.26}$$

and expand f up to linear terms in the $\mathbf{\Lambda}$'s, i.e.

$$f = f_E\left(1 - \frac{m}{k}\lambda_\alpha c_\alpha\right). \tag{1.27}$$

Hence follows by (1.23)

$$u_\beta - u_\beta^E = -\frac{m^2}{k}\int c_\beta\, c_\alpha f_E d\mathbf{c}\lambda_\alpha, \quad \text{where} \quad u_\beta^E = m\int c_\beta\, f_E d\mathbf{c}. \tag{1.28}$$

This is a linear algebraic system of equations from which the λ's may be calculated as functions of the $\mathbf{u}$'s.

Once the λ's are known, the "last" flux and all nonvanishing productions may be found by integration:

$$\begin{aligned} F_{i\beta} &= m\int c_i c_\beta \left(1 - \frac{m}{k}\lambda_\alpha c_\alpha\right) f_E d\mathbf{c}, && \text{for} \quad \beta = N \\ \pi_\beta &= m\int c_\beta S\left(\left(1 - \frac{m}{k}\lambda_\alpha c_\alpha\right) f_E\right) d\mathbf{c}\,, && \text{for} \quad \beta = 2,\dots,N \end{aligned} \tag{1.29}$$

Thus the system (1.6) is closed. We proceed to demonstrate the closing procedure in the linear approximation in nonequilibrium variables for the relatively simple case $N = 3$.

[1]Actually though, a septic will say that one knows just as little about the moment $F_{<ij>kk}$ as one does about the Lagrange multipliers. There arises the question of how to use a system whose fields we cannot interpret, and we shall come back to this; see Paragraph 2.3.

1.5 The Case N=3: An Illustration

For $N = 3$ we have as densities, fluxes, and productions

$$\mathbf{u} = \mathbf{F}^0 = \begin{bmatrix} F \\ F_i \\ F_{ij} \\ F_{ijk} \end{bmatrix}, \quad \mathbf{F}_i = \begin{bmatrix} F_i \\ F_{ij} \\ F_{ijk} \\ F_{ijkl} \end{bmatrix}, \quad \mathbf{\Pi} = \begin{bmatrix} 0 \\ 0 \\ \Pi_{<ij>} \\ \Pi_{ijk} \end{bmatrix}. \tag{1.30}$$

The productions Π, Π_i, Π_{ii} vanish because of conservation of mass, momentum, and energy. There are 20 densities, all moments of the type (1.2).

The field equations read

$$\frac{\partial F_{i_1 \ldots i_l}}{\partial t} + \frac{\partial F_{i i_1 \ldots i_l}}{\partial x_i} = \Pi_{i_1 \ldots i_l} \qquad (l = 0, 1, 2, 3). \tag{1.31}$$

The approximate phase density reads (cf. (1.27))

$$f = f_E \left(1 - \frac{m}{k}\left(\lambda + \lambda^i c_i + \lambda^{ij} c_i c_j + \lambda^{ijk} c_i c_j c_k\right)\right). \tag{1.32}$$

For the determination of the $\mathbf{\Lambda}$'s we rely on the linear system (1.28), which now has the explicit form

$$\begin{aligned}
0 &= -\frac{m^2}{k}\left\{\int f_E d\mathbf{c}\, \lambda + \int c_l f_E d\mathbf{c}\, \lambda^l + \int c_l c_m f_E d\mathbf{c}\, \lambda^{lm} + \int c_l c_m c_n f_E d\mathbf{c}\, \lambda^{lmn}\right\} \\
0 &= -\frac{m^2}{k}\left\{\int c_i f_E d\mathbf{c}\, \lambda + \int c_i c_l f_E d\mathbf{c}\, \lambda^l + \int c_i c_l c_m f_E d\mathbf{c}\, \lambda^{lm} + \int c_i c_l c_m c_n f_E d\mathbf{c}\, \lambda^{lmn}\right\} \\
F_{<ij>} - F^E_{<ij>} &= -\frac{m^2}{k}\left\{\int c_i c_j f_E d\mathbf{c}\, \lambda + \int c_i c_j c_l f_E d\mathbf{c}\, \lambda^l + \int c_i c_j c_l c_m f_E d\mathbf{c}\, \lambda^{lm} + \int c_i c_j c_l c_m c_n f_E d\mathbf{c}\, \lambda^{lmn}\right\} \\
F_{ijk} - F^E_{ijk} &= -\frac{m^2}{k}\left\{\int c_i c_j c_k f_E d\mathbf{c}\, \lambda + \int c_i c_j c_k c_l f_E d\mathbf{c}\, \lambda^l + \int c_i c_j c_k c_l c_m f_E d\mathbf{c}\, \lambda^{lm} + \int c_i c_j c_k c_l c_m c_n f_E d\mathbf{c}\, \lambda^{lmn}\right\}.
\end{aligned} \tag{1.33}$$

The left-hand sides of the first and second equation and of the trace of the third one are equal to zero, since the densities of mass, momentum, and energy are identical in equilibrium and nonequilibrium.

It is convenient to solve the linear system (1.33) in the rest frame, because in that case f_E is even in $\mathbf{c}$ and the integrals are solved easily—half of them vanish. Of course, in that case the $\mathbf{F}$'s on the left-hand side of (1.33) are the internal densities which we have previously denoted by ϱ; see Chapter 4. It is then appropriate to rewrite the system (1.31) in the reduced form—for the internal quantities—which reads (cf. (3.4.24))

$$\begin{array}{llllll}
\frac{d\rho}{dt} & +\rho & \frac{\partial v_n}{\partial x_n} & & & =0 \\
 & & & \frac{\partial \rho_{in}}{\partial x_n} & +\left\{\rho\frac{dv_i}{dt} \qquad \right\} & =0 \\
\frac{d\rho_{ij}}{dt} & +\rho_{ij} & \frac{\partial v_n}{\partial x_n}+ & \frac{\partial \rho_{ijn}}{\partial x_n} & +\left\{ \qquad +\rho_{n(i}\frac{\partial v_{j)}}{\partial x_n}\right\} & =\hat{\Pi}_{<ij>} \\
\frac{d\rho_{ijk}}{dt} & +\rho_{ijk} & \frac{\partial v_n}{\partial x_n}+ & \frac{\partial \rho_{ijkn}}{\partial x_n} & +\left\{3\rho_{(ij}\frac{dv_k}{dt} +\rho_{n(ij}\frac{\partial v_{k)}}{\partial x_n}\right\} & =\hat{\Pi}_{ijk}.
\end{array} \tag{1.34}$$

Thus we see the quantities explicitly exhibited which require closure, viz. $\rho_{ijkl}, \hat{\Pi}_{<ij>}, \hat{\Pi}_{ijk}$.

If the system (1.33) is solved in the rest frame, it provides the internal parts of the Lagrange multipliers, i.e. the velocity-independent parts. We denote these by $\hat{\boldsymbol{\lambda}}$ and obtain

$$\hat{\lambda}=0, \quad \hat{\lambda}_i=\frac{1}{2}\frac{1}{\rho\frac{k}{m}T^2}\rho_{ijj}, \quad \hat{\lambda}_{ij}=-\frac{1}{2}\frac{1}{\rho\frac{k}{m}T^2}\rho_{<ij>}, \quad \hat{\lambda}_{ijk}=-\frac{1}{6}\frac{1}{\rho\left(\frac{k}{m}\right)^2T^3}\rho_{ijk}. \tag{1.35}$$

Hence follows the phase density in terms of $\rho_{<ij>}, \rho_{ijj}$ and $\rho_{<ijk>}$ by insertion into (1.32)

$$\begin{aligned} f \;=\; f_E\Bigg(1 & -\tfrac{1}{2}\tfrac{1}{\rho\left(\frac{k}{m}T\right)^2}\rho_{<ij>}C_iC_j-\tfrac{1}{2}\tfrac{1}{\rho\left(\frac{k}{m}T\right)^2}C_i\left(1-\tfrac{1}{5}\tfrac{C^2}{\frac{k}{m}T}\right)\rho_{ijj}- \\ & \tfrac{1}{6}\tfrac{1}{\rho\left(\frac{k}{m}T\right)^3}\rho_{<ijk>}C_iC_jC_k\Bigg)\,. \end{aligned} \tag{1.36}$$

Here $C_i = c_i - v_i$ is the velocity of the atoms in the local rest frame of the gas.

Now that the phase density is known we may use (1.29) to determine the "last" flux, here F_{ijkl}, or its internal part ρ_{ijkl}. We obtain

$$\rho_{ijkl}=\rho\left(\frac{k}{m}T\right)^2 3\delta_{(ij}\delta_{kl)}-\frac{k}{m}T6\delta_{(ij}\rho_{kl)}. \tag{1.37}$$

The internal productions may be calculated from the phase density (1.36) by insertion into $(1.3)_3$

$$\begin{aligned} \hat{\Pi}_{<ij>} &= m\int C_{<i}C_{j>}\left(f'f^{1\prime}-ff^1\right)sg\sin\theta d\theta d\varepsilon d\mathbf{c}^1d\mathbf{c} \\ \hat{\Pi}_{ijk} &= m\int C_iC_jC_k\left(f'f^{1\prime}-ff'\right)sg\sin\theta d\theta d\varepsilon d\mathbf{c}^1d\mathbf{c}. \end{aligned} \tag{1.38}$$

For the evaluation of the integrals we need to assume an interaction potential between the atoms and we choose to consider Maxwell atoms, whose interaction potential is given by $\Phi(r)=\frac{\kappa}{4}\frac{1}{r^4}$ as a function of the distance r. Thus we obtain

$$\begin{aligned} \hat{\Pi}_{<ij>} &= \sigma\rho_{<ij>} \\ \hat{\Pi}_{ijk} &= \sigma\left(\tfrac{3}{2}\rho_{<ijk>}+\tfrac{2}{5}\delta_{(ij}\rho_{k)ll}\right) \end{aligned} \qquad \text{where} \quad \sigma = -3\frac{\rho}{m}\sqrt{\frac{\kappa}{2m}}\,1.370347... \tag{1.39}$$

Insertion of ρ_{ijkl},$\hat{\Pi}_{<ij>}$,$\hat{\Pi}_{ijk}$ from (1.37), (1.39) into the reduced balance equations (1.34) closes the system; we thus obtain 20 explicit field equations for the 20 variables $\rho, v_i, \rho_{ij}, \rho_{ijk}$.

1.6 Field Equations for n=13, 14, 20, 21, 26, 35

The derivation of the closed system (1.34), (1.37), (1.39) for 20 moments was typical for the procedure of extended thermodynamics of moments. We can apply this method to an arbitrary finite number of moments, and in this paragraph we write the result for several values of n. In doing so we drop the synthetic notation and, wherever possible, we introduce the conventional notation by which we have

$$\rho_{ii} = 2\rho\varepsilon = 3kT, \quad \rho_{<ij>} = -t_{<ij>}, \quad \rho_{ijj} = 2q_i$$

with ρ and T as mass density and temperature and $t_{<ij>}$ and q_i as stress deviator and heat flux. Of course, there is no conventional notation for the higher moments, but among the people who work in this field, a quantity Δ is used which is equal to the nonequilibrium part of the double-trace of ρ_{ijkl}, viz.

$$\rho_{iijj} = 15\rho\left(\frac{k}{m}T\right)^2 + \Delta. \tag{1.40}$$

The following equations are all linearized in the nonequilibrium parts of the moments, in velocity, and in the derivatives of ρ and T. The uniform and time-independent fields of ρ and T underlying the process will be called $\tilde{\rho}$ and $\tilde{T}$.

n = 13 Variables: F, F_i, F_{ij}, F_{ijj} or $\varrho, v_i, T, t_{\langle ij\rangle}, q_i$

$$\begin{aligned}
&\frac{\partial \varrho}{\partial t} + \tilde{\varrho}\frac{\partial v_i}{\partial x_i} = 0\\
&\frac{\partial v_i}{\partial t} + \frac{\left(\frac{k}{m}\tilde{T}\right)}{\varrho}\frac{\partial \varrho}{\partial x_i} + \frac{\partial}{\partial x_i}\left(\frac{k}{m}T\right) - \frac{1}{\tilde{\varrho}}\frac{\partial t_{\langle ij\rangle}}{\partial x_j} = 0\\
&\frac{\partial\left(\frac{k}{m}T\right)}{\partial t} + \frac{2}{3}\left(\frac{k}{m}\tilde{T}\right)\frac{\partial v_k}{\partial x_k} + \frac{2}{3}\frac{1}{\tilde{\varrho}}\frac{\partial q_k}{\partial x_k} = 0.\\
&\frac{\partial t_{\langle ij\rangle}}{\partial t} - \frac{4}{5}\frac{\partial q_{\langle i}}{\partial x_{j\rangle}} - 2\tilde{\varrho}\left(\frac{k}{m}\tilde{T}\right)\frac{\partial v_{\langle i}}{\partial x_{j\rangle}} = \sigma t_{\langle ij\rangle}\\
&\frac{\partial q_i}{\partial t} - \left(\frac{k}{m}\tilde{T}\right)\frac{\partial t_{\langle ik\rangle}}{\partial x_k} + \frac{5}{2}\tilde{\varrho}\left(\frac{k}{m}\tilde{T}\right)\frac{\partial\left(\frac{k}{m}T\right)}{\partial x_i} = \frac{2}{3}\sigma q_i.
\end{aligned} \tag{1.41}$$

n = 14 Variables: $F, F_i, F_{ij}, F_{ijj}F_{iijj}$ or $\varrho, v_i, T, t_{\langle ij\rangle}, q_i, \Delta$

$$\begin{aligned}
&\frac{\partial \varrho}{\partial t} + \tilde{\varrho}\frac{\partial v_i}{\partial x_i} - 0\\
&\frac{\partial v_i}{\partial t} + \frac{\left(\frac{k}{m}\tilde{T}\right)}{\tilde{\varrho}}\frac{\partial \varrho}{\partial x_i} + \frac{\partial}{\partial x_i}\left(\frac{k}{m}T\right) - \frac{1}{\tilde{\varrho}}\frac{\partial t_{\langle ij\rangle}}{\partial x_j} = 0
\end{aligned}$$

$$\frac{\partial\left(\frac{k}{m}T\right)}{\partial t}+\frac{2}{3}\left(\frac{k}{m}\tilde{T}\right)\frac{\partial v_k}{\partial x_k}+\frac{2}{3}\frac{1}{\tilde{\varrho}}\frac{\partial q_k}{\partial x_k}=0. \tag{1.42}$$
$$\frac{\partial t_{\langle ij\rangle}}{\partial t}-\frac{4}{5}\frac{\partial q_{\langle i}}{\partial x_{j\rangle}}-2\tilde{\varrho}\left(\frac{k}{m}\tilde{T}\right)\frac{\partial v_{\langle i}}{\partial x_{j\rangle}}=\sigma t_{\langle ij\rangle}$$
$$\frac{\partial q_i}{\partial t}-\left(\frac{k}{m}\tilde{T}\right)\frac{\partial t_{\langle ik\rangle}}{\partial x_k}+\frac{5}{2}\tilde{\varrho}\left(\frac{k}{m}\tilde{T}\right)\frac{\partial\left(\frac{k}{m}T\right)}{\partial x_i}+\frac{1}{6}\frac{\partial\Delta}{\partial x_i}=\frac{2}{3}\sigma q_i$$
$$\frac{\partial\Delta}{\partial t}+8\left(\frac{k}{m}\tilde{T}\right)\frac{\partial q_i}{\partial x_i}=\frac{2}{3}\sigma\Delta.$$

n = 20 Variables: F, F_i, F_{ij}, F_{ijk} or $\varrho, v_i, T, t_{\langle ij\rangle}, q_i, \varrho_{\langle ijk\rangle}$

$$\frac{\partial\varrho}{\partial t}+\tilde{\varrho}\frac{\partial v_i}{\partial x_i}=0$$
$$\frac{\partial v_i}{\partial t}+\frac{\left(\frac{k}{m}\tilde{T}\right)}{\tilde{\varrho}}\frac{\partial\varrho}{\partial x_i}+\frac{\partial}{\partial x_i}\left(\frac{k}{m}T\right)-\frac{1}{\tilde{\varrho}}\frac{\partial t_{\langle ij\rangle}}{\partial x_j}=0$$
$$\frac{\partial\left(\frac{k}{m}T\right)}{\partial t}+\frac{2}{3}\left(\frac{k}{m}\tilde{T}\right)\frac{\partial v_k}{\partial x_k}+\frac{2}{3}\frac{1}{\tilde{\varrho}}\frac{\partial q_k}{\partial x_k}=0. \tag{1.43}$$
$$\frac{\partial t_{\langle ij\rangle}}{\partial t}-\frac{4}{5}\frac{\partial q_{\langle i}}{\partial x_{j\rangle}}-2\tilde{\varrho}\left(\frac{k}{m}\tilde{T}\right)\frac{\partial v_{\langle i}}{\partial x_{j\rangle}}-\frac{\partial\varrho_{\langle ijk\rangle}}{\partial x_k}=\sigma t_{\langle ij\rangle}$$
$$\frac{\partial q_i}{\partial t}-\left(\frac{k}{m}\tilde{T}\right)\frac{\partial t_{\langle ik\rangle}}{\partial x_k}+\frac{5}{2}\tilde{\varrho}\left(\frac{k}{m}\tilde{T}\right)\frac{\partial\frac{k}{m}T}{\partial x_i}=\frac{2}{3}\sigma q_i$$
$$\frac{\partial\varrho_{\langle ijk\rangle}}{\partial t}-3\frac{k}{m}\tilde{T}\left\{\frac{\partial t_{(\langle ij\rangle}}{\partial x_{k)}}-\frac{2}{5}\frac{\partial t_{\langle r(i\rangle}}{\partial x_r}\delta_{jk)}\right\}=\frac{3}{2}\sigma\varrho_{\langle ijk\rangle}$$

n = 21 Variables: $F, F_i, F_{ij}, F_{ijk}F_{iijj}$ or $\varrho, v_i, T, t_{\langle ij\rangle}, q_i, \varrho_{\langle ijk\rangle}, \Delta$

$$\frac{\partial\varrho}{\partial t}+\tilde{\varrho}\frac{\partial v_i}{\partial x_i}=0$$
$$\frac{\partial v_i}{\partial t}+\frac{\left(\frac{k}{m}\tilde{T}\right)}{\tilde{\varrho}}\frac{\partial\varrho}{\partial x_i}+\frac{\partial}{\partial x_i}\left(\frac{k}{m}T\right)-\frac{1}{\tilde{\varrho}}\frac{\partial t_{\langle ij\rangle}}{\partial x_j}=0$$
$$\frac{\partial\left(\frac{k}{m}T\right)}{\partial t}+\frac{2}{3}\left(\frac{k}{m}\tilde{T}\right)\frac{\partial v_k}{\partial x_k}+\frac{2}{3}\frac{1}{\tilde{\varrho}}\frac{\partial q_k}{\partial x_k}=0.$$
$$\frac{\partial t_{\langle ij\rangle}}{\partial t}-\frac{4}{5}\frac{\partial q_{\langle i}}{\partial x_{j\rangle}}-2\tilde{\varrho}\left(\frac{k}{m}\tilde{T}\right)\frac{\partial v_{\langle i}}{\partial x_{j\rangle}}-\frac{\partial\varrho_{\langle ijk\rangle}}{\partial x_k}=\sigma t_{\langle ij\rangle} \tag{1.44}$$
$$\frac{\partial q_i}{\partial t}-\left(\frac{k}{m}\tilde{T}\right)\frac{\partial t_{\langle ik\rangle}}{\partial x_k}+\frac{5}{2}\tilde{\varrho}\left(\frac{k}{m}\tilde{T}\right)\frac{\partial\left(\frac{k}{m}T\right)}{\partial x_i}+\frac{1}{6}\frac{\partial\Delta}{\partial x_i}=\frac{2}{3}\sigma q_i$$
$$\frac{\partial\varrho_{\langle ijk\rangle}}{\partial t}-3\frac{k}{m}\tilde{T}\left\{\frac{\partial t_{(\langle ij\rangle}}{\partial x_{k)}}-\frac{2}{5}\frac{\partial t_{\langle r(i\rangle}}{\partial x_r}\delta_{jk)}\right\}=\frac{3}{2}\sigma\varrho_{\langle ijk\rangle}$$
$$\frac{\partial\Delta}{\partial t}+8\frac{k}{m}\tilde{T}\frac{\partial q_i}{\partial x_i}=\frac{2}{3}\sigma\Delta.$$

n = 26 Variables: $F, F_i, F_{ij}, F_{ijk} F_{ijkk}$ or $\varrho, v_i, T, t_{\langle ij\rangle}, q_i, \varrho_{\langle ijk\rangle}, \Delta, \varrho_{kk\langle ij\rangle}$

$$\frac{\partial\varrho}{\partial t}+\tilde{\varrho}\frac{\partial v_i}{\partial x_i}=0$$

$$\frac{\partial v_i}{\partial t}+\frac{\left(\frac{k}{m}\tilde{T}\right)}{\tilde{\varrho}}\frac{\partial\varrho}{\partial x_i}+\frac{\partial}{\partial x_i}\left(\frac{k}{m}T\right)-\frac{1}{\tilde{\varrho}}\frac{\partial t_{\langle ij\rangle}}{\partial x_j}=0$$

$$\frac{\partial\left(\frac{k}{m}T\right)}{\partial t}+\frac{2}{3}\left(\frac{k}{m}\tilde{T}\right)\frac{\partial v_k}{\partial x_k}+\frac{2}{3}\frac{1}{\tilde{\varrho}}\frac{\partial q_k}{\partial x_k}=0.$$

$$\frac{\partial t_{\langle ij\rangle}}{\partial t}-\frac{4}{5}\frac{\partial q_{\langle i}}{\partial x_{j\rangle}}-2\tilde{\varrho}\left(\frac{k}{m}\tilde{T}\right)\frac{\partial v_{\langle i}}{\partial x_{j\rangle}}-\frac{\partial\varrho_{\langle ijk\rangle}}{\partial x_k}=\sigma t_{\langle ij\rangle} \tag{1.45}$$

$$\frac{\partial q_i}{\partial t}-\frac{5}{2}\left(\frac{k}{m}\tilde{T}\right)\frac{\partial t_{\langle ik\rangle}}{\partial x_k}+\frac{5}{2}\tilde{\varrho}\left(\frac{k}{m}\tilde{T}\right)\frac{\partial\frac{k}{m}T}{\partial x_i}+\frac{1}{6}\frac{\partial\Delta}{\partial x_i}+\frac{1}{2}\frac{\partial\varrho_{kk\langle ij\rangle}}{\partial x_r}=\frac{2}{3}\sigma q_i$$

$$\frac{\partial\varrho_{\langle ijk\rangle}}{\partial t}+\frac{3}{7}\left\{\frac{\partial\varrho_{rr(\langle ij\rangle}}{\partial x_{k)}}-\frac{2}{5}\frac{\partial\varrho_{nn\langle r(i\rangle}}{\partial x_r}\delta_{jk)}\right\}=\frac{3}{2}\sigma\varrho_{\langle ijk\rangle}$$

$$\frac{\partial\Delta}{\partial t}+8\left(\frac{k}{m}\tilde{T}\right)\frac{\partial q_i}{\partial x_i}=\frac{2}{3}\sigma\Delta$$

$$\frac{\partial\varrho_{kk\langle ij\rangle}}{\partial t}+14\tilde{\varrho}\left(\frac{k}{m}\tilde{T}\right)^2\frac{\partial v_{\langle i}}{\partial x_{j\rangle}}+\frac{56}{6}\frac{k}{m}\tilde{T}\frac{\partial q_{\langle i}}{\partial x_{j\rangle}}+9\frac{k}{m}\tilde{T}\frac{\partial\varrho_{\langle ijk\rangle}}{\partial x_k}=\frac{7}{6}\sigma\left(\varrho_{kk\langle ij\rangle}+t_{\langle ij\rangle}\right)$$

n = 35 Variables: $F, F_i, F_{ij}, F_{ijk} F_{ijkl}$ or $\varrho, v_i, T, t_{\langle ij\rangle}, q_i, \varrho_{\langle ijk\rangle}, \Delta, \varrho_{kk\langle ij\rangle}, \varrho_{\langle ijkl\rangle}$

$$\frac{\partial\varrho}{\partial t}+\tilde{\varrho}\frac{\partial v_i}{\partial x_i}=0$$

$$\frac{\partial v_i}{\partial t}+\frac{\left(\frac{k}{m}\tilde{T}\right)}{\tilde{\varrho}}\frac{\partial\varrho}{\partial x_i}+\frac{\partial}{\partial x_i}\left(\frac{k}{m}T\right)-\frac{1}{\tilde{\varrho}}\frac{\partial t_{\langle ij\rangle}}{\partial x_j}=0$$

$$\frac{\partial\left(\frac{k}{m}T\right)}{\partial t}+\frac{2}{3}\left(\frac{k}{m}\tilde{T}\right)\frac{\partial v_k}{\partial x_k}+\frac{2}{3}\frac{1}{\tilde{\varrho}}\frac{\partial q_k}{\partial x_k}=0.$$

$$\frac{\partial t_{\langle ij\rangle}}{\partial t}-\frac{4}{5}\frac{\partial q_{\langle i}}{\partial x_{j\rangle}}-2\tilde{\varrho}\left(\frac{k}{m}\tilde{T}\right)\frac{\partial v_{\langle i}}{\partial x_{j\rangle}}-\frac{\partial\varrho_{\langle ijk\rangle}}{\partial x_k}=\sigma t_{\langle ij\rangle}$$

$$\frac{\partial q_i}{\partial t}-\frac{5}{2}\frac{k}{m}\tilde{T}\frac{\partial t_{\langle ik\rangle}}{\partial x_k}+\frac{5}{2}\tilde{\varrho}\left(\frac{k}{m}\tilde{T}\right)\frac{\partial\frac{k}{m}T}{\partial x_i}+\frac{1}{6}\frac{\partial\Delta}{\partial x_i}+\frac{1}{2}\frac{\partial\varrho_{kk\langle ij\rangle}}{\partial x_r}=\frac{2}{3}\sigma q_i \tag{1.46}$$

$$\frac{\partial\varrho_{\langle ijk\rangle}}{\partial t}+\frac{\partial\varrho_{\langle ijkr\rangle}}{\partial x_r}+\frac{3}{7}\left\{\frac{\partial\varrho_{rr(\langle ij\rangle}}{\partial x_{k)}}-\frac{2}{5}\frac{\partial\varrho_{nn\langle r(i\rangle}}{\partial x_r}\delta_{jk)}\right\}=\frac{3}{2}\sigma\varrho_{\langle ijk\rangle}$$

$$\frac{\partial\Delta}{\partial t}+8\left(\frac{k}{m}\tilde{T}\right)\frac{\partial q_i}{\partial x_i}=\frac{2}{3}\sigma\Delta$$

$$\frac{\partial\varrho_{kk\langle ij\rangle}}{\partial t}+14\tilde{\varrho}\left(\frac{k}{m}\tilde{T}\right)^2\frac{\partial v_{\langle i}}{\partial x_{j\rangle}}+\frac{56}{5}\left(\frac{k}{m}\tilde{T}\right)\frac{\partial q_{\langle i}}{\partial x_{j\rangle}}+9\left(\frac{k}{m}\tilde{T}\right)\frac{\partial\varrho_{\langle ijk\rangle}}{\partial x_k}=\frac{7}{6}\sigma\left(\varrho_{kk\langle ij\rangle}+t_{\langle ij\rangle}\right)$$

$$\frac{\partial\varrho_{\langle ijkl\rangle}}{\partial t}+4\frac{k}{m}\tilde{T}\left\{\frac{\partial\varrho_{(\langle ijk\rangle}}{\partial x_{l)}}-\frac{3}{7}\frac{\partial\varrho_{\langle r(ij\rangle}}{\partial x_r}\delta_{kl)}\right\}=1.8731\ldots\sigma\varrho_{\langle ijkl\rangle}.$$

The derivation of these equation involves the repeated solution of the linear algebraic system (1.28) for an increasing number of equations.

There is an awkward feature in the equations (1.41) through (1.46) that eventually frustrates even the most diligent student: As n—the number of equations—is increased, it is not safe to assume that the equations that have been calculated for lower n's remain unchanged. Thus for instance, as we proceed from $n = 20$ to $n = 21$, we do not merely add another equation to the system, but we change the equation for $\partial q_i/\partial t$ in an unforeseeable manner.

Because of this feature, the method of moments is not the most useful manner for the improved description of a gas. In Section 3 we shall learn about a better procedure.

2 Characteristic Speeds

We may ask whether a system of field equations for Lagrange multipliers – like (1.15) or (1.22) – can be useful at all seeing that these quantities are without a suggestive interpretation and certainly not measurable nor controllable on the boundary. Yet they *are* useful, at least in a near-equilibrium theory, because in that case the coefficient matrices of the system may be calculated in equilibrium. And the equilibrium Lagrange multipliers *are* known in terms of density, velocity and temperature.

Therefore it is possible to calculate the characteristic speeds in equilibrium states and in particular the pulse speed, – the largest characteristic speed. This pulse speed grows to infinity as the number of moments increases.

2.1 Field Equations near Equilibrium

We refer to the closed system (1.22) and recall that there is *one* case for which we know the Lagrange multipliers in terms of suggestive variables, and that is the case of equilibrium where, by (1.25), we have

$$\Lambda^E = \frac{v^2}{2T} - \frac{k}{m}\ln\left(\frac{\rho}{m}\frac{1}{y}\sqrt{\frac{m}{2\pi kT}}^{3}\right), \quad \Lambda_i^E = -\frac{v_i}{T}, \quad \Lambda_{ii}^E = \frac{1}{2T}. \tag{2.1}$$

All other Lagrange multipliers vanish, so that $e^{-\frac{1}{k}\chi}$ is equal to the Maxwell distribution.

When the system (1.22) is linearized with respect to the nonequilibrium variables, the coefficient matrices

$$\int c_\alpha c_\beta e^{-\chi/k} d\mathbf{c} \quad \text{and} \quad \int c_i c_\alpha c_\beta e^{-\chi/k} d\mathbf{c} \tag{2.2}$$

can be evaluated in equilibrium—with the Maxwell distribution: Each entry is a simple Gauss integral which is listed in mathematical tables.

2.2 Pulse Speed

Since the moments $u_{i_1 \ldots i_l}$ and the corresponding Lagrange multipliers $\Lambda_{i_1 \ldots i_l}$ are symmetric, there are $(l+1)(l+2)/2$ independent ones among them—say the ones with $i_1 \leq i_2 \leq \ldots \leq i_l$. All of the indices may assume the numbers $1, 2, 3$ and therefore the independent moments may be mapped into the variables u_{pqr} or Λ_{pqr} with $p + q + r = l$, where p, q, and r are the number of indices among $i_1 \ldots i_l$ which are equal to $1, 2$, or 3 respectively.

We recall from Section 3.4 that the matrices (3.3.9) are semi-definite and for the present case of moments as variables we conclude that the matrix

$$\left(-\frac{\partial^2 h'^a}{\partial\Lambda^{pqr}\partial\Lambda^{stu}}n_a + V_{\max}\frac{\partial^2 h'^0}{\partial\Lambda^{pqr}\partial\Lambda^{stu}}\right) \quad \text{is negative semi-definite.} \tag{2.3}$$

By use of (1.13) this condition may be rewritten in the form

$$\int F''(\chi)(-c^i n_i + V_{\max})c_1^{p+s}c_2^{q+t}c_3^{r+u}d\mathbf{c} \quad \text{- negative semi-definite.} \tag{2.4}$$

The elements a_{ij} of a semi-definite matrix satisfy the inequalities

$$a_{ii}\ a_{jj} \geq a_{ij}^2$$

and therefore (2.4) implies

$$\left(\int F''(\chi)(-c^i n_i + V_{\max})c_1^{2p}c_2^{2q}c_3^{2r}d\mathbf{c}\right)\left(\int F''(\chi)(-c^i n_i + V_{\max})c_1^{2s}c_2^{2t}c_3^{2u}d\mathbf{c}\right) \geq$$

$$\geq \left(\int F''(\chi)(-c^i n_i + V_{\max})c_1^{p+s}c_2^{q+t}c_3^{r+u}d\mathbf{c}\right)^2 \tag{2.5}$$

With $F'' = -\frac{1}{k}f$ and with $f = f_E$ this inequality reduces to a simpler form,[2] viz.

$$\begin{aligned} &V_{\max}^2\left(\int f_E\ c_1^{2p}c_2^{2q}c_3^{2r}d\mathbf{c}\right)\left(\int f_E\ c_1^{2s}c_2^{2t}c_3^{2u}d\mathbf{c}\right) \geq \\ &\geq \left(\int f_E\left(-c^i n_i + V_{\max}\right)\ c_1^{p+s}c_2^{q+t}c_3^{r+u}d\mathbf{c}\right)^2. \end{aligned} \tag{2.6}$$

This estimate depends on the choice of the exponents p through u and we choose, rather arbitrarily, $p = N$, $s = N-1$, and all others zero. Also we set $n^i = (1,0,0)$. In that case (2.6) implies

$$V_{\max}^2 \geq \frac{\int f_E c_1^{2N}dc_1}{\int f_E c_1^{2(N-1)}dc_1} = \frac{6}{5}\sqrt{\frac{5}{3}\frac{k}{m}T}\left(N-\frac{1}{2}\right) \tag{2.7}$$

and we conclude that the maximal characteristic speed *tends to infinity* with the number of moments; recall that N, by (7.1), is the biggest tensorial rank in the ordered set of variables $\mathbf{u}$.

Of course, (2.7) is only an estimate for a special choice of the exponents p through u and there may be better choices. In this context it is interesting to note (see Chapter 10, Paragraph 4.7) that the true speeds V have been calculated in theories with very many moments. Weiss [27] has tabulated these values; they are represented in Table 10.4, and Figure 9.1 shows that they increase with N, albeit slightly faster than the estimate.

In this context it is appropriate to recall Section 3.3 about subsystems. In that section we have explained that the characteristic speed $V_{\max}$ cannot decrease when the number of equations is increased. This observation is born out by the present result on extended thermodynamics of an increasing number of moments.

---

[2]Note that the "paradox" of heat conduction arose in a *linear* theory.

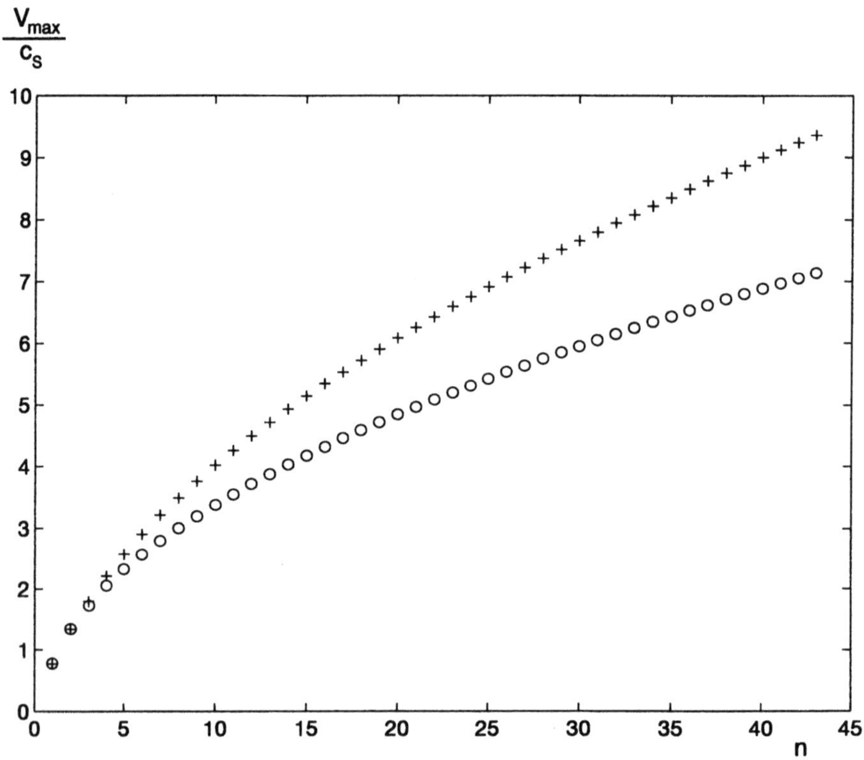

Figure 9.1: Maximum characteristic speeds of a monatomic gas in equilibrium.
Ordinate: In units of the adiabatic speed of sound.
Abscissa: Rank of highest moment.
Crosses: Exact values by Weiss[27].
Circles: The estimate (2.7) by Boillat and Ruggeri [25].

2.3 Discussion

Not knowing the interpretation of the Lagrange multipliers, we cannot possibly solve initial and boundary value problems for the system (1.15) or (1.22). Indeed, we cannot prescribe and control the data for the Λ^{α}'s—excepting the first five of them. But we are able to calculate characteristic speeds in equilibrium and we have found explicit lower bounds for the pulse speeds. Thus we were able to resolve the *original question* of extended thermodynamics and to show that the pulse speed tends to infinity for more and more moments.

The phenomenon of the characteristic speeds is the easiest one to deal with in a near-equilibrium theory, because it does not require a physical interpretation of the main field—the field of Lagrange multipliers—nor does it require the knowledge of the production terms. Other phenomena that do not require the interpretability of the fields include

- dispersion and absorption of sound, and
- light scattering.

To be sure, these phenomena do require the knowledge of the productions and we shall prepare the equations for their treatment in the next section.

One should think that all boundary value problems would require interpretable fields, and generally this is indeed the case. But for shock waves we may be satisfied with Lagrange multipliers or other fields that cannot be controlled at the boundary. Indeed, for shock waves we know that all non-equilibrium variables vanish in the limiting states at $\pm\infty$. And this is enough of a boundary condition to make it possible to deal with the case.

2.4 The Relativistic Case; Speeds Smaller than c

Extended thermodynamics of moments—the theory presented in Paragraph 1.2.—has a counterpart in the relativistic case. In that case the two equations (1.13) combine to form a single four-vector equation, and the same is true for the two equations (1.14). We obtain

$$h'^A = \int F(\chi) p^A \, dP, \qquad \text{and} \qquad h^A = \int \left(\chi F'(\chi) - F(\chi)\right) p^A \, dP. \tag{2.8}$$

p^A is the four-momentum of a particle and dP stands for the invariant element of momentum space, so that $dP = \sqrt{(-g)}\, dp^1 dp^2 dp^3 / p_0$ holds. χ is now equal to $\Lambda_\alpha p^\alpha$, where α is a multi-index; we have $p^0 = 1$ and $p^\alpha = p^{A_1} p^{A_2} ... p^{A_\alpha}$.

The symmetric system (1.15) becomes

$$H^{A\alpha\beta}(\Lambda_\gamma)\Lambda_{\beta},_A = \Pi^\alpha(\Lambda_\gamma), \quad \text{where} \quad H^{A\alpha\beta}(\Lambda_\gamma) = \frac{\partial^2 h'^A}{\partial\Lambda_\alpha \partial\Lambda_\beta} = \int F''(\chi) p^A p^\alpha p^\beta \, dP. \tag{2.9}$$

Since p^A is time-like, the matrix $\mathbf{H}^A \xi_A$ is positive definite for all time-like vectors ξ_A provided that $F''(\chi) > 0$ holds; indeed we have

$$H^{A\alpha\beta}\xi_A X_\alpha X_\beta = \int F'' p^A \xi_A (p^\alpha X_\alpha)^2 \, dP > 0 \quad \forall X \neq 0. \tag{2.10}$$

Now the four-vector $\varphi_{,A}$ normal to a wave front satisfies the characteristic eigenvalue problem

$$\det\left(H^{A\alpha\beta}\varphi_{,A}\right) = 0, \tag{2.11}$$

so that, by (2.10), the gradient $\varphi_{,A}$ cannot be time-like. Therefore we have

$$g^{AB}\varphi_{,A}\varphi_{,B} \leq 0. \tag{2.12}$$

By (8.4.27) and (8.4.28) this means that the characteristic speeds cannot exceed the speed of light.

These results by Boillat & Ruggeri—the ones just presented and those of Paragraph 2.2.—agree with those obtained from the study of infinitesimal disturbances for the Boltzmann equation. Indeed, the maximal phase speed in the high-frequency limit is unbounded in the classical case and less than c in the relativistic case; see Sirovich & Thurber [104], Wang Chang & Uhlenbeck [105], Cercignani & Majorana [106], and Cercignani [107].

3 Mean Eigenfunctions

We have seen that the fields need not be interpretable and yet a theory can provide interesting results, witness the estimate on the pulse speed in Paragraph 2.2. In that calculation we did not even need the production terms but with other phenomena - even near-equilibrium ones - we are not so lucky and we have to calculate productions. This is most easily done, if we choose variables that represent the mean eigenfunctions of the Boltzmann collision operator. We shall introduce these eigenfunctions here and show that they are orthogonal combinations of the irreducible parts of moments.

The mean eigenfunctions have essentially four advantages

- The calculation of the productions is easy.
- The coefficient matrix of the time derivatives is the identity matrix.
- The coefficient matrix of the space derivatives is only sparsely occupied.

All these properties make the mean eigenfunctions attractive alternatives to the moments when it comes to the choice of fields.

3.1 Eigenfunctions and Eigenvalues

Close to an equilibrium of a body at rest with $\tilde{\rho}$ and $\tilde{T}$ we may write

$$f = \tilde{f}(1+\varphi) \quad \text{where} \quad \tilde{f} = \frac{\tilde{\rho}}{m}\frac{1}{\sqrt{2\pi\frac{k}{m}\tilde{T}}^{3}}\, e^{-\frac{mc^2}{2k\tilde{T}}} \tag{3.1}$$

and linearize the Boltzmann equation in φ. Thus we obtain a linear integro-differential equation for φ, viz.

$$\frac{\partial \varphi}{\partial t} + c_i \frac{\partial \varphi}{\partial x_i} = I(\varphi) \quad \text{with} \quad I(\varphi) = \int \tilde{f}^1 (\varphi' + \varphi^{1'} - \varphi - \varphi^1) sg \sin \vartheta d\vartheta d\varphi d\mathbf{c}^1 . \tag{3.2}$$

The solutions φ may be represented in terms of the orthogonal eigenfunctions $\psi^{(r)}_{\langle i_1 \dots i_l \rangle}$ of the linear collision operator $I(\varphi)$ so that we may write

$$\varphi(\mathbf{x}, \mathbf{c}, t) = \sum_{r,l} b^{(r)}_{\langle i_1 \dots i_l \rangle}(\mathbf{x}, t)\ \psi^{(r)}_{\langle i_1 \dots i_l \rangle}, \quad \text{where } I\left(\psi^{(r)}_{\langle i_1 \dots i_l \rangle}(\mathbf{c})\right) = -\zeta_{rl} \psi^{(r)}_{\langle i_1 \dots i_l \rangle}(\mathbf{c}). \tag{3.3}$$

ζ_{rl} are the eigenvalues.

According to Waldmann [108] the eigenfunctions and eigenvalues for Maxwell atoms are given as follows.

$$\psi^{(r)}_{\langle i_1 \dots i_l \rangle}(\mathbf{c}) = \psi_{lr}(c) \underbrace{\frac{c_{\langle i_1} \cdots c_{i_l \rangle}}{c^l}}_{= \frac{(-1)^l}{1 \cdot 3 \cdot 5 \dots (2l-1)} c^{l+1} \frac{\partial^l}{\partial c_{i_1} \dots \partial c_{i_l}} \left(\frac{1}{c}\right)} , \tag{3.4}$$

where

$$\psi_{lr}(c) = \sqrt{\frac{\sqrt{\pi} r!}{2(l + \frac{1}{2} + r)!}} \underbrace{S^{(r)}_{l+\frac{1}{2}}\left(\frac{c^2}{2\frac{k}{m}\tilde{T}}\right)}_{= \sum_{p=0}^{r} \frac{(l+\frac{1}{2})(l+\frac{3}{2})\dots(l+\frac{1}{2}+r)}{p!(r-p)!(l+\frac{1}{2})(l+\frac{3}{2})\dots(l+\frac{1}{2}+p)} \left(-\frac{c^2}{2\frac{k}{m}\tilde{T}}\right)^p} \left(\frac{c}{\sqrt{2\pi\frac{k}{m}\tilde{T}}}\right)^l$$

$$\zeta_{rl} = \frac{\tilde{\rho}}{m} 2\pi \int_0^{\pi/2} [1 - \sin^{l+2r}\theta \underbrace{P_l(\sin\theta)}_{\text{Legendre polynomials}} - \cos^{l+2r}\theta P_l(\cos\theta)] \underbrace{\Gamma(\theta)}_{= 4\cos(\theta) sg} \sin\theta d\theta . \tag{3.5}$$

These functions satisfy orthogonality conditions as follows [3]

$$a_{\langle i_1 .. i_l \rangle} \int \frac{c_{\langle i_1 .. i_l \rangle}}{c^l} \frac{c_{\langle j_1 - j_n \rangle}}{c^n} \sin\vartheta d\vartheta d\varphi = \frac{\sqrt{\pi} l!}{2^{l+1} \Gamma\left(l + \frac{3}{2}\right)} \delta_{l\,n}\ a_{\langle j_1 .. j_l \rangle} \tag{3.6}$$

$$\int \psi_{lr} \psi_{ls} \frac{m}{\tilde{\rho}} \tilde{f} dc = \delta_{rs}.$$

[3] $\alpha_{\langle i_1 \dots i_l \rangle}$ is any symmetric traceless tensor.

Therefore the coefficients $b^{(r)}_{\langle i_1..i_l\rangle}$ in (3.4) may be given in terms of the mean values of the non-equilibrium parts of the eigenfunctions

$$b^{(r)}_{\langle i_1\ldots i_l\rangle} = \frac{2^{l+1}\Gamma\left(l+\frac{3}{2}\right)}{\sqrt{\pi}\,l!}\,\overline{\psi^{(r)}_{\langle i_1\ldots i_l\rangle}} \qquad \text{with} \qquad \overline{\psi^{(r)}_{\langle i_1\ldots i_l\rangle}} = \frac{m}{\tilde{\rho}}\int \psi^{(r)}_{\langle i_1\ldots i_l\rangle}(f-\tilde{f})d\mathbf{c}, \tag{3.7}$$

so that the expansion $(3.3)_1$ reads

$$\varphi(\mathbf{x},\mathbf{c},t) = \sum_{r,l} \frac{2^{l+1}\Gamma(l+\frac{3}{2})}{\sqrt{\pi}\,l!}\,\overline{\psi^{(r)}_{\langle i_1\ldots i_l\rangle}(\mathbf{x},t)}\;\psi^{(r)}_{\langle i_1\ldots i_l\rangle}. \tag{3.8}$$

Tables 9.1 and 9.2 give the first few eigenvalues and eigenfunctions and in Table 9.3 we have listed some of the mean eigenfunctions $\overline{\psi^{(r)}_{\langle i_1\ldots i_l\rangle}}$.

$l\backslash r$	0	1	2	3
0	0	0	2/3	1
1	0	2/3	1	1.228
2	1	7/6	1.342	1.491
3	3/2	1.570	1.667	1.763

Table 9.1 Some eigenvalues ζ_{rl}/ζ_{02}.

$l\backslash r$	0	1	2
0	1	$\sqrt{\frac{2}{3}}\left(\frac{3}{2}-(\beta c)^2\right)$	$\sqrt{\frac{8}{15}}\left(\frac{15}{8}-\frac{5}{2}(\beta c)^2+\frac{1}{2}(\beta c)^4\right)$
1	$\sqrt{\frac{2}{3}\beta c_i}$	$\frac{2}{\sqrt{15}}\left(\frac{5}{2}-(\beta c)^2\right)\beta c_i$	
2	$\frac{2}{\sqrt{15}}\beta^2 c_{\langle\, i}c_{j\rangle}$	$\sqrt{\frac{8}{105}}\left(\frac{7}{2}-(\beta c)^2\right)\beta^2 c_{\langle i}c_{j\rangle}$	
3	$\sqrt{\frac{8}{105}}^{\,3}\,\beta^3 c_{\langle i}c_j c_{k\rangle}$		
4	$\frac{4}{\sqrt{945}}\beta^4 c_{\langle i}c_j c_k c_{e\rangle}$		

Table 9.2 Some eigenfunctions $\overline{\psi^{(r)}_{\langle i_1\ldots i_l\rangle}}$. $\left(\beta \text{ stands for} \dfrac{1}{\sqrt{2\frac{k}{m}\tilde{T}}}\right)$

$l\backslash r$	0	1	2
0	$\left(\frac{\rho}{\tilde{\rho}}-1\right)$	$\sqrt{\frac{3}{2}}\left(1-\frac{T}{\tilde{T}}\right)$	$\sqrt{\frac{15}{8}}\left(3-\frac{\rho}{\tilde{\rho}}-2\frac{T}{\tilde{T}}\right)+\sqrt{\frac{8}{15}}\frac{\beta^4}{2\tilde{\rho}}\Delta$
1	$\sqrt{\frac{2}{3}}\beta v_i$	$\frac{2}{\sqrt{15}}\left(\frac{5}{2}\beta v_i-\beta^3\frac{2}{\tilde{\rho}}q_i\right)$	
2	$-\frac{2}{\sqrt{15}}\beta^2\frac{1}{\tilde{\rho}}t_{\langle ij\rangle}$	$-\frac{8}{\sqrt{105}}\left(\beta^2\frac{1}{\tilde{\rho}}\frac{7}{2}t_{\langle ij\rangle}-\beta^4\frac{1}{\tilde{\rho}}\rho_{\langle ij\rangle kk}\right)$	
3	$\sqrt{\frac{8}{105}}\beta^3\frac{1}{\tilde{\rho}}\rho_{\langle ijk\rangle}$		

Table 9.3 Some mean values $\overline{\psi^{(r)}_{\langle i_1\ldots i_l\rangle}}$.

3.2 Mean Eigenfunctions as the Main Field

Upon reflection it is clear that the arguments of Section 1 can be repeated with the eigenfunctions $\psi^{(r)}_{\langle i_1\ldots i_l\rangle}$ replacing the polynomials $mc_{i_1}\ldots c_{i_l}$. The Lagrange multipliers will be denoted by $\Lambda^{(r)}_{\langle i_1\ldots i_l\rangle}$ in this case and we obtain

$$f=y\ e^{-\frac{1}{k}\Lambda^{(r)}_{\langle i_1\ldots i_l\rangle}\psi^{(r)}_{\langle i_1\ldots i_l\rangle}} \tag{3.9}$$

in the same manner in which previously we obtained (1.18).

In an equilibrium at rest with constant fields $\tilde{\rho}$ and $\tilde{T}$ the phase density is given by (3.1) and we may write in a linear expansion

$$f=\tilde{f}\left(1-\frac{1}{k}\ \lambda^{(r)}_{\langle i_1\ldots i_l\rangle}\ \psi^{(r)}_{\langle i_1\ldots i_l\rangle}\right), \tag{3.10}$$

where $\lambda^{(r)}_{\langle i_1\ldots i_l\rangle}$ are the nonequilibrium Lagrange multipliers, all of which vanish in equilibrium.

The densities, fluxes, and productions will be defined as

$$\begin{aligned} u^{(r)}_{\langle i_1\ldots i_l\rangle} &= \int\ \psi^{(r)}_{\langle i_1\ldots i_l\rangle}\ f d\mathbf{c} \\ F^{(r)}_{i\langle i_1\ldots i_l\rangle} &= \int c_i\psi^{(r)}_{\langle i_1\ldots i_l\rangle}\ f d\mathbf{c} \\ \Pi^{(r)}_{\langle i_1\ldots i_l\rangle} &= \int\ \psi^{(r)}_{\langle i_1\ldots i_l\rangle}\ I(\varphi)\tilde{f} dc. \end{aligned} \tag{3.11}$$

Therefore the new densities will be equal to the mean eigenfunctions to within a factor $\frac{\tilde{\rho}}{m}$. Inspection of the Tables 9.2 and 9.3 shows that the new densities are given by combinations of the irreducible parts of the moments so that—grosso modo—we may say that we are still working in extended thermodynamics of moments.

The properties of the eigenfunctions ensure that the productions are proportional to the non-equilibrium values of the variables. Indeed we have

$$\Pi^{(r)}_{\langle i_1 \dots i_l \rangle} = -\zeta_{rl} \left(u^{(r)}_{\langle i_1 \dots i_l \rangle} - \tilde{u}^{(r)}_{\langle i_1 \dots i_l \rangle} \right), \tag{3.12}$$

where $\tilde{\mathbf{u}}$ are equilibrium values. The linearized distribution function has the form (cf. (3.1), (3.8))

$$f = \tilde{f} \left(1 + \sum_{rl} \frac{2l+1\Gamma(l+\frac{3}{2})}{\sqrt{\pi}l!} \frac{m}{\tilde{\rho}} \left(u^{(r)}_{\langle i_1 \dots i_l \rangle} - \tilde{u}^{(r)}_{\langle i_1 \dots i_l \rangle} \right) \psi^{(r)}_{\langle i_1 \dots i_l \rangle} \right) \tag{3.13}$$

and comparison with (3.10) identifies the nonequilibrium parts of the Lagrange multipliers as

$$\lambda^{(r)}_{\langle i_1 \dots i_l \rangle} = -\frac{k}{\tilde{\rho}} \frac{2l+1\Gamma(l+\frac{3}{2})}{\sqrt{\pi}l!} \left(u^{(r)}_{\langle i_1 \dots i_l \rangle} - \tilde{u}^{(r)}_{\langle i_1 \dots i_l \rangle} \right). \tag{3.14}$$

Thus the nonequilibrium Lagrange multipliers are in fact equal to the nonequilibrium variables to within a factor; i.e. to within a factor the variables $(3.11)_1$ represent the main field.

While the moments have a natural order, the variables (3.11) do not, because each $u^{(r)}_{\langle i_1 \dots i_l \rangle}$ is characterized by the pair (r, l). We consider it natural to order the $u^{(r)}_{\langle i_1 \dots i_l \rangle}$ so that their index increases with the polynomial degree $2s+1$ in c_i, and within those $u^{(r)}_{\langle i_1 \dots i_l \rangle}$ which have the same degree we let the index increase with an increase in l. Thus the $u^{(r)}_{\langle i_1 \dots i_l \rangle}$ form a family with *one* increasing parameter α, whose value for a given *pair* (r, l) is determined by

$$\alpha(r,l) = \begin{cases} \frac{M}{2}\left(l+\frac{M}{2}\right) + \frac{l+2}{2} & \text{for} \quad M = 2r+l \quad \text{even} \\ \frac{(M+1)^2}{4} + \frac{l+1}{2} & \phantom{\text{for}} \quad M = 2r+l \quad \text{odd} \end{cases} \tag{3.15}$$

In order to facilitate the ordering, Table 9.4 provides the index α for the first few pairs (r, l).

$l \backslash r$	0	1	2	3	4
0	1	3	7	13	21
1	2	5	10	17	26
2	4	8	14	22	32
3	6	11	18	27	38
4	9	15	23	33	45

Table 9.4 Number of equation in terms of (r, l).

In this manner the first three equations are the conservation laws of mass, momentum, and energy and the fourth one—for $(r, l) = (0, 2)$—is the balance of the stress-deviator. Accordingly ζ_{02} is the inverse of the relaxation time of $t_{<ij>}$, a quantity which we have previously denoted by σ, e.g. in (2.3.34). Recall that σ—and now ζ_{02}—is related to the viscosity μ by $\sigma = p/\mu$. This quantity has been used in Table 9.1 to make the eigenvalues dimensionless.

3.3 Linear Field Equations for the Mean Eigenfunctions

We refer to the equations (1.14) and recall that now—with mean eigenfunctions as variables—we have to replace $mc_{i_1}...c_{i_l}$ by $\psi^{(r)}_{\langle i_1...i_l\rangle}$, or mc_α by ψ_α; recall (3.15) for the relation between (r,l) and α. For a classical gas we have $F'' = -\frac{1}{k}f$ and in the linearized case f can be replaced by $\tilde{f}$. Also the Lagrange multipliers Λ^β may be replaced by λ^β, since the difference between the two is a constant in the equilibrium characterized by (3.1). Therefore we obtain

$$-\frac{1}{k}\int \psi_\alpha\psi_\beta\ \tilde{f}dc\frac{\partial\lambda^\beta}{\partial t} - \frac{1}{k}\int c_i\psi_\alpha\psi_\beta\ \tilde{f}dc\frac{\partial\lambda^\beta}{\partial x_i} = \Pi_\alpha. \tag{3.16}$$

Note that this system is closed. It has been closed by the exploitation of the entropy principle. This was explained in Section 1. If ν is the number of fields, the index α and the summation index β run from 1 to ν.

The production Π_α is given by (3.12), so that with (3.14) we may write

$$\Pi_\alpha = \zeta_\alpha\frac{\tilde{\rho}}{k}\frac{\sqrt{\pi}l!}{2l+1T(l+\frac{3}{2})}\lambda_\alpha. \tag{3.17}$$

We recall from Paragraph 3.1 that

$$\int \psi_\alpha\psi_\beta\ \tilde{f}dc\frac{\partial\lambda_\beta}{\partial t} = \frac{\tilde{\rho}}{m}\frac{\sqrt{\pi}l!}{2l+1\Gamma(l+\frac{3}{2})}\frac{\partial\lambda_\alpha}{\partial t}. \tag{3.18}$$

and now it remains to calculate $\int c_i\psi_\alpha\psi_\beta\ \tilde{f}dc$. We use the identities

$$\frac{c_k}{c}\frac{c_{<i_1}\cdots c_{i_n>}}{c^n} = \frac{c_{<i_1}\cdots c_{i_n}c_{k>}}{c^{n+1}} + \frac{n}{2n+1}\frac{c_{<<i_1}\cdots c_{i_{n-1}>}}{c^{n-1}}\delta_{i_n>k} \tag{3.19}$$

and recurrence relations for the isotropic functions $\psi_{lr}(c)$, viz.

$$\begin{aligned}\int c\psi_{lr}\psi_{l+1,s}\frac{m}{\tilde{\rho}}\tilde{f}dc &= \sqrt{2\frac{k}{m}\tilde{T}}\left(\sqrt{l+\frac{3}{2}+s}\ \delta_{rs} - \sqrt{s+1}\ \delta_{r,s+1}\right)\\ \int c\psi_{lr}\psi_{l-1,s}\frac{m}{\tilde{\rho}}\tilde{f}dc &= \sqrt{2\frac{k}{m}\tilde{T}}\left(\sqrt{l+\frac{1}{2}+s}\ \delta_{rs} - \sqrt{s}\ \delta_{r,s-1}\right).\end{aligned} \tag{3.20}$$

A careful calculation shows that the explicit form of (3.16) reads

$$\frac{\partial\lambda^{(r)}_{\langle i_1...i_l\rangle}}{\partial t} + \frac{l+1}{2l+3}\sqrt{2\frac{k}{m}\tilde{T}}\left(\sqrt{l+\frac{3}{2}+r}\frac{\partial\lambda^{(r)}_{\langle i_1...i_l i\rangle}}{\partial x_i} - \sqrt{r}\frac{\partial\lambda^{(r-1)}_{\langle i_1...i_l i\rangle}}{\partial x_i}\right) + \tag{3.21}$$

$$+\sqrt{2\frac{k}{m}\tilde{T}}\left(\sqrt{l+\frac{1}{2}+r}\frac{\partial\lambda^{(r)}_{<<i_1...i_{l-1}>}}{\partial x_{i_l>}} - \sqrt{r+1}\frac{\partial\lambda^{(r+1)}_{<<i_1...i_{l-1}>}}{\partial x_{i_l>}}\right) = -\zeta_{rl}\lambda^{(r)}_{\langle i_1...i_l\rangle}.$$

Alternatively, in terms of the new variables we may write (cf. (3.14))

$$\frac{\partial u^{(r)}_{\langle i_1 \dots i_l \rangle}}{\partial t} + \sqrt{2\frac{k}{m}\tilde{T}} \left(\sqrt{l + \frac{3}{2} + r} \frac{\partial u^{(r)}_{\langle i_1 \dots i_l i \rangle}}{\partial x_i} - \sqrt{r} \frac{\partial u^{(r-1)}_{\langle i_1 \dots i_l i \rangle}}{\partial x_i} \right) + \tag{3.22}$$

$$+ \frac{l}{2l+1} \sqrt{2\frac{k}{m}\tilde{T}} \left(\sqrt{l + \frac{1}{2} + r} \frac{\partial u^{(r)}_{<<i_1 \dots i_{l-1}>}}{\partial x_{i_l>}} - \sqrt{r+1} \frac{\partial u^{(r+1)}_{<<i_1 \dots i_l i>}}{\partial x_i} \right) =$$

$$= -\zeta_{rl} \left(u^{(r)}_{\langle i_1 \dots i_l \rangle} - \tilde{u}^{(r)}_{\langle i_1 \dots i_l \rangle} \right).$$

Inspection seems to indicate that the equation for the field with the index pair (r, l) depends on the fields that correspond to the index pairs $(r, l+1)$, $(r-1, l+1)$, $(r, l-1)$, and $(r+1, l-1)$. This observation would seem to represent the well-known closure problem. *But the closure problem is already decided at this stage*, because the equations (3.16)—from which we derived (3.21) and (3.22)—form already a closed system; a system that was closed by the exploitation of the entropy principle.

Therefore the systems (3.20), (3.21) contain terms which are zero by virtue of the orthogonality of the eigenfunctions. Those spurious terms are the ones for which the single index α—which corresponds to the pair (r, l)—exceeds ν, the number of fields. Recall the relation (3.15) between α and (r, l) which is also illustrated by Table 9.4.

Formally we arrive at the closed system (3.16) from (3.21) or (3.22) by ignoring all terms for which the pair (s, n) corresponds to a value $\alpha > \nu$. Such terms do actually not appear in the equations after the closure by the entropy principle.

In the first edition of this book we suspected this state of affairs, but were unable to prove it, except in the special case of the 13-moment theory. In the mean-time the situation has become clear.

The systems (3.21), (3.22) are entirely specific and therefore they may be written for an arbitrary number of equations with the greatest ease. We recall that in Section 1 we have deplored the fact that the addition of another variable not only added another equation but also changed the previously derived set in subtle and nontransparent ways. This is no longer a difficulty with the new variables, nor with the main field $\lambda^{(r)}_{\langle i_1 \dots i_l \rangle}$. Therefore the equations (3.21) will form the basis for the treatment of light scattering and dispersion of sound in following chapters.

4 Maximization of Entropy

There is an alternative to extended thermodynamics of moments when it comes to the determination of the phase density. That alternative is the maximization of entropy under constraints.

The two methods are equivalent.

The maximization of entropy is a method often used in statistical mechanics for the calculation of the phase density, and over the years it has acquired a certain plausibility so that its logic seems convincing. Therefore it is important to prove the consistency – even equivalence – of extended thermodynamics with the maximization method.

4.1 Maximizing Entropy

The two realistic choices for the entropy as a functional of the phase density are

$$h^0 = -k\int \left[\ln \tfrac{f}{y} - 1\right] f dc \qquad \text{for a classical gas, or}$$

$$h^0 = -k\int \left[\ln \tfrac{f}{y} \pm \tfrac{y}{f}\left(1 \mp \tfrac{f}{y}\right)\ln\left(1 \mp \tfrac{f}{y}\right)\right] f dc \quad \text{for a degenerate gas} \tag{4.1}$$

In the case of the degenerate gas the upper sign refers to Fermions, the lower to Bosons. We treat the more general case

$$h^0 = \int \psi(f) dc \tag{4.2}$$

under which the forms (4.1) are subsumed as special cases.

We ask for the phase density that provides a maximum of h^0 under the constraints of fixed values for the moments u_α. With the Lagrange multipliers λ^α we form the expression

$$\int \psi(f) dc + \lambda^\alpha \left(u_\alpha - m\int c_\alpha f dc\right) \tag{4.3}$$

and obtain

$$\int \left(\frac{d\psi}{df} - \lambda^\alpha m c_\alpha\right)\delta f dc = 0, \quad \text{i.e.} \quad \frac{d\psi}{df} = \lambda^\alpha m c_\alpha \tag{4.4}$$

as a necessary condition for an extremum. Hence follows that f is a function of

$$\chi = \lambda^\alpha m c_\alpha \tag{4.5}$$

and that $\psi(f)$ has the form

$$\psi(f) = \chi f - \int f d\chi. \tag{4.6}$$

As a sufficient condition for a maximum (4.3) provides a restriction on the function $\psi(f)$, viz. $d^2\psi/df^2 < 0$. This condition is satisfied by the ψ's of (4.1) for all $f > 0$ in the case of Bosons and for $0 < f < y$ in the case of Fermions. These are the physically relevant intervals.

4.2 Maximizing Entropy is Equivalent to Extended Thermodynamics of Moments

Insertion of (4.6) into (4.2) gives exactly the same result as (1.13), since $F' = f$ and $F = \int f d\chi$ hold. Thus we conclude that *the maximization of entropy leads to the same result as extended thermodynamics of moments.* The reverse is also true, since $\int(\chi F' - F)dc$ with $F'' < 0$ provides a maximum of h^0 under the constraint. In particular, by (1.11) and (4.5), the Lagrange multipliers λ^α are identical to the main field components Λ^α which were introduced as Lagrange multipliers of a different nature in Paragraph 3.2.1.

The specific form of f—the one that maximizes h^0—has to be calculated from $(4.4)_2$ for particular choices of $\psi(f)$. Of course the choice of $\psi(f)$ is not arbitrary; the function ψ must be chosen in such a manner that it makes the entropy production nonnegative. $\psi = -k\left[\ln\frac{f}{y} - 1\right]$—implied by $(4.1)_1$—does that; see (2.3.19) and (2.3.20). With respect to the ψ implied by $(4.1)_2$ the situation is less clear, but there are simple cases in which it can be proved that the production is nonnegative.

The first author that applied the idea of maximization of entropy in extended thermodynamics was Dreyer [13] who started from the observation of Kogan [109] that the thirteen-moment phase density of Grad maximizes the entropy. The procedure of maximizing entropy was introduced in information theory and physics by Jaynes [110] and it is extensively used under the name of ***Maximum Entropy Principle***, e.g. see the book by Kapur [111].

Chapter 10

Extended Thermodynamics and Light Scattering

Light scattering is a powerful tool for the experimental determination of the thermal and caloric equations of state and of transport coefficients like thermal conductivity and viscosity. The great advantage of the method lies in the fact that it works in equilibrium; in fact, it makes use of the density fluctuations in equilibrium: By Onsager's hypothesis the auto-correlation function of such a fluctuation is related to the macroscopic field equations.

If the equations of state and the transport coefficients are known, – as they are for a monatomic ideal gas – light scattering is capable of distinguishing between theories for dilute gases, where the Navier-Stokes-Fourier theory is inappropriate. In that case extended thermodynamics of moments comes into its own. Indeed, for lower and lower pressures we need more and more moments and light scattering shows us how many.

The organization of this chapter and its contents follow closely the presentation of Weiss & Müller in their review article [28].

1 Basic Electrodynamics

1.1 Distant Field Approximation

Light is scattered by a gas in equilibrium, because unavoidable fluctuations $\delta\rho$ of density make the dielectric constant ε fluctuate. Thus we have $\varepsilon = \bar{\varepsilon} + \delta\varepsilon$ where $\bar{\varepsilon}$ is the mean value of ε.We proceed to describe the basic electrodynamics that governs the effect. For visualization we refer to Figure 10.1.

The incident electromagnetic field is given by

$$\begin{array}{llll} \mathbf{E}^i & \text{electric field,} & \mathbf{D}^i & \text{dielectric displacement,} \\ \mathbf{H}^i & \text{magnetic field,} & \mathbf{B}^i & \text{magnetic flux density.} \end{array} \tag{1.1}$$

Inside the scattering volume the electromagnetic field is a superposition of the incident field and the scattered field. We have

$$\begin{array}{ll} \mathbf{E} = \mathbf{E}^i + \mathbf{E}^s, & \mathbf{D} = \varepsilon\left(\mathbf{E}^i + \mathbf{E}^s\right) \approx \bar{\varepsilon}\left(\mathbf{E}^i + \mathbf{E}^s\right) + \delta\varepsilon\mathbf{E}^i, \\ \mathbf{H} = \mathbf{H}^i + \mathbf{H}^s, & \mathbf{B} = \mu_0\left(\mathbf{H}^i + \mathbf{H}^s\right), \end{array} \tag{1.2}$$

provided we neglect $\delta\varepsilon \cdot \mathbf{E}^s$ and if no magnetization occurs in the gas. The fields $\mathbf{E}$, $\mathbf{D}$, $\mathbf{H}$, $\mathbf{B}$ and $\mathbf{E}^i$, $\mathbf{D}^i$, $\mathbf{H}^i$, $\mathbf{B}^i$ satisfy the Maxwell equations; therefore, $\mathbf{E}^s$, $\mathbf{D}^s$, $\mathbf{H}^s$, and $\mathbf{B}^s$ must also satisfy these equations so that we have

$$\begin{array}{lll} \frac{\partial \mathbf{B}^s}{\partial t} + \text{curl } \mathbf{E}^s = 0, & \text{div } \mathbf{B}^s = 0, & \\ & & \text{or by elimination of } \mathbf{B}^s, \mathbf{E}^s \\ -\frac{\partial \mathbf{D}^s}{\partial t} + \text{curl } \mathbf{H}^s = 0, & \text{div } \mathbf{D}^s = 0 & \end{array}$$

$$\begin{array}{ll} \mu_0 \frac{\partial \mathbf{H}^s}{\partial t} + \frac{1}{\bar{\varepsilon}} \text{ curl } \mathbf{D}^s = \frac{1}{\bar{\varepsilon}} \text{ curl}\left(\delta\varepsilon\mathbf{E}^i\right), & \text{div } \mathbf{H}^s = 0, \\ -\frac{\partial \mathbf{D}^s}{\partial t} + \quad \text{curl } \mathbf{H}^s = 0, & \text{div } \mathbf{D}^s = 0. \end{array} \tag{1.3}$$

By elimination of $\mathbf{H}^s$ we obtain with $c^2 = 1/\bar{\varepsilon}\mu_0$

$$\frac{1}{c^2}\frac{\partial^2 \mathbf{D}^s}{\partial t^2} - \Delta\mathbf{D}^s = \text{curl curl }\left(\delta\varepsilon\mathbf{E}^i\right). \tag{1.4}$$

We introduce a potential $\boldsymbol{\pi}$ defined by $\mathbf{D}^s = \text{curl curl } \boldsymbol{\pi}$ and obtain the differential equation

$$\frac{1}{c^2}\frac{\partial^2 \boldsymbol{\pi}}{\partial t^2} - \Delta\boldsymbol{\pi} = \delta\varepsilon \cdot \mathbf{E}^{(i)} + \mathbf{F} \quad \text{with curl curl } \mathbf{F} = 0. \tag{1.5}$$

The solution that vanishes at infinity is the well-known retarded potential given by

$$\boldsymbol{\pi}(\mathbf{R}, t) = \frac{1}{4\pi}\int_V d\mathbf{r}\frac{\delta\varepsilon(\mathbf{r}, t')\mathbf{E}^i(\mathbf{r}, t')}{|\mathbf{R} - \mathbf{r}|}, \quad \text{where} \quad t' = t - \frac{|\mathbf{R} - \mathbf{r}|}{c} \tag{1.6}$$

is the time it takes the light to reach the detector from the scattering site in V. The function $\mathbf{F}$ in (1.5) can be set equal to zero without loss of generality.

We calculate $\mathbf{D}^s(\mathbf{R},t)$ under two simplifying assumptions, namely,

i. *slow fluctuation* of $\delta\varepsilon(\mathbf{r},t)$, and
ii. detector *far-away* from the scattering site.

Assumption *i.* allows us to set $\delta\varepsilon(\mathbf{r},t') \approx \delta\varepsilon(\mathbf{r},t)$; it implies that the size of the fluctuation $\delta\varepsilon = \varepsilon - \bar{\varepsilon}$ does not change appreciably while the scattered light travels to the detector. By assumption *ii.* we retain only *the* term in

$$\mathbf{D}^s = \text{curl curl } \boldsymbol{\pi}$$

that is of order $\frac{1}{R}$. Thus we obtain as the electric field of scattering on a "slow" fluctuation in the far-field approximation, i.e., at the site of the detector, where we assume $\mathbf{E}^s$ to be equal to $\frac{1}{\varepsilon_0}\mathbf{B}^s$

$$E_p^s(\mathbf{R},t) = \frac{1}{4\pi\varepsilon_0 c^2}\frac{1}{R}\left(\frac{R_p R_q}{R^2} - \delta_{pq}\right)\int_V d\mathbf{r}\,\delta\varepsilon(\mathbf{r},t)\frac{\partial^2 E_q^i(\mathbf{r},t')}{\partial t'^2}. \tag{1.7}$$

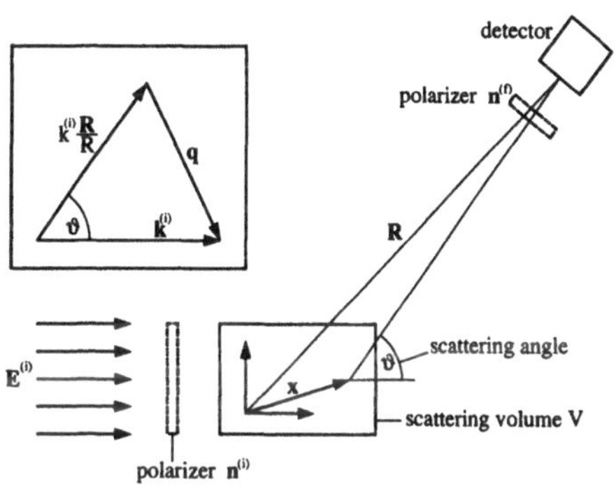

Figure 10.1 Incident light is scattered at point $\mathbf{r}$ in the scattering volume V.

1.2 Incident Plane Harmonic Wave

If the incident electromagnetic field is a plane harmonic light wave of frequency ω^i, wave length λ_i, and wave number k^i with the complex representation

$$\hat{E}_q^i(\mathbf{r},t') = n_q^i \hat{E}_0^i\, e^{i(k_n^i r_n - \omega^i t')}, \tag{1.8}$$

we obtain

$$\frac{\partial^2 \hat{E}_q^i(\mathbf{r},t')}{\partial t'^2} = -n_q^i \hat{E}_0^i \omega^{i2}\, e^{i\left(\left(k_e^i - \frac{\omega}{c}\frac{R_e}{R}\right)r_e + \frac{\omega^i}{c}R - \omega^i t\right)}. \tag{1.9}$$

In (1.9) $t' - t$ has been written as $\frac{1}{c}\frac{R_e}{R}r_e - \frac{1}{c}R$ as is appropriate for the far field. $\mathbf{n}^i$ is the unit vector in the direction of the incident electric field and $\hat{E}_0^i$ is the complex amplitude of the wave; the actual incident electric field is given by $\mathbf{E}^i = \Re e\hat{\mathbf{E}}^i$.

We insert (1.9) into (1.7) and obtain

$$\hat{E}_p^s(\mathbf{R},t) = -\frac{\hat{E}_0^i}{4\pi\varepsilon_0 c^2}\frac{1}{R}\omega^{i2}\left(\frac{R_p R_q}{R^2} - \delta_{pq}\right). \tag{1.10}$$

$$\left[\int_V d\mathbf{r}\, e^{i \underbrace{\left(k_e^i - \frac{\omega^i}{c}\frac{R_e}{R}\right)}_{\mathbf{q}_e} r_e} \delta\varepsilon(\mathbf{r}, t)\right] n_q^i e^{i\left(\frac{\omega^i}{c}R - \omega^i t\right)},$$

$$\hat{E}_p^s(\mathbf{R}, t) = -\tfrac{\hat{E}_0^i}{4\pi\varepsilon_0 c^2}\tfrac{1}{R}\omega^{i2}\left(\tfrac{R_p R_q}{R^2} - \delta_{pq}\right) \underbrace{\left[\int_V d\mathbf{r}\, e^{i q_e r_e} \delta\varepsilon(\mathbf{r}, t)\right]}_{\delta\varepsilon(\mathbf{q}, t)} n_q^i e^{i\left(\frac{\omega^i}{c}R - \omega^i t\right)}, \tag{1.11}$$

$$\hat{E}_p^s(\mathbf{R}, t) = -\tfrac{\hat{E}_0^i}{4\pi\varepsilon_0 c^2}\tfrac{1}{R}\omega^{i2}\left(\tfrac{R_p R_q}{R^2} - \delta_{pq}\right) \delta\varepsilon(\mathbf{q}, t) n_q^i e^{i\left(\frac{\omega^i}{c}R - \omega^i t\right)}. \tag{1.12}$$

If the detector receives the oscillations of the electric field in the direction n_p^s we record, since $\mathbf{R} \cdot \mathbf{n}^f \approx 0$

$$\hat{E}^s(\mathbf{R}, t) = \underbrace{\frac{\hat{E}_0^i}{4\pi}\mu_0 \frac{(\omega^i)^2}{R} \mathbf{n}^i \cdot \mathbf{n}^f\, e^{i\frac{\omega^i}{c}R} \delta\varepsilon(\mathbf{q}, t)}_{\hat{a}(t)}\, e^{-i\omega^i t}. \tag{1.13}$$

We note that $\delta\varepsilon(\mathbf{q}, t)$, defined by (1.11), is the spatial Fourier component of the fluctuation $\delta\varepsilon(\mathbf{r}, t)$ corresponding to the wave number $\mathbf{q}$, and we conclude that

- The scattered field at the position $\mathbf{R}$ is represented by an amplitude-modulated oscillation with the basic frequency ω^i and complex amplitudes $\hat{a}(t)$ (see Figure 10.2).
- The amplitude modulation at $\mathbf{R}$ is determined by the fluctuation $\delta\varepsilon(\mathbf{r}, t)$ in all points $\mathbf{r}\epsilon V$ but only through the combination $\delta\varepsilon(\mathbf{q}, t)$; i.e. the scattered field at the position $\mathbf{R}$ is determined by *the* spatial harmonic of the fluctuation field $\delta\varepsilon(\mathbf{r}, t)$, which has the wave number $\mathbf{q} = \mathbf{k}^i - \frac{\omega^i}{c}\frac{\mathbf{R}}{R}$ (see Figure 10.1).
- The magnitude of $\mathbf{q}$ is given by

$$q = 2k^i \sin\frac{\vartheta}{2} = \frac{4\pi}{\lambda^i}\sin\frac{\vartheta}{2} = 2\frac{\omega^i}{c}\sin\frac{\vartheta}{2}. \tag{1.14}$$

  That is to say, forward scattering with $\vartheta \ll \pi$, hence $q \ll 4\pi/\lambda^i$, represents the large scale fluctuations, and backward scattering with $\vartheta < \pi$, hence $q < 4\pi/\lambda^i$, represents small scale fluctuations of the order of magnitude of the wavelength of the incident light.

It follows that the position of the detector in the range $0 \leq \vartheta \leq \pi$ determines the scale of the fluctuation that has scattered the light which is being detected.

Typically wavelength and frequency of the incident laser light have the values

$$\lambda^i = 0.4 \cdot 10^{-6} m \qquad \omega^i = 4.7 \cdot 10^{15}\frac{1}{s}. \tag{1.15}$$

It is clear from (1.14) that $q \leq 4\pi/\lambda^i$ holds, which means that the scattered light contains no information about features of the fluctuation that are smaller than $2\pi/q = 2 \cdot 10^{-7} m$.

For more information about light scattering, theory and application, see the monograph by Berne & Pecora [112] or the review [113] by Fleury & Boon.

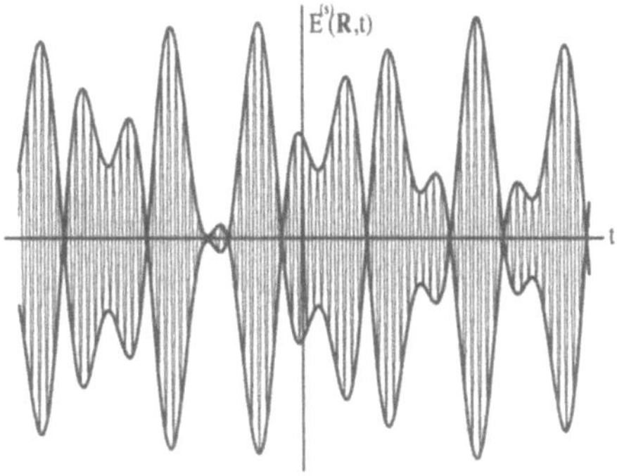

Figure 10.2 A qualitative picture of $\mathbf{E}^s(\mathbf{R},t) = \Re e\hat{\mathbf{E}}^s(\mathbf{R},t)$.

2 A Modicum of Fluctuation Theory

2.1 Expectation Values

2.1.1 Fluctuation Theory

We consider an adiabatic system of volume V_G containing the scattering volume V. The surface of V_G is at rest and the thermodynamic state of the system is given by $\rho(\mathbf{x},t), v_i(\mathbf{x},t), T(\mathbf{x},t)$ and by r fields of "internal" variables $I_\kappa(\mathbf{x},t)$ $(\kappa = 1,2,..r)$.[1] The random fluctuations from the constant and uniform equilibrium state $\tilde{\rho}, \tilde{v}_i = 0, \tilde{T}, \tilde{I}_\kappa = 0$ are

$$\delta\rho(\mathbf{x},t), \delta v_i(\mathbf{x},t), \delta T(\mathbf{x},t), \delta I_\kappa(\mathbf{x},t). \tag{2.1}$$

They are constrained by the requirement that the total mass and energy be constant and that the total momentum vanish. Thus

$$\int\limits_{V_G} \delta\rho \, d\mathbf{x} = 0, \quad \int\limits_{V_G} \delta(\rho v_i) d\mathbf{x} = 0, \quad \int\limits_{V_G} \delta\left(\rho\varepsilon + \frac{\rho}{2}v^2\right) d\mathbf{x} = 0, \tag{2.2}$$

where ε is the specific internal energy.

So as to simplify the mathematics we discretize V_G into a large number N of small cells ($\alpha = 1,2,...N$) of equal volume ΔV. The state of each cell is then given by

$$\rho^{(\alpha)}(t), v_i^{(\alpha)}(t), T^{(\alpha)}(t), I_\kappa^{(\alpha)}(t).$$

[1] In this book the internal variables will be the components of the heat flux or of the stress deviator, or of higher moments.

A fluctuation is characterized by

$$\delta\rho^{(\alpha)}, \delta v_i^{(\alpha)}, \delta T^{(\alpha)}, \delta I_\kappa^{(\alpha)}. \tag{2.3}$$

and the constraints (2.2) now read

$$\sum_{\alpha=1}^{N} \delta\rho^{(\alpha)} = 0, \quad \sum_{\alpha=1}^{N} \delta\left(\rho^{(\alpha)} v_i^{(\alpha)}\right) = 0, \quad \sum_{\alpha=1}^{N} \delta\left(\rho^{(\alpha)}\varepsilon^{(\alpha)} + \frac{\rho_{(\alpha)}}{2}\, v_{(\alpha)}^2\right) = 0. \tag{2.4}$$

We shall consider the case that the leading term in the entropy is given by $\sum_{\alpha=1}^{N} \rho^{(\alpha)}\, s(\rho^{(\alpha)}, \varepsilon^{(\alpha)})\Delta V$ where $s(\rho, \varepsilon)$ is the entropic state function. In addition to this there is a contribution from the internal variables of the form

$$-\sum_{\alpha=1}^{N}\sum_{\kappa=1}^{r} i_\kappa\left(\rho^{(\alpha)}, \varepsilon^{(\alpha)}\right) I_\kappa^{(\alpha)2}\Delta V,$$

so that the internal variables affect the entropy by a sum of squares. While this is not the most general case conceivable, it is the case provided by extended thermodynamics and by the kinetic theory—if the internal variables are chosen properly.

The entropy fluctuation thus reads

$$\delta S = \left[\sum_{\alpha=1}^{N} \delta\left(\rho^{(\alpha)} s\left(\rho^{(\alpha)}, \varepsilon^{(\alpha)}\right)\right) - \sum_{\alpha=1}^{N}\sum_{\kappa=1}^{r} \delta\left(i_\kappa\left(\rho^{(\alpha)}\varepsilon^{(\alpha)}\right) I_\kappa^{(\alpha)2}\right)\right]\Delta V; \tag{2.5}$$

it determines the probability density p to find the system with a fluctuation in the range

$$d\Xi = \prod_{\alpha=1}^{N} d\left(\delta\rho^{(\alpha)}\right) d\left(\delta v_i^{(\alpha)}\right) d\left(\delta T^{(\alpha)}\right) \prod_{\kappa=1}^{r} d\left(\delta I_\kappa^{(\alpha)}\right).$$

Specifically we have with k as the Boltzmann constant

$$pd\Xi = \frac{e^{\frac{\delta S}{k}}\hat{\delta}\left(\sum_{\alpha=1}^{N}\delta\rho^{(\alpha)}\right)\hat{\delta}\left(\sum_{\alpha=1}^{N}\delta\left(\rho^{(\alpha)}v_i^{(\alpha)}\right)\right)\hat{\delta}\left(\sum_{\alpha=1}^{N}\delta\left(\rho^{(\alpha)}\left(\varepsilon^{(\alpha)}+\frac{1}{2}v_i^{(\alpha)}v_i^{(\alpha)}\right)\right)\right)}{\int\ldots\int\left(-------------------------\right)d\Xi}d\Xi. \tag{2.6}$$

This formula is an adaptation of Boltzmann's formula $S = klnW$ appropriate to the present case. $\hat{\delta}$ denotes the Dirac delta function. The $\hat{\delta}$'s acknowledge the constraints (2.2). The broken line stands for the expression which already appears in the numerator.

We expand δS in a Taylor series, including second-order terms in $\delta\rho^{(\alpha)}, \delta\left(\rho^{(\alpha)}v_i^{(\alpha)}\right), \delta\left(\rho^{(\alpha)}\varepsilon^{(\alpha)}\right)$, and obtain

$$\begin{aligned}\frac{\delta S}{\Delta V} &= \frac{\partial\rho s}{\partial(\rho\varepsilon)}\sum_{\alpha=1}^{N}\delta\left(\rho^{(\alpha)}\varepsilon^{(\alpha)}\right) + \frac{\partial\rho s}{\partial\rho}\sum_{\alpha=1}^{N}\delta\rho^{(\alpha)} + \\ &+\frac{1}{2}\frac{\partial^2\rho s}{\partial(\rho\varepsilon)^2}\sum_{\alpha=1}^{N}\delta\left(\rho^{(\alpha)}\varepsilon^{(\alpha)}\right)^2 + \frac{\partial^2\rho s}{\partial\rho\partial(\rho\varepsilon)}\sum_{\alpha=1}^{N}\delta\rho^{(\alpha)}\delta\left(\rho^{(\alpha)}\varepsilon^{(\alpha)}\right) + \\ &+\frac{1}{2}\frac{\partial^2\rho s}{\partial\rho^2}\sum_{\alpha=1}^{N}\delta\rho^{(\alpha)2} - \sum_{\kappa=1}^{r}\tilde{i}_\kappa\sum_{\alpha=1}^{N}\delta I_\kappa^{(\alpha)}\delta I_\kappa^{(\alpha)}.\end{aligned} \tag{2.7}$$

By the rules of classical thermodynamics we have

$$\frac{\partial(\rho s)}{\partial(\rho\varepsilon)} = \frac{1}{T}, \qquad \frac{\partial(\rho s)}{\partial\rho} = -\frac{g}{T}, \qquad \text{and}$$
$$\frac{\partial^2(\rho s)}{\partial(\rho\varepsilon)^2}(\partial\rho\varepsilon)^2 + 2\frac{\partial^2(\rho s)}{\partial\rho\partial(\rho\varepsilon)}\partial(\rho\varepsilon)\partial\rho + \frac{\partial^2(\rho s)}{\partial\rho^2}(\partial\rho)^2 = -\frac{\rho}{T^2}c_v\,(\delta T)^2 - \frac{1}{\rho^2 T}\frac{1}{\kappa_T}(\partial\rho)^2,$$

where g is the specific free enthalpy. Therefore we may write

$$\begin{aligned}\frac{\delta S}{k\Delta V} &= -\frac{1}{2\rho^2 kT}\frac{1}{\kappa_T}\sum_{\alpha=1}^{N}\left(\delta\rho^{(\alpha)}\right)^2 - \frac{\rho}{2kT}\sum_{\alpha=1}^{N}\left(\delta v_i^{(\alpha)}\delta v_i^{(\alpha)}\right) - \\ &\quad -\frac{\rho}{2kT^2}\,c_v\sum_{\alpha=1}^{N}\left(\delta T^{(\alpha)}\right)^2 - \sum_{\kappa=1}^{r} i_\kappa\,(\rho,T)\sum_{\alpha=1}^{N}\delta I_\kappa^{(\alpha)}\delta I_\kappa^{(\alpha)}.\end{aligned} \tag{2.8}$$

κ_T and c_v are the isothermal compressibility and specific heat at constant ρ, respectively.

With (2.8) the exponential in the fluctuation probability (2.6) becomes an explicit expression. Since δS, according to (2.8), contains no first-order terms and since we ignore higher than second-order terms, the constraints (2.4) may be written as

$$\sum_{\alpha=1}^{N}\delta\rho^{(\alpha)} = 0, \qquad \sum_{\alpha=1}^{N}\delta v_i^{(\alpha)} = 0, \qquad \sum_{\alpha=1}^{N}\delta T^{(\alpha)} = 0. \tag{2.9}$$

Thus the fluctuation probability reads

$$pd\Xi = \frac{\exp([\ldots]\Delta V)\hat\delta\left(\sum_{\alpha=1}^{N}\delta\rho^{(\alpha)}\right)\hat\delta\left(\sum_{\alpha=1}^{N}\delta v_i^{(\alpha)}\right)\hat\delta\left(\sum_{\alpha=1}^{N}\delta T^{(\alpha)}\right)d\Xi}{\int\ldots\int(-----------------------------)d\Xi}. \tag{2.10}$$

where the square bracket in the exponent is given by the right-hand side of (2.8). Using the Fourier representation of the $\hat\delta$-function we may write

$$pd\Xi = \frac{\exp([\ldots]\Delta V)\iiint_{-\infty}^{\infty}\exp\left(i\omega\sum_{\alpha=1}^{N}\delta\rho^{(\alpha)} + i\nu_j\sum_{\alpha=1}^{N}\delta v_j^{(\alpha)} + i\gamma\sum_{\alpha=1}^{N}\delta T^{(\alpha)}\right)d\Xi\;d\omega\;d\nu_j\;d\gamma}{\int\ldots\int\iiint_{-\infty}^{\infty}(-----------------------------------)d\Xi\;d\omega\;d\nu_j\;d\gamma}. \tag{2.11}$$

The denominator D of the expression is easily calculated by quadratic extension in the exponent. We obtain

$$D = \sqrt{\frac{2\pi\Delta V}{N\rho^2 kT\kappa_T}}\sqrt{\frac{2\pi\rho^2 kT\kappa_T}{\Delta V}}^{N}\sqrt{\frac{2\pi\Delta V}{NkT}}^{3}\sqrt{\frac{2\pi kT}{\rho\Delta V}}^{3N}\sqrt{\frac{2\pi 3\rho\Delta V}{2NmT^2}}\sqrt{\frac{2\pi kT^2}{\rho c_v\Delta V}}^{N}\cdot\prod_{\kappa=1}^{r}\sqrt{\frac{\pi}{i_\kappa\Delta V}}^{N}. \tag{2.12}$$

2.1.2 Expectation Values

Expectation values of the fluctuations $\delta\rho^{(\alpha)}, \delta v_i^{(\alpha)}, \delta T^{(\alpha)}$, and $\delta I^{\kappa(\alpha)}$ are equal to zero, because of the specific form (2.11) of the probability. But products of fluctuations, like $\delta\rho^{(\alpha)}\;\delta\rho^{(\alpha')}$, have nonvanishing expectation values. They may be written in the form

$$\overline{\delta\rho^{(\alpha)}\delta\rho^{(\alpha')}} = \int\delta\rho^{(\alpha)}\delta\rho^{(\alpha')}\;pd\Xi \tag{2.13}$$

and an easy calculation—by use of (2.11) and quadratic extension in the exponent—gives

$$\overline{\delta\rho^{(\alpha)}\delta\rho^{(\alpha')}} = \frac{\rho^2 kT\kappa_T}{\Delta V}\left(\delta_{\alpha\alpha'} - \frac{1}{N}\right). \tag{2.14}$$

Similarly we obtain for the expectation values of the products of all variables

$$\begin{bmatrix} \overline{\delta\rho^{(\alpha)}\delta\rho^{(\alpha')}} & \overline{\delta\rho^{(\alpha)}\delta v_i^{(\alpha')}} & \overline{\delta\rho^{(\alpha)}\delta T^{(\alpha')}} & \overline{\delta\rho^{(\alpha)}\delta I_\kappa^{(\alpha')}} \\ \overline{\delta v_j^{(\alpha)}\delta\rho^{(\alpha')}} & \overline{\delta v_j^{(\alpha)}\delta v_i^{(\alpha')}} & \overline{\delta v_j^{(\alpha)}\delta T^{(\alpha')}} & \overline{\delta v_j^{(\alpha)}\delta I_\kappa^{(\alpha')}} \\ \overline{\delta T^{(\alpha)}\delta\rho^{(\alpha')}} & \overline{\delta T^{(\alpha)}\delta v_i^{(\alpha')}} & \overline{\delta T^{(\alpha)}\delta T^{(\alpha')}} & \overline{\delta T^{(\alpha)}\delta I_\kappa^{(\alpha')}} \\ \overline{\delta I_{\kappa'}^{(\alpha)}\delta\rho^{(\alpha')}} & \overline{\delta I_{\kappa'}^{(\alpha)}\delta v_i^{(\alpha')}} & \overline{\delta I_{\kappa'}^{(\alpha)}\delta T^{(\alpha')}} & \overline{\delta I_{\kappa'}^{(\alpha)}\delta I_\kappa^{(\alpha')}} \end{bmatrix} = \tag{2.15}$$

$$\begin{bmatrix} \frac{\rho^2 kT\kappa_T}{\Delta V}\left(\delta_{\alpha\alpha'} - \frac{1}{N}\right) & 0 & 0 & 0 \\ 0 & \frac{kT}{\rho\Delta V}\left(\delta_{\alpha\alpha'} - \frac{1}{N}\right)\delta_{ij} & 0 & 0 \\ 0 & 0 & \frac{kT^2}{\rho c_v\Delta V}\left(\delta_{\alpha\alpha'} - \frac{1}{N}\right) & 0 \\ 0 & 0 & 0 & \frac{1}{2i_\kappa\Delta V}\delta_{\alpha\alpha'}\delta_{\kappa\kappa'} \end{bmatrix}$$

In subsequent parts of this book we shall be interested in the expectation values of the **q**-harmonics $\delta\rho(\mathbf{q},0), \delta v_i(\mathbf{q},0)$, $\delta T(\mathbf{q},0)$, $\delta I_\kappa(\mathbf{q},0)$ of the Fourier transforms of the fields $\delta\rho(\mathbf{x},0)$, $\delta v_i(\mathbf{x},0)$, $\delta T(\mathbf{x},0)$ and $\delta I_\kappa(\mathbf{x},0)$ at time $t=0$. More specifically, we shall have to determine quantities like $\overline{\delta\rho^*(\mathbf{q},0)\ \delta\rho(\mathbf{q},0)}$, $\overline{\delta\rho^*(\mathbf{q},0)\ \delta v_i(\mathbf{q},0)}$, etc. These are given by

$$\overline{\delta\rho^*(\mathbf{q},0)\ \delta\rho(\mathbf{q},0)} = \int_V\int_V \overline{\delta\rho(\mathbf{x},0)\ \delta\rho(\mathbf{x}',0)}\ e^{i\mathbf{q}\cdot(\mathbf{x}-\mathbf{x}')}d\mathbf{x}d\mathbf{x}', \tag{2.16}$$

where V is the scattering volume. Using the discretization into cells ΔV, introduced in paragraph 2.1.1, we may write (2.16) in the form

$$\overline{\delta\rho^*(\mathbf{q},0)\ \delta\rho(\mathbf{q},0)} = (\Delta V)^2\sum_{\alpha,\alpha'}^{M}\overline{\delta\rho^{(\alpha)}\delta\rho^{(\alpha')}}\ e^{i\mathbf{q}\cdot(\mathbf{x}^{(\alpha)}-\mathbf{x}^{(\alpha')})}\ , \tag{2.17}$$

where M is the number of cells in the scattering volume. With (2.15) and $V_G = N\Delta V$ we obtain

$$\overline{\delta\rho^*(\mathbf{q},0)\delta\rho(\mathbf{q},0)} = \rho^2 kT\kappa_T V\underbrace{\left(1 - \frac{(\Delta V)^2}{VV_G}\sum_{\alpha,\alpha'}^{M} e^{i\mathbf{q}\cdot(\mathbf{x}^{(\alpha)}-\mathbf{x}^{(\alpha')})}\right)}_{Q(\mathbf{q})}. \tag{2.18}$$

The other expectation values of Fourier harmonics are similarly calculated. In particular, of course, if the expectation values themselves vanish, so do the expectation values of Fourier harmonics. We summarize the results in the following matrix:

$$\begin{bmatrix} \overline{\delta\rho^*\delta\rho} & \overline{\delta\rho^*\delta v_i} & \overline{\delta\rho^*\delta T} & \overline{\delta\rho^*\delta I_\kappa} \\ \overline{\delta v_j^*\delta\rho} & \overline{\delta v_j^*\delta v_i} & \overline{\delta v_j^*\delta T} & \overline{\delta v_j^*\delta I_\kappa} \\ \overline{\delta T^*\delta\rho} & \overline{\delta T^*\delta v_i} & \overline{\delta T^*\delta T} & \overline{\delta T^*\delta I_\kappa} \\ \overline{\delta I_{\kappa'}^*\delta\rho} & \overline{\delta I_{\kappa'}^*\delta v_i} & \overline{\delta I_{\kappa'}^*\delta T} & \overline{\delta I_{\kappa'}^*\delta I_\kappa} \end{bmatrix} = \begin{bmatrix} \rho^2 kT\kappa_T VQ & 0 & 0 & 0 \\ 0 & \frac{kT}{\rho}VQ\delta_{ij} & 0 & 0 \\ 0 & 0 & \frac{kT^2}{\rho c_v}VQ & 0 \\ 0 & 0 & 0 & \frac{1}{2i_\kappa}V\delta_{\kappa\kappa'} \end{bmatrix}. \tag{2.19}$$

Each $\delta\rho^*$, $\delta\rho$, δv_j^*, δv_i etc. on the left-hand side has the argument $(\mathbf{q}, 0)$.

The definition of the function Q is indicated in (2.18). This function is best evaluated by reconverting the sums into integrals. Thus

$$Q = 1 - \frac{1}{VV_G}\int\limits_{V}\int\limits_{V} e^{-i\mathbf{q}\cdot(\mathbf{x}-\mathbf{x}')}d\mathbf{x}d\mathbf{x}' = 1 - \frac{1}{VV_G}\left(\int\limits_{V} e^{-i\mathbf{q}\cdot\mathbf{x}}d\mathbf{x}\right)^2. \tag{2.20}$$

The integral is easily solved if V is a cube of side length $2L$, and we obtain

$$Q = 1 - \frac{V}{V_G}\left(\frac{\sin(q_x L)}{q_x L}\right)^2\left(\frac{\sin(q_y L)}{q_y L}\right)^2\left(\frac{\sin(q_z L)}{q_z L}\right)^2. \tag{2.21}$$

Note that the $\mathbf{q}$-dependence of Q comes from the consideration of the constraints (2.2), or (2.4). These are inessential for our purposes, as we shall presently see. Therefore we shall set $Q = 1$ in what follows.

2.1.3 Length Scales

According to (2.21) $Q(\mathbf{q})$ is different from 1 essentially only if $q < \frac{\pi}{L}$ holds, i.e. if the length scale $\Lambda = \frac{2\pi}{q}$ of a fluctuation is bigger than the whole scattering volume. If we take L to be of the order of magnitude of $1cm$ we may use (1.14) and (1.15) to estimate that such values of q could be observed for scattering angles $\sin\frac{\vartheta}{2} < 10^{-5}$. This is quite impractical, because at such an angle the scattered light could not be separated from the transmitted part of the incident light. In other words, realistic detector positions will imply that $q > \frac{\pi}{L}$ holds and therefore $Q = 1$ is an excellent approximation for all observations of scattered light. This means that the length scale of observed fluctuations is smaller than $2L$.

While $\Lambda = 2L$ is a natural limitation for the length scale of a fluctuation, it is not the only one. There is also a lower bound for Λ which is provided by the size ΔV of the cells. And while we have not limited that size so far, there is an implicit limitation in our assumptions and results. Indeed, from (2.15) we conclude that the fluctuations of different cells—even neighbouring ones—are uncorrelated [provided that we ignore the $\frac{1}{N}$-terms, which we may do, according to the preceding discussion about $Q \approx 1$]. This lack of correlation must mean that the cells are bigger than the range of interaction L_{int} of the atoms. Therefore we have for the length scale $\Lambda = \frac{2\pi}{q}$ of a fluctuation

$$L_{int} < \Lambda < L \quad \text{or} \quad \frac{2\pi}{L} < q < \frac{2\pi}{L_{int}}. \tag{2.22}$$

All discussions—previous and subsequent—about q being small or large concern the range of q's defined by $(2.22)_2$.

Actually the upper bound of q according to (2.22) is illusory for light scattering, because we have already seen that light cannot carry information about fluctuations with $\Lambda < 2\cdot 10^{-7}m$ which is much bigger than L_{int} ; hence we must have $q < 3\cdot 10^{7}\frac{1}{m}$. There are other types of scattering—notably neutron scattering—which may reach beyond these limits.

It is a good thing that we find it possible to set $Q = 1$ and thus ignore the constraints (2.2). Indeed these conventional constraints are not the only ones when we come to extended thermodynamics. Let us consider: First, since invariably the heat flux q_i is among the variables I_κ in extended thermodynamics, adiabaticity of the system in V_G implies a constraint of the form

$$\int_{\partial V_G} q_i n_i dA = 0 \tag{2.23}$$

along with the common constraints (2.2). Since the latter did not affect the expectation value $\overline{\rho^*(q,0)\rho(q,0)}$ for $V << V_G$ we hope – and trust – that (2.23) does not either.

Also, in extended thermodynamics an adiabatic system is not necessarily a system in which the entropy grows. This is so, because the entropy flux contains many terms apart from the leading term which reads q_i/T. The conditions for a growing entropy – and hence the basis for the probability distribution (2.6) – depend on the degree of the extension which is chosen. They will surely introduce more constraints, apart from (2.2) and (2.23). None of those are taken into account here in the hope that they do not matter much.

2.2 Temporal Evolution of a Fluctuation

2.2.1 Mean Regression and Autocorrelation

We consider fluctuations in time only, like the fluctuation of the density $\delta\rho^{(\alpha)}(t)$ at one point, or—more appropriately—the fluctuation of the $\mathbf{q}$-harmonic $\delta\rho(\mathbf{q},t)$ of density. While the fluctuations seem to be entirely stochastic, averaging brings out quite regular features about their decay as follows:

Let a particular value A_0 of a fluctuating real function $A(t)$—with vanishing mean value—recur at times $t_i (i = 1, 2, ...N >> 1)$. We define the mean regression of A_0 at time τ as the mean value

$$R_A(\tau; A_0) = \frac{1}{N}\sum_{i=1}^{N} A(t_i + \tau), \quad \text{where} \quad A(t_i) = A_0 \tag{2.24}$$

over many recurrence times t_i. This is a smooth even function of τ. That function has the value A_0 at $\tau = 0$ and we expect it to fall off to zero for large $|\tau|$, because of the randomness of the fluctuation.

The autocorrelation of $A(t)$ represents an average of mean regressions $R_A(\tau; A_0)$ over all initial values A_0. Before averaging we multiply $R_A(\tau; A_0)$ by A_0 so as to avoid mutual cancellation of the functions for positive and negative values of A_0. We denote the autocorrelation by $\langle A(0)\ A(\tau)\rangle$ and from the definition we conclude that it may be calculated as the time average

$$\langle A(0)\ A(\tau)\rangle = \frac{1}{T}\int_0^T A(t)\ A(t+\tau)dt, \tag{2.25}$$

where T is large on the scale of fluctuation times.

2.3 Autocorrelation of $E^s(\mathbf{R}, t)$

2.3.1 Autocorrelation

According to (2.25) the autocorrelation of $E^s(\mathbf{R}, t)$ reads

$$\begin{aligned}
\langle E^s(\mathbf{R},0)\ E^s(\mathbf{R},\tau)\rangle &= \tfrac{1}{T}\int_0^T E^s(\mathbf{R},t)\ E^s(\mathbf{R},t+\tau)dt \\
&= \tfrac{1}{T}\int_0^T \Re e\hat{E}^s(\mathbf{R},t)\Re e\ \hat{E}^s(\mathbf{R},t+\tau)dt \quad \text{and with (1.13)} \\
&= \tfrac{1}{2T}\int_0^T \Re e\left[\hat{a}(t)\hat{a}(t+\tau)\ e^{-i\omega^i(2t+\tau)}\right] dt+ \\
&\quad +\tfrac{1}{2T}\int_0^T \Re e\left[\hat{a}(t)\hat{a}^*(t+\tau)\ e^{i\omega^i\tau}\right] dt \\
&= \tfrac{1}{2}\Re e\left[\left(\tfrac{1}{T}\int_0^T \hat{a}(t)\hat{a}(t+\tau)dt\right) e^{i\omega^i\tau}\right],
\end{aligned}$$

since the first integral vanishes for large T. Reintroducing $\hat{a}(t)$ from (1.13), we obtain

$$\begin{aligned}
\langle E^s(\mathbf{R},0)\ E^s(\mathbf{R},\tau)\rangle &= \frac{1}{2}\frac{E_0^{i2}}{(4\pi)^2}\mu_0^2\frac{\omega^{i4}}{R^2}\left(\mathbf{n}^i\cdot\mathbf{n}^f\right)^2\cdot \\
&\quad \cdot\Re e\left[\left(\frac{1}{T}\int_0^T \delta\varepsilon^*(\mathbf{q},t)\delta\varepsilon(\mathbf{q},t+\tau)dt\right) e^{i\omega^i\tau}\right].
\end{aligned} \tag{2.26}$$

It can be proved that the integral in (2.26) is real. We call the time-average in (2.26)—i.e. the expression in round brackets—the autocorrelation of $\delta\varepsilon(\mathbf{q}, t)$ and denote it by $\langle\delta\varepsilon^*(\mathbf{q}, 0)\ \delta\varepsilon(\mathbf{q}, \tau)\rangle$, often dropping $\mathbf{q}$.[2] Thus

$$\langle E^s(\mathbf{R},0)\ E^s(\mathbf{R},\tau)\rangle = \frac{1}{2}\frac{E_0^{i2}}{(4\pi)^2}\mu_0^2\frac{\omega^{i4}}{R^2}\left(\mathbf{n}^i\cdot\mathbf{n}^f\right)^2\langle\delta\varepsilon^*(\mathbf{q},0)\delta\varepsilon(\mathbf{q},\tau)\rangle\cos\omega^i\tau. \tag{2.27}$$

This autocorrelation of $E^s(\mathbf{R}, t)$ once again is a high frequency oscillation—with frequency ω^i—whose amplitude is modulated by the autocorrelation; see Figure 10.3 for a schematic picture.

[2] Since $\delta\varepsilon(\mathbf{q}, t)$ is complex, the autocorrelation $\langle\delta\varepsilon^*(0)\delta\varepsilon(\tau)\rangle$ is a combination of the two ingredients of the fluctuating q-harmonic $\delta\varepsilon(\mathbf{q}, t)$, e.g. of its amplitude and its phase. Note that $\langle\delta\varepsilon^*(0)\delta\varepsilon(\tau)\rangle$ is an average of the mean regressions $R_{\delta\varepsilon}(\tau; \delta\varepsilon_0)$ over all initial values $\delta\varepsilon(\mathbf{q}, 0) = \delta\varepsilon_0$. Before averaging, $R_{\delta\varepsilon}(\tau; \delta\varepsilon_0)$ is multiplied by $\delta\varepsilon^*(\mathbf{q}, 0)$ so as to ensure that the autocorrelation begins with a positive real value.

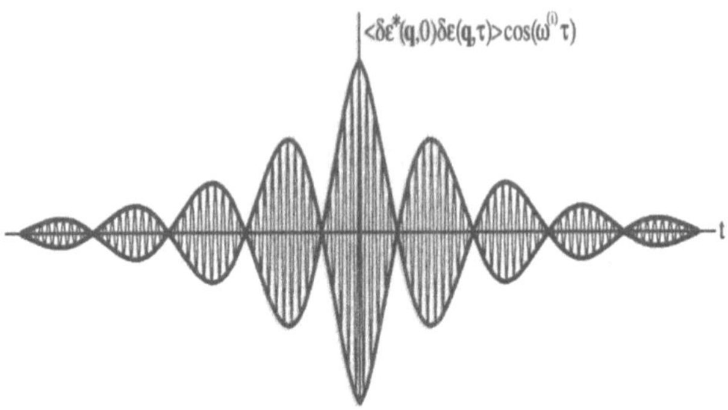

Figure 10.3 Time-dependent part of the autocorrelation of $E^s(\mathbf{R},t)$.

2.3.2 Spectral Density and Dynamic Form Factor

The temporal Fourier transform of the autocorrelation (2.26) is called the spectral density and denoted by $I(\mathbf{q},\omega)$. We have

$$\begin{aligned} I(\mathbf{q},\omega) &= \tfrac{1}{2\pi}\int_{-\infty}^{\infty} \langle E^s(0)\ E^s(\tau)\rangle\ e^{i\omega\tau}dt \quad \text{with (2.27)} \qquad (2.28) \\ &= \tfrac{1}{2}\tfrac{E_0^{i2}}{(4\pi)^2}\mu_0^2\tfrac{\omega^{i4}}{R^2}\left(\mathbf{n}^i\cdot\mathbf{n}^f\right)^2\tfrac{1}{2\pi}\int_{-\infty}^{\infty}\langle\delta\varepsilon^*(0)\delta\varepsilon(\tau)\rangle\cos(\omega^i\tau)\ e^{i\omega\tau}d\tau. \end{aligned}$$

The spectral density is important, because it is this function which is measured in an interferometer; see Section 3.

The essential part of $I(\mathbf{q},\omega)$ is the dynamic form factor $S(\mathbf{q},\omega)$ defined as

$$S(\mathbf{q},\omega) = \frac{1}{2\pi}\int_{-\infty}^{\infty}\langle\delta\varepsilon^*(\mathbf{q},0)\delta\varepsilon(\mathbf{q},\tau)\rangle\, e^{i\omega\tau}d\tau. \qquad (2.29)$$

Indeed we have by (2.28)

$$I(\mathbf{q},\omega) = \frac{1}{4}\frac{E_0^{i2}}{(4\pi)^2}\mu_0^2\frac{\omega^{i4}}{R^2}\left(\mathbf{n}^i\cdot\mathbf{n}^f\right)\left(S(\omega+\omega^i)+S(\omega-\omega^i)\right)$$

so that $I(\mathbf{q},\omega)$ is essentially the sum of two dynamic form factors centered around $\omega = \pm\omega^i$.

Since the autocorrelation is an even function of τ, we may write the dynamic form factor (2.29) as

$$S(\mathbf{q},\omega) = \tfrac{1}{\pi}\int_0^{\infty}\langle\delta\varepsilon^*(\mathbf{q},0)\delta\varepsilon(\mathbf{q},\tau)\rangle\cos\omega\tau\ d\tau$$

$$S(\mathbf{q},\omega) = \tfrac{1}{\pi}\Re e\left\langle\delta\varepsilon^*(\mathbf{q},0)\delta\hat{\varepsilon}(\mathbf{q},s)\right\rangle_{s=i\omega}\ , \qquad (2.30)$$

where $\langle \delta\varepsilon^*(\mathbf{q},0)\delta\hat{\varepsilon}(\mathbf{q},s)\rangle$ is the Laplace transform of the autocorrelation $\langle \delta\varepsilon^*(\mathbf{q},0)\delta\varepsilon(\mathbf{q},\tau)\rangle$.

ε fluctuates, because the density ρ and the temperature fluctuate and because ε depends on these values. The dependence of ε on T is usually weak and we ignore it. This leaves us with (2.29) in the form

$$S(\mathbf{q},\omega) = \frac{1}{\pi}\,\varepsilon_\rho^2\,\Re e\,\langle \delta\rho^*(\mathbf{q},0)\delta\hat{\rho}(\mathbf{q},s)\rangle_{s=i\omega}\,, \tag{2.31}$$

where ε_ρ is the derivative of the dielectric costant ε with respect to ρ.

2.3.3 Onsager Hypothesis

We have identified the dynamic form factor $S(\mathbf{q},\omega)$ (see (2.29) or (2.31)) as the essential part of the *measured* spectral density $I(\mathbf{q},\omega)$. Now we proceed to show how $S(\mathbf{q},\omega)$ may be *calculated*. The calculation is based on the Onsager hypothesis, which links fluctuations to the solutions of the macroscopic field equations. The hypothesis refers to rate-type systems of differential equations and therefore to the spatially Fourier-transformed field equations of thermodynamics.

Specifically the Onsager hypothesis equates the mean regression of the $\mathbf{q}$-harmonic of the density fluctuation $\delta\rho(\mathbf{x},t)$ to the solution $\rho(\mathbf{q},\tau)$ of the spatially Fourier-transformed field equations. [$\rho(\mathbf{q},\tau)$ is the Fourier transform of $\rho(\mathbf{x},t)-\tilde{\rho}$, where $\tilde{\rho}$ is the equilibrium value.] Thus

$$R_\rho(\tau,\rho_0) = \rho(\mathbf{q},\tau) \quad \text{with} \quad \rho(\mathbf{q},0)=\rho_0. \tag{2.32}$$

If this is true,[3] the autocorrelation of the fluctuating function $\delta\rho(\mathbf{q},\tau)$ equals the average of the solution $\rho(\mathbf{q},\tau)$ of the field equations averaged over the initial values $\rho^*(\mathbf{q},0)$ – after multiplication by $\rho^*(\mathbf{q},0)$. Therefore we have

$$\langle \delta\rho^*(\mathbf{q},0)\delta\rho(\mathbf{q},\tau)\rangle = \overline{\rho^*(\mathbf{q},0)\rho(\mathbf{q},\tau)}. \tag{2.33}$$

The right-hand side may be calculated from the linearized field equations. We proceed to do this in Section 4 for the prototypical case of a Navier-Stokes-Fourier fluid.

3 Measuring the Spectral Density—The Fabry-Perot Interferometer

3.1 Signal and Spectral Density

The scattered light passes through a Fabry-Perot interferometer where it is split into N beams by multiple partial reflection between parallel partially reflecting plates. In each reflection the electric field is weakened by a factor $r \approx 0.95$. Finally the N beams are focused on a point where they interfere; see Figure 10.4. A photomultiplier in the focus registers a signal $P(\mathbf{q},\tau)$ by averaging over the square of the electric field

[3]de Groot & Mazur [114], the authors of the standard textbook on irreversible thermodynamics, call this assumption "not altogether unreasonable."

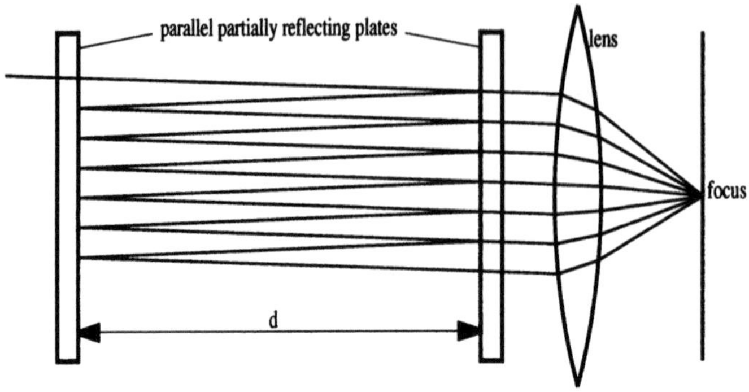

Figure 10.4 Fabry-Perot interferometer.

$$P(\mathbf{q},\tau) = \frac{1}{T}\int_0^T (1-r)\sum_{m=0}^{N-1} r^{2n} E^s(t+n\tau)\cdot(1-r)\sum_{m=0}^{N-1} r^{2m} E^s(t+m\tau)dt. \tag{3.1}$$

The averaging time T is large compared to the duration of a fluctuation. $\tau = 2d/c$ is the time it takes the light for a passage back and forth between the plates. Rearranging the sums in (3.1), we obtain

$$P(\mathbf{q},\tau) = (1-r)^2 \sum_{n=-(N-1)}^{N-1} r^{2|n|}\frac{1-r^{4(N-|n|)}}{1-r^4}\ \langle E^s(0)E^s(n\tau)\rangle \tag{3.2}$$

thus providing a relation between the signal and the autocorrelation function. We replace $\langle E^s(0)E^s(n\tau)\rangle$ in (3.2) by its Fourier transform, the spectral density $I(\mathbf{q},\omega)$; see (2.28). Thus with

$$\langle E^s(0)E^s(n\tau)\rangle = \int_{-\infty}^{\infty} I(\mathbf{q},\omega)e^{-in\tau\omega}d\omega. \tag{3.3}$$

we obtain

$$P(\mathbf{q},\tau) = \int_{-\infty}^{\infty} I(\mathbf{q},\omega)\ \underbrace{(1-r)^2\sum_{n=-(N-1)}^{N-1} r^{2|n|}\frac{1-r^{4(N-|n|)}}{1-r^4}e^{-in\tau\omega}}_{(1-r)^2\frac{1-r^{2N}e^{-iN\tau\omega}}{1-e^{-i\tau\omega}}\frac{1-r^{2N}e^{+iN\tau\omega}}{1-e^{+i\tau\omega}}\approx\frac{(1-r)^2}{1+r^4-2r^2\cos(\tau\omega)}}\ d\omega. \tag{3.4}$$

The sum in (3.4) can be calculated as indicated below the brace, and for a realistic N between 10^3 and 10^4—where r^N may be neglected—we get the explicit approximation indicated in (3.4). Thus the registered signal $P(\mathbf{q},\omega)$ and the spectral density $I(\mathbf{q},\omega)$ are related by

$$P(\mathbf{q},\tau) = \int_{-\infty}^{\infty} I(\mathbf{q},\omega)\ g(\omega,\tau)d\omega, \quad \text{where} \quad g(\omega,\tau) = \frac{(1-r)^2}{1+r^4-2r^2\cos(\tau\omega)}. \tag{3.5}$$

The two factors in the integrand are schematically plotted in Figure 10.5. The spectral density has a width of approximately $2 \cdot 10^{10} Hz$—as we shall see later—and is centered around $\omega = \omega^{(i)} \approx 5 \cdot 10^6 GHz$. The function $g(\omega, \tau)$ in the integrand consists of a series of narrow peaks around $\omega_m = m\frac{2\pi}{\tau} \approx m \cdot 4 \cdot 10^{10} Hz$. The width of the peaks depends on r; it is small for $r \lesssim 1$ and each one has the area $\frac{1}{\tau}\frac{2\pi}{1-r^4}$.

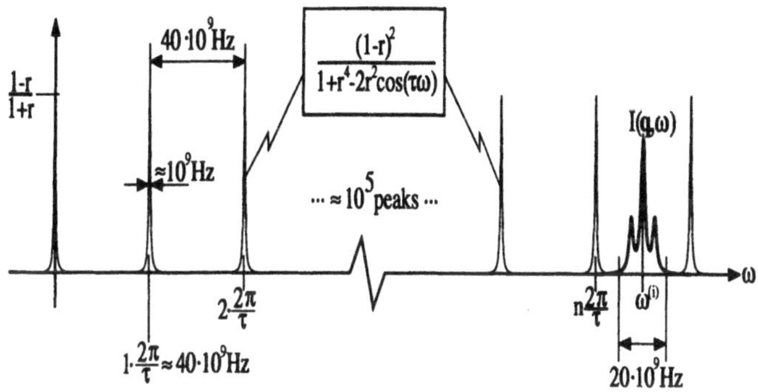

Figure 10.5 On the measurement of $I\left(\mathbf{q}, n\frac{2\pi}{\tau}\right)$.

Since—with the numbers given—$I(\mathbf{q}, \omega)$ intersects with at most one of the narrow peaks, we may write the signal $P(\mathbf{q}, \tau)$ according to $(3.5)_2$ and the figure as

$$P(\mathbf{q}, \tau) = \frac{1}{\tau} \frac{2\pi}{1 - r^4} I(\mathbf{q}, n\frac{2\pi}{\tau})$$

Once this signal has been registered, the Fabry-Perot is made to change the time $\tau = 2d/c$ by a tiny parallel shift of one of the plates. If the distance between the plates is decreased, so is τ, and the nth peak shifts to the right and thereby identifies another value of $I(\mathbf{q}, \omega)$.

In practice the movable plate makes the nth peak scan the curve $I(\mathbf{q}, \omega)$ continuously while the gap between the plates is smoothly decreased by a piezocrystal. In this sense it was correct to say—in Chapter 2—that the spectral density $I(\mathbf{q}, \omega)$ is measured by the Fabry-Perot interferometer—with a photomultiplier behind it.

The reality is not quite as clear-cut as the theory. Indeed, the peaks of the curve

$$g(\omega, \tau) = \frac{1 - r^2}{1 + r^4 - 2r^2 \cos(\tau\omega)}$$

are not quite only narrow lines; they have a certain width and this obviously affects the interpretation of the signal. The quality of a Fabry-Perot interferometer is determined by its *finesse*, the ratio of the distance of two peaks of the function g to the peak-width.

The practice of correcting for too small a finesse proceeds by simplifying the function $(3.5)_2$ so as to make it a function of the difference $\omega - n\frac{2\pi}{\tau}$ instead of a function of the product $\omega\tau$. Indeed, as long as $I(\mathbf{q},\omega)$ has nonzero values only in a small neighborhood of $\omega^{(i)}$ we may write

$$\cos(\tau\omega) \approx \cos\left(n2\pi\frac{\omega - n\frac{2\pi}{\tau}}{\omega^i}\right) \approx 1 - 2n^2\pi^2\left(\frac{\omega - n\frac{2\pi}{\tau}}{\omega^i}\right)^2.$$

The advantage is that thus (3.6) may be written as a convolution

$$P(\mathbf{q},\tau) = \int_{-\infty}^{\infty} I(\mathbf{q},\omega)\hat{g}\left(\omega - n\frac{2\pi}{\tau}\right)d\omega, \quad \text{where} \quad \hat{g} = \frac{(1-r^2)}{(1-r^2)^2 + 4r^2n^2\pi^2\left(\frac{\omega - n\frac{2\pi}{\tau}}{\omega^i}\right)^2}. \quad (3.6)$$

With the knowledge of the ***instrument function*** $\hat{g}$, the convolution may be "unfolded" so as to give $I(\mathbf{q},\omega)$ from the measured values $P(\mathbf{q},\tau)$.

More information about the Fabry-Perot interferometer can be obtained from the article [115] by Simonsohn.

3.2 Measured Data and Their Dependence on Pressure

Typical measured data for Argon—still tainted with noise and parasitic scattered light—are shown on the left-hand side of Figure 10.6. These data have been acquired by Wehr [116]; see also Au & Wehr [117].The individual curves refer to 13 different pressures in the scattering volume between $p = 14bar$ and $p = 0.019bar$. The low pressure curve and its central peak are due to parasitic scatter. This can be suppressed or compensated for, and thus we obtain curves like those on the right-hand side of Figure 10.6, in which the scatter is smoothed out. These correspond to pressures between $18.8bar$ and $1.7bar$. The curves with three peaks correspond to high pressures. As the pressure drops, the lateral peaks "melt" away and eventually the single peak in the center resembles a Gaussian.

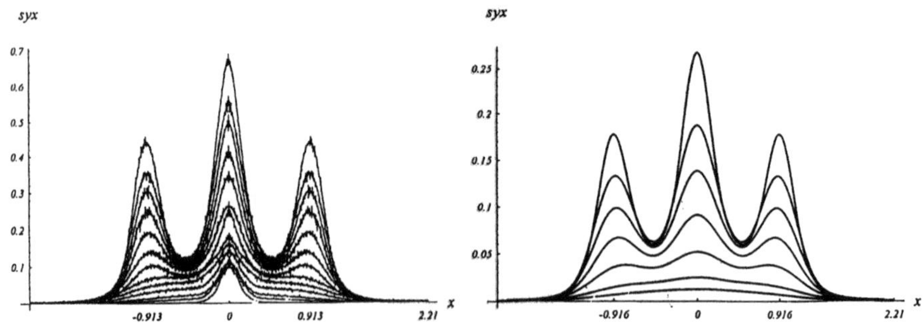

Figure 10.6 Measured curves for Argon for different pressures.
Left: With parasitic scatter and noise.
Right: Parasitic scatter compensated and smoothed.
x is a dimensionless frequency; see (4.16) below.

It turns out that the relevant dimensionless quantity y which determines the shape of the curves depends on the pressure and the scattering angle; see (4.16) below. Experimentally it is much easier to change the pressure than to move the detector. Therefore the data are acquired from a scattering chamber whose pressure may be controlled.

4 Navier-Stokes-Fourier Fluid

4.1 Dynamic Form Factor

In a viscous heat-conducting fluid of the Navier-Stokes-Fourier type we have only three fields, viz. ρ, v_i, T—no internal variables—and the field equations result from the equations of balance of mass, momentum, and (internal) energy and from constitutive relations for stress t_{ij}, heat flux q_i, and internal energy ε.

$$\begin{aligned} \frac{\partial \rho}{\partial t} &+ \frac{\partial \rho v_i}{\partial x_i} &&= 0 \\ \frac{\partial \rho v_j}{\partial t} &+ \frac{\partial (\rho v_j v_i - t_{ij})}{\partial x_i} &&= 0 \\ \frac{\partial \rho \varepsilon}{\partial t} &+ \frac{\partial (\rho \varepsilon v_i + q_i)}{\partial x_i} &&= t_{ij} \frac{\partial v_i}{\partial x_j} \end{aligned} \tag{4.1}$$

with $t_{ij} = -p(\rho, T)\delta_{ij} + \lambda \frac{\partial v_u}{\partial x_u} \delta_{ij} + 2\mu \frac{\partial v_{(i}}{\partial x_{j)}}$, $\varepsilon = \varepsilon(\rho, T)$ and $q_i = -\kappa \frac{\partial T}{\partial x_i}$.

We linearize the field equations about the constant and uniform fields
$\tilde{\rho}, \tilde{v}_i = 0, \tilde{T}$ and obtain

$$\begin{aligned} &\frac{\partial \rho}{\partial t} + \tilde{\rho} \frac{\partial v_i}{\partial x_i} = 0 \\ &\tilde{\rho} \frac{\partial v_i}{\partial t} + \frac{\partial p}{\partial x_i} - \tilde{\mu} \frac{\partial^2 v_i}{\partial x_n \partial x_n} - \left(\tilde{\lambda} + \frac{1}{3}\tilde{\mu}\right) \frac{\partial}{\partial x_i} \left(\frac{\partial v_n}{\partial x_n}\right) = 0, \\ &\tilde{\rho} \frac{\partial \varepsilon}{\partial t} - \tilde{\kappa} \frac{\partial^2 T}{\partial x_n \partial x_n} = -\tilde{p} \frac{\partial v_n}{\partial x_n} . \end{aligned} \tag{4.2}$$

Equivalent to the momentum balance $(4.2)_2$ is its div and curl, namely

$$\begin{aligned} \tilde{\rho} \frac{\partial}{\partial t} \left(\frac{\partial v_i}{\partial x_i}\right) + \Delta p - \left(\tilde{\lambda} + \frac{4}{3}\tilde{\mu}\right) \Delta \left(\frac{\partial v_i}{\partial x_i}\right) &= 0, \\ \tilde{\rho} \frac{\partial \; curl \; v}{\partial t} &= -\tilde{\mu} \Delta \; curl \; v. \end{aligned} \tag{4.3}$$

We note that the equation for curlv is not coupled to the other equations; hence curlv cannot affect the density fluctuations which determine the scattering. We may therefore ignore $(4.3)_2$ and replace the momentum balance by $(4.3)_1$. Inserting $p = p(\rho, T)$, $\varepsilon = \varepsilon(\rho, T)$, and $\psi \equiv \frac{\partial v_i}{\partial x_i}$, we thus obtain as the relevant system of field equations

$$\frac{\partial \rho}{\partial t} + \tilde{\rho}\psi = 0,$$

$$\tilde{\rho}\frac{\partial\psi}{\partial t} + \tilde{p}_\rho\Delta\rho + \tilde{p}_T\Delta T - \tilde{\rho}\tilde{D}_v\Delta\psi = 0,$$

$$\tilde{\rho}\tilde{c}_v\frac{\partial\tilde{T}}{\partial t} - \frac{T}{\tilde{\rho}}\tilde{p}_T\frac{\partial\rho}{\partial t} - \tilde{\rho}\tilde{c}_p\tilde{D}_T\Delta T = 0\,. \tag{4.4}$$

Here c_p and c_v are the specific heats whose ratio will be denoted by γ. D_v and D_T represent the viscosities λ and μ and the heat conductivity κ; they are defined by

$$D_v = \frac{1}{\rho}\left(\lambda + \frac{4}{3}\mu\right), \qquad D_T = \frac{\kappa}{\rho c_p}\,. \tag{4.5}$$

p_ρ and p_T denote the derivatives of p with respect to ρ and T.

The spatial Fourier transform of (4.4) provides us with a rate-type system of differential equations, viz.

$$\frac{\partial\rho(\mathbf{q},t)}{\partial t} + \tilde{\rho}\psi(\mathbf{q},t) = 0,$$

$$\frac{\partial\psi(\mathbf{q},t)}{\partial t} - \frac{\tilde{p}_\rho}{\tilde{\rho}}q^2\,\rho(\mathbf{q},t) - \frac{\tilde{p}_T}{\tilde{\rho}}q^2T(\mathbf{q},t) + \tilde{D}_v q^2\psi(\mathbf{q},t) = 0, \tag{4.6}$$

$$\frac{\partial T(\mathbf{q},t)}{\partial t} + \tilde{\rho}(\tilde{\gamma}-1)\frac{\tilde{p}_\rho}{\tilde{p}_T}\psi(\mathbf{q},t) + \tilde{\gamma}\,\tilde{D}_T q^2 T(\mathbf{q},t) = 0\,.$$

[Recall that $\rho(\mathbf{q},t), \psi(\mathbf{q},t), T(\mathbf{q},t)$ are the Fourier transforms of the differences $\rho(\mathbf{x},t)-\tilde{\rho}$, etc.]

The solution of (4.6) proceeds by Laplace transformation. Thus the equations (4.6) reduce to a linear algebraic system of equations for the Laplace transforms $\hat{\rho}(\mathbf{q},s)$, $\hat{\psi}(\mathbf{q},s)$, $\hat{T}(\mathbf{q},s)$, namely

$$\begin{bmatrix} s & \tilde{\rho} & 0 \\ -\frac{\tilde{p}_\rho}{\tilde{\rho}}q^2 & s+\tilde{D}_v q^2 & -\frac{\tilde{p}_T}{\tilde{\rho}}q^2 \\ 0 & \tilde{\rho}(\tilde{\gamma}-1)\frac{\tilde{p}_\rho}{\tilde{\rho}} & s+\tilde{\gamma}\tilde{D}_T q^2 \end{bmatrix} \begin{bmatrix} \hat{\rho}(\mathbf{q},s) \\ \hat{\psi}(\mathbf{q},s) \\ \hat{T}(\mathbf{q},s) \end{bmatrix} = \begin{bmatrix} \rho(\mathbf{q},0) \\ \psi(\mathbf{q},0) \\ T(\mathbf{q},0) \end{bmatrix}. \tag{4.7}$$

All we need from this system is the solution $\hat{\rho}(\mathbf{q},s)$, which reads

$$\begin{aligned}\hat{\rho}(\mathbf{q},s) \;=\; & \frac{s^2+\left(\tilde{\gamma}\;\tilde{D}_Tq^2+\tilde{D}_vq^2\right)s+\left(\tilde{\gamma}\;\tilde{D}_Tq^2\;\tilde{D}_vq^2+(\tilde{\gamma}-1)\;\tilde{p}_\rho q^2\right)}{s^3+\left(\tilde{\gamma}\;\tilde{D}_Tq^2+\tilde{D}_vq^2\right)s^2+\left(\tilde{\gamma}\;\tilde{p}_\rho q^2+\tilde{\gamma}\;\tilde{D}_Tq^2\;\tilde{D}_vq^2\right)s+\tilde{\gamma}\;\tilde{D}_Tq^2\;\tilde{p}_\rho q^2}\,\rho(\mathbf{q},0)- \\ & -\frac{\tilde{\rho}\left(s+\tilde{\gamma}\;\tilde{D}_Tq^2\right)}{s^3+\left(\tilde{\gamma}\;\tilde{D}_Tq^2+\tilde{D}_vq^2\right)s^2+\left(\tilde{\gamma}\;\tilde{p}_\rho q^2+\tilde{\gamma}\;\tilde{D}_Tq^2\;\tilde{D}_vq^2\right)s+\tilde{\gamma}\;\tilde{D}_Tq^2\;\tilde{p}_\rho q^2}\,\psi(\mathbf{q},0)- \\ & -\frac{\tilde{p}_Tq^2}{s^3+\left(\tilde{\gamma}\;\tilde{D}_Tq^2+\tilde{D}_vq^2\right)s^2+\left(\tilde{\gamma}\;\tilde{p}_\rho q^2+\tilde{\gamma}\;\tilde{D}_Tq^2\;\tilde{D}_vq^2\right)s+\tilde{\gamma}\;\tilde{D}_Tq^2\;\tilde{p}_\rho q^2}\,T(\mathbf{q},0).\end{aligned} \tag{4.8}$$

From this equation we calculate $\overline{\rho^*(q,0)\;\hat{\rho}(q,s)}$ by reference to (2.19)[4] and obtain, dropping the tildas (recall that $Q=1$ holds),

$$\overline{\rho^*(\mathbf{q},0)\;\hat{\rho}(\mathbf{q},s)} = \rho^2kT\kappa_TV\,\frac{s^2+\left(\gamma\;D_Tq^2+D_vq^2\right)s+\left(\gamma\;D_Tq^2\;D_vq^2+(\gamma-1)\;p_\rho q^2\right)}{s^3+(\gamma\;D_Tq^2+D_vq^2)s^2+(\gamma\;p_\rho q^2+\gamma\;D_Tq^2\;D_vq^2)s+\gamma\;D_Tq^2\;p_\rho q^2}. \tag{4.9}$$

[4]Note that $\overline{\rho^*(\mathbf{q},0)\;\psi(\mathbf{q},0)} = q_i\;\overline{\rho^*(\mathbf{q},0)\;v_i(\mathbf{q},0)}$, and this vanishes according to (2.19).

According to the Onsager hypothesis (2.33) this quantity is equal to $\langle\delta\rho^*(\mathbf{q},\mathbf{0})\delta\hat{\rho}(\mathbf{q},s)\rangle$, which determines the dynamic form factor $S(\mathbf{q},\omega)$. Thus we have by reference to (2.31)

$$S(\mathbf{q},\omega)=\frac{1}{\pi}\varepsilon_{\rho}^{2}\rho^{2}kT\kappa_T V\Re e\left\{\frac{s^2+\left(\gamma\ D_Tq^2+D_vq^2\right)s+\left(\gamma\ D_Tq^2\ D_vq^2+(\gamma-1)\ p_\rho q^2\right)}{s^3+(\gamma\ D_Tq^2+D_vq^2)s^2+\left(\gamma\ p_\rho q^2+\gamma\ D_Tq^2\ D_vq^2\right)s+\gamma\ D_Tq^2\ p_\rho q^2}\right\}_{s=i\omega} \tag{4.10}$$

$$S(\mathbf{q},\omega)=\frac{1}{\pi}\varepsilon_{\rho}^{2}\rho^{2}kT\kappa_T V\Re e\left[\frac{\left\{\gamma D_Tq^2D_vq^2+(\gamma-1)\ p_\rho q^2-\omega^2\right\}+i\left\{\ \gamma D_Tq^2+D_vq^2\right\}\omega}{\left\{\gamma\ D_Tq^2\ p_\rho q^2-(\gamma\ D_Tq^2+D_vq^2)\omega^2\right\}+i\left\{\gamma\ p_\rho q^2+\gamma D_Tq^2+\ D_vq^2\right\}\omega-\omega^3}\right]. \tag{4.11}$$

The next and last step consists in choosing a value for $\mathbf{q}$—by positioning the detector—and adjusting the values γ, p_ρ, D_T, D_v so that the right-hand side as a function of ω agrees with the measured function $S(\mathbf{q},\omega)$ to within a constant factor. In this way we "measure" γ, p_ρ, D_T, D_v.

Or else, if no amount of adjusting of γ, p_ρ, D_T, D_v will produce agreement, we conclude that the Navier-Stokes-Fourier theory is insufficient for describing the fluid for the chosen value of $\mathbf{q}$.

4.2 An Alternative Form of the Dynamic Form Factor. Also: An Approximate Form for Forward Scattering.

The algebraic fraction on the right-hand side of (4.10) may be decomposed into partial fractions such that

$$S(\mathbf{q},\omega)=\frac{1}{\pi}\varepsilon_{\rho}^{2}\rho^{2}kT\kappa_T V\Re e\left\{\sum_{i=-1}^{1}\frac{N(s_i)}{\dot{D}(s_i)}\frac{1}{s-s_i}\right\}_{s=i\omega},\ \text{where } \dot{D}=\frac{dD}{ds}. \tag{4.12}$$

N and D denote the numerator and denominator respectively of the fraction in (4.10). s_i are the roots of the cubic polynomial of the denominator of which one—we take it to be s_0—is real while the other two are complex conjugates. More explicitly therefore (4.12) may be written in the form

$$\begin{aligned} S(\mathbf{q},\omega) \;=\; & \frac{1}{\pi}\varepsilon_{\rho}^{2}\rho^{2}kT\kappa_T V\left\{\frac{N(s_0)}{\dot{D}(s_0)}\frac{-s_0}{s_0^2+\omega^2}+\right. \\ & +\Re e\left[\frac{N(s_1)}{\dot{D}(s_1)}\right]\left(\frac{-s_1'}{s_1'^2+(\omega+s_1'')^2}+\frac{-s_1'}{s_1'^2+(\omega-s_1'')^2}\right)- \\ & \left.-Im\left[\frac{N(s_1)}{\dot{D}(s_1)}\right]\left(\frac{\omega+s_1''}{s_1'^2+(\omega+s_1'')^2}+\frac{\omega-s_1''}{s_1'^2+(\omega-s_1'')^2}\right)\right\}. \end{aligned} \tag{4.13}$$

We conclude that the dynamic form factor consists of three Lorentz curves with maxima at $\omega=0$ and $\omega=\pm\ s_1''$; see Figure 10.7. The central peak is called the Rayleigh peak and the lateral ones are the Brillouin peaks. We also speak of the heat mode and sound modes respectively for a reason to be discussed later.

In forward scattering—according to Figure 10.1—we have small values of q. In that case the roots of the denominator function in (4.10) are easily determined to be

$$s_0 = -D_T q^2, \; s_{\pm 1} = -\Gamma q^2 \pm i\sqrt{\gamma p_\rho}\, q \quad \text{where} \quad \Gamma \equiv \frac{1}{2}\left[(\gamma - 1) D_T + D_v\right] \tag{4.14}$$

so that the dynamic form factor reads

$$\begin{aligned} S(\mathbf{q},\omega) \underset{q\to 0}{\longrightarrow} & \frac{1}{\pi}\varepsilon_\rho^2 \rho^2 kT\kappa_T V \left\{\left(1-\frac{1}{\gamma}\right)\frac{D_T q^2}{(D_T q^2)^2+\omega^2}\right. \\ & +\frac{1}{2\gamma}\frac{\Gamma q^2}{(\Gamma q^2)^2+(\omega-\sqrt{\gamma p_\rho}q)^2}+\frac{1}{2\gamma}\frac{\Gamma q^2}{(\Gamma q^2)^2+(\omega-\sqrt{\gamma p_\rho}q)^2} \\ & \left.-\frac{1}{2\gamma\sqrt{\gamma p_\rho}}\frac{(3\Gamma-D_v)q(\omega-\sqrt{\gamma p_\rho}q)}{(\Gamma q^2)^2+(\omega-\sqrt{\gamma p_\rho}q)^2}+\frac{1}{2\gamma\sqrt{\gamma p_\rho}}\frac{(3\Gamma-D_v)q(\omega+\sqrt{\gamma p_\rho}q)}{(\Gamma q^2)^2+(\omega+\sqrt{\gamma p_\rho}q)^2}\right\}. \end{aligned} \tag{4.15}$$

The first two lines within the curly bracket represent the Lorentz peaks with characteristics as summarized in Table 10.1. The corrective terms in the last line of (4.15) play no important role in forward scattering.

	central peak	lateral peaks
site	$\omega = 0$	$\omega = \pm\sqrt{\gamma p_\rho}\, q$
height	$\left(1-\frac{1}{\gamma}\right)\frac{1}{D_T q^2}$	$\frac{1}{2\gamma}\frac{1}{\Gamma q^2}$
half-width	$D_T q^2$	Γq^2

Table 10.1 Lorentz curves in forward scattering.

We conclude from Table 10.1 that γ, p_ρ, D_T, and D_v may be read off from the characteristics of the three peaks. Thus the heat conductivity κ and the viscosity term $\lambda + \frac{4}{3}\mu$ may be calculated. Note also that the lateral peaks lie where $\frac{\omega}{q}$ equals the sound speed.

4.3 Graphical Representation of the Dynamic Form Factor for a Monatomic Ideal Gas

In a monatomic gas we have $c_p = \frac{5}{2}\frac{k}{m}$, $\gamma = \frac{5}{3}$, $\lambda = 0$, $p = \rho\frac{k}{m}T$, and $\frac{\kappa}{\mu} \approx \frac{15}{4}\frac{k}{m}$. We introduce the dimensionless quantities

$$\hat{s} = \frac{s}{\sqrt{2\frac{k}{m}T}q} \qquad x = \frac{\omega}{\sqrt{2\frac{k}{m}T}q} \qquad y = \frac{\sqrt{2}}{3}\frac{p}{\sqrt{\frac{k}{m}T}}\frac{1}{\mu}\frac{1}{q}\,; \tag{4.16}$$

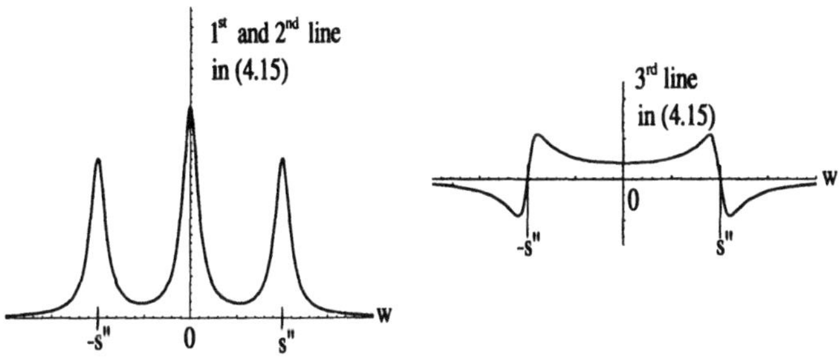

Figure 10.7 Lorentz curves and correction terms.

y represents—in order of magnitude—the ratio of the fluctuation length to the mean free path of the atoms. We obtain

$$\frac{S(q,\omega)}{\frac{1}{\pi}\varepsilon_\rho^2\rho^2 kT\kappa_T V\frac{1}{\sqrt{2\frac{k}{m}T}q}} = \Re e\left\{\frac{\hat{s}^2+\frac{23}{18}\frac{1}{y}\hat{s}+\left(\frac{1}{3}+\frac{10}{27}\right)\frac{1}{y^2}}{\hat{s}^3+\frac{23}{18}\frac{1}{y}\hat{s}^2+\left(\frac{5}{6}+\frac{10}{27}\frac{1}{y^2}\right)\hat{s}+\frac{5}{12}\frac{1}{y}}\right\}_{\hat{s}=ix} \tag{4.17}$$

Note that in terms of x and y we have a universal expression for the essential—i.e. ω-dependent—part of the dynamic form factor. In Figure 10.8 we have plotted that function versus x for various values of y.

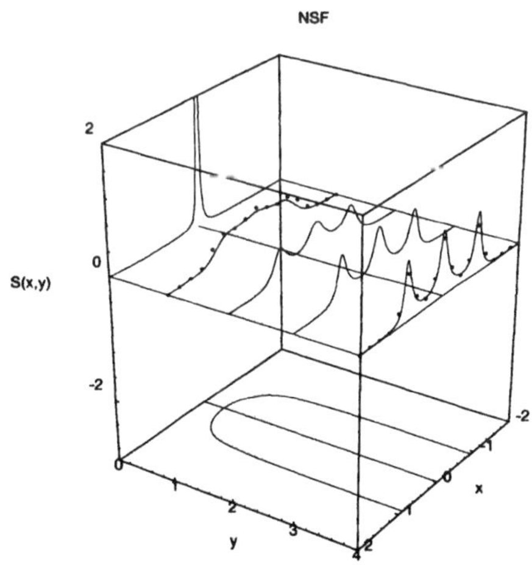

Figure 10.8 Top: The dynamic form factor as a function of x and y in a Navier-Stokes-Fourier theory.
Bottom: The site of the peaks of the Lorentz curves.

We observe that for large y, i.e. small q, i.e. in forward scattering, the curves are characterized by a central maximum and two lateral ones, as expected from Paragraph 4.2. These are called Rayleigh and Brillouin peaks respectively. The Brillouin peaks occur at $x = 0.913$ for large y—their position is determined by the sound speed—and they move to smaller values of x, becoming less distinct as y decreases, i.e. as the scattering angle increases. Eventually the Brillouin peaks vanish and from there on, for still smaller values of y, there is only one central peak. In the (x, y)-plane on the bottom of the three-dimensional plot of Figure 10.8 we have drawn the imaginary part $\pm s_1''$ of the complex root of the cubic polynomical in the denominator of (4.17). The point where it vanishes marks the value of y below which all three Lorentz curves in (4.13) are centered at the origin of x, or ω.

4.4 Comparison with Experimental Data

N. A. Clark [118], [119] has carefully measured the dynamic form factor in xenon for different values of y. The dots in Figure 10.8 represent his values for $y = 4$ and $y = 1$. Neither fit well to the calculated curves. A perfect agreement between theory and experiment appears only for $y \geq 15$, but this is not shown in the figure because we shall be interested in this book mostly in small values of y. Anyway, however, Figure 10.8 shows that the fit is better for $y = 4$ than for $y = 1$.

This can be understood: Indeed, we recall from (4.16) that large y means forward scattering and—from Section 1—that forward scattering contains information about large-scale features of the fluctuations. And it is reasonable to expect that such large-scale, macroscopic fluctuations are well described by the macroscopic Navier-Stokes-Fourier theory.

On the other hand small y means backward scattering and backward scattering contains information about small-scale features of the fluctuations, down to $10^{-7}m$. This is smaller even than a mean free path at normal conditions and therefore it is not surprising that the Navier-Stokes-Fourier theory fails.

It is true that Clark has measured at *one* angle only, corresponding to one q. He changed y by changing the pressure in the scattering volume; cf. (4.16). Thus large pressure makes y big and the q-fluctuations large compared to the mean free path. Or else, small pressure makes y small and lets the scale of the q-fluctuations become comparable to the mean free path.

We recall the discussion of Section 3 on the registered signal $P(q, \tau)$, the spectral density $I(q, \omega)$ and the instrument function $\hat{g}(\omega - n\frac{2\pi}{\tau})$. These functions are related by (3.6). Once the instrument function is known, there are two ways to proceed in order to compare theory and experiment:

- Either we unfold the convolution (3.6) so as to give $I(q, \omega)$ from the *measured* values $P(q, \tau)$
- Or we fold the *calculated* function $I(q, \omega)$ with the instrument function so as to obtain values that may be compared to the measured signals.

Clark has proceeded in the latter way but we have chosen the former one: Our dots in Figure 8 are taken from Clark's $P(q, \tau)$ unfolded with Clark's instrument function $\hat{g}$ which – fortunately – was published in tabular form in [118]. This way is the more natural procedure for us in this book, since we emphasize the theoretical curves and do not wish to taint them with the properties of a particular device, namely Clark's.

4.5 Autocorrelation

It is instructive to invert the temporal Fourier or Laplace transform in the dynamic form factor to obtain the autocorrelation itself, rather than its Fourier transform. By (4.12) we thus have

$$\frac{\langle \rho^*(\mathbf{q},0)\rho(\mathbf{q},\tau)\rangle}{\frac{1}{\pi}\varepsilon_\rho^2\rho^2 kT\kappa_T V} = \sum_{i=-1}^{1} \frac{N(s_i)}{\dot{D}(s_i)}\frac{1}{2\pi i}\int_{x-i\infty}^{x+i\infty}\frac{1}{s-s_i}\; e^{|\tau|s}ds \tag{4.18}$$

$$= \sum_{i=-1}^{1} \frac{N(s_i)}{\dot{D}(s_i)}\; e^{|\tau|s_i}$$

and since s_0 is real while $s_{\pm 1}$ are conjugate complex

$$\frac{\langle \rho^*(\mathbf{q},0)\rho(\mathbf{q},\tau)\rangle}{\frac{1}{\pi}\varepsilon_\rho^2\rho^2 kT\kappa_T V} = \frac{N(s_0)}{\dot{D}(s_0)}\, e^{s_0|\tau|} + \tag{4.19}$$

$$+2\Re e\left[\frac{N(s_1)}{\dot{D}(s_1)}\right]\; e^{s_1'|\tau|}\;\cos(s_1''\tau) -$$

$$-2Im\;\left[\frac{N(s_1)}{\dot{D}(s_1)}\right]\; e^{s_1'|\tau|}\sin(s_1''\,|\tau|).$$

In order to obtain an even more specific form we use the forward-scattering approximation (4.14) and (4.15) and write

$$\frac{\langle \rho^*(\mathbf{q},0)\rho(\mathbf{q},\tau)\rangle}{\frac{1}{\pi}\varepsilon_\rho^2\rho^2 kT\kappa_T V} = \left(1-\frac{1}{\gamma}\right)e^{-D_T\; q^2|\tau|}+\frac{1}{\gamma}e^{-\Gamma\; q^2|\tau|}\cos(\sqrt{\gamma p_\rho}\,q\tau) \tag{4.20}$$

$$+\frac{1}{\gamma}\,(3\Gamma - D_v)\,\frac{q}{\sqrt{\gamma p_\rho}}e^{-\Gamma\; q^2|\tau|}\sin(\sqrt{\gamma p_\rho}\,q\,|\tau|).$$

We note that the autocorrelation has the form of a damped oscillation. Also it is an even function with a zero slope at $\tau = 0$ as it should be.

4.6 Heat and Sound Modes

Fourier and Laplace transforms are not easily interpreted—except by an expert—and certainly a combined Fourier and Laplace transform such as (4.8) is hard to understand in its physical significance. In this paragraph we attempt to provide an intuitive interpretation of the Rayleigh and Brillouin peaks in the dynamic form factor; this will explain why these peaks are referred to as heat mode and sound modes respectively.

We recall the Onsager hypothesis and the specific form (2.19) of the expectation values, and conclude that the right-hand sides of (4.19) and (4.20) are the solutions of an initial value problem—in the Fourier space—for which $\rho(\mathbf{q},0) \neq 0$ and $\mathbf{v}(\mathbf{q},\mathbf{0}) = \mathbf{0}$, $T(\mathbf{q},0) = 0$. What is the form of the solution in space-time? We answer that question for the particular choice

$$\rho(\mathbf{q},0) = \rho_0\; e^{i\mathbf{q}\mathbf{x}_0} \text{ i.e. } \rho(\mathbf{x},0) - \tilde{\rho}(\mathbf{x},0) = \rho_0\delta(\mathbf{x}-\mathbf{x}_0)$$

i.e. an initial point-disturbance of density at $\mathbf{x}_0$ when initially the fluid is at rest and has a uniform temperature. Also we restrict the attention—somewhat arbitrarily—to one dimension only and obtain by back-transformation of the right-hand side of (4.20)

$$\begin{aligned}
\rho(x,\tau)-\tilde{\rho} &= \frac{\rho_0}{2\pi}\Bigg\{\left(1-\frac{1}{\gamma}\right)\int_{-\infty}^{\infty} e^{-D_T\, q^2|\tau|}e^{-i(x-x_0)q}dq \\
&\quad +\frac{1}{\gamma}\int_{-\infty}^{\infty} e^{-\Gamma q^2|\tau|}\cos(\sqrt{\gamma p_\rho}\,q\tau)e^{-i(x-x_0)q}dq \\
&\quad +\frac{1}{\gamma}(3\Gamma-D_v)\int_{-\infty}^{\infty}\frac{q}{\sqrt{\gamma p_\rho}}e^{-\Gamma q^2|\tau|}\sin\sqrt{\gamma p_\rho}q\,|\tau|\,e^{-i(x-x_0)q}dq\Bigg\}, \\
&= \frac{\rho_0}{2\pi}\left(1-\frac{1}{\gamma}\right)\sqrt{\frac{\pi}{\Gamma|\tau|}}e^{-\frac{(x-x_0)^2}{4D_T|\tau|}} \\
&\quad +\frac{1}{\gamma}\sqrt{\frac{\pi}{\Gamma|\tau|}}\left(e^{-\frac{(x-x_0-\sqrt{\gamma p_\rho}|\tau|^2}{4\Gamma|\tau|}}+e^{-\frac{(x-x_0+\sqrt{\gamma p_\rho}|\tau|)^2}{4\Gamma|\tau|}}\right) \\
&\quad +\frac{1}{\gamma}(3\Gamma-D_v)\sqrt{\frac{\pi}{\Gamma|\tau|}}\left(-\frac{x-x_0-\sqrt{\gamma p_\rho}|\tau|}{4\Gamma}e^{-\frac{(x-x_0-\sqrt{\gamma p_\rho}|\tau|)^2}{4\Gamma|\tau|}}\right. \\
&\quad \left.+\frac{x-x_0+\sqrt{\gamma p_\rho}|\tau|}{4\Gamma\sqrt{\gamma p_\rho}|\tau|}e^{-\frac{(x-x_0+\sqrt{\gamma p_\rho}|\tau|)^2}{4\Gamma|\tau|}}\right)\Bigg\}.
\end{aligned} \tag{4.21}$$

[The argument of this paragraph is meant to be heuristic only, and therefore we are not disturbed by the fact that we have integrated over all q, in order to obtain (4.21) when the integrand holds true only for small values of q.]

Inspection shows that the solution has a central peak at the site of the initial dense spot. This peak diffuses according to the linear laws of heat diffusion; it is the heat peak. Also there are diffusing peaks propagating to the right and left with the sound speed $\sqrt{\gamma p_\rho}$; these are the sound peaks.

We notice that the heat and sound peaks originate from the Rayleigh and Brillouin peaks in the Fourier-Laplace transform which represents the dynamic form factor. Therefore we refer to those as heat mode and sound modes respectively.

While in the Navier-Stokes-Fourier fluid there are but three peaks, extended thermodynamics has many. And we shall always refer to the central ones as heat modes and to the lateral ones as sound modes. The latter ones correspond to sound waves propagating outwards with the many speeds appropriate to extended thermodynamics.

5 Extended Thermodynamics

5.1 Introducing Extended Thermodynamics. The Case of 13 Moments.

We have seen that the Navier-Stokes-Fourier theory is not good enough for the description of light scattering data. An alternative of that theory—for monatomic ideal gases—is provided by the 13-moment theory of Grad which recognizes the 13 fields $\rho, v_i, T, t_{\langle ij\rangle}, q_i$ as defining the thermodynamic state of the gas.[5] The field equations are provided by the conservation laws of mass, momentum, and energy and by balance equations for the stress deviator and the heat flux; e.g. see Grad [11], Waldmann [108], or Müller [7],[18]. Its form, linearized about constant fields $\tilde{\rho}, \tilde{v}_i, \tilde{T}, \tilde{t}_{\langle ij\rangle}, \tilde{q}_i$, reads (cf. (9.1.26))

$$\begin{aligned}
\frac{\partial \rho}{\partial t}+\tilde{\rho}\frac{\partial v_i}{\partial x_i} &= 0,\\
\frac{\partial v_i}{\partial t}+\frac{\left(\frac{k}{m}\tilde{T}\right)}{\tilde{\rho}}\frac{\partial \rho}{\partial x_i}+\frac{\partial}{\partial x_i}\left(\frac{k}{m}T\right)-\frac{1}{\tilde{\rho}}\frac{\partial t_{\langle ij\rangle}}{\partial x_j} &= 0,\\
\frac{\partial\left(\frac{k}{m}T\right)}{\partial t}+\frac{2}{3}\left(\frac{k}{m}\tilde{T}\right)\frac{\partial v_k}{\partial x_k}+\frac{2}{3}\frac{1}{\tilde{\rho}}\frac{\partial q_k}{\partial x_k} &= 0, \qquad (5.1)\\
\frac{\partial t_{\langle ij\rangle}}{\partial t}-\frac{4}{5}\frac{\partial q_{\langle i}}{\partial x_{j\rangle}}-2\tilde{\rho}\left(\frac{k}{m}\tilde{T}\right)\frac{\partial v_{\langle i}}{\partial x_{j\rangle}} &= -\sigma t\langle ij\rangle,\\
\frac{\partial q_i}{\partial t}-\left(\frac{k}{m}\tilde{T}\right)\frac{\partial t_{\langle ik\rangle}}{\partial x_k}+\frac{5}{2}\tilde{\rho}\left(\frac{k}{m}\tilde{T}\right)\frac{\partial\left(\frac{k}{m}T\right)}{\partial x_i} &= -\frac{2}{3}\sigma q_i,
\end{aligned}$$

where $1/\sigma$ is a relaxation time related to the viscosity μ and the thermal conductivity κ. Indeed we have

$$\sigma=\frac{\rho\left(\frac{k}{m}T\right)}{\mu}=\frac{15}{4}\left(\frac{k}{m}\right)^2\frac{\rho T}{\kappa} \qquad (5.2)$$

This system may be decomposed into two uncoupled systems for

$$\rho, \psi=\frac{\partial v_i}{\partial x_i}, T, \tau=\frac{\partial^2 t_{\langle ij\rangle}}{\partial x_i\partial x_j}, Q=\frac{\partial q_i}{\partial x_i} \text{ and curl } \mathbf{v}, \varepsilon_{ijk}\frac{\partial^2 t_{\langle jn\rangle}}{\partial x_k\partial x_n}, \text{curl}\mathbf{q}\,. \qquad (5.3)$$

Since only the former affects the density, for our purposes the system (5.1) is equivalent to the system

$$\begin{aligned}
\frac{\partial \rho}{\partial t}+\tilde{\rho}\psi &= 0,\\
\tilde{\rho}\frac{\partial \psi}{\partial t}+\frac{k}{m}\tilde{T}\Delta\rho+\tilde{\rho}\Delta\left(\frac{k}{m}T\right)-\tau &= 0,\\
\frac{3}{2}\tilde{\rho}\frac{\partial\frac{k}{m}T}{\partial t}+\tilde{\rho}\frac{k}{m}\tilde{T}\psi+Q &= 0, \qquad (5.4)\\
\frac{\partial \tau}{\partial t}-\frac{8}{15}\Delta Q-\frac{4}{3}\tilde{\rho}\frac{k}{m}\tilde{T}\Delta\psi &= -\sigma\tau,
\end{aligned}$$

[5] $\mathbf{t}_{\langle ij\rangle}$ is the stress deviator, i.e. the traceless part of the stress.

$$\frac{\partial Q}{\partial t} - \frac{k}{m}\tilde{T}\tau + \frac{5}{2}\tilde{\rho}\frac{k}{m}\tilde{T}\Delta\left(\frac{k}{m}T\right) = -\frac{2}{3}\sigma Q\,.$$

Its Fourier and Laplace transform provides an algebraic system, viz.

$$\begin{bmatrix} s & \tilde{\rho} & 0 & 0 & 0 \\ \frac{k}{m}\hat{T}q^2 & \tilde{\rho}s & -\tilde{\rho}q^2 & -1 & 0 \\ 0 & \tilde{\rho}\frac{k}{m}\hat{T} & \frac{3}{2}\tilde{\rho}s & 0 & 1 \\ 0 & \frac{4}{3}\tilde{\rho}\frac{k}{m}\tilde{T} & 0 & s-\sigma & \frac{8}{15}q^q \\ 0 & 0 & -\frac{5}{2}\tilde{\rho}\frac{k}{m}\tilde{T}q^2 & -\frac{k}{m}\hat{T} & s-\frac{2}{3}\sigma \end{bmatrix} \begin{bmatrix} \hat{\rho}(\mathbf{q},s) \\ \hat{\psi}(\mathbf{q},s) \\ \frac{k}{m}\hat{T}(\mathbf{q},s) \\ \hat{\tau}(\mathbf{q},s) \\ \hat{Q}(\mathbf{q},s) \end{bmatrix} = \begin{bmatrix} \rho(\mathbf{q},0) \\ \psi(\mathbf{q},0) \\ \frac{k}{m}T(\mathbf{q},0) \\ \tau(\mathbf{q},0) \\ Q(\mathbf{q},0) \end{bmatrix} \tag{5.5}$$

where the right-hand side denotes the initial values of the spatial Fourier transform of the fluctuations, in analogy to (4.7). From this equation we calculate $\overline{\rho^*(\mathbf{q},0)\hat{\rho}(\mathbf{q},s)}$ by reference to (2.19)[6] and obtain, dropping the tildas

$$\overline{\rho^*(\mathbf{q},0)\hat{\rho}(\mathbf{q},s)} = \rho^2 kT\kappa_T V \frac{1}{\sqrt{2\frac{k}{m}Tq}} \frac{\hat{s}^4 - \frac{5}{2}\,y\,\hat{s}^3 + \left(\frac{9}{4}\,y^2 + \frac{21}{10}\right)\hat{s}^2 - \frac{11}{2}\,y\,\hat{s} + \left(\frac{1}{2}\,y^2 + \frac{1}{5}\right)}{\hat{s}^5 - \frac{5}{2}\,y\,\hat{s}^4 + \left(\frac{9}{4}\,y^2 + \frac{13}{5}\right)\hat{s}^3 - 8y\,\hat{s}^2 + \left(\frac{5}{4}\,y^2 + \frac{3}{4}\right)\hat{s} - \frac{5}{8}\,y}. \tag{5.6}$$

Here we have introduced the same nondimensional quantities $\hat{s}$ and y that were used in the Navier-Stokes-Fourier theory; see (4.16). The essential part of the dynamic form factor is given by

$$\frac{S(\mathbf{q},\omega)}{\frac{1}{\pi}\varepsilon_\rho^2\rho^2 kT\kappa_T V\frac{1}{\sqrt{2\frac{k}{m}Tq}}} = $$

$$= \Re e\left\{\frac{\hat{s}^4 - \frac{5}{2}\,y\,\hat{s}^3 + \left(\frac{9}{4}\,y^2 + \frac{21}{10}\right)\hat{s}^2 - \frac{11}{2}y\,\hat{s} + \left(\frac{1}{2}\,y^2 + \frac{1}{5}\right)}{\hat{s}^5 - \frac{5}{2}\,y\,\hat{s}^4 + \left(\frac{9}{4}\,y^2 + \frac{13}{5}\right)\hat{s}^3 - 8y\,\hat{s}^2 + \left(\frac{5}{4}\,y^2 + \frac{3}{4}\right)\hat{s} - \frac{5}{8}\,y}\right\}_{\hat{s}\,=\,ix} \tag{5.7}$$

In Figure 10.9 this function is plotted. The plot has to be compared with the one of Figure 10.8. We observe that for the large value $y = 4$ the curve $S(\mathbf{q},\omega)$ coincides with the dots to the same degree approximately as the Navier-Stokes theory does. Apart from that there is little agreement. Indeed:

- There are five roots of the denominator function now instead of three. For $y = 4$ three are real and two are conjugate complex. Therefore there are still only two lateral Brillouin peaks whose position is determined by the sound speed. The three real roots correspond to three Lorentz curves at $\omega = 0$ which together form the Rayleigh peak.

---

[6]Note that $t_{\langle ij\rangle}$ and q_i—or τ and Q—are the internal variables of this theory.

- At a smaller y two more Brillouin peaks are bifurcated off from the line $\omega = 0$, while the previous Brillouin peaks are pushed outwards. The position of the two sets of Brillouin peaks are determined by the speeds of the two propagating modes in the 13-moment theory. In Figure 10.9 the imaginary parts of the four complex roots are plotted in the (x, y)-plane so that the dependence of the phase speeds on y can easily be followed. All four Brillouin peaks persist down to the smallest y. For $y \to 0$ the values of ω/q for the four of them are

$$\pm 0.63\sqrt{\gamma p_\rho} \qquad \text{and} \qquad \pm 1.65\sqrt{\gamma p_\rho} \tag{5.8}$$

which are the characteristic speeds of the hyperbolic system (5.1); see earlier chapters of this book, e.g. Chapter 8.

- While there is strong qualitative disagreement between the Navier-Stokes-Fourier theory and the 13-moment theory for small values y, neither describes the experiments well for $y = 1$. This fact is put in evidence by the disagreement between the appropriate calculated curves and the measured dots; see Figures 10.8 and 10.9.

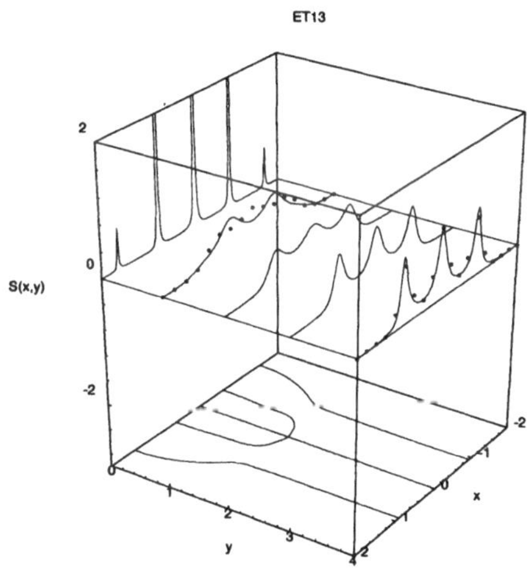

Figure 10.9 The dynamic form factor as a function of x and y in a 13-moment fluid.

5.2 Dynamic Form Factors for n=20, 35, 84

We proceed on the assumption that more moments are needed, and therefore we make use of the systems (9.1.43) and (9.1.46), which are appropriate for 20 and 35 moments respectively. Also we consider the case of 84 moments for which the system was not written down, but it is available to us. In all these systems the vectorial and tensorial equations can be decomposed—by differentiation—into sets of uncoupled equations of which only *one* affects the density field. The case of a Navier-Stokes-Fourier fluid and the 13-moment case have offered explicit examples of that decomposition. The subsystem linked to the density field is solved by Fourier and Laplace transform. All this is similar to the previous cases, except that the determinants to be calculated become bigger and bigger.

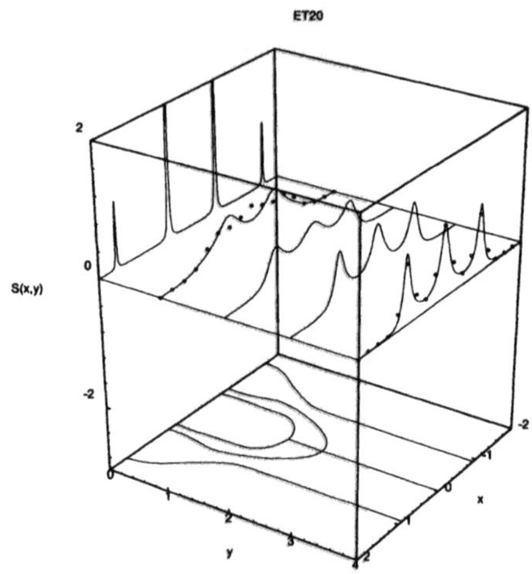

Figure 10.10 Dynamic form factor in a 20-moment theory.

The Figures 10.10 through 10.12 demonstrate the results in the same plot that was used for Figures 10.8 and 10.9. The new results apply to the cases $n = 20$, 35, and 84 and we proceed to discuss them.

First of all, all theories agree in their dynamic form factor for large y, and for those values of y there is also agreement with the measured values. Otherwise there is considerable disagreement. Let us consider:

It may seem from Figure 10.12 that the case $n = 84$ furnishes good results at $y = 1$, but this is an illusion created by the—otherwise illustrative—three-dimensional plot. The true situation is much better evaluated in the (S, x)-projection of Figure 10.14 below. There we see that neither of the theories for $n = 20, 35, 56$, and 84 agree with any other one, nor does any one of them agree with the experimental dots. This is most irritating, since it means that even 84 moments do not suffice. Therefore we proceed to higher moments; this will be done by the method of eigenfunctions that was described and evaluated in Section 9.3.

Before doing that, however, we discuss the complicated situation around the heat and sound modes which may be studied in the (x, y)-projection of the Figures 10.8 through 10.12.

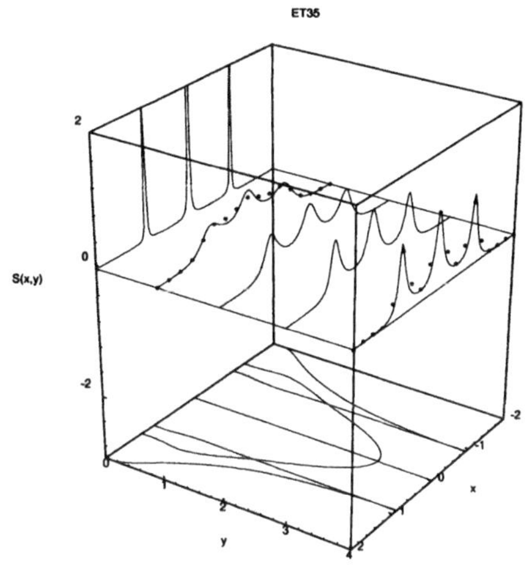

Figure 10.11 Dynamic form factor in a 35-moment theory.

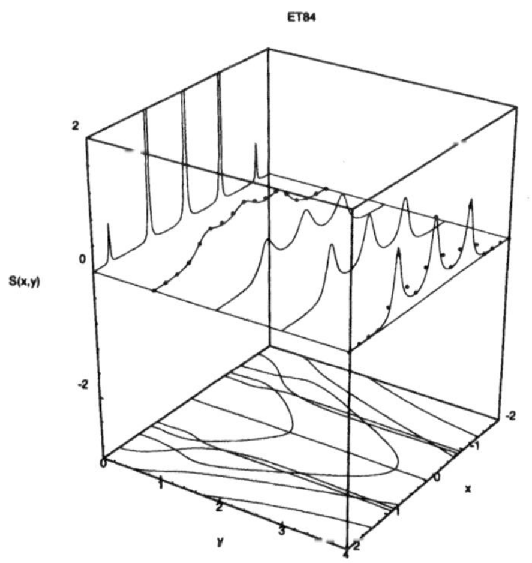

Figure 10.12 Dynamic form factor in an 84-moment theory.

5.3 Heat and Sound Modes in Extended Thermodynamics

Inspection of the Figures 10.9 through 10.12 shows an increasing richness of the bottom plane, which exhibits the imaginary parts of the complex roots of the denominator function $D(s)$ in the dynamic form factor

$$\frac{S(\mathbf{q},\omega)}{\frac{1}{\pi}\varepsilon_\rho^2\rho^2 kT\kappa_T V\frac{1}{\sqrt{2\frac{k}{m}T}q}} = \Re e\left\{\frac{D_1(s)}{D(s)}\right\}_{\hat{s}=ix}. \tag{5.9}$$

The greater richness is of course due to the fact that the denominator polynomial increases in order for more and more equations and therefore has more roots. Thus for $n = 20, 35$, and 84 the number of roots are 6, 9, and 16 respectively.

Not all of these complex roots correspond to sound modes however, as we can see from Figure 10.10. Here we have three complex roots in the (x, y)-plane for small y, but only two sound modes in the (S, x)-plane, i.e. four peaks symmetric to $x = 0$. This is due to the fact that some of the roots of $D(s)$ are also roots of the numerator function $D_1(s)$ in (5.9). Figure 10.13 exhibits this fact by showing all roots $x(y)$ of D for $n = 84$ on the right-hand side and the roots of D_1 on the left-hand side. A sound mode appears only when the right curves are not mirror images of the left curves. That is the case for three pairs of curves.[7] Table 10.2 shows the number of sound and heat modes for $y \to 0$ for theories of up to 286 moments, thus anticipating some results of Paragraph 10.5.4. We conclude that there are many sounds in a many-moment theory, albeit less than there are complex roots of D.

Then n's in the first column of Table 10.2—excepting $n = 13$—are chosen so as to represent the number of equations in a system which takes full moments into account.[8] A close inspection of the table exhibits a certain regularity:

- If the order $\mathcal{O}$ of the highest (full) moment is odd, there are no heat modes for $y \to 0$, and the number of sound modes, always being even, equals $(\mathcal{O} + 1)$.
- If the order $\mathcal{O}$ of the highest (full) moment is even, the number of heat modes equals $(\frac{1}{2}\mathcal{O}+1)$, while the number of sound modes equals $\mathcal{O}$.

Something must also be said for the limit $y \to \infty$: Indeed, starting with $n = 21$ there are more than two complex roots of D_1. For $n = 84$ in fact there are six such roots. All but two of them are also roots of D_1, so that we have only two sound modes in this case, as expected. These are the two "first sounds"—running to the right and left. We expect this situation to persist to any number of moments.

5.4 Higher Moments by Method of Eigenfunctions

We recall from Chapter 9 that for more and more moments the actual moment method becomes cumbersome and impractical. In Chapter 9 we have therefore chosen eigenfunctions, which we

[7] The discrepancy of the uppermost curves on the right and left does not show up on this scale, but it is there, as a magnification would show.

[8] $n = 13$ accounts only for part of the third moment, but we include it, because of the significant role which 13 moments play in the literature—and in this book.

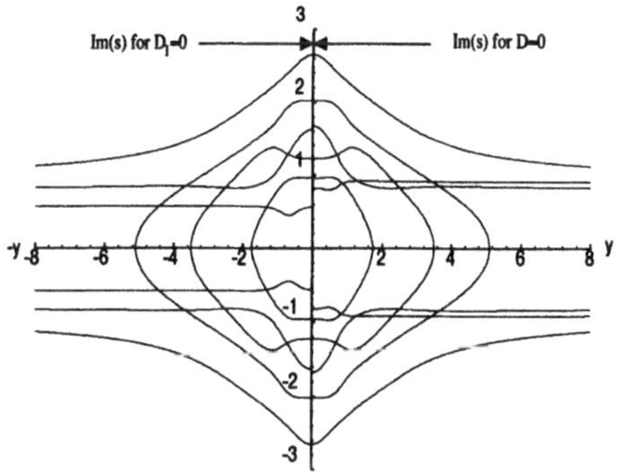

Figure 10.13 The roots $x(y)$ of D (right) and D_1 (left) for $n = 84$.

shall now use in the effort to proceed to theories of very many moments.

The mean eigenfunctions are combinations of moments (cf. Table 9.3) and their governing system of equations is given by (9.3.23).

n	Highest order of moments	SM	HM
13		4	1
20	3	4	0
35	4	4	3
56	5	6	0
84	6	6	4
120	7	8	0
165	8	8	5
220	9	10	0
286	10	10	6

Table 10.2 Number of sound and heat modes for $y \to 0$.

Actually we shall be interested in the subsystem of equations that results from the application of the operator $\partial^l / \partial x_{i_1} ... \partial x_{i_l}$ and obtain with

$$\frac{\partial^l \overline{u^{(s)}_{\langle i_1 ... i_l \rangle}}}{\partial x_{i_1} ... \partial x_{i_l}} \equiv B^{(s)}_l \quad \text{and} \quad \frac{\partial^l}{\partial x_{i_1} ... \partial x_{i_l}} \frac{\partial u^{(s)}_{\langle i_1 ... i_{l-1}}}{\partial x_{l \rangle}} = \frac{l}{2l-1} \Delta B^{(s)}_{l-1} : \tag{5.10}$$

$$\frac{\partial B^{(s)}_l}{\partial t} + \sqrt{2 \frac{k}{m} \tilde{T}} \left\{ \frac{l^2}{(2l+1)(2l-1)} \left(\sqrt{l + \frac{1}{2} + s}\, \Delta B^{(s)}_{l-1} - \sqrt{s+1} \Delta B^{(s+1)}_{l-1} \right) \right.$$

$$+\sqrt{l+\frac{3}{2}+s}\,B_{l+1}^{(s)}-\sqrt{s}\,B_{l+1}^{(s-1)}\Big\}=-\zeta_{sl}B_l^{(s)}\,. \tag{5.11}$$

Δ is the Laplace operator. This system is obviously uncoupled from the tensorial systems that follow from (9.3.23) by multiple application of the operators $\frac{\partial}{\partial x_i}$ and $\epsilon_{ijk}\frac{\partial}{\partial x_k}$. It governs longitudinal waves and density changes, and that is why we consider it.[9]

From the considerations of Chapter 9, specifically Paragraph 9.3.3, we know that certain terms must be omitted from the system (5.11). If the number of equations is ν, the terms to be omitted are those in which the index pair (s,l) corresponds to a value $\alpha > \nu$. [Recall the relation 9.3.15 between α and the pair (r,l), which is also spelled out in Table 9.4.]

For illustration we have explicitly written the first nine equations of the system (5.11) on the next page—as equation (5.12). The system is written in such a manner that the spurious terms—those that must be omitted—are the ones outside the appropriate frames: Thus in a 13-moment theory we ignore the terms with $B_3^{(0)}$, $B_0^{(2)}$, and $B_2^{(1)}$ in the fourth and fifth equation. In a 35-moment theory we ignore the terms with $B_1^{(2)}$, $B_3^{(1)}$, and $B_5^{(0)}$ in the seventh through ninth equation. The remainder represents a closed system in each case.

The closure procedure for 13 moments sets $B_3^{(0)}$, $B_0^{(2)}$ and $B_2^{(1)}$ equal to zero which – by Table 9.3 – amounts to

$$\rho_{\langle ijk\rangle}=0,\;\rho_{iikk}=5\frac{1}{\beta^2}\left(\rho_{ii}-\tilde{\rho}_{ii}\right)-\frac{15}{4}\frac{1}{\beta^4}(\rho-\tilde{\rho})\qquad \rho_{\langle ij\rangle kk}=\frac{7}{2}\frac{1}{\beta^2}\rho_{\langle ij\rangle}\,. \tag{5.13}$$

Since according to Chapter 9 this closure results from the exploitation of the entropy inequality, it is not surprising that the conditions (5.12) are identical to those that were obtained by the direct exploitation explained in Chapter 4, viz. cf. (4.2.45) with $F(z)=\frac{k}{m}z$

$$\rho_{\langle ijk\rangle}=0,\;\rho_{iikk}=5\frac{1}{\beta^2}\rho_{ii}-\frac{15}{4}\frac{1}{\beta^4}\rho,\qquad \rho_{\langle ij\rangle kk}=\frac{7}{2}\frac{1}{\beta^2}\rho_{\langle ij\rangle}\,. \tag{5.14}$$

The constant terms, by which (5.13) and (5.14) differ, are unimportant.

The hierarchy of equations (5.11) can be written explicitly with the greatest ease; moreover the existing equations at one stage are not altered in the next step, which is of course a great advantage of the method of eigenfuctions over extended thermodynamics of moments.

There is also a disadvantage. Indeed, the method of eigenfunctions is inherently and pervasively linear, whereas extended thermodynamics of moments permits the formulation of nonlinear theories simply by letting the expansion (9.1.27) run to higher orders. This difference does not matter much for light scattering in gases in equilibrium. It might matter though in other applications, like shock waves, or even in light scattering in nonequilibrium gases.

[9]The restriction to the subsystem (5.11) is akin to leaving out the equation for *curl***v** from the Navier-Stokes-Fourier theory (see Paragraph 4.1.) or the equations for *curl***v**, $\epsilon_{ijk}\partial t_{\langle jn\rangle}/\partial x_k\partial x_n$ and *curl***q** from the 13-moment equations (5.1).

$$
\begin{aligned}
&\partial_t B_0^{(0)} + \frac{1}{\beta}\left\{ \sqrt{\frac{3}{2}}\, B_1^{(0)} \right\} = 0 \\
&\partial_t B_1^{(0)} + \frac{1}{\beta}\left\{ \frac{1}{\sqrt{6}}\Delta B_0^{(0)} - \frac{1}{3}\Delta\, B_0^{(1)} + \sqrt{\frac{5}{2}}\, B_2^{(0)} \right\} = 0 \\
&\partial_t B_0^{(1)} + \frac{1}{\beta}\left\{ -B_1^{(0)} + \sqrt{\frac{5}{2}}\, B_1^{(1)} \right\} = 0 \\
&\partial_t B_2^{(0)} + \frac{1}{\beta}\left\{ +\frac{1}{3\sqrt{10}}\Delta\, B_1^{(0)} - \frac{4}{15}\Delta\, B_1^{(1)} + \sqrt{\frac{7}{2}}\, B_3^{(0)} \right\} = -\zeta_{02} B_2^{(0)} \\
&\partial_t B_1^{(1)} + \frac{1}{\beta}\left\{ \sqrt{\frac{5}{18}}\Delta\, B_0^{(1)} - B_2^{(0)} - \frac{\sqrt{2}}{3}\Delta\, B_0^{(2)} + \frac{7}{2}\, B_2^{(1)} \right\} = -\zeta_{11} B_1^{(1)} \qquad (13) \\
&\partial_t B_3^{(0)} + \frac{1}{\beta}\left\{ +\frac{9}{5\sqrt{14}}\Delta\, B_2^{(0)} - \frac{9}{35}\Delta\, B_2^{(1)} + \sqrt{\frac{9}{2}}\Delta B_4^{(0)} \right\} = -\zeta_{03} B_3^{(0)} \qquad (20) \\
&\partial_t B_0^{(2)} + \frac{1}{\beta}\left\{ -\sqrt{2}\, B_1^{(1)} + \frac{7}{2}\, B_1^{(2)} \right\} = -\zeta_{20} B_0^{(2)} \qquad (21) \\
&\partial_t B_2^{(1)} + \frac{1}{\beta}\left\{ \frac{4\sqrt{7}}{15\sqrt{2}}\Delta\, B_1^{(1)} - B_3^{(0)} - \frac{4\sqrt{2}}{15}\Delta\, B_1^{(2)} + \sqrt{\frac{9}{2}}\Delta B_3^{(1)} \right\} = -\zeta_{12} B_2^{(1)} \qquad (26) \\
&\partial_t B_4^{(0)} + \frac{1}{\beta}\left\{ +\frac{16}{7\sqrt{18}}\Delta\, B_3^{(0)} - \frac{16}{63}\Delta B_3^{(1)} + \sqrt{\frac{11}{2}}\, B_5^{(0)} \right\} = -\zeta_{04} B_4^{(0)} \qquad (35)
\end{aligned}
$$

(5.12)

5.5 Dynamic Form Factors for Many Moments

Once the system (5.11) has been curtailed and closed in the manner described the procedure is clear: We take the spatial Fourier transform and the temporal Laplace transform and calculate

$$\frac{S(\mathbf{q},\omega)}{\frac{1}{\pi}\varepsilon_{\rho}^{2}\rho^{2}kT\kappa_{T}V\frac{1}{\sqrt{2\frac{k}{m}T}q}} = \Re e\left\{\text{------}\right\}_{\hat{s}=ix} .$$

The bracket contains a polynomial of $n$the order (say) in $\hat{s}$ in the numerator and of order $(n+1)$ in the denominator, whose coefficients depend on y. Once these polynomials are known we may calculate and plot $S(\mathbf{q},\omega)$.

This method of calculation of $S(\mathbf{q},\omega)$—or $S(y,x)$—was followed by Sugawara, Yip & Sirovich [120], who took 21 equations into account. Kamgar-Parsi & Cohen [121] improved on that with 55 equations, and Weiss [27] has implemented the scheme using up to 506 equations. The plots of Figures 10.14 through 10.16 show his results. Let us consider:

We are already convinced by the results of Sections 4 and 5 that for large y all moment theories agree with the Navier-Stokes-Fourier theory, which in turn agrees with experiments.

This statement is confirmed by the theories of many moments, and there is no need to document it visually. On the other hand, for small y we have so far not seen curves that follow the experimental dots. Even the 84-moment theory was not good enough for $y = 1$. Moreover, all theories up to 84 moments disagree widely among themselves. This is put in evidence by Figure 10.14.

However, going further up in the number of moments provides curves that agree among themselves *and with the experimental dots.* This fact is illustrated by Figure 10.15 for several high-moment theories.

It is thus clear, however, that there is not one particular satisfactory theory. A theory may be sufficient down to a value y, *but not beyond.* Indeed, if we plot $S(\mathbf{q},\omega)$ for 210 through 250 moments and for the very small value $y = 0.1$, we see that all these theories once again disagree among themselves—and hence with the experiments—even though for $y = 1$ all of them did agree. See Figures 10.15 and 10.16.

5.6 Evaluation of Moment Theories

It is interesting to note that the hierarchy of moment equations, which we now have at our disposal, provides a check of validity of a particular theory at a given value y *even without experimental values.* Indeed, if $S(\mathbf{q},\omega)$ for one set (A) of moments is not appreciably changed for another set (B) with more moments, we take it for granted that the smaller set is sufficient and satisfactory.

We define the difference of theories A and B at a given ω_i by

$$\Delta S_i = \left|\frac{S_A(\mathbf{q},\omega_i) - S_B(\mathbf{q},\omega_i)}{S_A(\mathbf{q},\omega_i)}\right|$$

and consider the theories equivalent if $\Delta S_i \leq 0.01$ holds for ten equally spaced values x_i, or ω_i between $x = 0$ and $x = 1.5$.

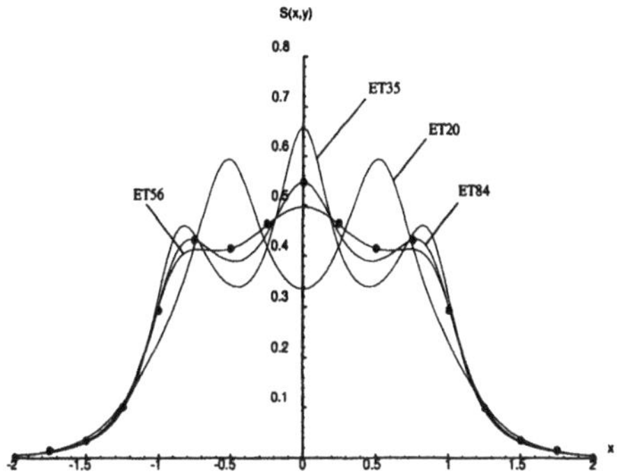

Figure 10.14 $S(\mathbf{q},\omega)$ at $y = 1.0$ for $20, 35, 56$, and 84 moments.

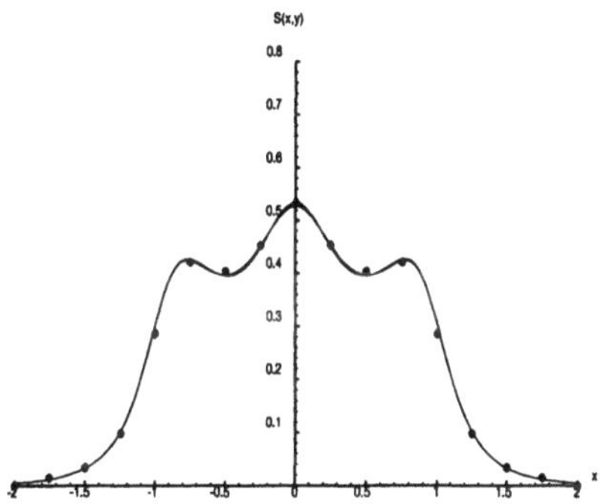

Figure 10.15 $S(\mathbf{q},\omega)$ at $y = 1.0$ for $120, 165, 220$, and 286 moments.

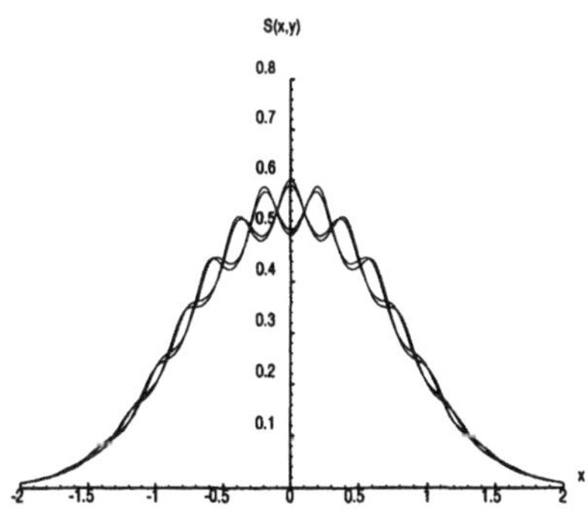

Figure 10.16 $S(\mathbf{q},\omega)$ at $y = 0.1$ for 210, 225, 240, and 256 moments.

Value of y	No. of moments	No. of equ's (5.11)	Highest order of moments
10.884	20	6	3
3.153	35	9	4
1.851	56	12	5
1.333	84	16	6
1.040	120	20	7
0.859	165	25	8
0.730	220	30	9
0.638	286	36	10
0.564	364	42	11
0.509	455	49	12
0.462	560	56	13
0.424	680	64	14
0.391	816	72	15
0.365	969	81	16
0.340	1140	90	17
0.320	1330	100	18
0.302	1540	110	19
0.286	1771	121	20
0.271	2024	132	21
0.259	2300	144	22
0.247	2600	156	23
0.236	2925	169	24
0.226	3276	182	25
0.218	3654	196	26
0.209	4060	210	27
0.202	4495	225	28
0.195	4960	240	29
0.189	5456	256	30
0.182	5984	272	31
0.177	6545	289	32
0.171	7140	306	33
0.167	7770	324	34
0.162	8436	342	35

Table 10.3 No. of moments and equations needed for a satisfactory theory at a given value y.

According to this criterion low-moment theories are satisfactory only for large values of y, while the range of validity of high-moment theories extends down to small values of y. In Table 10.3 we have collected this information quantitatively.

From that table it is seen that for a satisfactory theory at $y = 1.0$ we need about 120 moments. In the system (5.11) this corresponds to 20 equations, and these press into service components of moments up to order 7.

[Note that the number of moments is greater than the number of equations, because the former is given by the number of $\overline{\psi^{(s)}_{\langle i_1 \ldots i_l \rangle}}$, while the latter is given by the number of $B^{(s)}_l = \frac{\partial^l \overline{\psi^{(s)}_{\langle i_1 \ldots i_l \rangle}}}{\partial x_{i_1} \ldots \partial x_{i_l}}$.]

Let us further evaluate the case $y = 1$, which requires 120 moments—or maybe a few more, but less than 165. We recall from (4.16) that

$$y = \frac{\sqrt{2}}{3} \frac{p}{\sqrt{\frac{k}{m}T}} \frac{1}{\mu} \frac{1}{q} \tag{5.15}$$

holds. For $T = 293K$ and $p = 1bar$ in xenon with $m = 131.3 \cdot 1.67 \cdot 10^{-27} kg$ and $\mu = 2.26 \cdot 10^{-5} \frac{Ns}{m^2}$ (see Kestin et al. [122]) we have

$$y = 1.53 \cdot 10^7 \frac{1}{m} \frac{1}{q}, \qquad \text{hence for} \quad y = 1 : q = 1.53 \cdot 10^7 \frac{1}{m} \ . \tag{5.16}$$

Therefore the length scale $\Lambda = \frac{2\pi}{q}$ of the fluctuation at $y = 1$ in xenon is equal to $4.11 \cdot 10^{-7} m$. This is just within the range of detectability for the laser light according to the consideration of Sections 1 and 2. Also this length scale is still ten times as big as the mean free path of the atoms.

The results of this chapter are therefore both satisfactory and disappointing. The deeply satisfying feature is the fact that the moment methods are capable of quantitatively describing the experimental observations of the gas which the classical theories—including the 13-moment theory—definitely cannot do. Disappointment, however, comes from the observation that a sufficient description for length scales still well above the free-path scale requires so *many* moments: It is not good enough to go up to 13 or 14 moments; rather we have to proceed as far as 120 moments or beyond.

5.7 Characteristic speeds

As the number of moments increases, so does the number of modes, either heat modes or sound modes. The pattern is the same as in Figures 10.10 through 10.12 for $n = 35$ and $n = 84$, except that the number and complexity of lines increase. Not only are there more sound modes for $y \to 0$ as n grows but also the range of their values widens and, in particular, the maximum value $V_{\max}$ of all sound speeds increases. This is put in evidence by Table 10.4 which lists the maximum speed—referred to the "normal" sound speed $c_0 = \sqrt{\frac{5}{3}\frac{k}{m}T}$—for different theories up to 15180 moments as calculated by Weiss [27]. Recently Boillat & Ruggeri [25] have proved that $V_{\max}$ increases to infinity if the number of moments tends to infinity; see Chapter 9, Section 2.

Note that Table 10.4 starts with 4 moments, i.e. ρ and v_i. The corresponding equations are those of an isothermal fluid and the sound propagates with the "Newtonian" speed $\sqrt{\frac{k}{m}T}$, which equals $0.77c_0$. This is the only case with only two sound modes.

The fact that $V_{\max}$ tends to infinity for infinitely many moments is somewhat of an anticlimax for extended thermodynamics. This seemingly paradoxial result, which—for different reasons—is also a prediction of the Navier-Stokes-Fourier theory, has originally motivated the formulation of extended thermodynamics; see Chapter 2. It may be understood as an artifact of the classical, i.e. nonrelativistic kinetic theory which, after all, permits infinite velocities in its phase density.

5.8 More Experimental and Theoretical Evidence

In order to emphasize the power of extended thermodynamics in simulating measured values of the dynamic form factor, we show recent results by Wehr [116] and Au & Wehr [117] in Figure 10.17. All diagrams refer to xenon at $20°C$ and each one to a different pressure, starting with $p = 16bar$ ($y = 8$) down to $p = 1bar$ ($y = 0,5$). The dashed curves are the measured ones and the solid curves are the calculated ones; the latter ones are folded with the instrument function; see (3.6).

We see that the fit between theory and experiment is not perfect, but the figure shows very definitely qualitative agreement of calculation and measurement over a wide range of pressures, i.e. parameters y. [The quantitative disagreement is probably due to stray light which enters the interferometer without having been scattered by the gas. A more expensive equipment would presumably improve the fit.]

It is worth noting that the fit—such as it is—requires more and more moments as y drops. Thus the eight form factors of Figure 10.17 were obtained respectively for the theories of

35	moments for	$y > 3$
56	moments for	$y = 2$
120	moments for	$y = 1$
560	moments for	$y = 0,5$.

These numbers confirm the previous observation that many moments are needed when the fluctuation length scale $\Lambda = \frac{2\pi}{q} \approx 2 \cdot 10^{-7}$m approaches the order of magnitude of the mean free path of the atoms. For $p = 1bar$ we estimate the mean path to have the length $2 \cdot 10^{-8}m$.

6 Extrapolation of $S(q, \omega)$ for $y \to 0$

6.1 The Problem

From the calculation presented in Paragraph 5.6. for xenon at $T = 293K$ and $p = 1bar$ we conclude that the smallest value of y for which data are reliable with a laser of $\lambda^{(i)} = 0.4 \cdot 10^{-6}m$ is equal to $y \approx 0.5$. And yet, smaller values of y are still relevant, because y is proportional to p (see (4.16) or (5.15)) and, if we measure in a dilute gas at $p = 0.1$ bar, the interesting range of y extends down to $y \approx 0.05$. Therefore the curves of Figure 10.16 for $y = 0.1$ are relevant predictions of the many-moment theories, albeit for gases under low pressures.

We proceed to discuss the curves of Figure 10.16. While they disagree—and thus indicate that more moments should be taken into account—there is a plait-like structure in these curves: They appear to crowd around a monotone function for $x > 0$. This feature becomes more pronounced for even more moments, and the plait becomes thinner. This fact suggests that for small y the dynamic form factor tends to a limiting curve. In this chapter we proceed to determine the shape of that curve, which will turn out to be a Gaussian of the form $exp(-x^2)$.

No. of moments	No. of equ's (5.11)	Highest order of moments	$V_{\max}/c_0$
4	2	1	0.77459667
10	4	2	1.34164079
20	6	3	1.80822948
35	9	4	2.21299946
56	12	5	2.57495874
84	16	6	2.90507811
120	20	7	3.21035245
165	25	8	3.49555791
220	30	9	3.76412372
286	36	10	4.01860847
364	42	11	4.26098014
455	49	12	4.49279023
560	56	13	4.71528716
680	64	14	4.92949284
816	72	15	5.13625617
969	81	16	5.3362913
1140	90	17	5.53020569
1330	100	18	5.71852112
1540	110	19	5.90168962
1771	121	20	6.08010585
2024	132	21	6.25411673
2300	144	22	6.42402919
2600	156	23	6.59011627
2925	169	24	6.75262213
3276	182	25	6.91176615
3654	196	26	7.06774631
4060	210	27	7.22074198
4495	225	28	7.37091629
4960	240	29	7.51841807
5456	256	30	7.66338362
5984	272	31	7.80593804
6545	289	32	7.94619654
7140	306	33	8.08426549
7770	324	34	8.22024331
8436	342	35	8.35422129
9139	361	36	8.48628432
9880	380	37	8.61651144
10660	400	38	8.74497644
11480	420	39	8.87174833
12341	441	40	8.99689171
13244	462	41	9.12046722
14190	484	42	9.24253184
15180	506	43	9.36313918

Table 10.4 Highest characteristic speed.

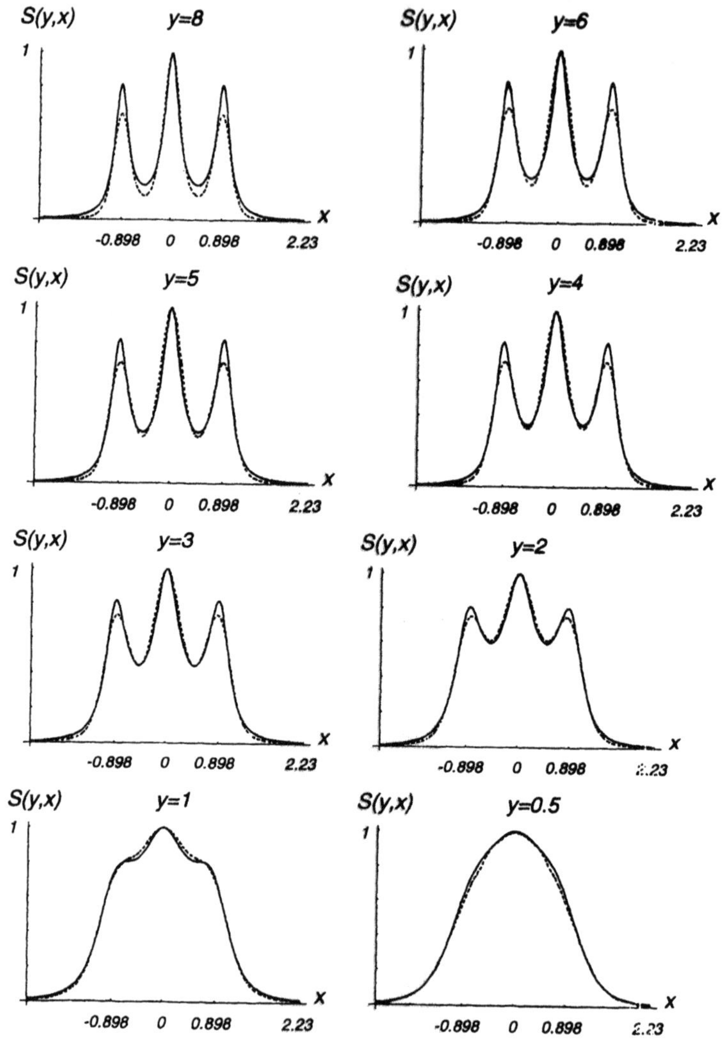

Figure 10.17 Dynamic form factor for decreasing pressures $p = 8bar$ to $p = 0.5bar$ and $T = 293K$.
Dashed curves: measurement
Solid curves: calculation
The heights of the central peak of the calculated curves is fitted to the measured heights.

6.2 The Boltzmann Equation in the Krook Approximation

Since we are interested now in small values of y, i.e. small pressures, collisions are rare and far apart. Therefore it may suffice to approximate the Boltzmann collision integral by a simple relaxation Ansatz. This is the so-called Krook approximation [123] of the Boltzmann equation. It reads

$$\frac{\partial f}{\partial t} + c_i \frac{\partial f}{\partial x_i} = -\frac{1}{\tau}(f - \tilde{f}) \quad \text{where} \quad \tilde{f} = \frac{\tilde{\rho}}{m} \frac{1}{\sqrt{2\pi \frac{k}{m} \tilde{T}}^{3}} e^{-\frac{mc^2}{2k\tilde{T}}}. \tag{6.1}$$

At every phase point $(\mathbf{x}, \mathbf{c})$ the fluctuation of f around the equilibrium distribution $\tilde{f}$ is denoted by $\delta f = f - \tilde{f}$.

Accordingly the mass density is decomposed into an equilibrium part $\tilde{\rho}$ and a fluctuation $\delta\rho$ by

$$\rho = \int m f dc = \underbrace{\int m \tilde{f} dc}_{\tilde{\rho}} + \underbrace{\int m \delta f(\mathbf{x}, t, \mathbf{c}) dc}_{\delta\rho(\mathbf{x}, t)} . \tag{6.2}$$

We proceed to calculate $\hat{\rho}(\mathbf{q}, s)$, the Fourier and Laplace transform of $\delta\rho$ $(\mathbf{x}, t)$, which we need for the calculation of the dynamic form factor.

6.3 The Dynamic Form Factor $S(\mathbf{q}, \omega)$; General Formula

From (6.2) and (6.1) we obtain by a Fourier-Laplace transform

$$\hat{\rho}(\mathbf{q}, s) = \int m \hat{f}(\mathbf{q}, s, \mathbf{c}) dc \quad \text{where} \quad \hat{f}(\mathbf{q}, s, \mathbf{c}) = \frac{1}{s - i c_i q_i + \frac{1}{\tau}} f(\mathbf{q}, 0, \mathbf{c})$$

and $f(\mathbf{q}, 0, \mathbf{c})$ is the initial value of the Fourier transform $f(\mathbf{q}, t, \mathbf{c})$ of $\delta f(\mathbf{x}, t, \mathbf{c})$. Thus we have

$$\hat{\rho}(\mathbf{q}, s) = m \int \frac{f(\mathbf{q}, 0, \mathbf{c})}{s - i c_i q_i + \frac{1}{\tau}} dc.$$

We multiply by $\rho^*(\mathbf{q}, 0) = m \int f^*(\mathbf{q}, 0, \mathbf{c}') dc'$ and average over all initial values $f(\mathbf{q}, 0, \mathbf{c})$ to obtain

$$\overline{\rho^*(\mathbf{q}, 0)\, \hat{\rho}(\mathbf{q}, s)} = m^2 \iint \frac{\overline{f^*(\mathbf{q}, 0, \mathbf{c}') f(\mathbf{q}, 0, \mathbf{c})}}{s - i c_i q_i + \frac{1}{\tau}} dc dc'. \tag{6.3}$$

This, by (2.31) is the important part of the dynamic form factor. It remains to calculate the expectation value in the integrand. For that purpose we revisit the fluctuation theory of Section 2, which we extend to the phase space spanned by $\mathbf{x}$ and $\mathbf{c}$.

6.4 Fluctuations in Phase Space

We consider an adiabatic system of volume V_G containing the scattering volume V and decomposed in N small cells $(\alpha = 1, 2, ...N)$ of equal volume ΔV. In each cell there are atoms of all velocities $\mathbf{c}$ and—for simplicity—we also discretize the velocity space into M equal cells $(\beta = 1, 2, ...M)$ of "volume" ΔC. The state of each six-dimensional cell is then given by $f_{(\alpha\beta)}(t)$, and a fluctuation $\delta f_{(\alpha\beta)}$ from the Maxwellian

$$\tilde{f}(\mathbf{c}_\beta) = \frac{\tilde{\rho}}{m} \frac{1}{\sqrt{2\pi \frac{k}{m} \tilde{T}}^3} e^{-\frac{mc_\beta^2}{2k\tilde{T}}} \tag{6.4}$$

is constrained by the requirement that total mass and energy are constant and that total momentum vanishes:

$$\sum_\alpha \sum_\beta m\delta f_{(\alpha\beta)} = 0, \quad \sum_\alpha \sum_\beta m\mathbf{c}_\beta \delta f_{(\alpha\beta)} = 0, \quad \sum_\alpha \sum_\beta \frac{m}{2} c_\beta^2 \delta f_{(\alpha\beta)} = 0. \tag{6.5}$$

Since the entropy of the gas is given by $S = -k \int\limits_{V_G} \int \ln(f)\, f\, d\mathbf{c}\, d\mathbf{x}$, the entropy fluctuation is given by

$$\delta S = -k \sum_\alpha \sum_\beta \delta \left(\ln(f_{(\alpha\beta)}) f_{(\alpha\beta)} \right) \Delta V \Delta C. \tag{6.6}$$

We expand it into a Taylor series around the Maxwellian, including second-order terms in $\delta f_{(\alpha\beta)}$

$$\frac{\delta S}{k\Delta V \Delta C} = -\underbrace{\sum_{\alpha\beta} \left(1 + \ln \tilde{f}(\mathbf{c}_\beta)\right) \delta f_{(\alpha\beta)}}_{0} - \frac{1}{2} \sum_{\alpha\beta} \frac{1}{\tilde{f}(\mathbf{c}_\beta)} \delta f_{(\alpha\beta)} \delta f_{(\alpha\beta)}. \tag{6.7}$$

The first term vanishes because of the constraints. δS determines the probability density p to find the system with the fluctuation $\delta f_{(\alpha\beta)}$ in the range $d\Xi = \prod_\alpha \prod_\beta d(\delta f_{(\alpha\beta)})$. Specifically we have

$$p d\Xi = \frac{\exp\left(\frac{\delta S}{k}\right) d\Xi}{\int \int (\quad) d\Xi} = \frac{\exp\left(-\frac{1}{2} \sum_{\alpha\beta} \frac{1}{\tilde{f}(\mathbf{c}_\beta)} \delta f_{(\alpha\beta)} \delta f_{(\alpha\beta)} \Delta V \Delta C \right) d\Xi}{\int \int (\qquad) d\Xi}.$$

For the expectation values $\overline{\delta f_{(\alpha\beta)}\; \delta f_{(\alpha'\beta')}}$ we thus obtain

$$\overline{\delta f_{(\alpha\beta)}\; \delta f_{(\alpha'\beta')}} = \int \delta f_{(\alpha\beta)}\; \delta f_{(\alpha'\beta')}\; p d\Xi = \frac{\tilde{f}(\mathbf{c}_\beta)}{\Delta V \Delta C} \delta_{\alpha\alpha'} \delta_{\beta\beta'}\;. \tag{6.8}$$

The desired expectation value $\overline{f^*(\mathbf{q}, 0, \mathbf{c}')\; f(\mathbf{q}, 0, \mathbf{c})}$ in (6.3) is determined from

$$\overline{f^*(\mathbf{q}, 0, \mathbf{c}')\; f(\mathbf{q}, 0, \mathbf{c})} = \int\limits_V \int\limits_V \overline{\delta f(\mathbf{x}', 0, \mathbf{c}')\; \delta f(\mathbf{x}, 0, \mathbf{c})}\; e^{i\mathbf{q}(\mathbf{x}-\mathbf{x}')} d\mathbf{x} d\mathbf{x}'$$

or, with the discretization

$$\overline{f^*(\mathbf{q}, 0, \mathbf{c}_{\beta'})\; f(\mathbf{q}, 0, \mathbf{c}_\beta)} = (\Delta V)^2 \sum_{\alpha\alpha'} \overline{\delta f_{(\alpha'\beta')}\; \delta f_{(\alpha\beta)}}\; e^{i\mathbf{q}(\mathbf{x}^{(\alpha)} - \mathbf{x}^{(\beta)})},$$

where V is the scattering volume. The summation now extends over the scattering volume V. By (6.8) we obtain

$$\overline{f^*(\mathbf{q}, 0, \mathbf{c}_{\beta'})\; f(\mathbf{q}, 0, \mathbf{c}_\beta)} = \frac{V}{\Delta C} \tilde{f}(\mathbf{c}_\beta) \delta_{\beta\beta'}. \tag{6.9}$$

Hence follows $\overline{\rho^*(\mathbf{q},0)\ \hat{\rho}(\mathbf{q},s)}$ from (6.3). We have

$$\begin{aligned}\overline{\rho^*(\mathbf{q},0)\ \hat{\rho}(\mathbf{q},s)} &= m^2 \sum_{\beta\beta'} \frac{\overline{f^*(\mathbf{q},0,\mathbf{c}_{\beta'})\ f(\mathbf{q},0,\mathbf{c}_\beta)}}{s - i\mathbf{c}_\beta \mathbf{q} + \frac{1}{\tau}} (\Delta C)^2 \text{ or, by (6.9)} \\ &= m^2 V \sum_\beta \frac{\tilde{f}(\mathbf{c}_\beta)}{s - i\mathbf{c}_\beta \mathbf{q} + \frac{1}{\tau}} \Delta C \\ &= m^2 V \int \frac{\tilde{f}(\mathbf{c})}{s - i\mathbf{c}\mathbf{q} + \frac{1}{\tau}} d\mathbf{c}. \end{aligned} \tag{6.10}$$

It remains to evaluate the integral in (6.10).

Note that in this section we have not been quite as conscientious in the calculation of expectation values as we were in Section 2. Indeed we have not accounted for the three constraints (6.5) in $pd\Xi$. We feel that the error thus introduced is slight and are confirmed in this view by the results of Paragraph 2.1.

6.5 The Dynamic Form Factor $S(\mathbf{q},\omega)$; Specific Form

By (2.31) and (2.33) the dynamic form factor $S(\mathbf{q},\omega)$ results from (6.10). We have by use of (6.4)

$$S(\mathbf{q},\omega) = \frac{1}{\pi}\varepsilon_\rho^2\ m\tilde{\rho} \frac{V}{\sqrt{2\pi \frac{k}{m}\tilde{T}}^3} \Re e \left(\int \frac{e^{-\frac{mc^2}{2k\tilde{T}}}}{s - i\mathbf{c}\mathbf{q} + \frac{1}{\tau}} d\mathbf{c} \right)_{s=i\omega} \tag{6.11}$$

As before we introduce the dimensionless quantities (4.16). Of course there was no relaxation time τ in previous sections, but we shall identify this coefficient with $\frac{3}{2}\mu/\rho\frac{k}{m}T$, the relaxation time of the heat flux in Section 5; e.g. see (5.2). In this manner we obtain

$$\begin{aligned}\frac{S(\mathbf{q},\omega)}{\frac{1}{\pi}\varepsilon_\rho^2\ m\tilde{\rho}V\frac{1}{\sqrt{2\frac{k}{m}T}}} &= \Re e \left\{ \frac{1}{\sqrt{\pi}^3} \int \frac{e^{-\hat{c}^2}}{\hat{s} - i\frac{\mathbf{q}}{q}\hat{\mathbf{c}} + y} d\hat{\mathbf{c}} \right\}_{\hat{s}=ix} \\ &= \frac{1}{\sqrt{\pi}^3} \int \frac{y}{y^2 + (x - \frac{\mathbf{q}}{q}\hat{\mathbf{c}})^2} e^{-\hat{c}^2} d\hat{\mathbf{c}} \\ &\underset{y\to 0}{\longrightarrow} \sqrt{\pi}\, e^{-x^2}. \end{aligned} \tag{6.12}$$

The last line follows from the observation that a rotation of one of the $\hat{c}$-coordinates, for instance $\hat{c}_1$, into the $\mathbf{q}$-direction provides a right-hand side of the form

$$\frac{1}{\sqrt{\pi}} \int \frac{y}{y^2 + (x - \hat{c}_1)^2} e^{-\hat{c}_1^2} d\hat{c}_1$$

and that for $y \to 0$ only the value $\hat{c}_1 = x$ contributes to the integral and provides the value πe^{-x^2}.

6.6 Discussion

We have proved that the limiting form of the dynamic form factor for $y \to 0$, i.e. low pressure and/or large q is a Gaussian distribution; see (6.12). This is consistent with the calculations presented in Figure 10.16 and with observations made in very dilute gases; see Copley & Lovesey [124]. It would have been nice to have this result for the Boltzmann equation proper, rather than for the Krook approximation, but that seems impossible. Other authors, notably Yip & Nelkin [125], have obtained the result (6.12) as well. These authors have used the local Maxwell distribution in $(6.1)_1$, where we have—for simplicity—employed the Maxwell distribution of global equilibrium.

Chapter 11

Testing Extended Thermodynamics by Sound

Experimentalists measure phase speeds and attenuations in acoustic resonators. On the other hand, any thermodynamic theory will provide a mathematical expression for the phase speeds and attenuations. If the theory contains unknown parameters a comparison of the empirical data with the theoretical prediction will provide values for these parameters. But, if a theory is specific – as is the case with extended thermodynamics of moments – the experiments provide an estimate of the range of validity of the theory.

The measurements of dispersion of sound and of scattering of light complement each other. They show that a given theory of n moments is satisfactory up to a frequency ω_n and down to a wave length λ_n respectively. Thus we obtain information about the number of moments we need for dealing with a given initial-boundary value problem: If ω_n and λ_n are significantly represented in the Fourier spectra of the initial and boundary values, we shall have to use the n-moment theory.

The chapter begins with a short description of how an acoustic resonator works. Our misgivings about the interpretation of the acquired data are detailed and explained.

1 Basic Acoustics

1.1 How the Acoustic Resonator Measures Phase Speeds in Principle

The acoustic resonator basically consists of a gas-filled tube with a harmonic sound source at one end and an adjustable piston at the other end; see Figure 11.1. The resonator is stimulated by an alternating voltage of amplitude U_0 and adjustable frequency ω. On the left end, at $x = 0$, this creates a velocity $v(0,t) = V_0 \cos\omega t$, while on the right end, at $x = L$, the velocity is zero.

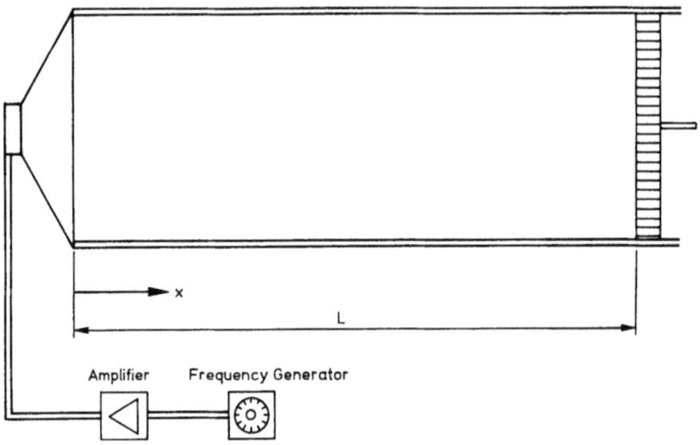

Figure 11.1 Acoustic resonator (schematic)

So as to explain the principle of measuring phase speeds, we first concentrate on the idealized case of an Eulerian fluid in which the viscosity and the heat conductivity may be neglected. The linearized equations of balance of mass, momentum, and energy read

$$\begin{aligned} \frac{\partial \varrho}{\partial t} + \tilde{\varrho}\frac{\partial v_i}{\partial x_i} &= 0 \\ \tilde{\varrho}\frac{\partial v_i}{\partial t} + \tilde{p}_T\frac{\partial T}{\partial x_i} + \tilde{p}_\varrho\frac{\partial \varrho}{\partial x_i} &= 0 \\ \tilde{\varrho}\tilde{\varepsilon}_T\frac{\partial T}{\partial t} + \tilde{T}\tilde{p}_T\frac{\partial v_i}{\partial x_i} &= 0. \end{aligned} \tag{1.1}$$

We set $\varrho = \varrho(x,t), \mathbf{v} = (v(x,t),0,0), T = T(x,t)$ and eliminate ϱ and T to obtain.

$$\frac{\partial^2 v}{\partial t^2} - c_0^2\frac{\partial^2 v}{\partial x^2} = 0 \quad \text{where} \quad c_0^2 = p_\varrho + \frac{T p_T^2}{\varrho^2 \varepsilon_T}. \tag{1.2}$$

The solution appropriate to the boundary conditions

$$v(0,t) = V_0 \cos\omega t, \qquad v(L,t) = 0 \tag{1.3}$$

is a superposition of two harmonic waves of wavelength $\lambda = 2\pi \frac{c_0}{\omega}$ running to the right and left, viz.

$$v(x,t) = -\frac{1}{2}\frac{V_0}{\sin\frac{\omega L}{c_0}}\left[\sin\left\{(x+c_0 t)\frac{\omega}{c_0} - \frac{\omega L}{c_0}\right\} + \sin\left\{(x-c_0 t)\frac{\omega}{c_0} - \frac{\omega L}{c_0}\right\}\right] \quad (1.4)$$

$$v(x,t) = \frac{V_0}{\sin\frac{\omega L}{c_0}}\cos\omega t \sin\left\{(L-x)\frac{\omega}{c_0}\right\}. \quad (1.5)$$

This represents a standing wave with knots at the points

$$x_n = L - n\pi\frac{c_0}{\omega} = L - n\frac{\lambda}{2} \quad (1.6)$$

and singular bulges if the length L is adjusted to the value

$$L = m\pi\frac{c_0}{\omega} = m\frac{\lambda}{2}. \quad (1.7)$$

The bulges indicate that even a tiny amplitude V_0 will create strong velocity oscillations between the knots, provided that L has the values (1.7). In that case we say that the gas is in resonance, or that the resonator is well-tuned.

An old fashioned method of measuring the sound speed in a gas uses sand on the bottom of the tube. The strong oscillations at the bulges displace the sand and sweep it to the motionless regions at the knots. Thus inspection determines λ, and hence $c_0 = \lambda\frac{\omega}{2\pi}$, since ω is known.

1.2 Piezoelectric Transducer and the Mechanical Impedance

In modern days the source of sound is a piezoelectric crystal with a thin layer of metal on opposite sides. An alternating voltage of frequency ω is applied to those sides. It creates a current I and induces oscillations in the crystal which are transmitted to the gas. Thus we have what is called a piezoelectric transducer. We proceed to describe the properties of this instrument.

The metal layers provide a capacity C so that the current I is split as follows

$$I = I_C + I_P, \quad \text{where} \quad I_C = C\frac{dU}{dt}. \quad (1.8)$$

I_C is the current through the capacitor and I_P is the current through the piezocrystal. The power UI_P equals the power of the piezo force F_P on the crystal, which is $F_P v(0,t)$. Thus we conclude with

$$F_P = \alpha U \qquad \text{that} \qquad I_P = \alpha v(0,t) \quad (1.9)$$

holds, where α is a parameter that characterizes the transducer.

The crystal may be considered to be a linearly viscoelastic body. To fix the ideas, we represent it by a mass m which is connected to a support by an elastic spring and a dashpot; see Figure 11.2. In addition to the piezo force the crystal is subject to the force

$$F_G = At_{xx}(0,t) \quad (1.10)$$

exerted by the gas. A is the cross–sectional area and t_{xx} is the normal stress in x-direction. The equation of motion reads

$$m\ddot{l} + D\dot{l} + \lambda l = F_P + At_{xx}(0,t). \tag{1.11}$$

We note that $\dot{l}$ equals $v(0,t)$ and differentiate equation (1.11). Thus we obtain, by use of (1.9),

$$m\ddot{I}_P + D\dot{I}_P + \lambda I_P - \alpha A\dot{t}_{xx}(0,t) = \alpha^2\dot{U}. \tag{1.12}$$

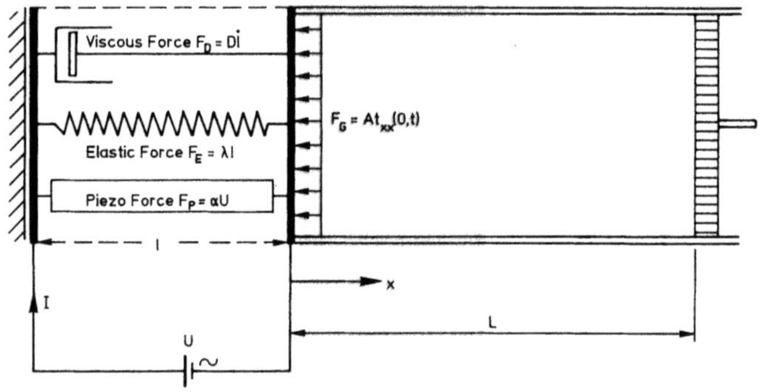

Figure 11.2 Acoustic resonator and schematic representation of the piezocrystal

A sinusoidal input $U = U_0 \mathcal{R}e\, e^{i\omega t}$ will create sinusoidal functions

$$I_C = \mathcal{R}e\left(\bar{I}_C e^{i\omega t}\right),\ I_P = \mathcal{R}e\left(\bar{I}_P e^{i\omega t}\right),\ t_{xx}(0,t) = \mathcal{R}e\left(\bar{t}_{xx} e^{i\omega t}\right),\ v(0,t) = \mathcal{R}e\left(\bar{v} e^{i\omega t}\right) \tag{1.13}$$

where $\bar{I}_C, \bar{I}_P, \bar{t}_{xx}$, and $\bar{v}$ are complex amplitudes. Insertion into (1.8) and (1.12) provides

$$\bar{I} = \left(i\omega C + \frac{\alpha^2}{i\omega m + D - i\frac{\lambda}{\omega} - \alpha A\frac{\bar{t}_{xx}}{\bar{I}_P}} \right) U_0 \qquad \text{or, by} \quad (1.9)_2$$

$$\bar{I} = \underbrace{\left(i\omega C + \frac{\alpha^2}{i\omega m + D - i\frac{\lambda}{\omega} - A\frac{\bar{t}_{xx}}{\bar{v}}} \right)}_{1/Z} U_0. \tag{1.14}$$

We define the *impedance* Z of the transducer as indicated in (1.14). The expression

$$Z_m = A\frac{\bar{t}_{xx}}{\bar{v}} \tag{1.15}$$

is called the *external mechanical impedance.*

For a given piezoelectric transducer the quantities C, α, m, D, λ, and A are known. Thus measuring $\bar{I}$ for a given U_0 we may calculate Z_m. Now the question arises how to use that knowledge to calculate the wavelength λ in the gas. We proceed to answer that question, first once again for an Eulerian fluid.

1.3 External Mechanical Impedance and Wavelength

We recall from (1.5) the form of $v(x,t)$ for an Eulerian fluid, viz.

$$v(x,t) = \frac{V_0}{\sin 2\pi\frac{L}{\lambda}} \sin\left\{(L-x)\frac{2\pi}{\lambda}\right\} \cos\omega t = \frac{V_0}{\sin 2\pi\frac{L}{\lambda}} \mathcal{R}e\left[\sin\left\{(L-x)\frac{2\pi}{\lambda}\right\} e^{i\omega t}\right] \tag{1.16}$$

Also for an Eulerian fluid $t_{xx}(x,t) = -p(x,t)$, and it follows from $(1.1)_2$ that we have

$$-p(x,t) = \tilde{\varrho}c_0 \frac{V_0}{\sin 2\pi\frac{L}{\lambda}} \cos\left\{(L-x)\frac{2\pi}{\lambda}\right\} \sin\omega t = \tilde{\varrho}c_0 \frac{V_0}{\sin 2\pi\frac{L}{\lambda}} \mathcal{R}e\left[-i\cos\left\{(L-x)\frac{2\pi}{\lambda}\right\} e^{i\omega t}\right]. \tag{1.17}$$

Therefore we obtain

$$Z_m = A\frac{\bar{t}_{xx}(0,t)}{\bar{v}(0,t)} = -i\tilde{\varrho}c_0 ctg\frac{2\pi L}{\lambda}. \tag{1.18}$$

Inspection shows that Z_m is singular for

$$L = n\frac{\lambda}{2} \tag{1.19}$$

i.e. when the resonator is well-tuned. In that case the impedance Z has a maximum and the current I is minimal. Also Z_m vanishes for

$$L = \left(n + \frac{1}{2}\right)\frac{\lambda}{2} \tag{1.20}$$

and I assumes a maximum. Since I may be measured as a function of L, the wavelength—hence the speed c_0—may be read off from the distance of the maxima or minima of I on the L-axis.

All this is perfectly simple and convincing—for an Eulerian fluid!

1.4 Difficulties with Many Modes and Damping

Things are less clear-cut when damping occurs, as it does in a Navier-Stokes-Fourier fluid or in the extended theories of various number of moments. The problems that occur have been discussed by Weiss [27] in some detail. They arise from the following observations.

i.) In a Navier-Stokes-Fourier fluid we have two modes of propagation, i.e. waves with different phase speeds and damping, and each one runs in both directions in the resonator tube. The fields in the tube are superpositions of those four waves and the solutions cannot be determined from the boundary values $v(0,t)$ and $v(L,t)$ alone. Rather we need additional boundary conditions on $T(x,t)$ (say), e.g. conditions of adiabaticity, or fixed temperatures at the ends.

In extended thermodynamics of many moments that problem is compounded: There are generally n modes of propagation and the field in the tube is therefore a superposition of $2n$ waves. Thus we need $2n$ boundary conditions and nobody knows how to fix them.

ii.) Even if that problem were solved, we immediately face another one. Each one of the n modes has its own phase speed and its own attenuation. And all of these influence the external mechanical impedance.

However—from what we were able to determine—the experimentalists extract only a single phase speed and a single attenuation from the measured impedance. How can that be? Is it the case that only *one* mode dominates the impedance, because the others are damped too strongly? And which mode is that?

iii.) Let us suppose that we know the answer to this question and that indeed one wave prevails. But this single wave will be *damped* and, as soon as there is damping, there are no standing waves any more and the maxima and minima of the impedance have different distances along the L-axis. So, where do we measure that distance in order to obtain the wavelength of that privileged wave?

iv.) Calculations have shown that at low frequencies indeed all modes but one are strongly damped and this one is the *sound mode*, i.e. the mode whose phase speed tends to $\sqrt{\frac{5}{3}\frac{k}{m}T}$ for $\omega \to 0$.

At higher frequencies, however, even this mode picks up considerable damping and eventually—according to extended thermodynamics—it looses its singular role to the *mode* V_{max}, so called by Weiss [27], because it has the maximum phase speed. It stands to reason that where the two modes interchange their role as the least damped one, both have equal influence on the impedance. Therefore it must be hard to interpret that function.

We are bothered and confused by these uncertainties and we are unable to remove them. This being so, we stick to a simple strategy: We calculate the phase speeds of only two modes, the sound mode and the mode V_{max}, the latter in the range where its attenuation is smaller than that of the sound mode.

2 Dispersion Relations

2.1 Navier-Stokes-Fourier Theory

According to the Navier-Stokes-Fourier theory the constitutive relations for a monatomic ideal gas have the forms

$$t_{ij} = -\varrho\frac{\mathrm{k}}{m}T\delta_{ij} + 2\mu\frac{\partial v_{\langle i}}{\partial x_{j\rangle}}, \quad q_i = -\kappa\frac{\partial T}{\partial x_i}, \quad u = \frac{3}{2}\frac{\mathrm{k}}{m}T. \tag{2.1}$$

We introduce these into the equations of balance of mass, momentum, and energy and linearize about a uniform and constant state $\tilde{\varrho}, \tilde{\mathbf{v}} = 0, \tilde{T}$. Furthermore we assume that the vector $\mathbf{v}$ points in the x-direction and that all fields depend only on x and t. Thus we obtain

$$\begin{bmatrix} \frac{\partial}{\partial t} & \tilde{\varrho}\frac{\partial}{\partial x} & 0 \\ \frac{\mathrm{k}}{m}\tilde{T}\frac{\partial}{\partial x} & \tilde{\varrho}\frac{\partial}{\partial t} - \frac{4}{3}\mu\frac{\partial^2}{\partial x^2} & \tilde{\varrho}\frac{\partial}{\partial x} \\ 0 & \frac{2}{3}\tilde{\varrho}\frac{\mathrm{k}}{m}\tilde{T}\frac{\partial}{\partial x} & \tilde{\varrho}\frac{\partial}{\partial t} - \frac{5}{2}\mu\frac{\partial^2}{\partial x^2} \end{bmatrix} \begin{bmatrix} \varrho \\ v \\ \frac{\mathrm{k}}{m}T \end{bmatrix} = 0 \tag{2.2}$$

where we have assumed that $\kappa = \frac{15}{4}\frac{\mathrm{k}}{m}\mu$ holds which is exact for Maxwellian molecules.

We consider harmonic waves of frequency ω, complex wave number k, and complex amplitudes $\bar{\varrho}, \bar{v}$ and $\bar{T}$. These amplitudes may be calculated from the algebraic equations

$$\begin{bmatrix} \omega & -\tilde{\varrho}k & 0 \\ -\frac{\mathrm{k}}{m}\tilde{T}k & \tilde{\varrho}\omega - i\frac{4}{3}\mu k^2 & -\tilde{\varrho}k \\ 0 & -\frac{2}{3}\tilde{\varrho}\frac{\mathrm{k}}{m}\tilde{T}k & \tilde{\varrho}\omega - i\frac{5}{2}\mu k^2 \end{bmatrix} \begin{bmatrix} \bar{\varrho} \\ \bar{v} \\ \frac{\mathrm{k}}{m}\bar{T} \end{bmatrix} = 0. \tag{2.3}$$

The dispersion relation results from the requirement that the system (2.3) have a nontrivial solution. We introduce the dimensionless inverse frequency

$$z = \frac{\tilde{\varrho}c_0^2}{\omega\mu}, \qquad \text{where} \qquad c_0 = \sqrt{\frac{5}{3}\frac{\mathrm{k}}{m}\tilde{T}} \tag{2.4}$$

and obtain

$$\left(\frac{10}{3}\frac{1}{z^2} + i\frac{3}{2}\frac{1}{z}\right)\left(\frac{k}{\omega/c_0}\right)^4 + \left(1 - i\frac{23}{6}\frac{1}{z}\right)\left(\frac{k}{\omega/c_0}\right)^2 - 1 = 0. \tag{2.5}$$

This is a biquadratic equation which has two relevant solutions, called modes 1 and 2. Both $\mathcal{R}e\, k$ and $\mathcal{I}m\, k$ are functions of ω, and the phase speeds and attenuations are given by

$$v_{Ph} = \frac{\omega}{\mathcal{R}e\, k} \qquad \text{and} \qquad \alpha = -\mathcal{I}m\, k. \tag{2.6}$$

Figure 11.3 shows the phase speeds and attenuations of both modes as functions of z. The figure also shows experimental points measured by Meyer & Sessler [126]. We see that for small ω, i.e. large z, these agree with the calculated values of v_{Ph} and α of mode 1. There is no agreement for large frequencies and—in particular—the phase speeds tend to ∞ for $\omega \to \infty$. We have previously—in Chapter 2—commented on this fact, which is known as the paradox of heat conduction; it has motivated the development of extended thermodynamics.

2.2 Extended Thermodynamics of 13 Fields

Extended thermodynamics in its simplest form is the 13-field theory whose equation we have written down on several occasions in this book, e.g. see (9.1.41). For a longitudinal harmonic wave we obtain the dispersion relation

$$\left(i\frac{3}{2}-\frac{9}{2}\frac{1}{z}\right)\left(\frac{k}{\omega/c_0}\right)^4-\left(z+8i-13\frac{1}{z}\right)\left(\frac{k}{\omega/c_0}\right)^2+\left(z+i\frac{25}{2}-\frac{25}{6}\frac{1}{z}\right)=0. \tag{2.7}$$

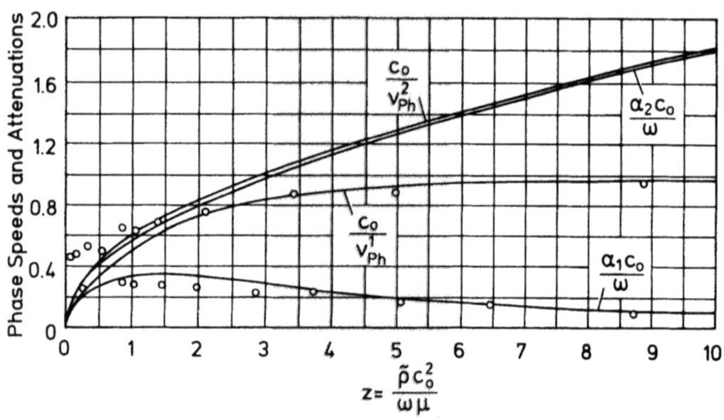

Figure 11.3 Phase speeds and attenuations in the NSF theory. Experimental points by Meyer & Sessler [126]

This is a biquadratic equation, so that there are again two essential modes, i.e. two phase speeds and attenuations. These are shown in Figure 11.4. We conclude that in extended thermodynamics the phase speeds v_{Ph} are finite for all values of ω and, in particular, for $\omega \to \infty$. Therefore, as explained in Chapter 2, the paradox seems to be resolved by the transition to extended thermodynamics.

However, although the phase speed v^1_{Ph} is finite for $\omega \to \infty$, it does not agree with the measured values for high frequencies. This fact motivates us to use further extensions of thermodynamics.

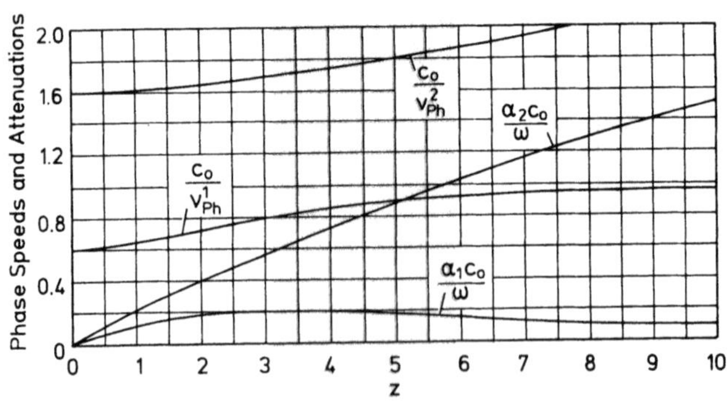

Figure 11.4 Phase speeds and attenuations of extended thermodynamics of 13 variables.

2.3 Extended Thermodynamics with Many Variables

In Chapter 9, when we explained extended thermodynamics of moments, we have written the systems of field equations for up to 35 variables and given the recipe for an arbitrary number of variables. All these systems imply dispersion relations that allow us to calculate the phase speed v_{Ph}^1 of the sound mode as a function of ω. Figure 11.5 shows the corresponding graphs. Inspection shows that the range of agreement between theories of differently many moments extends to higher values of ω if we increase the number of moments. We also observe that in the range where different theories agree they also all agree with the experimental dots. Once again we thus see that extended thermodynamics is capable of determining its own range of validity, because it is a "theory of theories"—those with differently many moments: Whenever the addition of more moments does not change the values of the dispersion curve in a particular range of $\omega' s$, the theory is good enough for those values of ω.

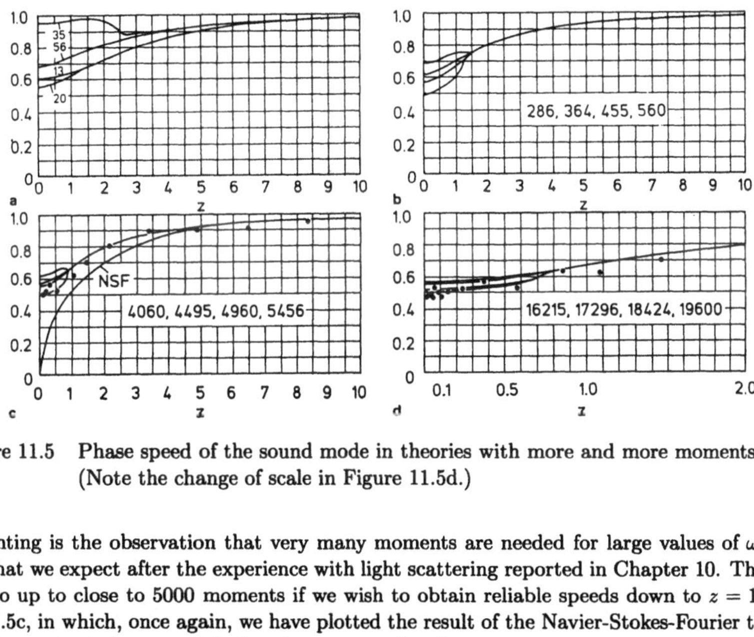

Figure 11.5 Phase speed of the sound mode in theories with more and more moments. (Note the change of scale in Figure 11.5d.)

Disappointing is the observation that very many moments are needed for large values of ω, but that is what we expect after the experience with light scattering reported in Chapter 10. Thus we have to go up to close to 5000 moments if we wish to obtain reliable speeds down to $z = 1$. See Figure 11.5c, in which, once again, we have plotted the result of the Navier-Stokes-Fourier theory so as to demonstrate how much better the moment theories are.

Figure 11.5d shows that there is no point in going beyond 20,000 moments, because between 16,000 and 19,000 the theories disagree only as much as the experimental points scatter.

2.4 Conclusion and Estimate

The conclusion from these findings is this: The smaller the value z is, i.e. the bigger the frequency ω is, the more moments are needed. This conclusion complements the conclusion of Chapter 9 which stated that many moments are needed for the proper description of the scattering from fluctuations of small wavelength. Here and there 13 or 14 moments bring little improvement over the Navier-Stokes-Fourier theory. Big improvements require the consideration of hundreds of moments. We make an estimate:

According to Figure 11.5b we need 300 moments for an adequate description of the phase speed v^1_{Ph} for $z = 3$. From (2.4) we have

$$z = \frac{\tilde{\varrho} c_0^2}{\omega \mu} = \frac{5}{3} \tilde{p} \frac{1}{\omega \mu}.$$

With $z = 3$ and $\tilde{p} = 1 bar$, and $\mu = 2.26 \cdot 10^{-5} \frac{Ns}{m^2}$ for Xe according to Kestin et al. [122] we obtain $\omega \approx 2.4\ 10^9 Hz$. This is of the order of magnitude of the collision frequency of the atoms under the given circumstances. Therefore more moments provide macroscopic theories for frequencies *well below the collision frequency.*

And once again z depends on p, so that lowering p has the same effect as increasing ω. A dilute gas requires many moments for its description sooner than a dense gas.

3 Maximum Speed

3.1 Modes of Least Damping

The reason for the concentration on the sound mode is the belief that this mode experiences least damping. This belief is surely satisfied in the Navier-Stokes-Fourier theory and in the 13-moment theory as an inspection of Figures 11.3 and 11.4 shows. But then, when more and more moments are taken into account, we obtain more modes and the question arises whether it is always and for all frequencies the sound mode that has minimal damping. In fact Weiss [27] found that this is not so.

He instructed the computer to plot the phase speed of *the* mode with least damping and obtained graphs like those in Figure 11.6: For small frequencies the least damped mode is always the sound mode, but starting with the 57-moment theory the sound mode looses that privileged role for high frequencies to the mode of maximum phase speed, which Weiss calls the mode V_{max}.

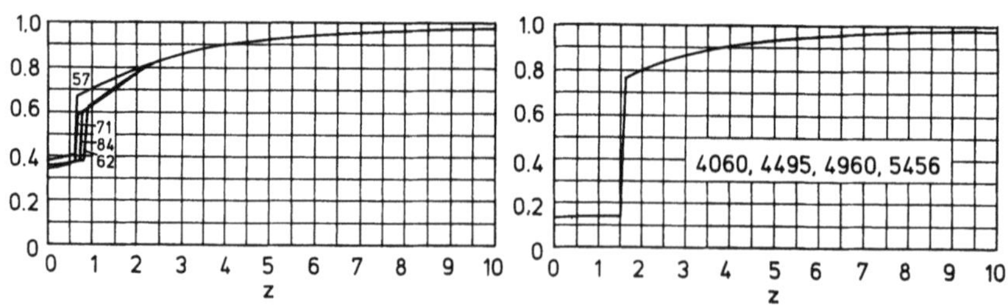

Figure 11.6 Phase speeds of modes with least damping.

This phenomenon is obviously of some concern for the proper interpretation of the resonance experiments. Indeed, at least at the switching point at about $z = 1.5$ we expect the two modes to contribute to the experimental findings, since they are bound to be equally weakly damped there. This consideration seems to be unknown to the experimentalists.

3.2 The Maximum Speed

If we only consider the sound mode, inspection of Figure 11.5 shows that the phase speed remains finite for $\omega \to \infty$. Thus the paradox of infinite pulse speeds would not seem to arise and, indeed, that was the original motivation for formulating extended thermodynamics.

However, there are the other modes, and in particular there is the mode V_{max}. We see from Figure 11.6 that V_{max} becomes bigger as the theory is improved by taking more and more moments into account. In fact, Weiss has calculated V_{max} and come up with the values reported in Table 10.4, in which V_{max} shows a steady growth as the number of moments is raised up to 15180. There seems little doubt that V_{max} keeps on increasing to infinity as the number of moments is raised to infinity.[1] Nor is the mode V_{max} irrelevant; after all, it is least damped.

This is somewhat of an anticlimax because, after all, extended thermodynamics originally started with the desire to have finite pulse speeds. But then, it has already been indicated by Cercignani and Majorana [106] that the nonrelativistic kinetic theory—and hence extended thermodynamics of infinitely many moments—permits infinite speeds, e.g. in the Maxwell distribution. Therefore the present result does not come as a surprise.

Anyway extended thermodynamics has outgrown its original motif of determining a finite speed of heat propagation. Indeed, we may now say that it provides a hierarchy of symmetric hyperbolic systems which are relevant when rapid changes occur or steep gradients. "Rapid" means that the fields contain high frequencies in their Fourier spectra, and the values of these frequencies will determine the number of necessary fields in the manner explained in Paragraph 2.4 above.

[1] Recall the general proof given in Paragraph 9.2.2 by which $V_{max} \to \infty$.

Chapter 12

Structure of Shock Waves

by **Wolf Weiss**, Thermodynamics, TU Berlin

Shock waves are treated as singular surfaces in gas dynamics, i.e. as abrupt jumps of the thermodynamic fields. Experiments show, however that the thermodynamic fields are smooth in a shock, although quite steep. Thus the pressure may increase by a large factor on the short distance of a few mean free paths. The purpose of this chapter is the calculation of ***shock structure***, that is to say the calculation of the thermodynamic fields across the shock in monatomic gases.

This is a long-standing problem in gas dynamics and it has been "proved" several times that the method of moments cannot solve it, because the smooth shock structure breaks down at finite Mach numbers. In this context Israel [127] has raised the question: "Internal Structure of Strong Shocks: A Crisis of Extended Thermodynamics?".

No Crisis! If there ever was one, it has been resolved by the recognition that the objections are based on misunderstandings. Weiss [16], [128] has begun to prove that the smooth shock structure persists to higher and higher Mach numbers when more and more moments are considered.

1 Experimental Evidence

The density field across the shock wave is measured in sophisticated and highly accurate experiments, essentially by sending an electron beam through the gas which is absorbed or scattered depending on the density. Figure 12.1 shows a typical result for argon: On the distance of a few mean free paths Λ_0 the density ρ increases from ρ_0 to ρ_1, its values before and behind the wave. And the shape of the curve depends on the speed of the wave, or the Mach number M_0 before the shock.

It is common in the field to define the shock thickness δ as

$$\delta = \frac{\varrho_1 - \varrho_0}{\left.\frac{d\varrho}{dx}\right|_{\max}}. \tag{1.1}$$

Inspection of Figure 12.1 shows that δ will depend on M_0. The shock thickness, defined by (1.1), is $\delta \approx 4\Lambda_0$ for $M_0 = 3.38$ and $\delta \approx 8\Lambda_0$ for $M_0 = 1.55$. The thickness δ depends on M_0, as shown by the circles in Figure 12.2. We see that δ first decreases rapidly with M_0; then it reaches a minimum and proceeds to increase slowly as M_0 grows.

There is considerable disagreement between experiment and the Navier-Stokes-Fourier theory and this fact has prompted Weiss [16] to apply extended thermodynamics to the problem.

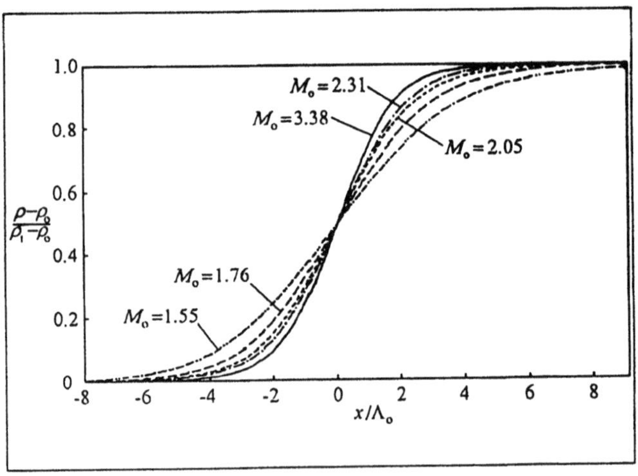

Figure 12.1 Shock wave structure in argon. Measurements by Alsmeyer [129] at a pressure of $p \approx 0.6$ Pa.

2 Review of Previous Work

2.1 Rankine-Hugoniot Relations

We rewrite the equations of balance of mass, momentum, and energy for a monatomic gas with $t_{<ij>}$ and q_i as stress deviator and heat flux

$$\begin{aligned}
&\frac{\partial\rho}{\partial t}+\frac{\partial\rho v_j}{\partial x_j}=0\\
&\frac{\partial\rho v_i}{\partial t}+\frac{\partial(\rho v_i v_j-t_{ij})}{\partial x_j}=0 \qquad\qquad \text{with}\quad t_{ij}=-\rho\frac{k}{m}T\delta_{ij}+t_{<ij>} \qquad (2.1)\\
&\frac{\partial\rho\left(\varepsilon+\frac{1}{2}v^2\right)}{\partial t}+\frac{\partial\left(\rho\left(\varepsilon+\frac{1}{2}v^2\right)v_j-t_{ij}v_i+q_j\right)}{\partial x_j}=0. \quad \varepsilon=\frac{3}{2}\frac{k}{m}T
\end{aligned}$$

For a stationary and one-dimensional wave these equations reduce to

$$\begin{aligned}
\frac{d}{dx}(\varrho v) &= 0,\\
\frac{d}{dx}\left(\varrho v^2+\varrho\frac{k}{m}T+\sigma\right) &= 0, \qquad \text{with } \sigma=-t_{11}\\
\frac{d}{dx}\left(\varrho v^3+5\varrho\frac{k}{m}Tv+2\sigma v+2q\right) &= 0.
\end{aligned} \qquad (2.2)$$

Integration provides

$$\begin{aligned}
\varrho v &= C_1,\\
\varrho v^2+\varrho\frac{k}{m}T+\sigma &= C_2,\\
\varrho v^3+5\varrho\frac{k}{m}Tv+2\sigma v+2q &= C_3,
\end{aligned} \qquad (2.3)$$

so that the expressions on the left-hand sides are constant along the whole x-axis and, in particular, they have the same values in the equilibria at the points x_0 and x_1, far before and far behind the wave, where σ and q vanish.

$$\begin{aligned}
C_1 &= \varrho_0 v_0 &&= \varrho_1 v_1\\
C_2 &= \varrho_0 v_0^2+\varrho_0\frac{k}{m}T_0 &&= \varrho_1 v_1^2+\varrho_1\frac{k}{m}T_1\\
C_3 &= \varrho_0 v_0^3+5\varrho_0\frac{k}{m}T_0 v_0 &&= \varrho_1 v_1^3+5\varrho_1\frac{k}{m}T_1 v_1
\end{aligned} \qquad (2.4)$$

We may use these equations to relate the values of ϱ, v, T before and behind the wave

$$\begin{aligned}
\varrho_1 &= \frac{4\varrho_0 v_0^2}{5\frac{k}{m}T_0+v_0^2},\\
v_1 &= \frac{5\frac{k}{m}T_0+v_0^2}{4v_0},\\
T_1 &= \frac{3v_0^4+14\frac{k}{m}T_0v_0^2-5\frac{k^2}{m^2}T_0^2}{16\frac{k}{m}v_0^2}.
\end{aligned} \qquad (2.5)$$

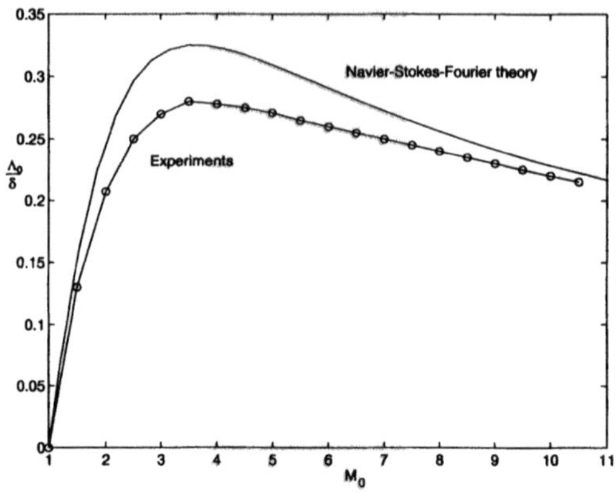

Figure 12.2 Shock wave thickness as function of the Mach number. Navier-Stokes-theory and experiments. The figure is taken from Ruggeri [130].

These are the Rankine-Hugoniot equations. It is common in gas dynamics to introduce dimensionless velocities, the Mach numbers

$$M_0 = \frac{v_0}{a_0} = \frac{v_0}{\sqrt{\frac{5}{3}\frac{k}{m}T_0}} \quad \text{and} \quad M_1 = \frac{v_1}{a_1} = \frac{v_1}{\sqrt{\frac{5}{3}\frac{k}{m}T_1}}. \tag{2.6}$$

With these the Rankine-Hugoniot equations read

$$\begin{aligned} \varrho_1 &= \frac{4M_0^2}{M_0^2+3}\varrho_0, \\ v_1 &= \frac{M_0^2+3}{4M_0^2}v_0, \\ T_1 &= \frac{5M_0^4+14M_0^2-3}{16M_0^2}T_0, \\ M_1 &= \frac{M_0^2+3}{\sqrt{5M_0^4+14M_0^2-3}}. \end{aligned} \tag{2.7}$$

In Figures 12.3 and 12.4 these functions are plotted. We itemize a few salient features which are well known in gas dynamics:

- $\rho_1 > \rho_0$ holds for all M_0, i.e. the wave is compressive. The maximal value of ρ_1 is $4\rho_0$.
- v_1/v_0 is non-monotone as a function of M_0; it reaches its minimal value $\sqrt{3}/2$ for $M_0 = \sqrt{3}$.
- T_1 grows monotonically with M_0 and is unbounded for $M_0 \to \infty$.

- $M_1 < M_0$ holds for all $M_0 > 1$, so that the wave leads from supersonic to subsonic flow. For $M_0 \to \infty$ we obtain $M_1 \to 1/\sqrt{5}$.

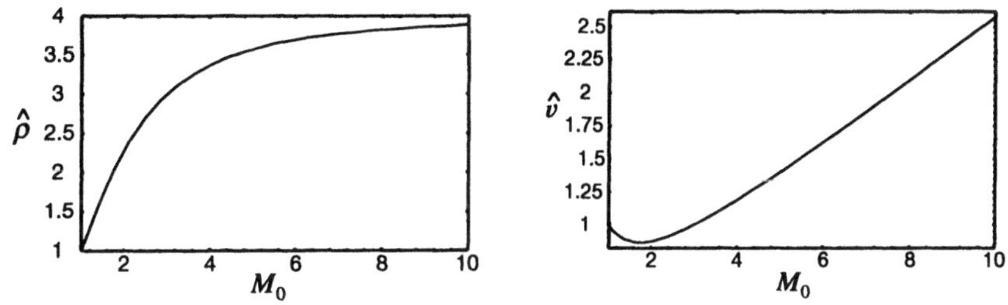

Figure 12.3 $\hat{\rho} = \rho_1/\rho_0$ and $\hat{v} = v_1/v_0$ as functions of M_0.

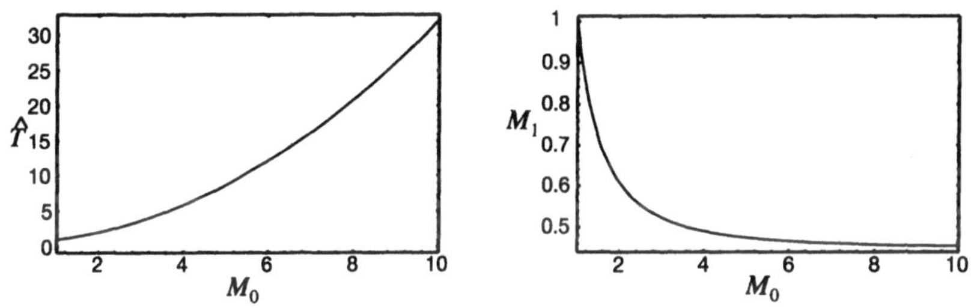

Figure 12.4 $\hat{T} = T_1/T_0$ and M_1 as functions of M_0.

2.2 Becker's Solutions

As early as 1922 Becker [131] attempted to calculate the shock structure for a Navier-Stokes-Fourier fluid. In that case we have

$$t_{<ij>} = 2\mu \frac{\partial v_{<i}}{\partial x_{j>}} \qquad \text{and} \qquad q_i = -\kappa \frac{\partial T}{\partial x_i},$$

so that the equations (2.3) read

$$\begin{aligned} \rho v &= C_1, \\ \rho v^2 + \rho \frac{k}{m} T - \frac{4}{3}\mu \frac{dv}{dx} &= C_2, \\ \rho v^3 + 5\rho \frac{k}{m} T - \frac{8}{3}\mu \frac{dv}{dx} v - 2\kappa \frac{dT}{dx} &= C_2. \end{aligned} \tag{2.8}$$

By elimination of ρ and some easy manipulation we may rewrite this system in the form

$$\begin{aligned} \frac{4}{3}\mu \frac{dv}{dx} &= C_1 \left(v + \frac{k}{m} T \frac{1}{v} \right) - C_2, \\ 2\kappa \frac{dT}{dx} &= C_1 \left(3\frac{k}{m} T - v^2 \right) + 2C_2 v - C_3. \end{aligned} \tag{2.9}$$

The system is a coupled quasi-linear system of ordinary differential equations. It would be quite explicit,[1] if it were not for the fact that the viscosity μ and the thermal conductivity κ are temperature dependent in a nontrivial manner.

For the simple—and unrealistic—cases that either μ or κ are zero Becker was able to obtain solutions by quadrature. And in the equally unrealistic case $\kappa/\mu = \frac{10}{3}\frac{k}{m}$ with both κ and μ constants he obtained the algebraic relation

$$5\frac{k}{m}T = C_3/C_1 - v^2$$

between T and v; and then he was able to find an exact solution for the remaining differential equation, viz.

$$\begin{aligned} x - C &= \frac{5}{6}\frac{\mu}{\rho_0\sqrt{\frac{5}{3}\frac{k}{m}T_0}}\frac{1}{M_0} \\ &\left[\frac{\frac{5}{3}M_0^2+1}{\sqrt{\left(\frac{5}{3}M_0^2+1\right)^2 - \frac{16}{15}M_0^2\left(\frac{5}{3}M_0^2+5\right)}}\ln\frac{1-\frac{v}{v_0}}{\frac{v}{v_1}-1} + \ln\left(1-\frac{v}{v_0}\right)\left(\frac{v}{v_1}-1\right)\right], \quad (2.10) \end{aligned}$$

where C is a constant of integration and $v_{\substack{0\\1}}$ are given by

$$v_{\substack{0\\1}} = \sqrt{\frac{k}{m}T_0}\left[\frac{5}{8}\frac{\frac{5}{3}M_0^2+1}{\frac{5}{3}M_0} \pm \sqrt{\frac{25}{64}\frac{\left(\frac{5}{3}M_0^2+1\right)^2}{\frac{5}{3}M_0^2} - \frac{1}{4}\left(\frac{5}{3}M_0^2+5\right)}\right]. \quad (2.11)$$

Figure 12.5 shows plots of the function (2.10) for different Mach numbers. We recognize that the thickness of the shock decreases monotonically—in contrast to the observed behaviour which is exhibited by the circles in Figure 12.2.

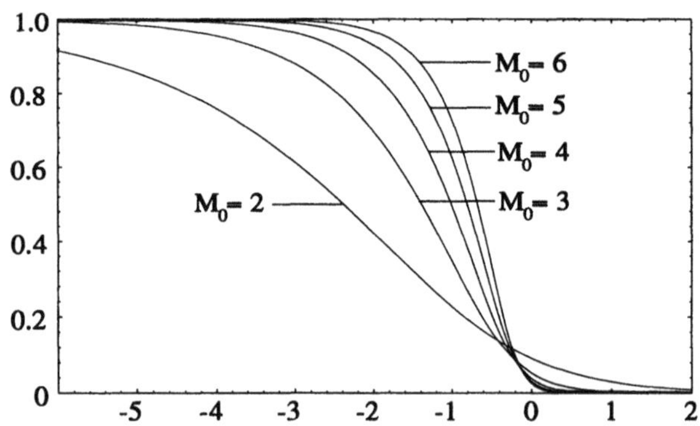

Figure 12.5 Shock wave structure for the exact solution by Becker. Ordinate: $\frac{v-v_1}{v_0-v_1}$, Abscissa: $x - C \Big/ \frac{5}{6}\frac{\mu}{\rho a_0}$.

[1]Note that, by (2.4), the C's are known as the fluxes of mass, momentum, and energy far ahead of the wave.

2.3 Singular Perturbation Analysis

Explicit exact solutions of the quasi-linear system (2.9) cannot be found except for unrealistic assumptions on κ and μ. But an explicit approximate solution is possible by a perturbation analysis, although only for small shocks. The smallness parameter is denoted by ε and it is chosen as

$$\varepsilon = M_0 - 1,$$

because for $M_0 = 1$ there is no shock; cf. Paragraph 2.1. We refer v and T to the constant solutions v_0 and T_0 and set

$$\begin{aligned} v &= v_0 + \varepsilon \overset{I}{v}(x) + \varepsilon^2 \overset{II}{v}(x) + .. \\ T &= T_0 + \varepsilon \overset{I}{T}(x) + \varepsilon^2 \overset{II}{T}(x) + ... \end{aligned}$$

Insertion into (2.9) provides polynomials in ε on the two sides of the equations, and their coefficients must match. This is not the place to describe the method of singular perturbations, which can be quite cumbersome in detail. Suffice it to say that

- the zero-order terms match identically,
- the first-order terms lead to a relation between $\overset{I}{T}$ and $\overset{I}{v}$, viz. $\frac{k}{m}\overset{I}{T} = -\frac{2}{5}a_0 \overset{I}{v}$,
- the second-order terms provide a simple differential equation for $\overset{I}{v}$, viz

$$\left(5\mu_0 + \frac{\kappa_0}{k/m}\right)\frac{d\overset{I}{v}}{dx} = 5\rho_0\left(\overset{I}{v}^2 - \frac{1}{25}a_0^2\right).$$

μ_0 and κ_0 are values appropriate to the temperature T_0 before the shock.

The solution $\overset{I}{v}(x)$ is a hyperbolic tangent and the velocity field itself is given by

$$v(x) = a_0\left(1 - (M_0 - 1)\frac{1}{5}\tanh\frac{x - C}{5\left(\frac{5\mu_0 + \frac{\kappa_0}{k/m}}{\rho_0 a_0}\right)}\right) \tag{2.12}$$

to within second-order terms in the smallness parameter ε. We do not exhibit this function in a figure, because it is a well-known hyperbolic tangent and because the shock structure is independent of the Mach number in this approximation. The only interesting remark about this solution is the observation that the shock becomes steeper, with decreasing viscosity and thermal conductivity. Incidentally, this could also have been concluded from Becker's solution in Figure 12.5.

Instructive as it is to do the first nontrivial perturbation, it is virtually useless nowadays, since the system (2.9) has been solved exactly, albeit by numerical means.

2.4 Numerical Solution by Gilbarg and Paolucci

An important contribution in this field is the 1953 paper by Gilbarg & Paolucci [132]. These authors have solved the system (2.9) numerically for the correct value $\kappa/\mu \approx \frac{15}{4}\frac{k}{m}$—valid for all monatomic gases—and they allowed μ to depend on T by a power law

$$\mu \sim T^s.$$

It turns out that the M_0-dependence of the shock thickness is strongly affected by s, even if s is limited to the range between $\frac{1}{2}$ and 1, which corresponds to hard sphere molecules and Maxwell molecules respectively. For argon Gilbarg & Paolucci chose $s = 0.816$—as recommended by Chapman & Cowling [43]—and calculated the shock thickness up to $M_0 = 3$. Thus they just missed the maximum of the curve. Ruggeri [130] has recalculated the curve proceeding to $M_0 = 11$ and beyond. His curve is shown in Figure 12.2. Although it runs roughly "parallel" to the experimental curve, the discrepancy is noticable and it is considered unacceptable.[2]

Despite the discrepancy between theory and experiment the Gilbarg & Paolucci paper is of lasting interest to the mathematical physicist, because the authors identified—and explained neatly—an important difference between the stationary states before and behind the wave:

- the state before the wave is a two-tangent nodal point, while
- the state behind the wave is a saddle.

Therefore, if we consider the problem of calculating the shock structure in a Navier-Stokes-Fourier fluid as an initial value problem, we have to choose the starting point *behind* the wave; otherwise we shall never obtain a solution. We postpone the discussion of this interesting point, because it will come up later in a more general context; see Paragraph 4.2.

2.5 The 13-Moment Theory by Grad

Right at the beginning of this book—in and around Figure 1.1—we have argued that the Navier-Stokes-Fourier theory may be unsatisfactory for shock waves where rapid changes and steep gradients occur, because the theory ignores the time derivatives $\dot{t}_{<ij>}$ and $\dot{q}_i$ of the viscous stress and the heat flux. And of course Grad knew that.

So when Grad had developed the 13-moment theory—which preserves the time derivatives of $t_{<ij>}$ and q_i—he tried to improve on the shock structure calculations; see Grad [99]. But the result was a surprise: It is *dramatically worse* than Navier-Stokes-Fourier. Indeed, there is no smooth shock structure for Mach numbers beyond $M_0 = 1.65$, which, of course, is in stark contrast with the experiments. Rather the calculations show that a "subshock" appears, so that the profile of the shock looks as shown in Figure 12.6.

We do not know whether Grad recognized the significance of the number 1.65, but, of course, the careful reader of this book will recognize that number as the largest characteristic speed

[2] Actually, according to Kestin et al. [122], $s = 0.64$ is a better value for argon than $s = 0.816$. And for $s = 0.64$ the discrepancy of theory and experiment is still bigger; see Kremer & Müller [133].

of a monatomic gas, the "pulse speed" in equilibrium of a 13-moment fluid; e.g. see equation (8.5.17) and many other places of this book. The fact that subshocks appear when the Mach number equals the pulse speed for 13-moments suggests that the same might happen when we have more moments where—by Table 10.4 and Figure 9.1—the pulse speed is bigger. So perhaps the appearance of subshocks could be pushed to higher and higher Mach numbers as more and more moments come into play. This was conjectured by Weiss, and he has begun to prove it as we shall see in Section 4.

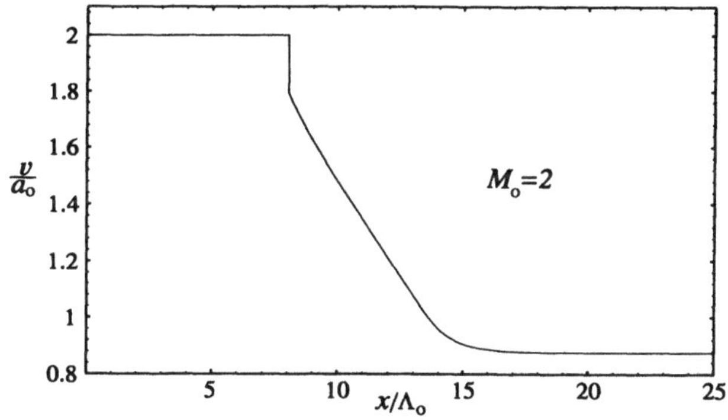

Figure 12.6 A subshock for $M_0 = 2$ in the 13-moment theory.

2.6 The 13-Moment Theory by Anile & Majorana

Anile & Majorana [100] were aware of Grad's subshock dilemma and they also knew the early version of extended TIP which we have presented in Chapter 2. It will be recalled that in that theory the pulse speed depends on the parameters a, b, and K (see (2.2.20)) of which not much is known so that there is some freedom. Hoping against hope, Anile & Majorana attempted to use this freedom to push the appearance of subshocks out of the observable range of Mach numbers. It did not work, at least not well enough to be convincing. The authors suggested that the fault might lie in the omission of nonlinear terms of which there are plenty; e.g. see (1.3.34). This conjecture was also adopted by Jou & Pavon [134]. But the conjecture is not valid, as we know now. Nevertheless the papers revived interest in the field.

2.7 Criticism of Moment Methods for Shock Structure

Holway [135] argued in 1964 that the expansion of the phase density into Hermite polynomials—the very essence of moment methods—does not allow smooth shock structures beyond $M_0 = 1.851$. This argument, if it were correct, would obviously be the death blow for a successful application of moment theories to shocks. But the argument was erroneous, as Weiss [136] showed recently: Holway had confused the front side and the backside of the shock.

Another objection to the application of high-moment methods was raised by Ruggeri [137]. This is more subtle. Ruggeri noticed that the singularity—which occurred at $M_0 = 1.65$ in the 13-moment theory—moves closer and closer to $M_0 = 1$ as the number of moments increases. Therefore the subshock should appear sooner, and no smooth shock structure could be expected or, at best, only very close to $M_0 = 1$. The fallacy in this case was a misjudgement of the nature of the singularity. We shall come back to this point in detail later. The question of why Ruggeri's objection is not valid represents one of the most intriguing problems at this time of shock structure analysis. Numerical calculations all disqualify the objection, but an analytical proof is yet to come.

2.8 Alternative Methods for Shock Structure Calculations

The moment method is just one possiblity for an approximate treatment of the Boltzmann equation. Another one is the Chapman-Enskog approximation. This method invariably leads to a parabolic system, so that subshocks—being the mark of hyperbolicity—do not occur. When the Chapman-Enskog method was applied to shocks by Zoller [138] and later by Foch [139], the authors obtained what one expects: qualitatively reasonable results but quantitatively not quite satisfactory.

Another ingenious attempt was invented by Mott-Smith [140], who published as early as 1951. This method is based on a superposition of two Maxwell distributions chosen in such a manner as to satisfy the Boltzmann equation approximately. The results are no better and no worse than those of the other attempts. Gilbarg & Paolucci [132] present comparative plots for all these attempts, and we refer the interested reader to their paper.

The best results and, as a matter of fact, perfect results are obtained by molecular dynamics via Monte Carlo calculations, which where first presented by Bird [141]. Therefore one may think that the matter is closed. But we are not satisfied, because we prefer analytical methods and we are confident that they will work.[3]

3 Preliminaries on Singular Points and Characteristic Speeds

3.1 Field Equations and Boundary Values

We recall the field equations of extended thermodynamics in their synthetic form as written in Chapter 3

$$\frac{\partial \mathbf{F}^0(\mathbf{u})}{\partial t} + \frac{\partial \mathbf{F}^k(\mathbf{u})}{\partial x^k} = \mathbf{\Pi}(\mathbf{u}). \tag{3.1}$$

In the stationary and one-dimensional case appropriate for shock waves we have

$$\frac{d\mathbf{F}(\mathbf{u})}{dx} = \mathbf{\Pi}(\mathbf{u}), \quad \text{or} \quad \mathbf{A}(\mathbf{u}) \cdot \frac{d\mathbf{u}}{dx} = \mathbf{\Pi}(\mathbf{u}), \quad \text{where} \quad \mathbf{A} = \frac{\partial \mathbf{F}}{\partial \mathbf{u}}. \tag{3.2}$$

These equations are supplemented by boundary conditions which are the values of $\mathbf{u}$ at x_0 far before the wave and x_1 far behind it. At those boundary points all components of the production density $\mathbf{\Pi}(\mathbf{u})$ vanish.

[3] Of course in the long run, when computers do everything, continuum mechanics and thermodynamics may become obsolete for natural science—as obsolete as alchemy and astrology. But not yet!

For the first three components of **u**, which are invariably the density of mass ρ, the velocity v, and the temperature T, the boundary values ρ_0, v_0, T_0 and ρ_1, v_1, T_1 are related by the Rankine-Hugoniot relations (2.5) or (2.7). The remaining components of **u** will be chosen to be the nonequilibrium parts of higher moments—e.g. stress deviator $t_{<ij>}$ and heat flux q_i. Therefore they vanish at x_0 and x_1, so that the boundary conditions for (3.2) read

$$(\rho_0, v_0, T_0, 0, 0, ...) \quad \text{and} \quad (\rho_1, v_1, T_1, 0, 0, ...) . \tag{3.3}$$

3.2 Singular Points and Stationary Points

It will turn out that the system (3.2) is replete with singularities, and we start with a characterization of such singularities, because there seems to be no standardized nomenclature in the literature. Our nomenclature is taken from the books of Kamke [142] and Bieberbach [143].

First a single ordinary differential equation: For the equation

$$\frac{du}{dx} = \frac{p(u,x)}{q(u,x)} \tag{3.4}$$

we have the following definitions:

- A point x_s, for which $q(u(x_s), x_s) = 0$ holds, is singular.
- A point x_s, for which $q(u(x_s), x_s) = 0$ and $p(u(x_s), x_s) = 0$ hold, but so that the right-hand side exists, is a regular singular point.
- A point x_s, for which $q(u(x_s), x_s) = 0$ and $p(u(x_s), x_s) \neq 0$ hold, is an irregular singular point.

For *systems* like $(3.2)_2$ things are less straightforward. Let us first solve the system for $\frac{du_B}{dx}$ by linear algebra to give

$$\frac{du_B}{dx} = \frac{z_B(\mathbf{u})}{D(\mathbf{u})}, \qquad (B = 1, 2, ...N) \tag{3.5}$$

where u_B $(B = 1, 2, ...N)$ are components of **u** and D is the determinant of **A**; see (3.2). z_B is the determinant of **A** with the Bth column replaced by **Π**. Unlike (3.4) the variable x does not appear explicitly on the right-hand side. But we may recover the form (3.4) by elimination of x between the first $N-1$ equations and the Nth one. Thus we obtain

$$\frac{du_B}{du_N} = \frac{z_B(\overline{\mathbf{u}}, u_N)}{z_N(\overline{\mathbf{u}}, u_N)} \qquad (B = 1, 2, ...N-1) \tag{3.6}$$

a system which looks more like (3.4), providing equations for the $N-1$ functions $\overline{\mathbf{u}}(u_N)$. It is now clear how to extrapolate to *systems* the above characterization of singular points.

- A point x_s is *singular* if $z_N(\overline{\mathbf{u}}(x_s), u_N(x_s)) = 0$ holds.
- A point x_s is a *regular singular* point if $z_N(\overline{\mathbf{u}}(x_s), u_N(x_s)) = 0$ and all $z_B(\overline{\mathbf{u}}(x_s), u_N(x_s)) = 0$ $(B = 1, 2, ...N-1)$ hold, but such that z_B/z_N is finite for all B.
- A point x_s is an *irregular singular* point if $z_N(\overline{\mathbf{u}}(x_s), u_N(x_s)) = 0$ holds and at least one $z_B(\overline{\mathbf{u}}(x_s), u_N(x_s)) \neq 0$.

These definitions are independent of the determinant D in (3.5). But of course D may also vanish and thereby create a singularity. In fact that will be the case which interests us most in the sequel. Therefore, motivated by the above definitions, we extrapolate further and—with reference to the system (3.5)—we define singularities as follows:

- A point x_s is singular if $D(\mathbf{u}(x_s)) = 0$ holds.
- A point x_s is a regular singular point if $D(\mathbf{u}(x_s)) = 0$ and all $z_B(\mathbf{u}(x_s)) = 0$ hold, but such that z_B/D is finite for all B.
- A point x_s is an irregular singular point if $D(\mathbf{u}(x_s)) = 0$ holds but at least one $z_B(\mathbf{u}(x_s))$ is unequal to 0.

Relevant to our study are also *stationary* points x_s of the system (3.5). These occur where $z_B(\mathbf{u}(x_s)) = 0$ holds for all $B = 1, 2, ...N$. For the wave this happens in the equilibrium points x_0 and x_1 far before and behind the wave, because $\mathbf{\Pi} = 0$ holds there. Stationary points of the system (3.5) are obviously also regular singular points of the system (3.6), which is why the two notions are sometimes confused.

3.3 The Singularities D = 0

The determinant D of $\mathbf{A}$ is a function of all variables including v. And if $D = 0$ holds, we may consider this as an equation for v. The N solutions are the characteristic speeds and they are denoted by $c^{(\alpha)} (\alpha = 1, 2, ...N)$. They depend on the other variables.

In equilibrium, where all variables vanish with the exception of ρ, v, T, the characteristic speeds in an ideal gas depend only on T and they are proportional to a, the adiabatic sound speed. Thus, in the equilibrium points x_0 and x_1 we have

$$c_0^{(\alpha)} = \hat{c}^{(\alpha)} a_0 \quad \text{and} \quad c_1^{(\alpha)} = \hat{c}^{(\alpha)} a_1. \tag{3.7}$$

In order to explain what happens with D as v_0—and M_0—increases we focus the attention on the schematic curves of Figure 12.7. This figure refers to the case when we have three positive characteristic speeds, two larger than a and one smaller.[4] These are marked—along with a—before and behind the wave and their rise in-between is schematically indicated by the dashed straight lines. Also the figure shows a schematic shock structure of the velocity field v.

Now let us first consider the situation before the wave when v_0 is increased; it will soon reach $c_0^{(2)}$, so that D becomes zero at x_0 and the system becomes singular before the wave. If we increase v_0 further—assuming that $v(x)$ remains smooth—the velocity curve $v(x)$ will intersect $c^{(2)}$ at a point $x > x_0$ so that the singularity moves into the wave. A second singularity before the wave develops when v_0 reaches $c_0^{(3)}$ and that singularity moves into the wave as well as v_0 is further increased.

A similar situation occurs behind the wave: As v_0 increases, v_1 goes down, further away from the sound speed a_1 and eventually reaches $c_1^{(1)}$, in which case the system becomes singular at x_1,

[4]This is the case for the 21-moment theory, as we shall see below.

because D becomes zero there. A further increase in v_0 will let that singularity move into the wave from behind.

We conclude that the occurrence of singularities before the wave is governed by the characteristic speeds $c_0^{(\alpha)} > a_0$, while the singularities behind the wave are governed by the characteristic speeds $c_0^{(\alpha)} < a_0$.

The former statement is evident and the latter follows from a little calculation: Elimination of $\hat{c}^{(\alpha)}$ and the use of $(2.7)_3$ provides the relation between the characteristic speeds before and behind the wave, viz.

$$c_1^{(\alpha)} = \sqrt{\frac{5M_0^4 + 14M_0^2 - 3}{16M_0^2}}\, c_0^{(\alpha)}.$$

If v_0 increases, v_1 changes as dictated by $(2.7)_2$ and v_1 reaches $c_1^{(\alpha)}$ for the initial velocity

$$v_0 = \sqrt{\frac{3a_0^2 + c_0^{(\alpha)2}}{5c_0^{(\alpha)2} - a_0^2}}\, a_0.$$

Since we must have $v_0 > a_0$, it follows that $c_0^{(\alpha)} < a_0$ must hold. q.e.d.

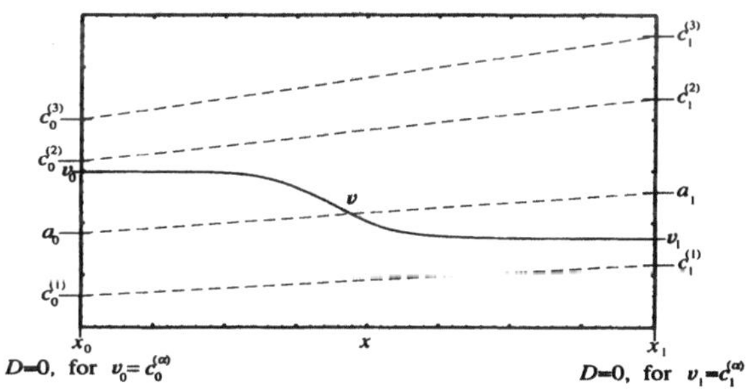

Figure 12.7 Schematic representation of flow velocity v, sound speed a, and the characteristic speeds $c^{(\alpha)}$, all as functions of x.

3.4 Regular and Irregular Singularities

Of course it might have been that all $c_0^{(\alpha)}$ are so far above or below a_0 that the singularities are never reached for reasonable Mach numbers $M_0 = v_0/a_0$. However, we already know that this hope is vain, since Grad saw the singularity appear at $M_0 = 1.65$. In fact, the smallest characteristic speed greater than v_0 tends towards a_0 as the number of moments increases; see Figure 12.8. Therefore the first singularity will appear right away, as soon as v_0 grows beyond a_0.

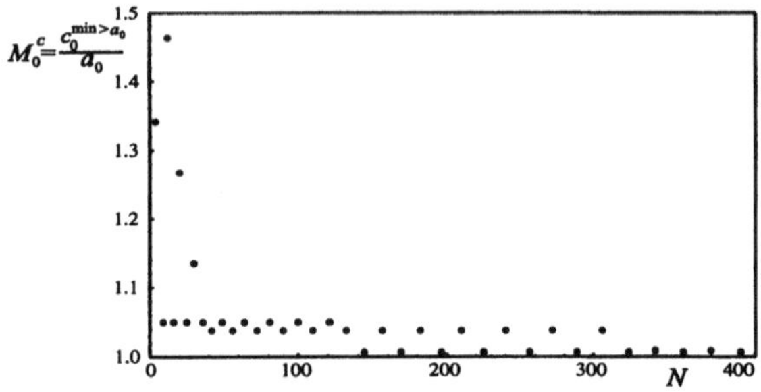

Figure 12.8 The smallest characteristic speed greater than a_0 versus the number of moments.

We may ask what happens at these singularities? Ruggeri [130] thought—in effect—that they were irregular singularities and a subshock would appear right away at $v_0 \geq a_0$. And after that all bets are off; Ruggeri did not speculate.

Weiss [16] showed—and we shall see the demonstration in subsequent sections—that all but one of these singularities are regular ones. The one that is not regular corresponds to the highest characteristic speed. Thus he calculates smooth shock structures up to $v_0 = c_0^{(\max)}$. Beyond that value of v_0 the subshock appears.

Obviously this result makes a lot of sense. After all, $c_0^{(\max)}$ is the largest characteristic speed in equilibrium; what we have called the pulse speed. And if the wave moves faster than that? Well, in that case the wave is truly *supersonic* and we are not surprised that a shock appears, albeit a subshock.

4 Numerical Calculation of the Shock Structure

4.1 Initial and Boundary Value Problems

We refer to the system (3.2) or, equivalently, to the forms (3.5) or (3.6), viz.

$$\frac{d\mathbf{F}(\mathbf{u})}{dx} = \mathbf{\Pi}(\mathbf{u}), \quad \frac{d\mathbf{u}}{dx} = \frac{\mathbf{z}(\mathbf{u})}{D(\mathbf{u})}, \quad \frac{du_B}{du_N} = \frac{z_B(\overline{\mathbf{u}}, u_N)}{z_N(\overline{\mathbf{u}}, u_N)} \quad (B = 1, 2, ...N-1). \tag{4.1}$$

These are N coupled equations. Far on the left, at x_0, the values $\mathbf{u}_0$ are given and at x_1, at the far right, the values $\mathbf{u}_1$ are given. $\mathbf{u}_0$ and $\mathbf{u}_1$ are related by the Rankine-Hugoniot relations.

The $2N$ data are a peculiarity. Since the system is of first order, we might have expected the solutions to be determined by N data. In the present case, however, the solution of the problem is not unique without $\mathbf{u}_0$ and $\mathbf{u}_1$. The reason for the non-uniqueness is the fact that both end-points are stationary points, see Section 3.2.

For the solution of the system we have two options: We may consider our problem as an

- initial value problem, in which case all values $\mathbf{u}$ are given at either x_0 or x_1, or a
- boundary value problem, in which all data are given at x_0 and at x_1.

We discuss these options consecutively.

4.2 Algorithm for the Initial Value Problem

Among the three equivalent formulations (4.1) of the problem, the last one seems to be most appropriate for the solution of the initial value problem at x_0 (or x_1). Therefore we consider u_N as the variable, and $u_N(x_0)$ is its initial value. The initial values are then given by

$$u_A(u_N(x_0)) = u_A(x_0) \qquad (A = 1, 2, ...N-1). \tag{4.2}$$

There is a difficulty though, because at x_0 all productions vanish, and so do all determinants $\mathbf{z}(\mathbf{u}_0)$, so that the right-hand side of $(4.1.)_3$ is indeterminate at the initial point; in the jargon of Section 3.2 the initial point is a regular singular point. So as to avoid the singular point we choose $u_N(x_0) + \Delta u_N$ as the initial value of the variable u_N—with an arbitrary, but small Δu_N—and formulate the initial value problem of the system as

$$u_A(u_N(x_0) + \Delta u_N) = u_A(x_0) + \Delta u_A. \tag{4.3}$$

It remains to determine Δu_A appropriately so that it fits to the chosen value of Δu_N. This can be done from $(4.1)_3$ by the L'Hospital rule. We write $(4.1.)_3$ at x_0 as

$$\left.\frac{du_A}{du_N}\right|_0 = \lim_{\Delta \mathbf{u}\to 0} \frac{\overbrace{z_A(\mathbf{u}_0)}^{=0} + \left.\frac{\partial z_A}{\partial u_B}\right|_0 \Delta u_B}{\underbrace{z_N(\mathbf{u}_0)}_{=0} + \left.\frac{\partial z_N}{\partial u_B}\right|_0 \Delta u_B} = \lim_{\Delta \mathbf{u}\to 0} \frac{\left.\frac{\partial z_A}{\partial u_B}\right|_0 \left.\frac{\Delta u_B}{\Delta u_N}\right|_0}{\left.\frac{\partial z_N}{\partial u_B}\right|_0 \left.\frac{\Delta u_B}{\Delta u_N}\right|_0} = \frac{\left.\frac{\partial z_A}{\partial u_B}\right|_0 \left.\frac{du_B}{du_N}\right|_0}{\left.\frac{\partial z_A}{\partial u_B}\right|_0 \left.\frac{du_B}{du_N}\right|_0}. \tag{4.4}$$

$\left.\frac{\partial z_A}{\partial u_B}\right|_0$ is known for all A and therefore (4.4) represents a system of algebraic equations for $\left.\frac{\partial u_A}{\partial u_N}\right|_0$ whence follow the required increments

$$\Delta u_A = \left.\frac{du_A}{du_N}\right|_0 \Delta u_N. \tag{4.5}$$

With these values the stepwise numerical solution can start and proceed. Once we have obtained $u_A(u_N)$, we may use $(4.1)_2$ to obtain x as a function of u_N. Indeed, we have

$$x - x_0 = \int\limits_{u_N(x_0)}^{u_N} \frac{D\left(\mathbf{u}\left(u_N'\right)\right)}{z_N\left(\mathbf{u}\left(u_N'\right)\right)} du_N'. \tag{4.6}$$

It is true that the equations (4.4) have more than one solution. The choice of the correct one must be made by ulterior arguments. After all, we know something about the required wave solution – in particular we know that it has to reach the values $\mathbf{u}_1$ at the other end. Thus the multiplicity of solutions is reduced. In the 13-moment theory there exists just *one* sensible solution.

Of course the numerical solution is only approximate and this raises another problem which we shall try to illuminate for the Navier-Stokes-Fourier theory, for which we have already put down the relevant equations in Section 2.2. By (2.9) the form equivalent to the system $(4.1)_3$ in this case is the single equation (with $\frac{\kappa}{\mu}=\frac{15}{4}\frac{k}{m}$, $\hat{v}=\frac{v}{a_0}$, $\hat{T}=\frac{T}{T_0}$.)

$$\frac{d\hat{T}}{d\hat{v}} = \frac{z_T(\hat{v},\hat{T})}{z_v(\hat{v},\hat{T})} = \frac{8}{9}\,\frac{-M_0\hat{v}^2 + 2\left(M_0^2+\frac{3}{5}\right)\hat{v} + M_0\left(\frac{9}{5}\hat{T} + \left(M_0^2+3\right)\right)}{M_0\hat{v} - \left(M_0^2+\frac{3}{5}\right) + \frac{3}{5}M_0\hat{T}/\hat{v}}. \tag{4.7}$$

The two curves $z_T(\hat{v},\hat{T})=0$ and $z_v(\hat{v},\hat{T})=0$ are both parabolae in a (T,v)-diagram, convex and concave respectively as shown – schematically – in Figure 9. The stationary points – or regular singular points – occur where the curves intersect. They are denoted by S_0 and S_1 in the figure signifying the stationary points before and behind the wave. We know that the left intersection is behind the wave, because $v_1 < v_0$ and $T_1 > T_0$ must hold according to the Rankine-Hugoniot conditions. The shock structure solution that connects S_0 and S_1 must lie entirely between the two parabolae, if T and v are to be monotone functions with finite slope.

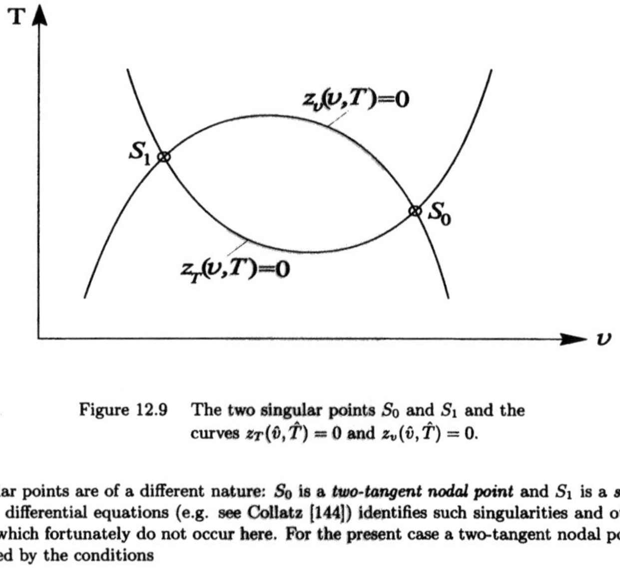

Figure 12.9 The two singular points S_0 and S_1 and the curves $z_T(\hat{v},\hat{T})=0$ and $z_v(\hat{v},\hat{T})=0$.

But the two singular points are of a different nature: S_0 is a ***two-tangent nodal point*** and S_1 is a ***saddle point***. The theory of ordinary differential equations (e.g. see Collatz [144]) identifies such singularities and others – like stars and whirlpools, – which fortunately do not occur here. For the present case a two-tangent nodal point and a saddle point are recognized by the conditions

$$\mathcal{D} = \left(\frac{\partial z_v}{\partial\hat{v}}\cdot\frac{\partial z_T}{\partial\hat{T}} - \frac{\partial z_v}{\partial\hat{T}}\cdot\frac{\partial z_T}{\partial\hat{v}}\right)\Bigg|_{x_0 \text{ or } x_1} \begin{matrix} > \\ < \end{matrix} 0 \begin{matrix}\text{node}\\ \text{saddle}\end{matrix} \tag{4.8}$$

From (4.7)—and with the Rankine Hugoniot conditions – we calculate for $M_0 > 1$

$$\mathcal{D}_0 = \tfrac{9}{5}M_0^2 > 0 \text{ and } \mathcal{D}_1 = -\tfrac{180}{25}\tfrac{M_0^2}{(M_0^2+3)^2}\left(M_0^4 + 2M_0^2 - 3\right) < 0,$$

so that we can confirm the above statement about S_0 and S_1 : S_0 is a node and S_1 is a saddle respectively.

Figure 12.10 shows an ensemble of solution curves of the differential equation (4.7), among them many that pass through the nodal point and two through the saddle—the saddle has only two! The solid line connecting S_0 and S_1 is the exact solution of (4.7), and the curves that lie close to it are the approximate solutions which we may obtain by the algorithm described above. But now consider: *No* approximate solution near S_0 will ever reach S_1, while *all* approximate solutions near S_1—in the right half plane—will reach S_0. Therefore it is imperative to start the numerical calculation of the shock structure *behind* the wave.

Interestingly enough the positions of nodal point and saddle are reversed in the 13-moment case, as Figure 12.11 shows, at least for $M_0 = 1.4$. Therefore here we may—and must—start an initial value problem *before* the wave.

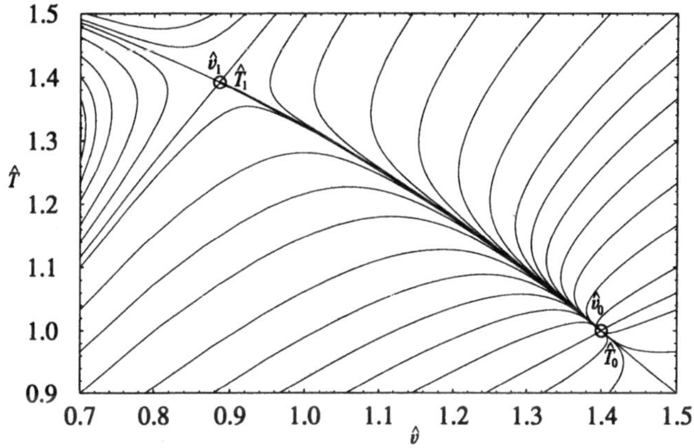

Figure 12.10 Solution curves of Navier-Stokes-Fourier theory with singular points 0 and 1 ($M_0 = 1.4$) $\left(\hat{v} = \frac{v}{a_0}, \hat{T} = \frac{T}{T_0}\right)$.

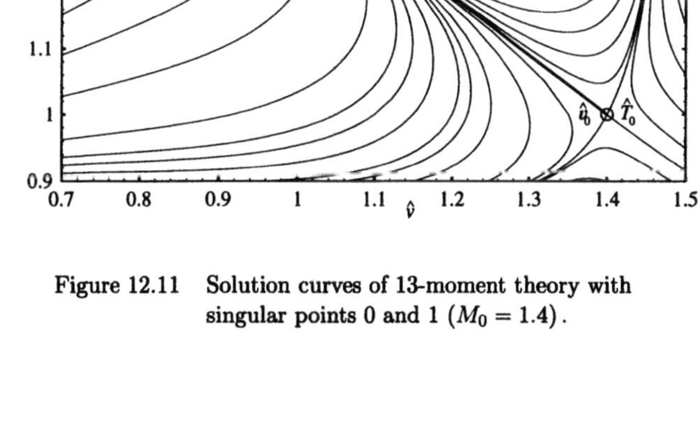

Figure 12.11 Solution curves of 13-moment theory with singular points 0 and 1 ($M_0 = 1.4$).

4.3 Algorithm for the Boundary Value Problem

The solution of boundary value problems for the system (4.1) is best started from the form $(4.1)_1$ of the system. We discretize a sufficiently large interval (x_0, x_1) by choosing S grid points

$$x^{(n)} = x_0 + (n-1)\Delta x, \quad (n = 1, 2, ...S), \quad \text{where} \quad x_0 + (S-1)\Delta x = x_1. \tag{4.9}$$

At the endpoints x_0 and x_1 the boundary values $\mathbf{u}_0$ and $\mathbf{u}_1$ are prescribed. The differential quotient in $(4.1)_1$ is replaced by the central difference quotient, so that we have

$$\begin{aligned} \mathbf{u}\left(x^{(1)}\right) &= \mathbf{u}_0 \\ \mathbf{F}\left(\mathbf{u}(x^{(n+2)})\right) - \mathbf{F}\left(\mathbf{u}(x^{(n)})\right) &= 2\Delta x\, \mathbf{\Pi}\left(\mathbf{u}(x^{(n+1)})\right), \quad (n = 1, 2...S-2) \\ \mathbf{u}\left(x^{(s)}\right) &= \mathbf{u}_1. \end{aligned} \tag{4.10}$$

This system consists of S algebraic equations for the S unknown vectors $\mathbf{u}(x^n)$. And since $\mathbf{u}$ itself has N components we have a system of $S \cdot N$ equations for the $S \cdot N$ unknowns $\mathbf{u}_A(x^{(n)})$. Thus for $S = 200$ and $N = 5$—as in the 13-moment case—there are 1000 nonlinear algebraic equations to be solved. There are standard solutions for such problems in the program libraries. They start from an initial guess $\mathbf{u}(x^n)$ and improve on it in repeated steps until the system (4.10) is satisfied. In the present case the Navier-Stokes-Fourier solution is a good candidate for the initial guess.

4.4 The 13-Moment Case

The 13-moment case has already been treated by Grad [99] and we discussed Grad's paper—in Paragraph 2.5—among the important ones leading up to our problem. The Grad equations appropriate to this stationary and one-dimensional case—and with the right-hand side calculated for Maxwell molecules—read

$$\begin{aligned} &\frac{d}{dx}\{\varrho v\} = 0, \\ &\frac{d}{dx}\left\{\varrho v^2 + \varrho\frac{k}{m}T + \sigma\right\} = 0, \\ &\frac{d}{dx}\left\{\varrho v^3 + 5\varrho\frac{k}{m}Tv + 2\sigma v + 2q\right\} = 0, \\ &\frac{d}{dx}\left\{\frac{2}{3}\varrho v^3 + \left(\frac{4}{3}\varrho\frac{k}{m}T + \frac{7}{3}\sigma\right)v + \frac{8}{15}q\right\} = -\alpha\varrho\sigma, \\ &\frac{d}{dx}\left\{\frac{1}{2}\varrho v^4 + \left(4\varrho\frac{k}{m}T + \frac{5}{2}\sigma\right)v^2 + \frac{16}{5}qv + \frac{5}{2}\varrho\left(\frac{k}{m}T\right)^2 + \frac{7}{2}\frac{k}{m}T\sigma\right\} = -\alpha\varrho\left(\sigma v + \frac{2}{3}q\right). \end{aligned} \tag{4.11}$$

σ is the (11)-component of the pressure deviator and α stands for $2\frac{k}{m}T/\mu$; note that α is a constant for Maxwellian molecules. We introduce dimensionless quantities by

$$\hat{\varrho} := \frac{\varrho}{\varrho_0}, \quad \hat{v} := \frac{v}{a_0}, \quad \hat{T} := \frac{T}{T_0}, \quad \hat{\sigma} := \frac{\sigma}{\varrho_0 {}^k/_m T_0}, \quad \hat{q} := \frac{q}{\varrho_0 {}^k/_m T_0 a_0}, \quad \hat{x} := \frac{\varrho_0\alpha}{a_0}x, \tag{4.12}$$

and obtain

$$\begin{aligned}
\hat{\varrho} &= \frac{M_0}{\hat{v}}, \\
\hat{\sigma} &= M_0\left(\frac{5}{3}(M_0-\hat{v})-\frac{\hat{T}}{\hat{v}}\right)+1, \\
\hat{q} &= \frac{M_0}{2}\left(\frac{5}{3}(M_0-\hat{v})^2-3\hat{T}+5\right)-\hat{v}
\end{aligned} \tag{4.13}$$

by integration of the first three equations of (4.11), and

$$\begin{aligned}
&\frac{d}{d\hat{x}}\left\{\frac{7}{3}M_0\hat{v}^2+3\left(M_0^2+\frac{3}{5}\right)\hat{v}-\frac{9}{5}M_0\hat{T}+\frac{4}{3}M_0\left(\frac{M_0^2}{3}+1\right)\right\}= \\
&\qquad -\frac{M_0}{\hat{v}}\left[M_0\left(\frac{5}{3}(M_0-\hat{v})-\frac{\hat{T}}{\hat{v}}\right)+1\right] \\
&\frac{d}{d\hat{x}}\left\{-\frac{2}{3}M_0\hat{v}^3\left(\frac{7}{6}M_0^2+\frac{7}{10}\right)\hat{v}^2+\left(-\frac{34}{5}M_0\hat{T}+\frac{8}{3}M_0^3+8M_0\right)\hat{v}-\right. \\
&\qquad \left.\frac{3}{5}M_0\frac{\hat{T}^2}{\hat{v}}+\frac{7}{2}\left(M_0^2+\frac{3}{5}\right)\hat{T}\right\}= \\
&-\frac{M_0}{9\hat{v}}\left\{-10M_0\hat{v}^2+\left(3+5M_0^2\right)\hat{v}+M_0\left(15+5M_0^2-18\hat{T}\right)\right\}
\end{aligned} \tag{4.14}$$

for the remaining two equations. We note that the conditions (4.13) are valid for all numbers of moments—not only for 13—and these relations may be used to eliminate $\hat{\rho}, \hat{\sigma}$, and $\hat{q}$ from the rest of the equations, which has been done in (4.14).

The system (4.14) is in the form $(4.1)_1$ appropriate for the solution of a boundary value problem as described in Section 4.3.

If the differentiations with respect to $\hat{x}$ are carried out, we obtain the system in the generic forms

$$\begin{bmatrix} \hat{A}_{11} & \hat{A}_{12} \\ \hat{A}_{21} & \hat{A}_{22} \end{bmatrix}\begin{bmatrix} \frac{d\hat{v}}{d\hat{x}} \\ \frac{d\hat{T}}{d\hat{x}} \end{bmatrix}=\begin{bmatrix} \hat{\Pi}_1 \\ \hat{\Pi}_2 \end{bmatrix} \quad \text{or} \quad \begin{aligned} \frac{d\hat{v}}{d\hat{x}} &= \frac{1}{\hat{D}}\begin{bmatrix} \hat{\Pi}_1 & A_{12} \\ \hat{\Pi}_2 & A_{22} \end{bmatrix}=\frac{\hat{z}_v}{\hat{D}} \\ \frac{d\hat{T}}{d\hat{x}} &= \frac{1}{\hat{D}}\begin{bmatrix} A_{11} & \hat{\Pi}_1 \\ A_{21} & \hat{\Pi}_2 \end{bmatrix}=\frac{\hat{z}_T}{\hat{D}}. \end{aligned} \tag{4.15}$$

The right-hand side of $(4.15)_1$ is the same as in (4.14) and the matrix elements $\hat{A}_{BC}$ are easily calculated from (4.14); they are too cumbersome to be written here. However, the computer knows the matrix and, of course, it also knows $\hat{D}$, its determinant. For x_0 and x_1 the values of $\hat{D}$ are given by

$$\hat{D}_0=\frac{25M_0^4-78M_0^2+27}{10} \quad \text{and} \quad \hat{D}_1=\frac{155M_0^4-606M_0^2+243}{80}. \tag{4.16}$$

The vanishing of these determinants identifies the Mach number M_0 for which the velocities are characteristic before and behind the wave respectively. They read

$$\begin{matrix} M_0=\pm 0.62972... & & M_0=\pm 0.67351... \\ & \text{and} & \\ M_0=\pm 1.65029... & & M_0=\pm 1.85905... \end{matrix} \tag{4.17}$$

respectively. Only $M_0 > 1$ is relevant to our discussion since no wave solution appears for $M_0 \leq 1$. Therefore we expect singularities before the wave at $M_0 = 1.650$ and behind the wave at $M_0 = 1.859$.

$\hat{D}_0$ and $\hat{D}_1$ are represented in Figure 12.12 as functions of M_0. Of course we can also plot $\hat{D}$ as a function of $\hat{x}$, but we need the solutions $\hat{v}(\hat{x})$ and $\hat{T}(\hat{x})$ in order to do that.

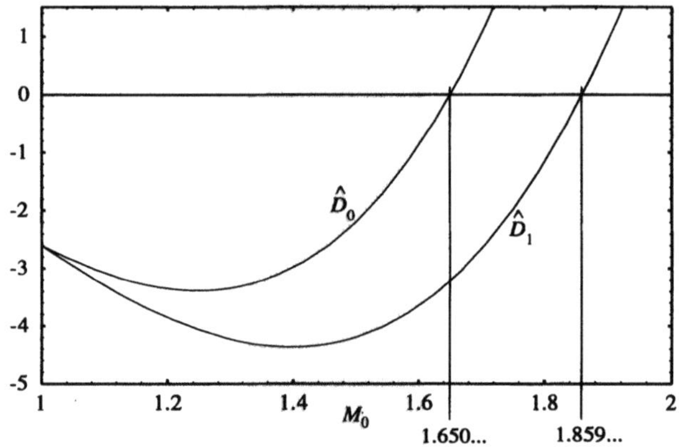

Figure 12.12 The determinants $\hat{D}_0$ and $\hat{D}_1$ for 13 moments.

Figure 12.13a shows a solution—calculated as a boundary value problem—for $M_0 = 1.5$. The velocity $\hat{v}$ decreases from 1.5 to 0.875 and the temperature $\hat{T}$ increases between 1 and 1.494. The shock structure is smooth: no singularity is seen nor is a singularity expected, since $\hat{D}_0$ and $\hat{D}_1$ are both negative for $M_0 = 1.5$, so that it is unlikely that $\hat{D}$ has changed sign along the way. In fact, the determinant $\hat{D}$—when it is calculated and plotted versus $\hat{x}$, as in Figure 12.14a—is nonzero everywhere along the way. The numerator determinants $\hat{z}_v$ and $\hat{z}_T$ are only zero at x_0 and x_1, while inside the shock structure Figure 12.14a shows that they appear with nonzero values. This is so because productions $\mathbf{\Pi}$ do occur within the shock structure; shock structures after all are dissipative.

More interesting is the case $M_0 = 1.65$ represented in Figures 12.13b and 12.14b. We recognize kinks in the curves in $\hat{v}(\hat{x})$ and $\hat{T}(\hat{x})$ which are due to the fact that M_0 is now equal to the characteristic speed $\hat{c}_0^{(2)}$, so that the system becomes singular at x_0 where $\hat{D}$ vanishes. $\hat{z}_T$ and $\hat{z}_v$ also vanish and the quotients $\hat{z}_v/\hat{D}$ and $\hat{z}_T/\hat{D}$ are finite; they determine the slopes of $\hat{v}$ and $\hat{T}$ at x_0. Therefore the singular point at x_0 is a regular singular point.

Since no kinks appear in the observations of shock waves, we must conclude that the 13-moment method fails at $M_0 = 1.65$—and probably before. This is our motivation for an increase in the number of moments. We postpone this, however, so as to make room for an additional instructive argument:

Despite the kink at $M_0 = 1.65$ we proceed to $M_0 = 2.0$, see Figures 12.13c and 12.14c. We see that the kink has developed into a subshock now. This is not surprising, since Grad [99] already saw this. But what *is* surprising is the fact that there is no second subshock behind the wave. After all, we are already past the Mach number $M_0 = 1.859..$, where a singularity should appear behind the wave according to (4.17). And a singularity does appear at that Mach number: it is again a regular singularity which, however—unlike the first one—does not develop into an irregular singularity but into a regular one. This fact is put in evidence by the inset in Figure 12.14c, which shows a magnification of the point where $\hat{D} = 0$ holds. We see that at the same point $\hat{z}_v$ and $\hat{z}_T$ vanish and the shock structure shows no sign of all this.

We shall soon see that this behavior is symptomatic for all singularities except the one connected with the largest characteristic speed before the wave. This is why we have discussed the point despite the fact that we must not trust 13 moments up to $M_0 = 2.0$.

4.5 The 14-Moment Case

For the stationary and one-dimensional shock wave the relevant equations of a 14-moment theory are

$$\frac{d}{dx}\{\varrho v\} = 0,$$

$$\frac{d}{dx}\left\{\varrho v^2 + \varrho\frac{k}{m}T + \sigma\right\} = 0,$$

$$\frac{d}{dx}\left\{\varrho v^3 + 5\varrho\frac{k}{m}Tv + 2\sigma v + 2q\right\} = 0, \tag{4.18}$$

$$\frac{d}{dx}\left\{\frac{2}{3}\varrho v^3 + \left(\frac{4}{3}\varrho\frac{k}{m}T + \frac{7}{3}\sigma\right)v + \frac{8}{15}q\right\} = -\alpha\varrho\sigma,$$

$$\frac{d}{dx}\left\{\frac{1}{2}\varrho v^4 + \left(4\varrho\frac{k}{m}T + \frac{5}{2}\sigma\right)v^2 + \frac{16}{5}qv + \frac{5}{2}\varrho\left(\frac{k}{m}T\right)^2 + \frac{7}{2}\frac{k}{m}T\sigma + \frac{1}{6}\Delta\right\} = -\alpha\varrho\left(\sigma v + \frac{2}{3}q\right),$$

$$\frac{d}{dx}\left\{\varrho v^5 + \left(14\varrho\frac{k}{m}T + 8\sigma\right)v^3 + \frac{84}{5}qv^2 + \left(35\varrho\left(\frac{k}{m}T\right)^2 + 28\frac{k}{m}T\sigma + \frac{7}{3}\Delta\right)v + 28\frac{k}{m}Tq\right\} =$$
$$-\alpha\varrho\left(4\sigma v^2 + \frac{16}{3}qv + \frac{2}{3}\Delta\right)$$

The quantity Δ, the nonequilibrium part of the trace ρ_{iijj} of the fourth moment, is the 14th variable in addition to the ones of the 13-moment theory. Δ is made nondimensional by the definition

$$\hat{\Delta} := \frac{\Delta}{\rho_0\frac{k}{m}T_0a_0^2}. \tag{4.19}$$

Our investigation proceeds much like before, except that everything is a little more cumbersome. In particular, there are now three remaining variables—after $\hat{\rho}, \hat{\sigma}, \hat{q}$ have been eliminated—namely $\hat{v}, \hat{T}, \hat{\Delta}$. Therefore we also have three equations—rather than the two of (4.15)—and there will be three determinants $\hat{z}_T(\hat{v}, \hat{T}, \hat{\Delta})$, $\hat{z}_v(\hat{v}, \hat{T}, \hat{\Delta})$, and $\hat{z}_\Delta(\hat{v}, \hat{T}, \hat{\Delta})$. Of course Δ vanishes in the equilibria at x_0 and x_1.

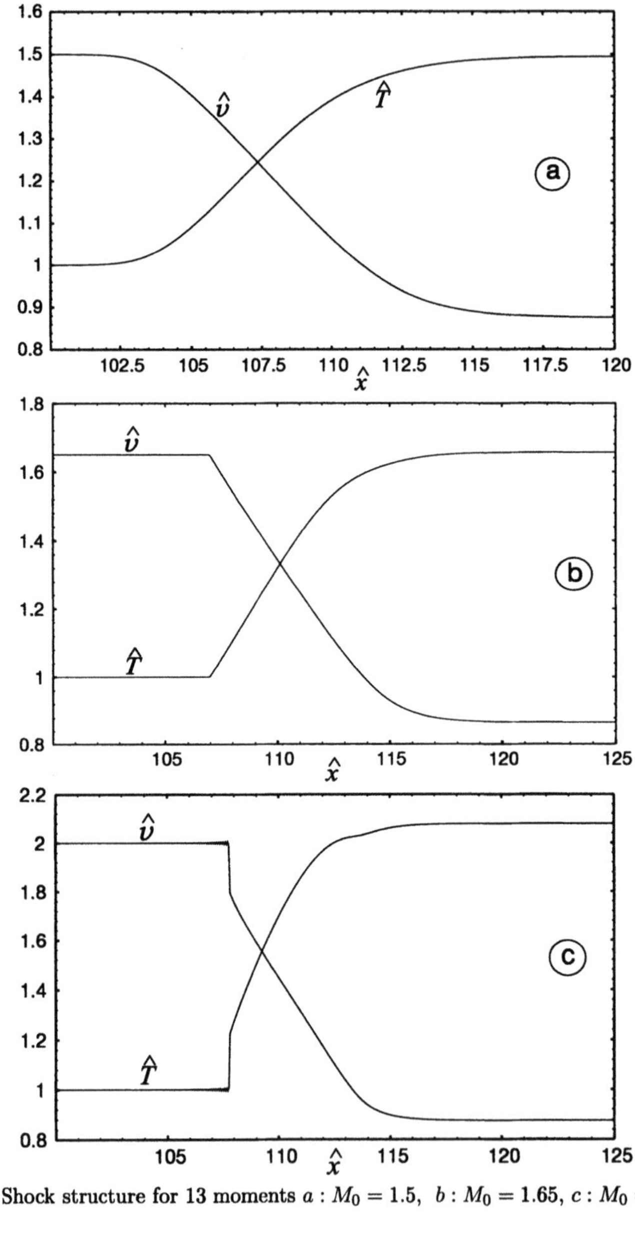

Figure 12.13 Shock structure for 13 moments $a : M_0 = 1.5$, $b : M_0 = 1.65$, $c : M_0 = 2.0$.

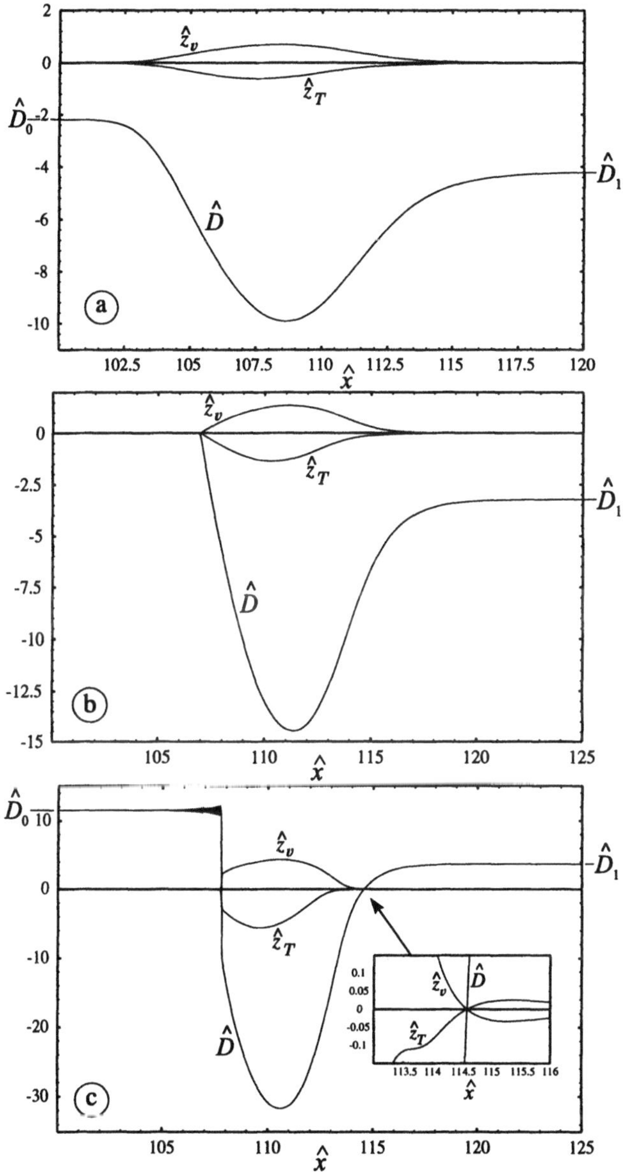

Figure 12.14 Determinants $\hat{z}_v$, $\hat{z}_T$ and $\hat{D}$ along the wave (13 moments) $a: M_0 = 1.5$, $b: M_0 = 1.65$, $c: M_0 = 2.0$.

We first write $\hat{D}_0$ and $\hat{D}_1$. They have the forms

$$\hat{D}_0 = M_0\left(\frac{25M_0^4 - 98M_0^2 + 63}{10}\right) \text{ and } \hat{D}_1 = \frac{555M_0^6 + 739M_0^4 - 2487M_0^2 + 873}{320M_0}, \tag{4.20}$$

and their real roots—apart from the trivial case $M_0 = 0$—are

$$\begin{matrix} M_0 = \pm 0.90021... \\ M_0 = \pm 1.76341... \end{matrix} \quad \text{and} \quad \begin{matrix} M_0 = \pm 0.64804... \\ M_0 = \pm 1.11737... \end{matrix} \tag{4.21}$$

respectively. $\hat{D}_0$ and $\hat{D}_1$ are shown in Figure 12.15, whence we conclude, as we do from (4.21), that for $M_0 = 1.117$ the system becomes singular behind the wave and for $M_0 = 1.763$ it becomes singular before the wave. Obvious $\hat{D}$ must therefore pass through zero inside the wave.

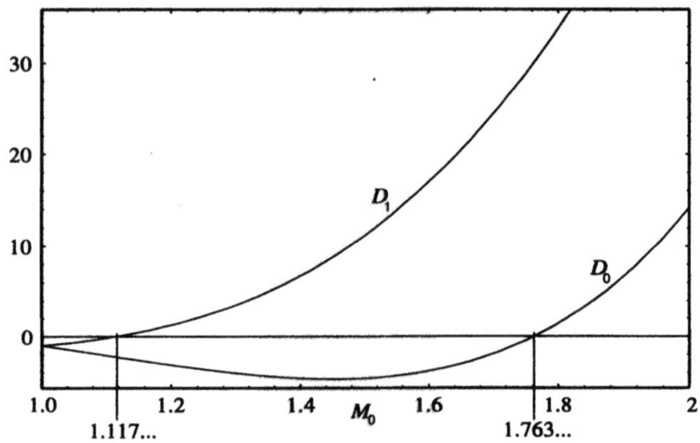

Figure 12.15 The determinants $\hat{D}_0$ and $\hat{D}_1$ for 14 moments.

The Figures 12.16 and 12.17 exhibit the solution of the boundary value problem for the 14-moment case. We itemize the salient points:

- For $M_0 = 1.1$ there is a smooth shock structure without any singularities.
- For $M_0 = 1.13$ a singularity—which first appeared behind the wave at $M_0 = 1.117$—has moved into the wave from behind. It does not disturb the smooth shock structure, however, because it is a regular singularity. Witness the fact that all determinants $\hat{z}_v, \hat{z}_T, \hat{z}_\Delta$, and $\hat{D}$ pass through zero at the same point.
- For $M_0 = 1.6$ the situation is the same: We have a smooth shock structure despite—or because of—the regular singularity, which has moved slightly forwards.
- For $M_0 = 1.76$ the previous regular singularity is still there but in the front the shock structure develops a kink as a second regular singularity appears for $M_0 = 1.76341$.

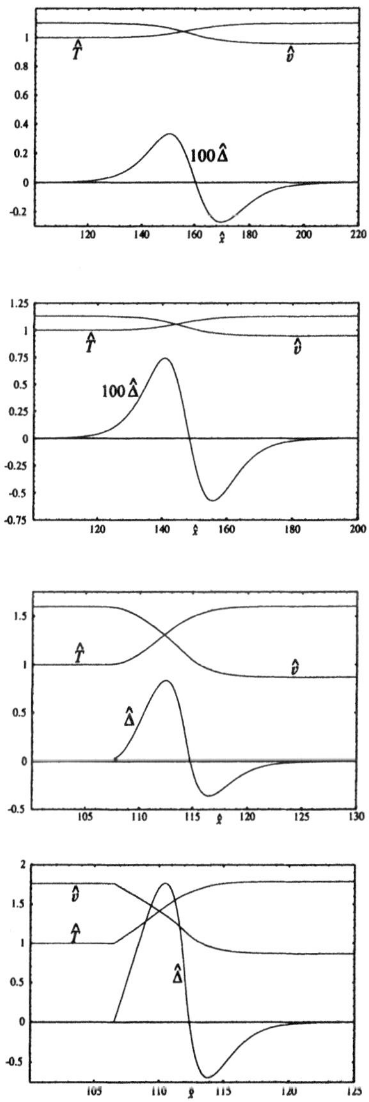

Figure 12.16 Shock structure for 14 moments
a: $M_0 = 1.1$, b: $M_0 = 1.13$,
c: $M_0 = 1.6$, d: $M_0 = 1.763$.

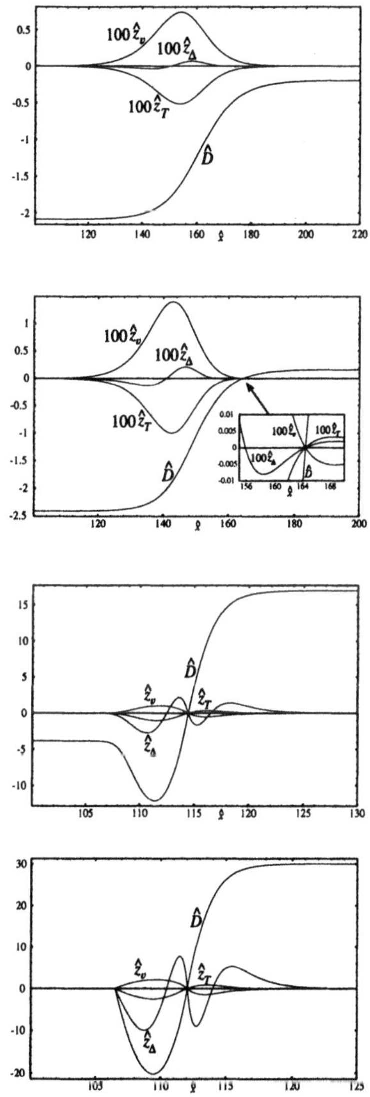

Figure 12.17 Determinants $\hat{z}_v$, $\hat{z}_T$, $\hat{z}_\Delta$ and $\hat{D}$ for 14 moments
a: $M_0 = 1.1$, b: $M_0 = 1.13$,
c: $M_0 = 1.6$, d: $M_0 = 1.76$.

The kink will become a subshock for higher values of M_0 but this is not shown here, because the 14-moment method has already disqualified itself for $M_0 = 1.76$ by the appearance of the kink.

But note that the kink—and subsequent subshocks—start at a slightly higher Mach number than in the 13-moment case. This fact encourages us to go beyond 14 moments, to 21 moments.

4.6 The 21-Moment Case

We follow the by now familiar procedure: the equations for the 21-moment case read

$$\begin{gathered}
\frac{d}{dx}\{\varrho v\} = 0, \\
\frac{d}{dx}\left\{\varrho v^2 + \varrho\frac{k}{m}T + \sigma\right\} = 0, \\
\frac{d}{dx}\left\{\varrho v^3 + 5\varrho\frac{k}{m}Tv + 2\sigma v + 2q\right\} = 0, \\
\frac{d}{dx}\left\{\frac{2}{3}\varrho v^3 + \left(\frac{4}{3}\varrho\frac{k}{m}T + \frac{7}{3}\sigma\right)v + \frac{8}{15}q\right\} = -\alpha\varrho\sigma, \\
\frac{d}{dx}\left\{\frac{1}{2}\varrho v^4 + \left(4\varrho\frac{k}{m}T + \frac{5}{2}\sigma\right)v^2 + \frac{16}{5}qv + \frac{5}{2}\varrho\left(\frac{k}{m}T\right)^2 + \frac{7}{2}\frac{k}{m}T\sigma + \frac{1}{6}\Delta\right\} = -\alpha\varrho\left(\sigma v + \frac{2}{3}q\right), \\
\frac{d}{dx}\left\{\frac{2}{5}\varrho v^4 + \left(\frac{6}{5}\varrho\frac{k}{m}T + 3\sigma\right)v^2 + \left(\frac{24}{25}q + \frac{14}{5}\Gamma\right)v + \frac{9}{5}\frac{k}{m}T\sigma\right\} = -\alpha\varrho\left(\frac{9}{5}\sigma v + \frac{3}{2}\Gamma\right), \\
\frac{d}{dx}\left\{\varrho v^5 + \left(14\varrho\frac{k}{m}T + 8\sigma\right)v^3 + \left(\frac{84}{5}q + 4\Gamma\right)v^2 + \left(35\varrho\left(\frac{k}{m}T\right)^2 + 28\frac{k}{m}T\sigma + \frac{7}{3}\Delta\right)v + 28\frac{k}{m}Tq\right\} \\
= -\alpha\varrho\left(4\sigma v^2 + \frac{16}{3}qv + \frac{2}{3}\Delta\right).
\end{gathered} \tag{4.22}$$

There is one new variable, denoted by Γ, which is the (111)-component of the traceless moment $\rho_{<ijk>}$. It is made nondimensional by the introduction of

$$\hat{\Gamma} := \frac{\Gamma}{\rho_0\frac{k}{m}T_0 a_0}. \tag{4.23}$$

The system (4.22) will reduce to four differential equations of the type $(4.1)_2$ after $\hat{\rho}, \hat{\sigma}, \hat{q}$ have been eliminated by use of (4.13). And the determinant $\hat{D}$ before and behind the wave is given by

$$\hat{D}_0 = \frac{125M_0^6 - 625M_0^4 + 693M_0^2 - 189}{50}. \qquad \text{and}$$
$$\hat{D}_1 = \frac{-2325M_0^8 + 3580M_0^6 + 20682M_0^4 - 30132M_0^2 + 8451}{3200M_0^2} \tag{4.24}$$

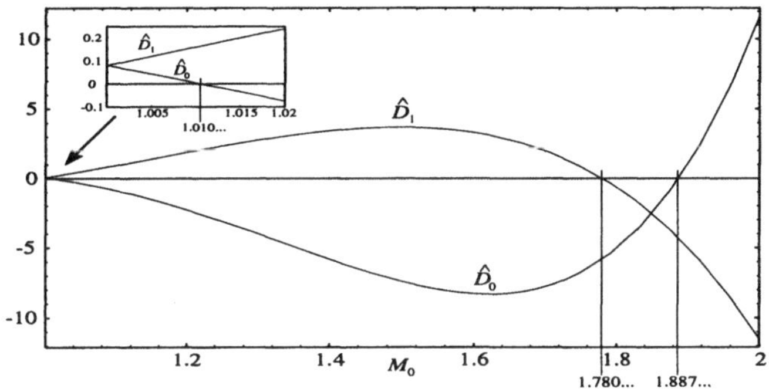

Figure 12.18 The determinants $\hat{D}_0$ and $\hat{D}_1$ for 21 moments.

whose real roots are

$$M_0 = \pm 0.64456... \qquad\qquad M_0 = \pm 0.62473...$$
$$M_0 = \pm 1.01062... \qquad \text{and} \qquad M_0 = \pm 0.98954... \;.$$
$$M_0 = \pm 1.88764... \qquad\qquad M_0 = \pm 1.78053...$$

respectively. Figure 12.18 shows $\hat{D}_0$ and $\hat{D}_1$ as functions of M_0 and we observe a rich structure, i.e. several switches between positive and negative values. From all we know so far, we expect singular points to appear before the wave at $M_0 = 1.010$ and $M_0 = 1.887$ and behind the wave at $M_0 = 1.780$.

The Figures 12.19 and 12.20 show shock structures as solutions of boundary value problems at relevant Mach numbers and the determinants $\hat{z}_v, \hat{z}_T, \hat{z}_\Delta, \hat{z}_\Gamma$, and $\hat{D}$ as functions of $\hat{x}$. Again we itemize the salient points.

- For $M_0 = 1.008$ there is a very weak wave (note the ordinate scale!) which is completely smooth and, indeed, no singularity does occur since $\hat{D}$ is nonzero everywhere.

- At $M_0 = 1.4$ the shock structure is still smooth even though a singularity has appeared before the wave at $M_0 = 1.010$. But that singularity is regular, as put in evidence by the simultaneous roots of all five determinants.

- For $M_0 = 1.8$ a second regular singularity has appeared—this time behind the wave. The first appearance was at $M_0 = 1.780$, of course. None of the singularities affects the smooth shock structure.

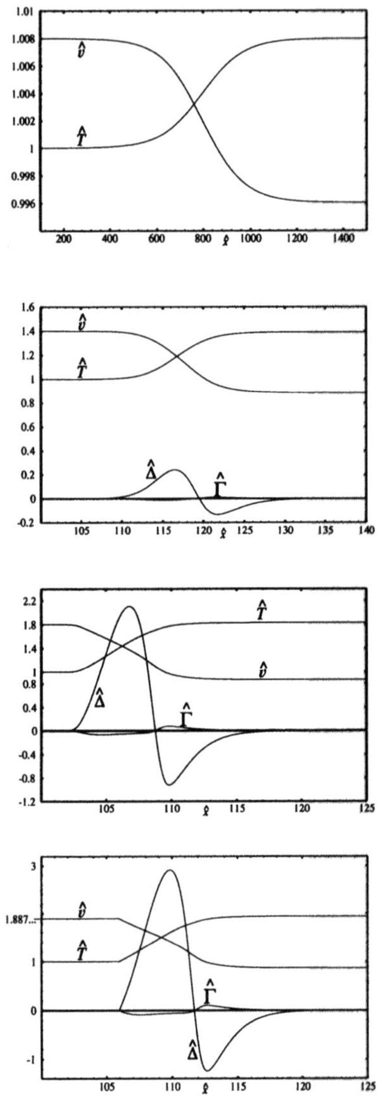

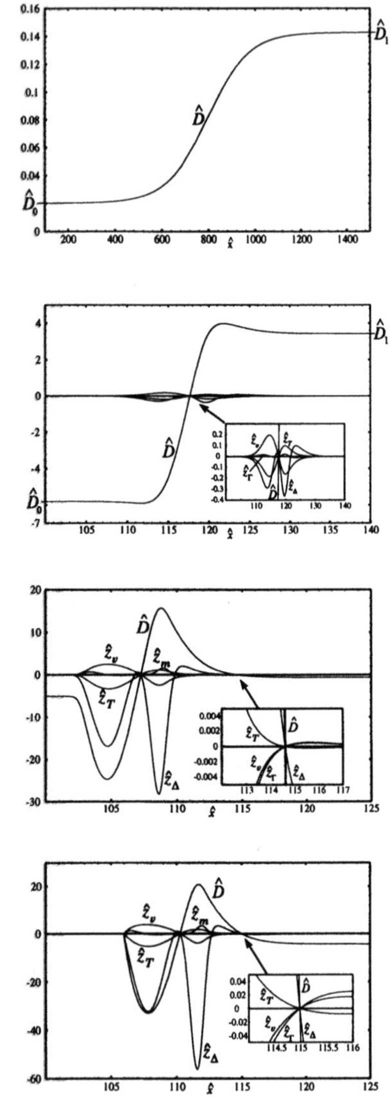

Figure 12.19 Shock structure for 21 moments
a: $M_0 = 1.008$, b: $M_0 = 1.4$,
c: $M_0 = 1.8$, d: $M_0 = 1.887$.

Figure 12.20 Determinants $\hat{z}_v$, $\hat{z}_T$, $\hat{z}_\Delta$, $\hat{z}_\Gamma$, and $\hat{D}$ for 21 moments
a: $M_0 = 1.008$, b: $M_0 = 1.4$,
c: $M_0 = 1.8$, d: $M_0 = 1.887$.

- The smooth shock structure *is* affected, however, by the singularity corresponding to the largest characteristic speed before the wave, i.e. at $M_0 = 1.887$. Indeed here we see a kink appear and, if we were to continue increasing M_0, this kink would develop into a subshock.

Note that again, by taking more moments into account, the initiation of kink and subshock has been pushed to a slightly higher Mach number.

5 Conclusion

It has turned out that the smooth shock structure was not affected by the occurrence of singular points in the system (4.1) *except when the singular point corresponds to the largest characteristic speed in front of the wave.*

This statement has been demonstrated for three cases where the number of moments was 13, 14, and 21. Weiss has confirmed that the statements holds true for the further cases of 20, 26, 34, and 35 fields. Thus he was able to push the first appearance of subshocks to a Mach number of about $M_0 = 2.2$; see Weiss [128]

Physically this is most satisfactory, because we may say that a flow calculated by the n-moment theory is *truly* supersonic only when its speed exceeds the largest characteristic speed implied by the theory.

Practically this is again disappointing, because we know from previous chapters that the largest characteristic speed creeps up very slowly as the number of moments increases. Thus, according to Table 10.4, for the largest speed to be equal to $M_0 = 9.36$ we need numbers of moments up to 15180 producing a system of 506 one-dimensional equations. Weiss is working on this project.

Mathematically the situation offers a neat challenge. We have seen numerically that all characteristic speeds—except the largest in front—lead to regular singular points. Up to now there is no *analytical* proof that this must be so. So the challenge is to find the proof, or in other words, to answer this question: what feature is it in the system that makes all but one of the singularities regular?

6 Addendum on Initial Value Problem for 13 Moments

We retrace our steps to the 13-moment case, because we want the simplest non-trivial moment case in order to investigate the appearance of sub-shocks a little more closely. For that purpose we solve the initial value problem, because it furnishes more information about the sub-shock than the boundary value solution.

We investigate the case $M_0 = 2.0$, well aware that this is in the range beyond $M_0 = 1.65$ where the sub-shock appears and where a regular singular point exists behind the wave, see Figure 14c. Before we can approach the initial value problem intelligently we must inquire which side, x_0 or x_1 represents a nodal point so as to be able to avoid it as the initial point.

For that purpose we have calculated $\mathcal{D}$ from (4.8) for the 13-moment case for x_0 and x_1 as functions of M_0 and we have plotted both functions in Figure 21. We conclude that x_0 is a saddle for $M_0 < 1.653$ (see Figure 12.11) and that x_1 is a saddle for $M_0 > 1.859$. Therefore, for $M_0 = 2$ we start the initial value problem behind the wave.

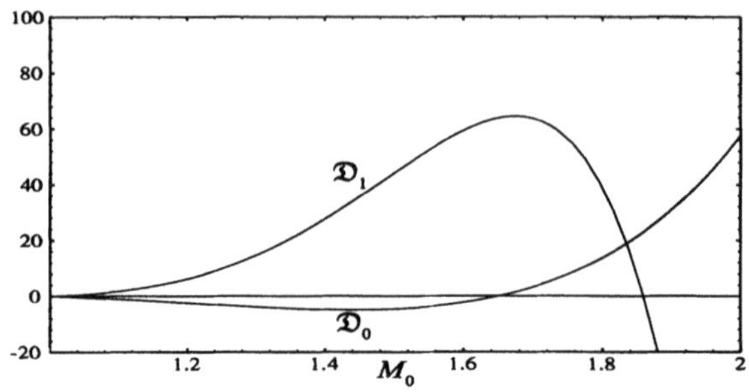

Figure 12.21 $\mathcal{D}_0$ and $\mathcal{D}_1$ as functions of M_0 for 13 moments.

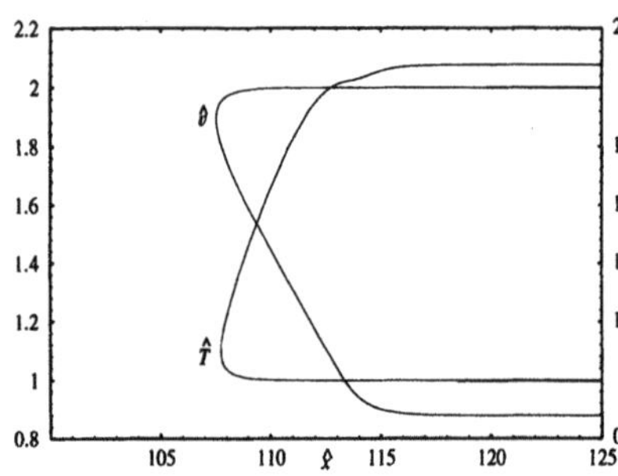

Figure 12.22 Nonunique solution for 13 moments at $M_0 = 2.0$.

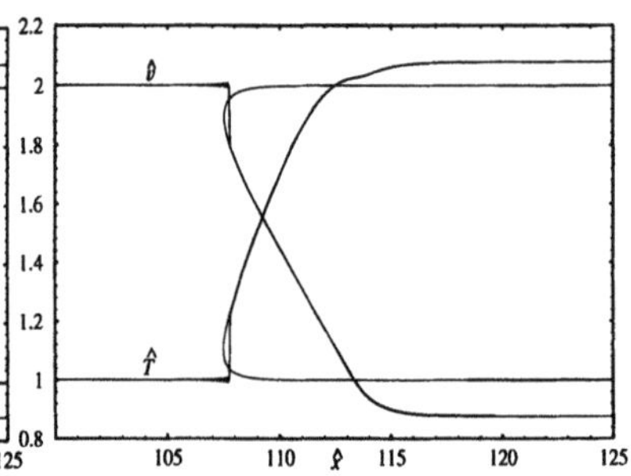

Figure 12.23 Superposition of the solutions of initial and boundary value problem for 13 moments at $M_0 = 2.0$.

The solutions $\hat{v}(\hat{x})$ and $\hat{T}(\hat{x})$ are shown in Figure 22. Both are non-unique with an infinite slope at $\hat{x} \approx 107.5$. In order to appreciate this figure better we have superposed upon it the solution of the boundary value problem from Figure 13c and obtained Figure 23.

Note that the solution of the initial value problem is not troubled at all by the regular singularity which we know to lie at $\hat{x} \approx 114.5$ cf. Figure 14.c.

Note also that the subshock of the boundary value solution cuts through the two branches of the initial value problem at $\hat{x} \approx 108$. It is interesting to remark that the sub-shock satisfies Rankine-Hugoniot conditions of its own between the values

$\rho_0, v_0, T_0, 0, 0$ before the shock and
$\rho_s, v_s, T_s, \sigma_s, q_s$ behind the shock.

For proof we refer to Weiss [128]. Here we close this subject. It is somewhat academic, since the occurence of the sub-shock is proof for us that the 13-moment theory is not appropriate for $M_0 = 2.0$.

7 Quantitative Results and Conclusions

The main objective of this chapter has been to prove that smooth shock structures can be calculated by the moment method beyond $M_0 = 1.65$, where the 13-moment method fails. This objective has been reached, we believe, because the break-down of the smooth solution has been linked to the largest characteristic speed. So what about quantitative results?

Figure 12.24 shows the shock structure calculated for differently many moments between 13 and 35 at $M_0 = 1.4$. There is not much disagreement between the curves and one might say that all the theories agree at this Mach number.

Figure 12.25 shows the results of the same theories—excluding only the 13-moment theory—for $M_0 = 1.8$, and we observe that now the theories *do not agree*. In fact the 20-moment case is just about ready to show the kink which should appear at $M_0 = 1.997$. The other curves are smoother, but the 26- and 35-moment cases exhibit an unbecoming bulge which is not observed in experiments.

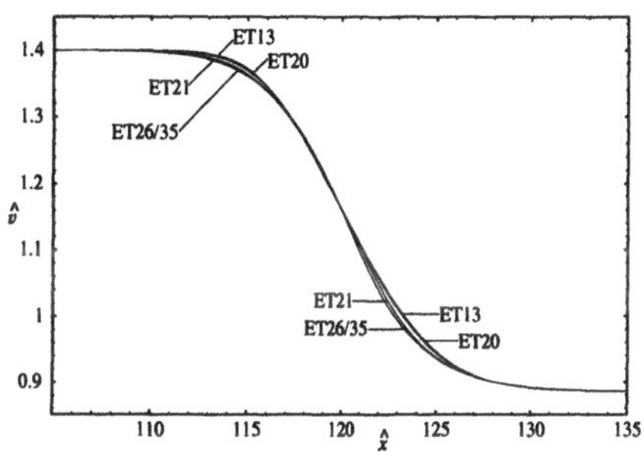

Figure 12.24 Calculated shock structures for $M_0 = 1.4$.

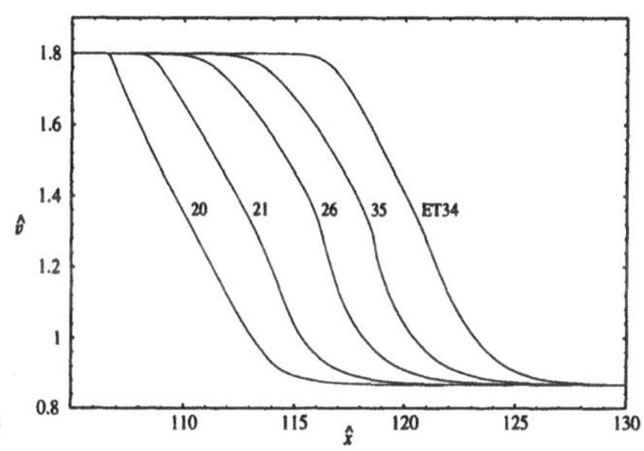

Figure 12.25 Calculated shock structures at $M_0 = 1.8$. (shifted arbitrarily along x-axis)

If we calculate the inverse shock thickness (1.1) for all these moment theories and compare with the Alsmeyer experiments, we obtain the dots shown in Figure 12.26. Inspection shows that the 34-moment theory achieves the best result. However, 35 moments are again bad, because of the bulge; see Figure 12.25. All we can say at this time is this:

- More variables lead to smooth shock structures for higher Mach numbers.
- The shock structure improves with more moments, but the improvement is not monotone with the number of moments.
- We expect the quantitative features of the shock structure to improve with more and more moments.

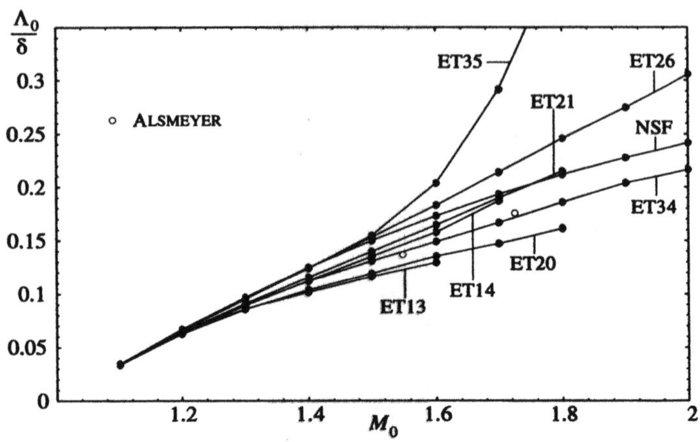

Figure 12.26 Shock wave thickness as function of Mach number.

The latter expectation is motivated by our experience with sound dispersion and light scattering that was presented in previous chapters. Unlike those chapters the transition to more moments is not so easy now, because we need to determine the form of the fully *nonlinear* moment equations in ρ, v, T, where previously the *linear* equations sufficed. The effect of the nonlinear parts of the constitutive quantities, however, can be neglected; Au [145] has shown this for the second-order terms.

Chapter 13

Extended Thermodynamics of Radiation

by **H. Struchtrup**, Thermodynamics, TU Berlin

It is useful to think of radiation as a photon gas which is governed by a transfer equation much like the Boltzmann equation for an atomic gas, albeit with a different production term. This makes it possible to develop an extended thermodynamics of photons.

In many respects this theory is similar to extended thermodynamics of moments of atoms. Thus we have moments and moment equations, we have a closure problem and this may be solved by maximizing the entropy. Given absorption and scattering mechanisms, we obtain an explicit set of field equations which can be applied to non-equilibrium situations.

The photonic moments lend themselves for an easy distinction of ***isotropic*** non-equilibrium, in which the frequency distribution of the photons deviates from the Planck distribution, and non-equilibrium through ***anisotropy*** like in a sun-beam. Both are generally coupled, of course, but there are illustrative phenomena where they are not.

As usual the essential question is where, at what stage, to close the hierarchy – or hierarchies – of moments and we answer that question for three suggestive problems: local equilibrium, sudden compression of a photon gas and penetration of a light beam into matter.

It is not very common in thermodynamics to deal with radiation as a part of thermodynamic systems; in particular the role which the photons play in the entropy, the entropy flux and the entropy production is not always well-understood. This problem is discussed in the last section of this chapter, albeit only for gray bodies.

While the bulk of this chapter is due to the work of Struchtrup [14],[15], the last section represents work by Müller [146].

1 Structure of Extended Thermodynamics of Photons

1.1 Energy and Momentum of Individual Photons

Radiation consists of electromagnetic waves which in some ways behave like particles. Thus a wave of frequency ω and wave vector $\mathbf{k}$ corresponds to a photon of energy $\hbar\omega$ and momentum $\hbar\mathbf{k}$.[1] The speed of the wave is c, the speed of light, so that the frequency ω and the wave number $k = |\mathbf{k}|$ are related by the dispersion relation $\omega = ck$. The unit vector $\mathbf{n}$ in the direction of propagation is given by $\mathbf{k}/k$.

A radiation field may be considered as a gas of photons. The number of photons in that gas with wave vectors between $\mathbf{k}$ and $\mathbf{k}+\mathbf{dk}$ is given by the phase density f. We have

$$density = f(\mathbf{x},\mathbf{k},t)\,d\mathbf{k} = f(\mathbf{x},\mathbf{k},t)\,k^2\,dk\,d\Omega. \tag{1.1}$$

$d\Omega$ is the solid angle element $\sin\vartheta\,d\vartheta\,d\varphi$ in the direction (ϑ,φ) of propagation.

We ignore the polarization of the radiation in this article. Moreover we deal only with radiation in interaction with matter at rest. The extension to interaction with matter of arbitrary velocities requires a relativistic treatment; see Thorne [147], or Struchtrup [15].

1.2 Radiative Transfer Equation

In the phase space of photons spanned by the coordinates of $\mathbf{x}$ and $\mathbf{k}$, the phase density f obeys a continuity equation with a production term, viz.

$$\frac{\partial f}{\partial t} + cn_k\frac{\partial f}{\partial x_k} = \mathcal{S}(f)\,; \tag{1.2}$$

$\mathcal{S}$ is the phase density of production of photons which is due to absorption, emission, and scattering. All three phenomena occur only by interaction of photons with matter, so that $\mathcal{S}$ is equal to zero in empty space.

The photon production density is calculated from models of the interaction of matter and radiation. We assume that its form is given by (cf. Chandrasekar [148], Mihalas & Mihalas [149] or Oxenius [150])

$$\mathcal{S}(f) = -\kappa\left(f - f_{|E}\right) - \zeta\left(f - \frac{1}{4\pi}\int f\left(k,\mathbf{n}'\right)d\Omega'\right); \tag{1.3}$$

The first term is due to absorption and emission of photons; these two phenomena—absorption and emission—are balanced when the phase density reaches its equilibrium value $f_{|E}$ to be discussed later. The effective absorption coefficient κ depends on k. The second term in (1.3) represents isotropic scattering due to nonisotropy of the phase density and ζ is called the scattering coefficient; it is a function of k in general. The form (1.2), (1.3) of the radiative transfer equation is valid only in the rest frame of the matter.

---

[1] $\hbar$ is the Planck constant divided by 2π.

1.3 Moments and Moment Equations. The Closure Problem.

The **k**-dependence of the phase density may often be approximated by a finite number of moments, viz.

$$u^r_{\langle i_1 i_2 \cdots i_n \rangle} = \int k^r n_{\langle i_1} \ldots n_{i_n \rangle} f \, d\mathbf{k}. \tag{1.4}$$

Naturally the $(\mathbf{x}, t)$-dependence of the moments is governed by the radiative transfer equation (1.2). And indeed, multiplication of (1.2) by $k^r n_{\langle i_1} \ldots n_{i_n \rangle}$ and integration over $\mathbf{k}$ yields

$$\frac{\partial u^r_{\langle i_1 i_2 \cdots i_n \rangle}}{\partial t} + \frac{n}{2n+1} c \frac{\partial u^r_{\langle\langle i_1 i_2 \cdots i_{n-1} \rangle}}{\partial x_{i_n \rangle}} + c \frac{\partial u^r_{\langle i_1 i_2 \cdots i_n k \rangle}}{\partial x_k} = P^r_{\langle i_1 i_2 \cdots i_n \rangle}, \tag{1.5}$$

where $P^r_{\langle i_1 i_2 \cdots i_n \rangle}$ is the moment production density, defined as

$$P^r_{\langle i_1 i_2 \cdots i_n \rangle} = \int k^r n_{\langle i_1} \ldots n_{i_n \rangle} S \, d\mathbf{k}. \tag{1.6}$$

The equations (1.5) are the basis for the field equations of extended thermodynamics of photons, but they do not form a closed system. Indeed, if we have reasons to believe that the moments

$$u^r_{\langle i_1 i_2 \cdots i_n \rangle} = \int k^r n_{\langle i_1} \ldots n_{i_n \rangle} f d\mathbf{k} \qquad \text{for} \qquad \begin{matrix} r = 1, ..., R \\ n = 0, 1, ..., N \end{matrix} \tag{1.7}$$

describe the state of the photon gas satisfactorily, we find that the system (1.5) is not appropriate as a system of field equations, because it is not closed on two counts:

- The left-hand side contains the moments $u^r_{\langle i_1 i_2 \cdots i_N k \rangle}$ which are r tensors of rank $N+1$,
- The right-hand side contains the productions $P^r_{\langle i_1 i_2 \cdots i_n \rangle}$ which are not related to the state variables (1.7) in any obvious manner.

Thus we encounter the familiar closure problem again, which is common to all kinetic theories based on balance equations of the type (1.2). The closure problem will be solved here by maximizing the entropy.

Note that (1.7) excludes moments for $r = 0$ in what seems to be an arbitrary manner. However, this exclusion anticipates a later result by which moments $u^0_{\langle i_1 i_2 \cdots i_N k \rangle}$ are singular. Actually most of the literature on the photon gas considers only one value of r, or R, namely $r = 1$.

We maintain the generic formal structure of the theory as long as possible. Therefore we do not at this stage specify values for R and N. Such values will be decided upon later and they depend on the phenomena under consideration.

1.4 Entropy and Maximization of Entropy

Since photons are bosons, the entropy density, entropy flux, and entropy production are given by

$$\begin{aligned} h &= - \; k_B \int \left[f \ln \tfrac{f}{y} - (y+f) \ln \left(1 + \tfrac{f}{y}\right)\right] d\mathbf{k} \\ h_i &= - \; k_B c \int n_i \left[f \ln \tfrac{f}{y} - (y+f) \ln \left(1 + \tfrac{f}{y}\right)\right] d\mathbf{k} \\ \sigma &= \; k_B \int \left[\ln \left(\tfrac{y}{f} + 1\right)\right] \mathcal{S} \, d\mathbf{k}. \end{aligned} \tag{1.8}$$

The Boltzmann constant, which is usually denoted by k, is called k_B here so as to distinguish it from the wave number. y is equal to $2/(2\pi)^3$; it represents the smallest element of the phase space that can accommodate a photon.

For the form (1.3) of the phase density of photon production an H-Theorem can be proved (cf. Struchtrup [14]); i.e. the total entropy production density of radiation and matter is non-negative.

In a case where the finite number of moments (1.7) represent the state of the photon gas, the phase density will be determined by this set of variables. We may calculate the appropriate phase density by maximizing the entropy $(1.8)_1$ under the constraints (1.7). This is not different from previous arguments about the maximum of entropy which we have presented in Chapter 9. Here we obtain

$$f = \frac{y}{\exp \Xi - 1}, \qquad \text{where} \qquad \Xi = \frac{1}{k_B} \sum_{r=1}^{R} \sum_{n=0}^{N} \Lambda^r_{\langle i_1 i_2 \cdots i_n \rangle} k^r n_{\langle i_1} \dots n_{i_n \rangle}. \tag{1.9}$$

The Λ's are Lagrange multipliers that take care of the constraints. They may be determined—in principle—by elimination of f between (1.7) and (1.9) and by solving the resulting equations to give $\Lambda^r_{\langle i_1 i_2 \cdots i_n \rangle}$ in terms of $u^s_{\langle i_1 i_2 \cdots i_l \rangle}$ for $s = 1, ..., R$, $l = 0, 1, ..., N$.

1.5 Closure

There are very few cases for which the Lagrange multipliers can be determined exactly, because the necessary integrations cannot be performed. Therefore we are usually forced to calculate the phase density as a function of $u^s_{\langle i_1 i_2 \cdots i_l \rangle} \; \begin{smallmatrix} s = 1, ..., R \\ l = 0, 1, ..., N \end{smallmatrix}$ by approximation which will in fact be a linearization about equilibrium.

However, once the phase density is known—exact or approximate—the system (1.5), (1.6) is closed by the equations

$$\begin{aligned} u^r_{\langle i_1 i_2 \cdots i_N k \rangle} &= \int k^r n_{\langle i_1} \dots n_{i_N} n_{k \rangle} f \, d\mathbf{k} && (r = 1, ..., R) \\ P^r_{\langle i_1 i_2 \cdots i_n \rangle} &= \int k^r n_{\langle i_1} \dots n_{i_N \rangle} \mathcal{S}(f) \, d\mathbf{k}. && \begin{pmatrix} r = 1, ..., R \\ n = 0, 1, ..., N \end{pmatrix} \end{aligned} \tag{1.10}$$

We shall now proceed to discuss equilibrium in Section 2, then the approximate closure in Section 3 and the explicit closed field equations in Section 4.

2 Equilibrium

2.1 The First Few Moments

It is obvious that the densities of energy and momentum of the photon gas and their fluxes are given by the relations

$$\begin{aligned} &\text{energy density} && e &&= \int \hbar ck && f\,d\mathbf{k} \\ &\text{energy flux} && Q_i &&= \int \hbar ckc\frac{k_i}{k} && f\,d\mathbf{k} \\ &\text{momentum density} && p_i &&= \int \hbar k_i && f\,d\mathbf{k} \\ &\text{momentum flux} && N_{ij} &&= \int \hbar k_i c\frac{k_j}{k} && f\,d\mathbf{k} \end{aligned} \tag{2.1}$$

In analogy to mechanics the momentum flux may also be called the pressure tensor—or the negative stress tensor of radiation. The radiation pressure itself is 1/3 of the trace of the pressure tensor, viz.

$$p = \frac{1}{3}\int \hbar ckf\,d\mathbf{k} \qquad \text{such that, by } (2.1)_1 \qquad p = \frac{1}{3}e, \tag{2.2}$$

i.e. the pressure and the energy density are related by a factor 1/3.

The quantities (2.1) are simple representatives of the class of moments of the phase density. Indeed, by comparison with (1.4) we have

$$e = \hbar cu^1, \qquad p_i = hu_i^1, \qquad Q_i = \hbar c^2 u_i^1, \qquad N_{\langle ij\rangle} = \hbar cu^1_{\langle ij\rangle}. \tag{2.3}$$

The moment equations (1.5) which concern these first few moments are the equations of balance of energy and momentum, viz.

$$\begin{aligned} &\frac{\partial e}{\partial t} && &&+\frac{\partial Q_j}{\partial x_j} &&= \hbar cP^1 \\ &\frac{\partial p_i}{\partial t} &&+\frac{1}{3}\frac{\partial e}{\partial x_i} &&+\frac{\partial N_{\langle ij\rangle}}{\partial x_j} &&= \hbar P_i^1. \end{aligned} \tag{2.4}$$

2.2 Equilibrium of Radiation with Matter

2.2.1 Matter in Motion[2]

Equilibrium of a photon gas means that it is in equilibrium with matter. Its state is then characterized by the densities of energy and momentum, e and p_i respectively. And according to (1.9)

[2]This subsection and subsection 2.2.3 are the only ones that consider moving matter.

the equilibrium phase density has the form

$$f_{|E}(\mathbf{x},t,\mathbf{k}) = \frac{y}{\exp\left[\frac{\Lambda^1 k + \Lambda_i^1 k n_i}{k_B}\right] - 1}. \tag{2.5}$$

Here we have one of the few cases where the Lagrange multipliers may be determined exactly from the variables e and p_i, because the integrals ((2.1) can be calculated. With $a = \frac{\pi^2}{15}\frac{k_B^4}{\hbar^3 c^3}$ we obtain

$$\begin{aligned} e &= \int \hbar c k f_{|E}\, d\mathbf{k} &= \frac{a(\hbar c)^4}{3}\frac{3(\Lambda^1)^2 + \Lambda_i^1\Lambda_i^1}{\left[(\Lambda^1)^2 - \Lambda_i^1\Lambda_i^1\right]^3} \\ p_i &= \int \hbar k_i f_{|E} d\mathbf{k} &= -\frac{1}{c}\frac{4a(\hbar c)^4}{3}\frac{\Lambda^1}{\left[(\Lambda^1)^2 - \Lambda_i^1\Lambda_i^1\right]^3}\Lambda_i^1. \end{aligned} \tag{2.6}$$

By inversion we obtain Λ^1 and Λ_i^1 in terms of e and p_i, viz.

$$\begin{aligned} \Lambda^1 &= \hbar c\left(\frac{a}{e}\right)^{1/4}\frac{(3-\chi)^{1/2}}{2^{3/4}3^{1/4}(1-\chi)^{3/4}} \\ \Lambda_i^1 &= -\hbar c\left(\frac{a}{e}\right)^{1/4}\frac{2^{1/4}}{3^{1/4}(1-\chi)^{3/4}(3-\chi)^{1/2}}\frac{cp_i}{e} \end{aligned} \quad \text{where } \chi = \frac{5}{3} - \frac{4}{3}\sqrt{1 - \frac{3}{4}\frac{c^2p^2}{e^2}}. \tag{2.7}$$

The integrations implied by (2.6) are not easy to perform, nor are the inversions trivial which lead from (2.6) to (2.7). To our knowledge these calculations were first done by Larecki [151] on the basis of the kinetic theory of phonons. Anile, Pennisi &Sammartino [152] have derived the results of this paragraph by extended thermodynamics and so did Kremer & Müller [153] in a slightly different manner.

The phase density $f_{|E}$ is thus explicitly related to e and p_i and we may now calculate the momentum flux, or pressure tensor N_{ij} from $(2.1)_4$. Also we may calculate the entropy density h and the entropy flux h_i from (1.8). We obtain, again by nontrivial integrations

$$N_{ij} = e\left(\frac{1-\chi}{2}\delta_{ij} + \frac{3\chi-1}{2}\frac{p_i p_j}{p^2}\right) \tag{2.8}$$

$$h = \frac{4}{3}\left(\frac{a}{e}\right)^{1/4} e\frac{3^{3/4}}{2^{7/4}}(3-\chi)^{1/2}(1-\chi)^{1/4} \tag{2.9}$$

$$h_i = \left(\frac{a}{e}\right)^{1/4}\frac{2^{5/4}}{3^{1/4}}\frac{(1-\chi)^{1/4}}{(3-\chi)^{1/2}}c^2 p_i. \tag{2.10}$$

χ is the Eddington factor, so called because Eddington first introduced it in connection with the radiative stress in stars. The specific form (2.7) was discussed by Levermore [154].

2.2.2 Matter at Rest

It stands to reason that the radiation has no momentum if it is in equilibrium with matter at rest. Therefore Λ_i is zero, and we have, by (2.5), $(2.7)_1$

$$f_{|E} = \frac{y}{\exp\left[\frac{\hbar c k}{k_B}\left(\frac{a}{e}\right)^{1/4}\right] - 1}. \tag{2.11}$$

This is the Planck distribution function if we set

$$e = aT^4, \tag{2.12}$$

where T is the temperature of matter. Equation (2.12) represents the Stefan-Boltzmann law which was established experimentally by Stefan and derived from thermodynamic principles by Boltzmann. a is called radiation constant.

The Eddington factor is $\chi = 1/3$ in this case and we obtain from (2.6) and (2.8) through (2.10)

$$p_i = 0, \qquad N_{ij} = \frac{e}{3}\delta_{ij}, \qquad h = \frac{4}{3}aT^3, \qquad h_i = 0. \tag{2.13}$$

This is the case of isotropic equilibrium.

2.2.3 Momentum of Radiation in a Moving Body

If the photon gas is in equilibrium with a body moving with velocity v_i with respect to the observer, we may ask how its momentum density p_i is related to the velocity v_i. The answer is most easily obtained if we realize that k is the fourth component of the Lorentz vector (kn_i, k) which transforms to a moving frame by, e.g. see Becker & Sauter [155]

$$k = \frac{\hat{k}}{\sqrt{1-\frac{v^2}{c^2}}}\left(1 - \frac{1}{c}v_k\hat{n}_k\right). \tag{2.14}$$

$\hat{k}$ and $\hat{n}_i$ are the wave number and direction vector in the observer frame respectively. Insertion of (2.14) into (2.11) provides

$$f_{|E}\left(\mathbf{x}, t, \hat{k}, \hat{n}_i\right) = \frac{y}{\exp\left[\frac{\hbar c\hat{k}}{k_B T}\frac{\left(1-\frac{1}{c}v_k\hat{n}_k\right)}{\sqrt{1-\frac{v^2}{c^2}}}\right] - 1}. \tag{2.15}$$

Comparison with (2.5) shows that we have

$$\Lambda^1 = \frac{\hbar c}{T}\frac{1}{\sqrt{1-\frac{v^2}{c^2}}}, \qquad \Lambda^1_i = -\frac{\hbar}{T}\frac{v_i}{\sqrt{1-\frac{v^2}{c^2}}}, \tag{2.16}$$

or, by (2.7)

$$T = \left(\frac{e}{a}\right)^{1/4}\frac{3^{1/4}}{2^{1/4}}(1-\chi)^{1/4}, \qquad \frac{v_i}{c} = \frac{2}{(3-\chi)}\frac{cp_i}{e} \tag{2.17}$$

or, inversely

$$e = aT^4\frac{1+\frac{1}{3}\frac{v^2}{c^2}}{1-\frac{v^2}{c^2}} \qquad \text{and} \qquad p_i = \frac{4}{3}\frac{aT^4}{c^2}\frac{v_i}{1-\frac{v^2}{c^2}}. \tag{2.18}$$

In particular, if the matter moves with the speed of light in the direction n_i, we have $p_i = \frac{e}{c}n_i$ as a relation between the densities of momentum and energy, although both are infinite in this limit. All photons move in the same direction; at least that is what the observer perceives.

Note: $f_{|E}$ in (2.5), or (2.15) is *not* a nonequilibrium density; rather it is the phase density of a photon gas in equilibrium with a body of temperature T that moves with velocity v_i. This fact was not always recognized before; see Anile, Pennisi & Sammartino [152] and Kremer & Müller [153].

3 Near Equilibrium

3.1 Phase Density in Near-Equilibrium

We recall the equations (1.9) and the closure problem described in Paragraph 1.5. Since the integrations prescribed by (1.10) are impossible, we approximate, and the method of approximation will be an expansion about equilibrium of the photon gas with matter at rest which has the temperature T. In that equilibrium we have by (2.11), (2.12)

$$f = \frac{y}{\exp \Xi_{|E} - 1} \quad , \text{ where} \qquad \Xi_{|E} = \frac{\hbar c k}{k_B T}. \tag{3.1}$$

so that the Lagrange multipliers in equilibrium are

$$\Lambda^r_{\langle i_1 i_2 \cdots i_n \rangle |E} = \begin{cases} \dfrac{\hbar c}{T}, & r = 1, n = 0 \\ 0, & \text{else.} \end{cases} \tag{3.2}$$

Therefore close to equilibrium it seems reasonable to define

$$\frac{1}{k_B} \Lambda^r_{\langle i_1 i_2 \cdots i_n \rangle} = \frac{\hbar c}{k_B T} \delta^{r1} \delta_{n0} + \left(\frac{\hbar c}{k_B T} \right)^r \lambda^r_{\langle i_1 i_2 \cdots i_n \rangle} \tag{3.3}$$

and to consider the λ's as small. The factor $\left(\frac{\hbar c}{k_B T} \right)^r$ in front of the λ's enables us to write the exponent Ξ, defined in (1.9), in the simple form

$$\Xi = \Xi_{|E} + \sum_{r,n} \lambda^r_{\langle i_1 i_2 \cdots i_n \rangle} \Xi^r_{|E} n_{\langle i_1} n_{i_2} \ldots n_{i_n \rangle}, \tag{3.4}$$

so that the k-dependence of Ξ is carried by $\Xi_{|E}$. We assume that the λ's are so small that we may write

$$\begin{aligned} f &\approx f_{|E}(k,T) + \frac{df}{d\Xi}_{|\Xi = \Xi_{|E}} (\Xi - \Xi_{|E}) \qquad \text{or else} \\ f &\approx f_{|E}(k,T) + \frac{df_{|E}}{d\Xi_{|E}} \sum_{r,n} \lambda^r_{\langle i_1 i_2 \cdots i_n \rangle} \Xi^r_{|E} n_{\langle i_1} \ldots n_{i_n \rangle}. \end{aligned} \tag{3.5}$$

This phase density is explicit in the $(k, \mathbf{n})$- dependence and it may be used to calculate the λ's.

3.2 Approximate Lagrange Multipliers

Insertion of the approximate phase density (3.5) into the moment equations (1.4) gives

$$\begin{aligned} u^t_{\langle j_1 j_2 \cdots j_m \rangle} - u^t_{\langle j_1 j_2 \cdots j_m \rangle |E} &= \sum_{r,n} \lambda^r_{\langle i_1 i_2 \cdots i_n \rangle} \left(\frac{\hbar c}{k_B T} \right)^r \int k^{t+r} n_{\langle j_1} \ldots n_{j_m \rangle} n_{\langle i_1} \ldots n_{i_n \rangle} \frac{df_{|E}}{d\Xi_{|E}} d\mathbf{k} \\ &= \sum_r \lambda^r_{\langle j_1 j_2 \cdots j_m \rangle} \left(\frac{\hbar c}{k_B T} \right)^r \frac{4\pi m!}{\prod_{j=0}^{m} (2j+1)} \int k^{t+r+2} \frac{df_{|E}}{d\Xi_{|E}} dk \end{aligned}$$

$$= -\frac{4\pi y m!}{\prod_{j=0}^{m}(2j+1)}\left(\frac{k_B T}{\hbar c}\right)^{t+3}\sum_{s=0}^{R} C_{tr}\lambda^r_{\langle j_1 j_2\cdots j_m\rangle}, \tag{3.6}$$

where $C_{tr} = \Gamma(t+r+3)\ \zeta(t+r+2)$ when ζ is Riemann's ζ–function.

Equation (3.6) is a linear system of algebraic equations from which the Lagrange multipliers can be calculated by inverting the matrix C_{tr}. We obtain

$$\lambda^r_{\langle i_1 i_2\cdots i_n\rangle} = -\frac{\prod_{j=0}^{n}(2j+1)}{4\pi y n!}\sum_{t=1}^{R} C^{-1}_{tr}\frac{u^t_{\langle i_1 i_2\cdots i_n\rangle} - u^t_{\langle i_1 i_2\cdots i_n\rangle|E}}{\left(\frac{k_B T}{\hbar c}\right)^{t+3}}. \tag{3.7}$$

$u^r_{\langle i_1 i_2\cdots i_n\rangle|E}$, the moments in equilibrium, are given by

$$u^t_{\langle i_1 i_2\cdots i_n\rangle|E} = 4\pi y\left(\frac{k_B T}{\hbar c}\right)^{r+3}\Gamma(r+3)\ \zeta(r+3)\,\delta_{n0}. \tag{3.8}$$

Thus we have an explicit phase density which results from inserting (3.7) into (3.5). The phase density depends on the moments $u^t_{\langle i_1 i_2\cdots i_n\rangle}$ for $r = 1,\ldots,R$ and $n = 0,1,\ldots,N$, which are the state variables of the photon gas.

4 Field Equations

4.1 Closure for Moments

Since, by (3.5), the approximate phase density f is a linear combination of the traceless tensors $n_{\langle i_1} n_{i_2}\ldots n_{i_n\rangle}$ for $n - 0,1,\ldots,N$, all moments $u^r_{\langle i_1 i_2\cdots i_l\rangle}$ with a higher rank than N vanish. This observation takes care right away of the first part of the closure problem, viz. $(1.10)_1$. Indeed we have

$$u^r_{\langle i_1 i_2\cdots i_{N+m}\rangle} = 0 \qquad \text{for} \qquad \begin{array}{l} r = 1,2,\ldots \\ m \geq 1. \end{array} \tag{4.1}$$

Nothing could be simpler.

4.2 Closure for Productions

We recall that the phase density of production S is given by (1.3) with—in general—coefficients of absorption and scattering that depend on k. Insertion of the approximate value (3.5) for f with $\lambda^r_{\langle i_1 i_2\cdots i_n\rangle}$ given by (3.7) provides explicit expressions for the productions $P^r_{\langle i_1 i_2\cdots i_n\rangle}$ in the moment equations (1.5). We obtain

$$P^r = -\sum_{s=1}^{R}\Theta_{rs}\left(u^s - u^s_{|E}(T)\right), \qquad P^r_{\langle i_1 i_2\cdots i_n\rangle} = -\sum_{s=1}^{R}\widehat{\Theta}_{rs}\,u^s_{\langle i_1 i_2\cdots i_n\rangle}, \tag{4.2}$$

where the Θ's depend on the k-dependence of the coefficients κ and ζ. We have

$$\Theta_{rs} = \left(\frac{k_B T}{\hbar c}\right)^{r-s} \sum_t C_{ts}^{-1} \int \kappa \frac{e^{\Xi}}{(e^{\Xi}-1)^2} \Xi^{t+r+2} d\Xi \tag{4.3}$$

$$\widehat{\Theta}_{rs} = \left(\frac{k_B T}{\hbar c}\right)^{r-s} \sum_t C_{ts}^{-1} \int \zeta \frac{e^{\Xi}}{(e^{\Xi}-1)^2} \Xi^{t+r+2} d\Xi + \Theta_{rs}. \tag{4.4}$$

The difference between the coefficients Θ in $(4.2)_{1,2}$ is due to the fact that the scalar productions P^r are unaffected by the scattering term in the production density S.

The specific form of the coefficients Θ in $(4.2)_{1,2}$ will depend on the k-dependence of κ and ζ. These relations embody the physics of the interaction of matter and radiation and we shall discuss them later. It is clear, however, that in general neither Θ_{rs} nor $\widehat{\Theta}_{rs}$ are diagonal matrices, and that fact has an important consequence: *The hierarchies of systems* (1.5) *for different values of* r *are coupled.* The coupling occurs through the productions on the right-hand side of the system, because the tensors $P^r_{\langle i_1 i_2 \cdots i_n \rangle}$ depend—in general—on all moments $u^s_{\langle i_1 i_2 \cdots i_n \rangle}$ for $s = 1, \ldots, R$.

4.3 The Hierarchies of Field Equations

In summary we may write the full system of field equations in the form of only three equations formally, viz.

$$\begin{aligned}
&\frac{\partial u^r}{\partial t} && && +c\frac{\partial u^r_k}{\partial x_k} && = -\sum_q \Theta_{rq}\left(u^q - u^q_{|E}\right) \\
&\frac{\partial u^r_{\langle i_1 i_2 \cdots i_n \rangle}}{\partial t} && +\frac{n}{2n+1}c\frac{\partial u^r_{\langle\langle i_1 i_2 \cdots i_{n-1}\rangle}}{\partial x_{i_n\rangle}} && +c\frac{\partial u^r_{\langle i_1 i_2 \cdots i_n k\rangle}}{\partial x_k} && = -\sum_q \widehat{\Theta}_{rq} u^q_{\langle i_1 i_2 \cdots i_n \rangle} \\
&\frac{\partial u^r_{\langle i_1 i_2 \cdots i_N \rangle}}{\partial t} && +\frac{N}{2N+1}c\frac{\partial u^r_{\langle\langle i_1 i_2 \cdots i_{N-1}\rangle}}{\partial x_{i_N\rangle}} && && = -\sum_q \widehat{\Theta}_{rq} u^q_{\langle i_1 i_2 \cdots i_N \rangle}.
\end{aligned} \tag{4.5}$$

In each of these r runs from 1 through R and n in $(4.5)_2$ runs from 1 to $N-1$. Thus we have R hierarchies of moment equations, each one consisting of tensor equations up to rank N. The main problem after this is the determination of R and N appropriate to the phenomenon under consideration. We address that problem in the remaining Sections 5 through 7.

4.4 Absorption and Emission of Bremsstrahlung. Thomson Scattering.

In a hot plasma the bremsstrahlung of free electrons in the neighborhood of heavy ions is the most important mechanism for absorption and emission of photons. In that case the absorption coefficient has the form (e.g. Bethe & Salpeter [156] or Chandrasekar [148])

$$\kappa = D\frac{1-\exp\left(-\Xi_{|E}\right)}{\Xi^3_{|E}}, \qquad \text{where, as before } \Xi_{|E} = \frac{\hbar c k}{k_B T}. \tag{4.6}$$

This shows a strong k-dependence of κ. The factor D is a constant as far as k goes, but it depends on the density of the matter. By use of (4.6) we may calculate Θ_{rs} from (4.3) and obtain

$$\Theta_{rs} = D\left(\frac{k_B T}{\hbar c}\right)^{r-s} \sum_{t=1}^{R} \Gamma(t+r)\,\zeta(t+r)\,C_{ts}^{-1}. \tag{4.7}$$

If we had moments with $r = 0$ in the theory, we should find Θ_{0s} to be infinite, because $\zeta(0) = \infty$. For this reason we have started our hierarchies with $r = 1$.

There are different mechanisms for scattering as well as for absorption and emission but here again we focus the attention on one of these: Thomson scattering on free electrons, where the scattering coefficient is constant; cf. Bethe & Salpeter [156] or Chandrasekar [148]. In that case, by (4.4), $\widehat{\Theta}_{rs}$ differs from Θ_{rs} by a unit matrix and we have

$$\widehat{\Theta}_{rs} = \Theta_{rs} + \zeta\delta_{rs}. \tag{4.8}$$

Thus, if there is only scattering and no absorption and emission, the systems (4.5) for each r are uncoupled.

Another case in which no coupling occurs is the artificial case of gray matter, for which κ is independent of k. In that case we have

$$\Theta_{rs}^{gray} = \kappa\delta_{rs}. \tag{4.9}$$

This case is not realistic but it may serve engineers and physicists as a model in order to make quick estimates easily. Section 8 below will make use of this simplification.

5 Local Radiative Equilibrium

5.1 The Rosseland Mean Value of the Absorption Coefficient

One of the most important problems in stellar physics is the reliable calculation of the energy flux Q_i of radiation in terms of the gradient of the energy density. If there is no scattering, this calculation may be done approximately by the assumption of radiative equilibrium. Local radiative equilibrium means that the phase density on the left-hand side of the radiative transfer equation may be replaced by $f_{|E}$ even though its derivatives do not vanish so that there is a production density. We have

$$\frac{\partial f_{|E}}{\partial t} + cn_k\frac{\partial f_{|E}}{\partial x_k} = -\kappa\left(f - f_{|E}\right). \tag{5.1}$$

Hence follows an approximate expression for the phase density f. This may be used in the energy flux

$$Q_i = \int \hbar c^2 k_i f d\mathbf{k} \tag{5.2}$$

and we obtain

$$Q_i = -\int \frac{1}{\kappa}\hbar c^3 k n_i n_k \frac{\partial f_{|E}}{\partial x_k} d\mathbf{k}. \tag{5.3}$$

$f_{|E}$ is given by (2.11) and hence follows after a little calculation

$$Q_i = -\frac{c^2}{3}\frac{1}{\breve{\kappa}}\frac{\partial e_{|E}}{\partial x_i}, \qquad \text{where} \qquad \frac{1}{\breve{\kappa}} = \frac{\displaystyle\int \frac{1}{\kappa} k^4 \frac{\partial f_{|E}}{\partial k} dk}{\displaystyle\int k^4 \frac{\partial f_{|E}}{\partial k} dk} \tag{5.4}$$

with $e_{|E} = aT^4$. $\breve{\kappa}$ is the so-called *Rosseland mean* of the absorption coefficient. The integrals in $(5.4)_2$ may be calculated for κ as given in (4.6) and we obtain

$$\frac{1}{\breve{\kappa}} = \frac{1}{D}\frac{1}{\Gamma(5)\,\zeta(4)}\int \frac{\Xi^7 d\Xi}{(1-e^{-\Xi})^2\,(e^{\Xi}-1)}. \tag{5.5}$$

We consider the Rosseland mean as a reliable approximate value—close to equilibrium, of course—and we use it as a benchmark for the reliability of the moment method, which we proceed to discuss now.

5.2 Maxwell Iteration

The equivalent of the assumption of local radiative equilibrium for the moment equations (4.5) requires equilibrium values of all moments on the left-hand sides. Thus, without scattering, we obtain for $n = 1$

$$\frac{1}{3}c\frac{\partial u^r_{|E}(T)}{\partial x_i} = -\sum_q \widehat{\Theta}_{rq} u^q_i \tag{5.6}$$

We solve this for u^1_i, or $Q_i = \hbar c^2 u^1_i$, and obtain with (3.8) and $e_{|E} = aT^4$

$$Q_i = -\frac{c^2}{3}\frac{1}{\hat{\kappa}}\frac{\partial e_{|E}(T)}{\partial x_i}, \qquad \text{where} \qquad \frac{1}{\hat{\kappa}} = \sum_r^R \widehat{\Theta}^{-1}_{1r}\frac{\Gamma(r+4)\,\zeta(r+3)}{\Gamma(5)\,\zeta(4)}\left(\frac{k_B T}{\hbar c}\right)^{r-1}. \tag{5.7}$$

$\hat{\kappa}$ may be called the *moment-mean-value* of the absorption coefficient. If we consider absorption and emission of bremsstrahlung (4.6), (4.7) we obtain

$$\frac{1}{\hat{\kappa}} = \frac{1}{D}\sum_{q,r=1}^R \frac{\Gamma(q+4)\,\zeta(q+3)}{\Gamma(5)\,\zeta(4)} A^{-1}_{qr}\,\Gamma(r+4)\,\zeta(r+3), \tag{5.8}$$

where $A_{rs} = \Gamma(r+s)\,\zeta(r+s)$.

$\hat{\kappa}$ depends on R, the number of systems (4.5) and therefore we now have a first possibility to determine how many systems of moments we need. We form $\hat{\kappa}/\breve{\kappa}$, the ratio of the moment mean value and the Rosseland mean, and plot this value for different choices of R. Figure 13.1 shows the result: From $R = 4$ onwards we find that both values agree. And since we consider the Rosseland mean as reliable we must conclude that an equally reliable moment method requires $R \geq 4$.[3]

[3]Of course, this number depends on our choice of the absorption coefficient. With different absorption coefficients $\kappa(k)$ a similar behaviour is observed, but the number $R_{\min}$, for which $\hat{\kappa}/\breve{\kappa} \simeq 1$ holds, will change. For gray matter we have $R_{\min} = 1$.

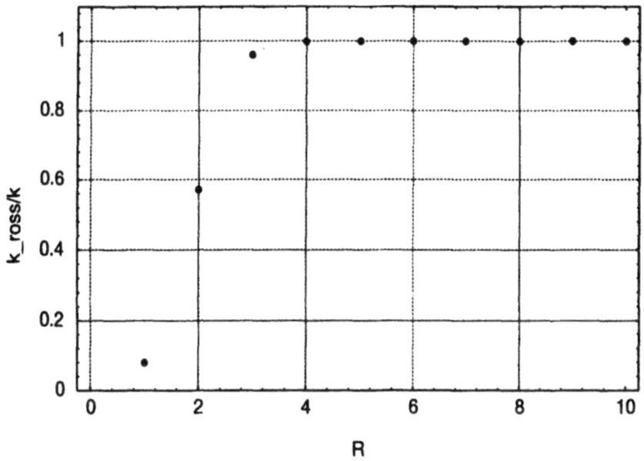

Figure 13.1 Comparison of moment mean and Rosseland mean.

5.3 Conclusion

The above observation implies that in a moment theory of radiation it is not sufficient to consider the balance laws of momentum and energy which are the first two equations of the $r = 1$ hierarchy. Nor is it enough to add higher moment equations of that hierarchy. Indeed, we now recognize that the coupling of the $r = 1$ hierarchy to at least three more hierarchies—the ones with $r = 2$ through $r = 4$—is essential for reliable results.

We shall presently confirm this observation.

6 Compression of Radiation

6.1 A Thought Experiment

We proceed with the investigation of the hierarchies of moment equations (4.5). For that purpose we consider a process of equilibration between radiation and matter. Initially we have a monatomic gas of temperature T_0 in a cylinder of volume V_0 which is closed off at the top by a piston. There is also a photon gas in the cylinder which is in equilibrium with the atomic gas so that its phase density is given by $f_{|E}(T_0)$. The interior walls of cylinder and piston are reflecting the photons perfectly so that the only thermally effective interaction of matter and radiation is between atoms and photons.

Now the piston is pushed into the cylinder so that the new volume is V_1. The push is fast enough so that no significant absorption and emission of photons occurs in its duration and that the compression of the gas is adiabatic; but the push is also slow enough that the compression may be considered as reversible.[4]

[4]There may be problems with all this; that is why it is a thought experiment.

After the compression the temperature of the gas is

$$T_1 = T_0 \left(\frac{V_0}{V_1}\right)^{2/3}, \tag{6.1}$$

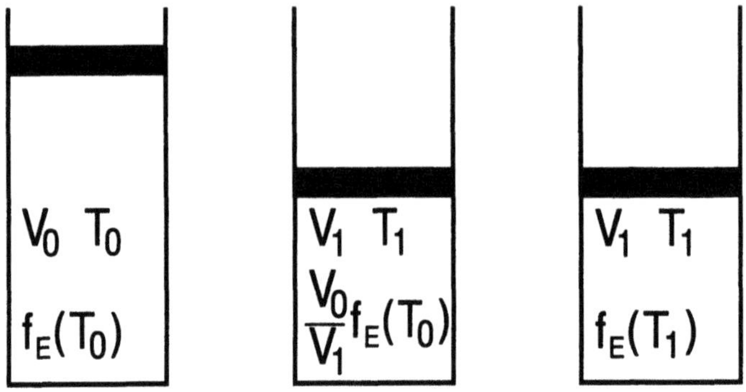

Figure 13.2 Compression of gas and radiation and subsequent equilibration.

and this temperature is henceforth kept constant by having the cylinder in a heat bath. The number of photons is unchanged during the compression but, since their number ***density*** is increased, the new phase density is given by $V_0/V_1 f_{|E}(T_0)$. This is our new initial condition and we shall proceed to calculate how the final equilibrium phase density $f_{|E}(T_1)$ is established. Figure 13.2 shows the different stages of the experiment. We exclude scattering, and since all fields are assumed homogeneous and isotropic during the whole process we obtain a radiative transfer equation of the form

$$\frac{df}{dt} = -\kappa \left(f - f_{|E}(T_1)\right). \tag{6.2}$$

Also the hierarchies of moment equations reduce to a single scalar equation each and we have

$$\frac{du^r}{dt} = -\sum_q^R \Theta_{rq}(T_1) \left(u^q - u^q_{|E}(T_1)\right) \qquad (r = 1, 2, \ldots, R). \tag{6.3}$$

We proceed to solve both the equations (6.2) and (6.3) and ask for which number R the energy density $\hbar c u^1(t)$ agrees in the two cases.

6.2 Solution of the Radiative Transfer Equation

According to Paragraph 6.1 the initial condition of the equilibration process is given by

$$f_1 = \frac{V_0}{V_1} f_{|E}(T_0)$$

and during the process, by (6.2), we have an exponential decay to the new equilibrium $f_{|E}(T_1)$, viz.

$$f(t) = f_{|E}(T_1) + \left(\frac{V_0}{V_1} f_{|E}(T_0) - f_{|E}(T_1)\right) e^{-\kappa t}. \tag{6.4}$$

From this phase density we calculate

$$u^r(t) = \int k^r f(t)\, d\mathbf{k}, \qquad \text{hence} \qquad u^r(0) = \frac{V_0}{V_1} u^r_{|E}(T_0)\,.$$

For easy comparison we introduce the dimensionless moments

$$v^r = \frac{u^r(t) - u^r_{|E}(T_1)}{u^r(0) - u^r_{|E}(T_1)} = \frac{u^r - u^r_{|E}(T_1)}{\frac{V_0}{V_1} u^r_{|E}(T_0) - u^r_{|E}(T_1)} \tag{6.5}$$

which—for all values r—start at 1 and approach 0 as t tends to infinity. We obtain

$$v^r(t) = \frac{\frac{V_0}{V_1}\left(\frac{T_0}{T_1}\right)^{r+3} \int \frac{\Xi^{r+2}}{e^{\Xi}-1} e^{-\kappa\left(\frac{T_0}{T_1}\Xi\right)t} d\Xi - \int \frac{\Xi^{r+2}}{e^{\Xi}-1} e^{-\kappa(\Xi)t} d\Xi}{\left[\frac{V_0}{V_1}\left(\frac{T_0}{T_1}\right)^{r+3} - 1\right] \Gamma(r+3)\,\zeta(r+3)\,.} \tag{6.6}$$

The function $\kappa(\Xi)$ is given by (4.6). The dashed curve in Figure 13.3 shows the function (6.6) for $r = 1$, i.e. the equilibration of the energy density. V_0/V_1 has been chosen as 4.

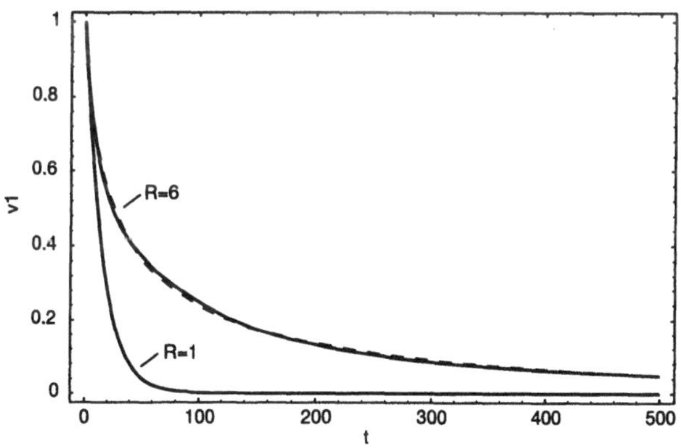

Figure 13.3 Equilibration of energy according to the radiative transfer equation (dashed) and according to moment theories for $R = 1$ and $R = 6$ (solid).

6.3 Solution of Moment Equations

We proceed to solve the moment equations (6.3), again introducing the dimensionless moments (6.5) appropriate to this problem. For these dimensionless quantities v^r the equations (6.3) assume the form

$$\frac{dv^r}{d(Dt)} = -\sum_q^R \tilde{\Theta}_{rq} v^q, \qquad \text{where}$$

$$\tilde{\Theta}_{rq} = \frac{\Gamma(q+3)\,\zeta(q+3)}{\Gamma(r+3)\,\zeta(r+3)} \frac{\frac{V_0}{V_1}\left(\frac{T_0}{T_1}\right)^{q+3} - 1}{\frac{V_0}{V_1}\left(\frac{T_0}{T_1}\right)^{r+3} - 1} \sum_s^R \Gamma(s+r)\,\zeta(s+r)\,C_{sq}^{-1}. \tag{6.7}$$

For the solution of $(6.7)_1$ we decouple the equations by a linear transformation $\hat{v}^s = \sum_{r=1}^R V_{sr}^{-1} v^r$, where V_{sr}^{-1} is the matrix of left eigenvectors of $\tilde{\Theta}_{rq}$. If α_q are the eigenvalues of $\tilde{\Theta}_{rq}$ we obtain as the required solution [note that $v^s(0) = 1$ for all s]

$$v^r(t) = \sum_{q,s}^R V_{rq} V_{qs}^{-1} e^{-\alpha_q Dt}. \tag{6.8}$$

The solid curves of Figure 13.3 represent $v^1(t)$, for the indicated choices of R, i.e. the equilibration of the energy density according to different r-hierarchies in the moment method.

6.4 Conclusion

Inspection of Figure 13.3 shows that—once again—it is not enough to use only one representative of the r-hierarchies. Indeed full agreement between the moment method and the exact solution of the radiative transfer equation requires that no less than 6 hierarchies be taken into account. This confirms the previous conclusion of Section 5; even though here we need an even higher value of R; presumably this is the case, because the present problem is further away from equilibrium.

7 Penetration of a Beam of Radiation into Matter

7.1 Field Equations

We consider the propagation of a one-dimensional beam of radiation into matter. A typical example occurs when a sunbeam penetrates the earth's atmosphere, which here we suppose has the uniform and constant temperature T.

The appropriate system of equations is again (4.5) except that now we specialize the system to one dimension, to wit the direction of the 1-axis or x-axis. We obtain

$$\begin{aligned}
&\frac{\partial u^r_{\langle 0\rangle}}{\partial t} && && +c\frac{\partial u^r_{\langle 1\rangle}}{\partial x} &&= -\sum_q \Theta_{rq}\left(u^q_{\langle 0\rangle} - u^q_{\langle 0\rangle|E}\right) \\
&\frac{\partial u^r_{\langle n\rangle}}{\partial t} && +\frac{n^2}{4n^2-1}c\frac{\partial u^r_{\langle n-1\rangle}}{\partial x} && +c\frac{\partial u^r_{\langle n+1\rangle}}{\partial x} &&= -\sum_q \widehat{\Theta}_{rq} u^q_{\langle n\rangle} \\
&\frac{\partial u^r_{\langle N\rangle}}{\partial t} && +\frac{N^2}{4N^2-1}c\frac{\partial u^r_{\langle N-1\rangle}}{\partial x} && &&= -\sum_q \widehat{\Theta}_{rq} u^q_{\langle N\rangle}.
\end{aligned} \tag{7.1}$$

Here we have introduced the abbreviation

$$u^r_{\langle n \rangle} = u^r_{\langle \underbrace{11 \cdots 1}_{n-times} \rangle}. \tag{7.2}$$

Also in the sequel we shall abbreviate $\frac{n^2}{4n^2-1}$ by α_n.

7.2 Characteristic Speeds and Amplitudes of the Propagating Beam

We know from electrodynamics that the beam propagates with the speed of light c, whereas the fastest propagation speed implied by the system of moments (7.1) is equal to the biggest characteristic speed of the system. This knowledge provides a criterion for the break-off of the system: *The number N of moments $u^r_{\langle n \rangle}$*—needed for the representation of a beam—*is such that the biggest characteristic speed equals c.*

The characteristic speeds of the system (7.1) are the eigenvalues of the matrix[5]

$$\mathcal{A}_{AB} = c \begin{bmatrix} 0 & 1 & & & \\ \alpha_1 & 0 & 1 & & \\ & \ddots & \ddots & \ddots & \\ & & & \alpha_N & 0 \end{bmatrix} \tag{7.3}$$

and since the matrix $\mathcal{A}$ has this simple form, its eigenvalues are easy to determine numerically. The value of the biggest eigenvalue—the largest characteristic speed—is plotted in Figure 13.4 for different values of N. Inspection shows that the value c is rapidly approached as N grows. For $N \geq 30$ we may say that the largest characteristic speed is equal to c.

Thus for a proper description of a beam of radiation we need as many as 30 moments—at least according to the speed criterion. We are not surprised, since the beam is a very nonisotropic phenomenon indeed. The characteristic speeds—i.e. eigenvalues of $\mathcal{A}$—are the speeds of propagation of acceleration waves. The amplitude of these waves are the right eigenvectors of $\mathcal{A}$. In particular, the amplitudes $\left[\left[u^r_{\langle n \rangle}\right]\right]$ of the fastest wave are the components of the eigenvector of the matrix $\mathcal{A}$ corresponding to its largest eigenvalue $V_{\max}$. For the calculation of the eigenvector it is useful to know that the determinant

$$D_{n+1} = \begin{vmatrix} -V_{\max}/c & 1 & & & & \\ \alpha_1 & -V_{\max}/c & 1 & & & \\ & \alpha_2 & -V_{\max}/c & 1 & & \\ & \ddots & \ddots & \ddots & & \\ & & & \alpha_{n-1} & -V_{\max}/c & 1 \\ & & & & \alpha_n & -V_{\max}/c \end{vmatrix} \quad n \geq 2 \tag{7.4}$$

obeys the recurrence relation

$$D_{n+1} = -V_{\max}/c\, D_n - \alpha_n D_{n-1}, \qquad \text{with } D_0 = 1, D_1 = -V_{\max}/c. \tag{7.5}$$

[5]Note that the matrix is the same for all of the R systems.

It is then easy to show that the eigenvector is given by the relation,

$$\left[\left[u^r_{\langle n\rangle}\right]\right] = (-1)^n D_n \left[\left[u^r_{\langle 0\rangle}\right]\right] . \tag{7.6}$$

or, if the beam propagates into an equilibrium state so that $u^r_{\langle n\rangle} = 0$ holds before the wave for $n \geq 1$

$$u^r_{\langle n\rangle} = (-1)^n D_n \left(u^r_{\langle 0\rangle} - u^r_{\langle 0\rangle|E}\right) . \tag{7.7}$$

What about the eigenvectors pertaining to the eigenvalues other than the largest one? We take all these to be zero in the beam problem, since we know that the beam propagates with just one speed, viz. $V_{\max} \simeq c$.

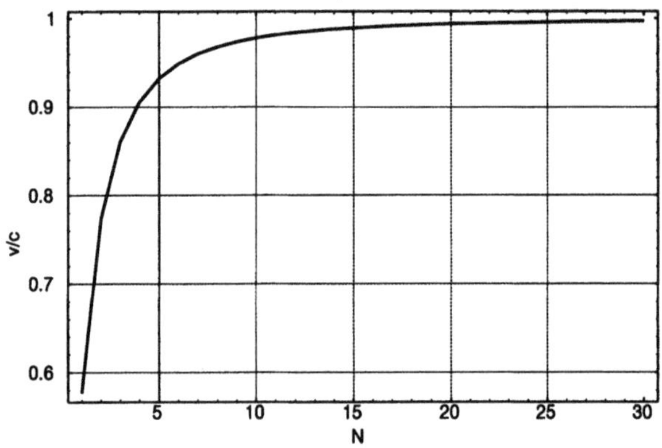

Figure 13.4 The largest characteristic speed of the moment equations approaches the speed of light rapidly as N goes up.

7.3 Plane Harmonic Waves and Dispersion Relation (General)

The hierarchies (7.1) are coupled through the right-hand sides. But they may easily be decoupled by a linear transformation of the form

$$u^r_{\langle n\rangle} - u^r_{\langle n\rangle|E} = \sum_s M_{rs} w^s_{\langle n\rangle} \tag{7.8}$$

such that the matrix $\mathbf{M}$ is the matrix of right eigenvectors of the matrix $\mathbf{\Theta}$. We have

$$\sum_{r,q} M^{-1}_{sr} \Theta_{rq} M_{qp} = \mu_s \delta_{sp} ,$$

where μ_s are the eigenvalues of Θ_{rq}. The decoupling works in this manner, because $\mathbf{\Theta}$ and $\widehat{\mathbf{\Theta}}$ differ only by a unit matrix; see (4.8). It follows that we may rewrite the R coupled systems (7.1) as R uncoupled systems for $w^s_{\langle n\rangle}$, viz.

$$\begin{aligned}
&\frac{\partial w^s_{\langle 0\rangle}}{\partial t} && && +c\frac{\partial w^s_{\langle 1\rangle}}{\partial x} &&= -\mu_s w^s_{\langle 0\rangle}\\
&\frac{\partial w^s_{\langle n\rangle}}{\partial t} && +\alpha_n c\frac{\partial w^s_{\langle n-1\rangle}}{\partial x} && +c\frac{\partial w^s_{\langle n+1\rangle}}{\partial x} &&= -\hat{\mu}_s w^s_{\langle n\rangle}\\
&\frac{\partial w^s_{\langle N\rangle}}{\partial t} && +\alpha_N c\frac{\partial w^s_{\langle N-1\rangle}}{\partial x} && &&= -\hat{\mu}_s w^s_{\langle N\rangle}
\end{aligned} \tag{7.9}$$

$\hat{\mu}_s$ equals $\mu_s + \zeta$.

We investigate plane harmonic waves of the form

$$\tilde{w}^{s,\Omega}_{\langle n\rangle} \exp i\left(\Omega t - q_s x\right) \tag{7.10}$$

as solutions of the s-system (7.9). $\tilde{w}^{s,\Omega}_{\langle n\rangle}$ are complex amplitudes and Ω and q_s are frequency and wave numbers respectively. The phase speeds and attenuation coefficients are given by

$$v_{Ph} = \frac{\Omega}{\mathrm{Re}\,(q)}, \qquad \alpha = -\,\mathrm{Im}\,(q)\,. \tag{7.11}$$

Insertion of (7.10) into (7.9) provides homogenous algebraic systems of the form

$$\begin{bmatrix}
-\gamma_s\beta_s & 1 & & & & & \\
\alpha_1 & -\beta_s & 1 & & & & \\
 & \alpha_2 & -\beta_s & 1 & & & \\
 & & \alpha_3 & -\beta_s & 1 & & \\
 & & & \ddots & \ddots & \ddots & \\
 & & & & \alpha_{N-1} & -\beta_s & 1 \\
 & & & & & \alpha_N & -\beta_s
\end{bmatrix}
\begin{bmatrix}
\tilde{w}^{s,\Omega}_{\langle 0\rangle} \\ \tilde{w}^{s,\Omega}_{\langle 1\rangle} \\ \\ \\ \\ \\ \tilde{w}^{s,\Omega}_{\langle N\rangle}
\end{bmatrix} = 0, \tag{7.12}$$

where

$$\beta_s = \frac{\hat{\mu}_s + i\Omega}{icq_s}, \qquad \gamma_s = \frac{\mu_s + i\Omega}{\hat{\mu}_s + i\Omega}. \tag{7.13}$$

These systems have nontrivial solutions if the determinant vanishes, a condition which defines the dispersion relation

$$\begin{vmatrix}
-\gamma_s\beta_s & 1 & & & & & \\
\alpha_1 & -\beta_s & 1 & & & & \\
 & \alpha_2 & -\beta_s & 1 & & & \\
 & & \alpha_3 & -\beta_s & 1 & & \\
 & & & \ddots & \ddots & \ddots & \\
 & & & & \alpha_{N-1} & -\beta_s & 1 \\
 & & & & & \alpha_N & -\beta_s
\end{vmatrix} = 0. \tag{7.14}$$

7.4 Intense Absorption. The Damped Wave Limit.

If there is much more absorption and emission than scattering, we may neglect ζ. Hence the difference between $\hat{\mu}_s$ and μ_s vanishes, so that $\gamma_s = 1$ holds.

In that case β_s are the eigenvalues of the matrix $\mathcal{A}$ in (7.3)—independent of s—or in other words the characteristic speeds of the system (7.1). Once again we assume that the eigenvectors—here $\tilde{w}^{s,\Omega}_{\langle n\rangle}$ by (7.12)—vanish for all but the largest β_s. That largest value is equal to $V_{\max}/c$ or, if N is large enough, equal to 1. Thus, by (7.13), we obtain

$$q_s = \frac{\Omega - i\mu_s}{c} \tag{7.15}$$

so that there is no dispersion and the damping is independent of Ω. We have

$$v_{Ph} = c \qquad \text{and} \qquad \alpha = \mu_s/c\,. \tag{7.16}$$

The $w^s_{\langle n\rangle}$-part of the beam may be represented by a superposition of harmonic waves—Fourier harmonics with different frequencies Ω—so that we may write

$$\begin{aligned} w^s_{\langle n\rangle} &= \int_{-\infty}^{\infty} \tilde{w}^{s,\Omega}_{\langle n\rangle}\, e^{i(\Omega t - q_s x)}\, d\Omega, \qquad \text{or by (7.15)} \\ w^s_{\langle n\rangle} &= \int_{-\infty}^{\infty} \tilde{w}^{s,\Omega}_{\langle n\rangle}\, e^{i\Omega(t - x/c)}\, d\Omega\; e^{-\frac{\mu_s x}{c}}\,. \end{aligned} \tag{7.17}$$

If we return to the moments $u^r_{\langle n\rangle}$ we thus obtain

$$u^r_{\langle n\rangle} - u^r_{\langle n\rangle|E} = \sum_s M_{rs} \int_{-\infty}^{\infty} \tilde{w}^{s,\Omega}_{\langle n\rangle}\, e^{i\Omega(t-x/c)}\, d\Omega\; e^{-\frac{\mu_s x}{c}}\,. \tag{7.18}$$

Therefore the beam penetrates the matter as a damped wave propagating with velocity c. Let the wave be created by a boundary stimulus at $x = 0$ that is given by the functions

$$\left(\overset{0}{u}{}^r_{\langle n\rangle} - \overset{0}{u}{}^r_{\langle n\rangle|E}\right)(t) = \sum_s M_{rs} \int_{-\infty}^{\infty} \tilde{w}^{s,\Omega}_{\langle n\rangle}\, e^{i\Omega t}\, d\Omega. \tag{7.19}$$

The propagating damped wave may thus be written in the form

$$\left(u^r_{\langle n\rangle} - u^r_{\langle n\rangle|E}\right)(x,t) = \sum_t \left(\sum_s M_{rs} e^{-\frac{\mu_s x}{c}} M^{-1}_{st}\right) \left(\overset{0}{u}{}^t_{\langle n\rangle} - \overset{0}{u}{}^t_{\langle n\rangle|E}\right)\left(t - \frac{x}{c}\right). \tag{7.20}$$

The boundary stimulus must be chosen carefully so as to create only the fastest mode. For this to be so we have to satisfy (7.6), and therefore (7.20) may be written in the form (recall that all $u^r_{\langle n\rangle|E} = 0$ for $n \geq 0$, since we let the beam propagate into an equilibrium state)

$$\left(u^r_{\langle n\rangle} - u^r_{\langle n\rangle|E}\right)(x,t) = \sum_t \left(\sum_s M_{rs} e^{-\frac{\mu_s x}{c}} M^{-1}_{st}\right) (-1)^n D_n \left(\overset{0}{u}{}^t_{\langle 0\rangle} - \overset{0}{u}{}^t_{\langle 0\rangle|E}\right)\left(t - \frac{x}{c}\right). \tag{7.21}$$

Note that starting from (7.15) the formulae are valid, if $N \geq 30$ holds, because otherwise $V_{\max}$ is not equal to c.

7.5 Intense Scattering. The Diffusion Limit.

If scattering is more important than absorption and emission we may set $\mu_s = 0$ and $\hat{\mu}_s = \zeta$. This makes the system (7.12) assume the form

$$\begin{bmatrix} -\frac{\Omega}{cq_s} & 1 & & & \\ \alpha_1 & -\frac{\zeta+i\Omega}{icq_s} & 1 & & \\ & \ddots & \ddots & \ddots & \\ & & \alpha_{N-1} & -\frac{\zeta+i\Omega}{icq_s} & 1 \\ & & & \alpha_N & -\frac{\zeta+i\Omega}{icq_s} \end{bmatrix} \begin{bmatrix} \tilde{w}^{s,\Omega}_{\langle 0\rangle} \\ \tilde{w}^{s,\Omega}_{\langle 1\rangle} \\ \vdots \\ \tilde{w}^{s,\Omega}_{\langle N-1\rangle} \\ \tilde{w}^{s,\Omega}_{\langle N\rangle} \end{bmatrix} = 0. \tag{7.22}$$

We proceed to solve this system for the special case of small frequency, i.e. $\Omega \ll \zeta$ and $(cq_s/\zeta)^2 \ll 1$. In that case the system (7.22) may be reduced to two recurrence equations, viz.

$$-\frac{\Omega}{cq}\tilde{w}^{s,\Omega}_{\langle 0\rangle} + \tilde{w}^{s,\Omega}_{\langle 1\rangle} = 0 \quad , \quad \frac{icq}{\zeta}\alpha_n \tilde{w}^{s,\Omega}_{\langle n-1\rangle} - \tilde{w}^{s,\Omega}_{\langle n\rangle} = 0 \qquad (n \geq 1)\ . \tag{7.23}$$

Hence follows that all amplitudes vanish—an obvious trivial result—unless we have

$$q_s = \pm\sqrt{-i\frac{3\zeta\Omega}{c^2}} \tag{7.24}$$

independent of s.

Just as the eigenvalue problem (7.12) belongs to the differential system (7.9), the new—approximate—eigenvalue problem (7.23) belongs to the differential system

$$\begin{aligned} \frac{\partial w^s_{\langle 0\rangle}}{\partial t} \qquad\qquad + c\frac{\partial w^s_{\langle 1\rangle}}{\partial x} &= 0 \\ + \alpha_n c \frac{\partial w^s_{\langle n-1\rangle}}{\partial x} &= -\zeta w^s_{\langle n\rangle} \qquad (n = 1, \ldots, N)\ . \end{aligned} \tag{7.25}$$

Thus we conclude that the approximations employed here have reduced the system in effect to two equations, one each for $n = 0$ and $n = 1$. These two equations combine to give the diffusion equation

$$\frac{\partial w^s_{\langle 0\rangle}}{\partial t} - \frac{c^2}{3\zeta}\frac{\partial^2 w^s_{\langle 0\rangle}}{\partial x^2} = 0\,; \tag{7.26}$$

all moments $w^s_{\langle n\rangle}$ with $n \geq 1$ may be obtained from the solution of (7.26) by differentiation; cf. $(7.25)_2$. We conclude that the predominance of scattering over absorption and emission makes radiation move diffusively into matter. Only $w^s_{\langle 0\rangle}$—hence $u^s_{\langle 0\rangle}$—matters now and, if we consider it as a superposition of harmonic waves, we may write

$$w^s_{\langle 0\rangle} = \int_{-\infty}^{\infty} \tilde{w}^{s,\Omega}_{\langle 0\rangle}\, e^{i(\Omega t - q_s x)}\, d\Omega.$$

To calculate the corresponding fields $\left(u^r_{\langle 0\rangle} - u^r_{\langle 0\rangle|E}\right)(x,t)$ is both easier and more difficult than in the case of intense absorption. Easier because q_s is independent of s now and more difficult since

q is not a linear function of Ω. We obtain

$$\left(u^r_{\langle 0\rangle} - u^r_{\langle 0\rangle|E}\right)(x,t) = \int_{-\infty}^{\infty} \left(\overset{0t}{u}_{\langle 0\rangle} - \overset{0t}{u}_{\langle 0\rangle|E}\right)(\tau) \left[\int_0^{\infty} e^{-\sqrt{\Omega/2}x} \cos\left(\Omega\,(t-\tau) - \sqrt{\Omega/2}x\right) d\Omega\right] d\tau, \tag{7.27}$$

where $\left(\overset{0t}{u}_{\langle 0\rangle} - \overset{0t}{u}_{\langle 0\rangle|E}\right)(t)$ are the boundary values as functions of time.

7.6 General Case and a Simple Example

In general we shall have both absorption and emission as well as scattering. The beam created by a boundary stimulus is then quite difficult to calculate and it would not be very instructive to describe the numerical method that must be followed. But the result is bound to be given—at least qualitatively—as shown in Figure 13.5, which refers to a short-pulse stimulus: The damped pulse arrives first and it drags a wake of scattered radiation behind it.

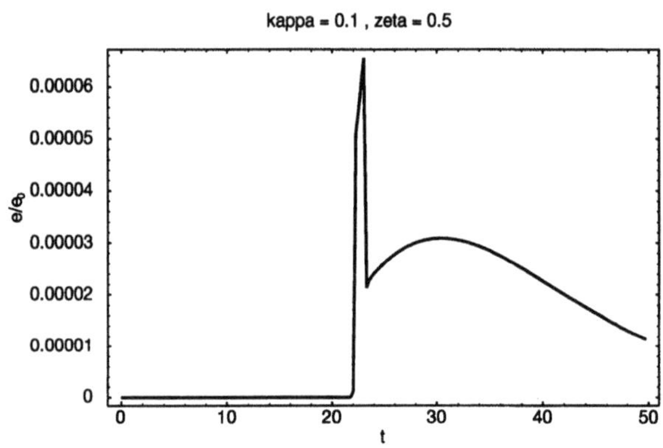

Figure 13.5 Propagation of a damped pulse followed by a wake of scattered radiation.

The curve of Figure 13.5 has been calculated for the case $R = 1$ and $N = 4$ and for κ and ζ independent of k. The duration Δt of the pulse is used to make t, x, κ, and ζ dimensionless:

$$\bar{t} = t/\Delta t, \qquad \bar{x} = x/c\Delta t, \qquad \bar{\kappa} = \kappa\Delta t, \qquad \bar{\zeta} = \zeta\Delta t.$$

The total energy supply $e = \hbar c\left(u^1 - u^1_{|E}\right)\Delta t$ is equal to e_0 at the place where the pulse is created and at $\bar{x} = 20$—in Figure 13.5—it has already dropped to less than $0.00081e_0$. The peak is due to photons that have passed the matter without any interaction, and we may call them ***ballistic photons***. The long wake behind the peak stems from scattered photons. [Note that the peak does not propagate with the speed of light, since we have chosen N equal to 4].

8 Radiative Entropy in Gray Bodies

8.1 Photon Gas and an Eulerian Fluid

Matter and radiation both carry an entropy and an entropy flux. And, if they interact, there is a production of entropy in both of them. Under most circumstances the entropy production of matter due to radiation equals r/T, where r is the material energy production and T the temperature of the matter. So, what is the entropy production of radiation due to matter and what is the total entropy production? We answer these questions. The answer involves a radiation temperature—or pseudo-temperature—which depends on both, the temperature of the source of radiation and the temperature of the target.

We recall the notions of previous sections and rewrite radiative quantities pertinent to this section in Table 13.1:

Radiation	Density	Flux	Production
Number	$n = \int f d\mathbf{k}$	$J_i = \int c\frac{k_i}{k} f d\mathbf{k}$	$\sigma^n = \int S d\mathbf{k}$
Momentum	$p_j = \int \hbar k_j f d\mathbf{k}$	$P_{ij} = \int \hbar c \frac{k_i}{k} k_j f d\mathbf{k}$	$\sigma_j^p = \int \hbar k_j S d\mathbf{k}$
Energy	$e = \int \hbar c k f d\mathbf{k}$	$Q_i = \int \hbar c^2 k_i f d\mathbf{k}$	$-r = \int \hbar c k S d\mathbf{k}$
Entropy	$h = -k_B \int [\star]\, d\mathbf{k}$	$\varphi_i = -k_B \int c \frac{k_i}{k} [\star]\, d\mathbf{k}$	$\sigma_{rm} = +k_B \int \ln\left(1 + \frac{y}{f}\right) S d\mathbf{k}$

$$\star = f \ln \frac{f}{y} - y\left(1 + \frac{f}{y}\right) \ln\left(1 + \frac{f}{y}\right)$$

Table 13.1 Thermodynamic fields of radiation.

Most of the entries in Table 13.1 are self-explanatory. y is equal to $2/(2\pi)^3$ as before. σ_j^p is the production of momentum of the photon gas due to the interaction with matter, and $-r$ and σ_{rm} are the productions of energy and entropy due to matter.

The matter we assume to be an Eulerian fluid and its thermodynamic quantities are listed in Table 13.2. Mass of course is conserved, but momentum is not, because a force is exerted on the matter if photons are absorbed or emitted. The power of that force contributes to the production of energy which otherwise is due to the absorption of photon energy. The entropy production σ_{mr} of matter due to radiation is equal to the energy absorption divided by the absolute temperature of the fluid; this is due to the second law in its classical form. Other contributions to the entropy production of matter may exist, but not in an Eulerian fluid.

Matter	Density	Flux	Production
Number	ρ	ρv_i	0
Momentum	ρv_j	$\rho v_i v_j + p\delta_{ij}$	$-\sigma_j^P$
Energy	$\rho\left(u+\frac{1}{2}v^2\right)$	$\rho\left(u+\frac{1}{2}v^2\right)v_i + pv_i$	$r - \sigma_j^P v_j$
Entropy	ρs	$\rho s v_i$	$\sigma_{mr} = \frac{r}{T}$

Table 13.2 Thermodynamic fields in an Eulerian fluid.

8.2 Equilibrium of Radiation with Matter at Rest

For equilibrium we maximize the entropy under the constraint of fixed energy and obtain the Planck distribution function

$$f_E(k,T) = \frac{y}{e^{\frac{\hbar ck}{k_B T}} - 1} . \tag{8.1}$$

For this case, if we calculate the entries in Table 13.1 we obtain the values given in Table 13.3 where

$$a = \frac{\pi^2}{15}\frac{k_B^4}{\hbar^3 c^3} \quad \text{and} \quad \zeta(3) = 1.202... .$$

a is called the radiation constant. All productions are zero and so are all fluxes except for the momentum flux which is isotropic. This reflects the isotropy of the equilibrium distribution (2.11). The energy density is the well-known T^4-law due to Stefan and Boltzmann, and the T^3-dependence of the entropy density was first derived by Boltzmann.

Radiation	Density	Flux	Production
Number	$2 \cdot \frac{15\zeta(3)}{\pi^4}\frac{a}{k_B}T^3$	0	0
Momentum	0	$\frac{1}{3}aT^4\delta_{ij}$	0
Energy	aT^4	0	0
Entropy	$\frac{4}{3}aT^3$	0	0

Table 13.3 Thermodynamic fields of radiation in equilibrium.

8.3 Entropy Production due to Matter-Photon Interaction

From the Tables 13.1 and 13.2 we may write for the total entropy production

$$\sigma_{rm} + \sigma_{mr} = k_B \int \ln\left(1+\frac{y}{f}\right) S d\mathbf{k} + \frac{r}{T} \tag{8.2}$$

which, with r from Table 13.1, can be written as

$$\sigma_{rm} + \sigma_{mr} = k_B \int \left(\ln\left(1+\frac{y}{f}\right) - \frac{\hbar c k}{k_B T} \right) S d\mathbf{k}$$

$$\sigma_{rm} + \sigma_{mr} = k_B \int \ln \frac{1+\frac{y}{f}}{1+\frac{y}{f_E}} S d\mathbf{k} \,. \tag{8.3}$$

This total entropy production ought to be nonnegative, and the simplest production density, which ensures that, is given by

$$S = \frac{1}{\tau}(f_E - f), \tag{8.4}$$

with τ being a constant relaxation time. This assumption may be unrealistic; it defines a "gray" fluid, and this is all we consider here.

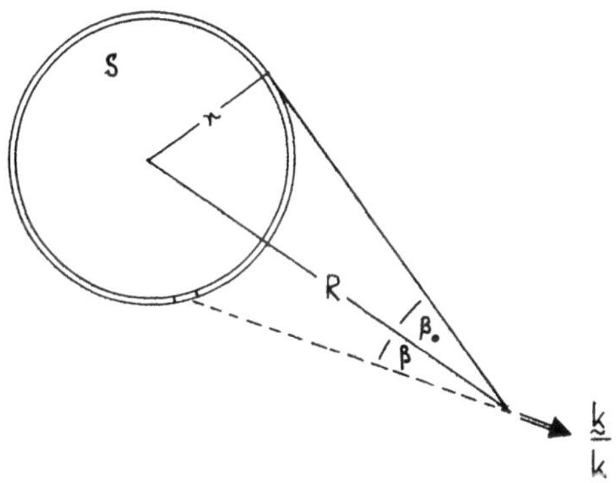

Figure 13.6 On radiation at the distance R from the center of a spherical source.

8.4 Thermodynamic Fields of Radiation in the Neighborhood of a Spherical Source

We refer to Figure 13.6, which shows a spherical source of radiation S. Inside the source the radiation is in equilibrium with the matter of the source, which has temperature T_s. Hence one concludes that in a point outside the source the distribution function is given by

$$f(x,t,k) = \begin{bmatrix} f_E(k,T_s) & 0 \le \varphi < 2\pi & \\ & 0 \le \beta \le \beta_0 & = arc\ sin\frac{r}{R} \\ 0 & \text{else}\,. & \end{bmatrix} \tag{8.5}$$

With this distribution function the entries of Table 13.1 may easily be calculated and we obtain the formulae summarized in Table 13.4. $\surd$ stands for $\left(1 - r^2/R^2\right)^{1/2}$, and the barred quantities stand for

$$\overline{\frac{k_i}{k}} = \frac{\int\limits_0^{2\pi}\int\limits_0^{\beta_0} \frac{k_i}{k}\sin\beta d\beta d\varphi}{\int\limits_0^{2\pi}\int\limits_0^{\beta_0} \sin\beta d\beta d\varphi} = \begin{bmatrix} 0 \\ 0 \\ \frac{1}{2}\frac{r^2/R^2}{1-\surd} \end{bmatrix} \begin{array}{l} \overrightarrow{r \to R} \begin{bmatrix} 0 \\ 0 \\ 1/2 \end{bmatrix} \\ \overrightarrow{r << R} \begin{bmatrix} 0 \\ 0 \\ 1 \end{bmatrix} \end{array} \tag{8.6}$$

$$\overline{\frac{k_i k_j}{k^2}} = \frac{\int\limits_0^{2\pi}\int\limits_0^{\beta_0} \frac{k_i k_j}{k^2}\sin\beta d\beta d\varphi}{\int\limits_0^{2\pi}\int\limits_0^{\beta_0} \sin\beta d\beta d\varphi} = \tag{8.7}$$

$$= \frac{1}{1-\surd}\begin{bmatrix} \frac{1}{6}\surd^3 - \frac{1}{2}\surd + \frac{1}{3} & & 0 \\ & \frac{1}{6}\surd^3 - \frac{1}{2}\surd + \frac{1}{3} & \\ 0 & & \frac{1}{3} - \frac{1}{3}\surd^3 \end{bmatrix} \begin{array}{l} \overrightarrow{r \to R} \frac{1}{3}\begin{bmatrix} 1 & & \\ & 1 & \\ & & 1 \end{bmatrix} \\ \overrightarrow{r << R} \begin{bmatrix} 0 & & \\ & 0 & \\ & & 1 \end{bmatrix} \end{array}$$

These formulae include the limiting cases when the point $\mathbf{x}$ is just outside the source and when it is far away.

These two limiting cases are separately registered in Tables 13.5 and 13.6. We note that close to the source the densities of the outgoing radiation are just equal to one half of the densities inside the source, and that, far away from the source, the fluxes are equal to the densities multiplied by c, the speed of light. Both conclusions are eminently sensible of course.

Radiation	Density	Flux	Production
Number	$\frac{1}{2}(1-\surd)\,2\cdot\frac{15\zeta(3)}{\pi^4}\frac{\sigma}{k_B}T_s^3$	$\left(1-\surd\right)c\cdot\frac{15\zeta(3)}{\pi^4}\frac{\sigma}{k_B}T_s^3\overline{\frac{k_i}{k}}$	0
Momentum	$\left(1-\surd\right)\frac{1}{2c}\sigma T_s^4\overline{\frac{k_i}{k}}$	$\left(1-\surd\right)\frac{1}{2c}\sigma T_s^4\overline{\frac{k_i k_j}{k^2}}$	0
Energy	$\frac{1}{2}(1-\surd)\,\sigma T_s^4$	$\left(1-\surd\right)\frac{1}{2c}cT_s^4\overline{\frac{k_i}{k}}$	0
Entropy	$\frac{1}{2}(1-\surd)\,\frac{4}{3}\sigma T_s^3$	$\left(1-\surd\right)c\frac{1}{2}\frac{4}{3}T_s^3\overline{\frac{k_i}{k}}$	0

Table 13.4 Thermodynamic fields of radiation from a spherical source.

$r \to R$	Density	Flux	Production
Number	$\frac{15\zeta(3)}{\pi^4}\frac{\sigma}{k_B}T_s^3$	$c\frac{15\zeta(3)}{\pi^4}\frac{\sigma}{k_B}T_s^3\begin{bmatrix}0\\0\\1/2\end{bmatrix}$	0
Momentum	$\frac{1}{2c}\sigma T_s^4\begin{bmatrix}0\\0\\1/2\end{bmatrix}$	$\frac{1}{2c}\sigma T_s^4\frac{1}{3}\begin{bmatrix}1&&\\&1&\\&&1\end{bmatrix}$	0
Energy	$\frac{1}{2}\sigma T_s^4$	$c\frac{1}{2}\sigma T_s^4\begin{bmatrix}0\\0\\1/2\end{bmatrix}$	0
Entropy	$\frac{1}{2}\frac{4}{3}\sigma T_s^3$	$c\frac{1}{2}\frac{4}{3}\sigma T_s^3\begin{bmatrix}0\\0\\1/2\end{bmatrix}$	0

Table 13.5 Thermodynamic fields of radiation close to a spherical source.

$r << R$	Density	Flux	Production
Number	$\frac{1}{2}\frac{r^2}{R^2}\frac{15\zeta(3)}{\pi^4}\frac{\sigma}{k_B}T_s^3$	$\frac{1}{2}\frac{r^2}{R^2}c\frac{15\zeta(3)}{\pi^4}\frac{\sigma}{k_B}T_s^3\begin{bmatrix}0\\0\\1\end{bmatrix}$	0
Momentum	$\frac{1}{2}\frac{r^2}{R^2}\frac{1}{2c}\sigma T_s^4\begin{bmatrix}0\\0\\1\end{bmatrix}$	$\frac{1}{2}\frac{r^2}{R^2}\frac{1}{2}\sigma T_s^4\begin{bmatrix}0&&\\&0&\\&&1\end{bmatrix}$	0
Energy	$\frac{1}{2}\frac{r^2}{R^2}\frac{1}{2}\sigma T_s^4$	$\frac{1}{2}\frac{r^2}{R^2}c\frac{1}{2}\sigma T_s^4\begin{bmatrix}0\\0\\1\end{bmatrix}$	0
Entropy	$\frac{1}{2}\frac{r^2}{R^2}\frac{1}{2}\frac{4}{3}\sigma T_s^3$	$\frac{1}{2}\frac{r^2}{R^2}c\frac{1}{2}\frac{4}{3}\sigma T_s^3\begin{bmatrix}0\\0\\1\end{bmatrix}$	0

Table 13.6 Thermodynamic fields of radiation far from a spherical source.

8.5 Absorption of Radiation from a Spherical Source in an Eulerian Fluid at Temperature T

We recall the radiative transfer equation and (8.4), which combine to give

$$\frac{\partial f}{\partial t} + c\cos\beta\frac{\partial f}{\partial z} = \frac{1}{\tau}\left(f_E(T) - f\right), \tag{8.8}$$

which is easily solved in the half-space $z > 0$ for the initial and boundary value problem

$$f(0,z) = f_E(T) \quad , \quad f(t,0) = \begin{cases} f_E(T) & t \leq 0 \\ f_E(T) + F(t,\mathbf{k}) & t > 0. \end{cases} \tag{8.9}$$

The solution reads

$$f(t,z,\mathbf{k}) = \begin{cases} f_E(T) & ct \leq \dfrac{z}{\cos\beta} \\ f_E(T) + F\left(t - \dfrac{z}{c\cos\beta}, \mathbf{k}\right) e^{-\frac{z}{c\,\tau\cos\beta}} & ct > \dfrac{z}{\cos\beta} \end{cases} \tag{8.10}$$

We let the disturbance $F(t,\mathbf{k})$ be furnished by incident radiation from a spherical source with $f_E(T_s)$; see Figure 13.7. In that case we have

$$F(t,\mathbf{k}) = \begin{cases} 0 \\ f_E(T_s) \end{cases} \quad \beta \gtrless \beta_0, \quad \sin\beta_0 = \frac{r}{R}. \tag{8.11}$$

For times $t > z \,/\, c\cos\beta_0$—when light of all directions $\beta \leq \beta_0$ has arrived at the depth z—the phase density is time-independent and we have

$$f(t,z,\mathbf{k}) = \begin{cases} f_E(T) & \beta > \beta_0 \\ f_E(T) + f_E(T_s)\, e^{-\frac{z}{c\,\tau\cos\beta}} & \beta < \beta_0 \end{cases} \tag{8.12}$$

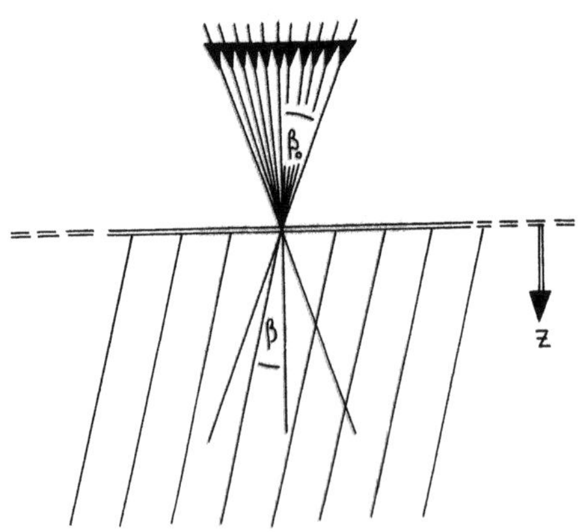

Figure 13.7 Incident radiation from the source.

Table 13.7 shows densities, fluxes, and productions for this distribution function, albeit only for the limiting case $r \ll R$, i.e. far away from the source, so that $\cos\beta_0 \approx 1$ can be assumed. Note that all nonequilibrium contributions fall off exponentially with increasing depth because of the accumulative effect of the absorption. Note also that Table 13.7 does not contain entropic terms. These will be separately calculated in the next section; they are more difficult than the entries of Table 13.7, because of their nonlinear dependence on f.

Radiation	Density	Flux
Number	$2\frac{15\zeta(3)}{\pi^4}\frac{\sigma}{k_B}\left[T^3+\frac{1}{4}\frac{r^2}{R^2}T_s^3\, e^{-\frac{z}{c\tau}}\right]$	$\left(0,0,c\frac{15\zeta(3)}{\pi^4}\frac{\sigma}{k_B}\frac{1}{2}\frac{r^2}{R^2}\, e^{-\frac{z}{c\tau}}\right)$
Momentum	$\left(0,0,\frac{1}{2c}\sigma T_s^4\frac{1}{2}\frac{r^2}{R^2}\, e^{-\frac{z}{c\tau}}\right)$	$\frac{1}{3}\sigma T^4\delta_{ij}+\frac{1}{4}\sigma T_s^4\begin{bmatrix}0 & & \\ & 0 & \\ & & \frac{r^2}{R^2}\, e^{-\frac{z}{c\tau}}\end{bmatrix}$
Energy	$\sigma\left[T^4+\frac{1}{4}\frac{r^2}{R^2}T_s^4 e^{-\frac{z}{c\tau}}\right]$	$\left(0,0,\frac{c}{2}\sigma T_s^4\frac{1}{2}\frac{r^2}{R^2}\, e^{-\frac{z}{c\tau}}\right)$

Production
$-\frac{1}{\tau}\frac{15\zeta(3)}{\pi^4}\frac{\sigma}{k_B}T_s^3\frac{r^2}{R^2}e^{-\frac{z}{c\tau}}$
$\left(0,0,-\frac{1}{\tau}\frac{1}{2c}\sigma T_s^4\frac{1}{2}\frac{r^2}{R^2}e^{-\frac{z}{c\tau}}\right)$
$-\frac{1}{\tau}\frac{1}{2c}\sigma T_s^4\frac{1}{2}\frac{r^2}{R^2}e^{-\frac{z}{c\tau}}$

Table 13.7 Thermodynamic fields of radiation of incident rays.

8.6 Entropic Production for Incident Rays

We write the total entropy production of matter and radiation as dictated by Table 13.1 and 13.2 and for the distribution function (8.12) of the incident rays. At the penetration depth z we obtain

$$\sigma_{mr}+\sigma_{rm}=r\left\{\frac{1}{T}-\frac{k_B\frac{1}{\tau}\int\limits_0^{\beta_0}\ln\left(1+\frac{y}{f_E(T)+f_E(T_s)\, e^{-\frac{z}{c\tau}\frac{1}{\cos\beta}}}\right)f_E(T_s)\, e^{-\frac{z}{c\tau}\frac{1}{\cos\beta}}d\mathbf{k}}{\frac{1}{\tau}\int\limits_0^{\beta_0}\hbar ck\, f_E(T_s)\, e^{-\frac{z}{c\tau}\frac{1}{\cos\beta}}d\mathbf{k}}\right\} \tag{8.13}$$

Note that the denominator of the second term is equal to $-r$ and thus cancels with the overall factor r. We have introduced it only to make the whole entropy production formally proportional to r. Far from the source the equation looks a little simpler, viz.

$$\sigma_{mr} + \sigma_{rm} \underset{r \ll R}{\longrightarrow} r \left\{ \underbrace{\frac{1}{T} - \frac{15}{\pi^4} \frac{T^3}{T_s^4} \int_0^\infty \ln \left(1 + \frac{1}{\frac{1}{e^\alpha - 1} + \frac{1}{e^{\alpha\, T/T_s} - 1} e^{-\frac{z}{cT}}} \right) \frac{\alpha^2}{e^{\alpha\, T/T_s} - 1} d\alpha}_{1/\Theta(z)} \right\} \tag{8.14}$$

$$\sigma_{mr} + \sigma_{rm} = r(z) \left(\frac{1}{T} - \frac{1}{\Theta(z)} \right), \tag{8.15}$$

where we have introduced a *pseudo-temperature* $\Theta(z)$. Also the z dependence of r (cf. Table 13.7) has been emphasized. The pseudo-temperature joins the temperature of the fluid in determining the total entropy production. That production is proportional to the difference of the reciprocals of the two temperatures.

8.7 Pseudo-Temperature

It is obvious from (8.14) that the pseudo-temperature depends on T and T_s—apart from z, of course. The integral cannot be solved analytically, but two limiting cases, which can be solved, are worthy of consideration—the cases $z \to 0$ and $z \to \infty$. We have

$$\begin{aligned} \frac{1}{\Theta} &\underset{z \to \infty}{\longrightarrow} \frac{15}{\pi^4} \frac{T^3}{T_s^4} \int_0^\infty \frac{\alpha^3 d\alpha}{e^{\alpha\, T/T_s} - 1} \\ \frac{1}{\Theta} &\underset{z \to 0}{\longrightarrow} \frac{15}{\pi^4} \frac{T^3}{T_s^4} \int_0^\infty \ln \left(1 + \frac{(e^\alpha - 1)\left(e^{\alpha\, T/T_s} - 1\right)}{e^\alpha + e^{\alpha\, T/T_s} - 2} \right) \frac{\alpha^2}{e^{\alpha\, T/T_s} - 1} d\alpha \, . \end{aligned} \tag{8.16}$$

Easy integrations provide

$$\begin{aligned} \Theta\left(z \to \infty; \frac{T}{T_s}\right) &= T \\ \Theta\left(z \to 0; \frac{T}{T_s}\right) &= \left[\begin{array}{ll} T_s & T \ll T_s \\ T + \left(4 - \frac{360}{\pi^4}\zeta(5)\right) T_s, & T \gg T_s \end{array} \right. \end{aligned} \tag{8.17}$$

where $\zeta(5) = 1.0369...$ All else must be calculated numerically and Figure 13.8 shows the results.

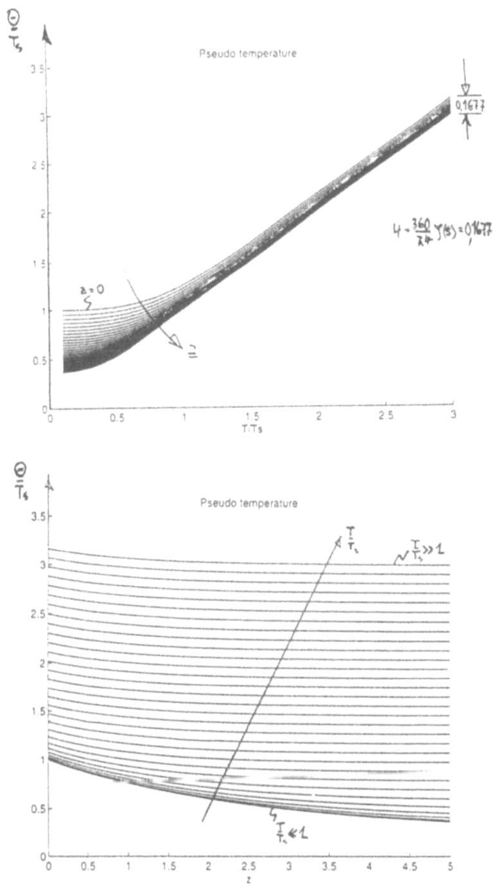

Figure 13.8 Pseudo-temperature as function of T/T_s and z.

We itemize some obvious observations.

- For the solar radiation absorbed by the earth—where $T \ll T_s$ holds—we have

$$\Theta = T_s \quad \text{on the surface}$$
$$\Theta = T \quad \text{at greater depth.}$$

- For a cold body's radiation absorbed by a hot body—where $T_s \ll T$ holds—we have

$$\begin{array}{ll} \Theta = T + 0.1677..T_s & \text{on the surface} \\ \Theta = T & \text{at greater depth.} \end{array}$$

- In both cases the decay of Θ from the surface into the interior is roughly exponential.

We may thus conclude that "hot radiation" to a cold body and "cold radiation" to a hot body have quite different entropic effects

Figure 13.9 shows the entropy production as a function of depth. The figure represents the case $r \ll R$ and $T \ll T_s$. It implies

$$\sigma_{rm} + \sigma_{mr} = \begin{bmatrix} r(0)\left(\frac{1}{T} - \frac{1}{T_s}\right) & \text{at the surface} \\ 0 & \text{for } z \to \infty \end{bmatrix} .$$

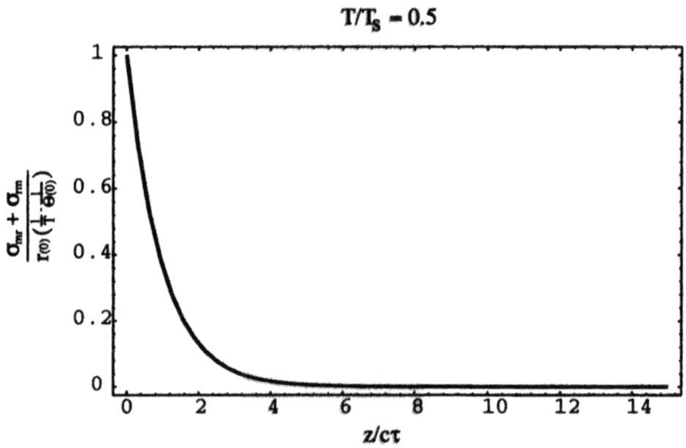

Figure 13.9 Entropy production as a function of depth.

8.8 Entropy Flux and Entropy

We continue to consider the situation represented in Figure 13.5 with (8.12) as the distribution function. We obtain from Table 13.1

$$\varphi_z = -k_B\, c \int \frac{k_z}{k}\left(\ln\frac{f}{y} - \left(\frac{y}{f} + 1\right)\ln\left(1 + \frac{f}{y}\right)\right) f d\mathbf{k} \tag{8.18}$$

$$= -2\pi k_B\, c \int_0^{\pi} \int_0^{\infty} k^2 \cos\beta \left(\ln\frac{f}{y} - \left(\frac{y}{f} - 1\right)\ln\left(1 + \frac{f}{y}\right)\right) f \sin\beta d\beta dk$$

which we solve—numerically—for $r \ll R$ and $T \ll T_s$; this is the case of absorption of solar radiation by the earth. The diagrams of Figure 13.10 show the results. In Figure 13.10a, we see φ_z as a function of z. In this limit, of course, φ_z is equal to the difference $h - h_E$ to within a factor c, and both drop off with increasing depth as must be expected. More interesting is the plot in Figure 13.10b, which represents the ratio of φ_z to q_z/Θ, which starts out at the value 4/3 at $z = 0$ and decreases to 1 with increasing depth. Together with the result (8.17) we conclude that

$$\varphi_z \underset{z\to 0}{\longrightarrow} \frac{4}{3}\frac{q_z}{T_s} \quad \text{and} \quad \varphi_z \underset{z\to\infty}{\longrightarrow} \frac{q_z}{T} . \tag{8.19}$$

Thus the entropy flux is not given by the well-known relation q/T except where the radiation approaches equilibrium, i.e. for $z \to \infty$.

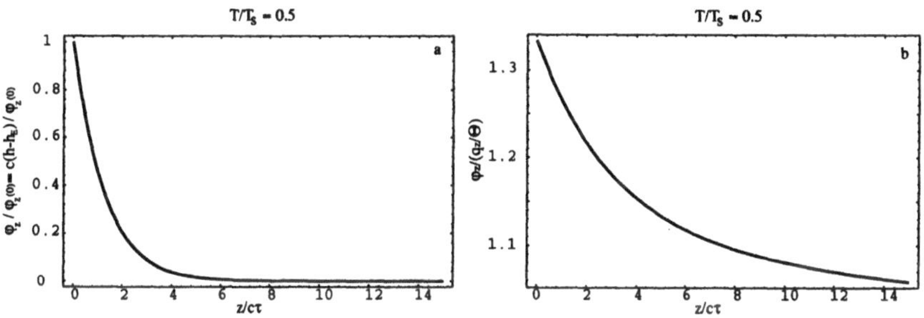

Figure 13.10 Entropy flux as a function of depth.

Chapter 14

Extended Thermodynamics of Phonons

by **H. Struchtrup**, Thermodynamics, TU Berlin

Phonons are much like photons which – as we have seen – are much like atoms of a gas. Infact phonons are much more like particles of a gas than photons, because they can exchange energy and momentum among themselves, which photons cannot do. Because of the similarity to gases it is possible to formulate a kinetic theory of the phonon gas, based on a phonon transfer equation. Peierls [157], [158] was the first to do that. And that equation may serve as the starting point for an extended thermodynamics of phononic moments. This theory was developed by Struchtrup [159] and Dreyer & Struchtrup [160].

Extended Thermodynamics of phonons is used here to describe the thermal propagation that occurs in low-temperature crystals, in particular in the so-called heat-pulse experiments. The theory covers most phenomena that are observed in that experiment, namely: ballistic phonons, second sound and thermal diffusion. All of these phenomena have their proper range and the ranges depend on the mean free paths of phonons between interactions among themselves and with the impurities of the body.

1 Phonon Transfer Equation

1.1 Energy and Momentum of Phonons

Many solid bodies consist of a crystalline lattice of atoms; this lattice is completely regular and uniform if the body is a single crystal. Nevertheless, in regard to the propagation of sound and heat such bodies may be viewed as a gas of phonons. Thus a wave of frequency ω and wave vector $\mathbf{k}$ is replaced by a phonon of energy $\hbar\omega$ and momentum $\hbar\mathbf{k}$. This is much like with photons representing electromagnetic radiation, but there are differences: For one thing the dispersion relation $\omega(k)$ is no longer a linear relation, except for small values of k. Also there are three types of phonons, corresponding to longitudinal and (two) transversal sound waves. And above all, phonons do interact among themselves.

In this chapter we ignore most of the intricacies of the phonon model. Thus we stick to small momenta and assume

$$\omega = ck \tag{1.1}$$

for all three types of phonons, where c is the Debye speed, i.e. a mean value of the longitudinal and transversal sound speeds; see Debye [161]. Thus we have for the phonon speed $\partial\omega/\partial k = c$.

In the physics of crystalline solids with lattice parameters a_i the components k_i of the wave vector range between $-a_i/\pi$ and a_i/π, the so-called Brillouin zone. Here, however, we let k_i range from $-\infty$ to $+\infty$. This is justified – approximately – at low temperatures, where we have no, or few phonons with large momenta.

1.2 Phonon Transfer Equation, Energy and Momentum

Since there are many phonons in the body, we introduce a phase density such that

$$f(x_i, t, k_i)\ d\mathbf{x}d\mathbf{k} \tag{1.2}$$

is the number density of phonons of wave vector $\mathbf{k}$ at the point $\mathbf{x}$ and time t. The phase density obeys a transfer equation with a production term, viz.

$$\frac{\partial f}{\partial t} + cn_k \frac{\partial f}{\partial x_k} = \mathcal{S}, \tag{1.3}$$

where $\mathbf{n}$ is a unit vector in the direction of $\mathbf{k}$. This equation is called the phonon transfer equation, and $\mathcal{S}$ is the phase density of production.

The densities of energy and momentum of the phonon gas are given as moments of the phase density. We have

$$e = \int \hbar c k f\, d\mathbf{k} \qquad \text{and} \qquad p_i = \int \hbar k_i f\, d\mathbf{k}. \tag{1.4}$$

1.3 The Phase Density of Production

If the crystalline lattice of the body is regular and undisturbed by impurities and grain boundaries and if there are few phonons, the phonons move through the crystal with the speed c and without any interaction. We then speak of the ballistic flight of the phonons or ballistic phonons.

In general a phonon will suffer collisions of two types:

- Collisions with other phonons which are similar to collisions between atoms in a gas. In particular energy and momentum are conserved in such collisions, which are called normal-processes or N-processes.[1]
- Collisions where energy is conserved, but momentum is not. These are collisions with impurities and grain boundaries of the body. They are called resistive processes or R-processes.

As a result of these collisions the phase density tends to an equilibrium. However, the N-processes favor a different equilibrium than the R-processes. Let us consider:

The body is at rest and, if the photon gas has equilibrated itself—through R-processes—it must have zero momentum so that the phase density reads (cf. Paragraph 13.2.2)[2]

$$f_R = \frac{y}{\exp\frac{\Lambda_R^1 k}{k_B} - 1}. \tag{1.5}$$

On the other hand, if the phonon gas has equilibrated itself by N-processes, it will in general have a momentum p_i and the phase density will assume the form (cf. Paragraph 13.2.2)

$$f_N = \frac{y}{\exp\left(\frac{\Lambda_N^1 k + \Lambda_{iN}^1 k n_i}{k_B}\right) - 1}. \tag{1.6}$$

Thus the R-processes will attempt to make the phonon gas approach f_R and the N-processes will try to force it toward f_N. The Callaway model combines the two tendencies and assumes a phase density of production of the form [162]

$$\mathcal{S} = -\frac{1}{\tau_R}(f - f_R) - \frac{1}{\tau_N}(f - f_N). \tag{1.7}$$

In general both relaxation times τ will depend on k, but here we consider them constants. τ_R may be interpreted as a mean time of free flight of phonons between succesive R-processes and τ_N as the mean time between N-processes. The values of the τ's and—in particular—their relative values will determine the behaviour of the phonon gas.

Since the R-processes conserve energy, while the N-processes conserve energy *and* momentum we must have

$$\int \hbar ck\,(f - f_R)\,d\mathbf{k} = 0, \qquad \int \hbar ck\,(f - f_N)\,d\mathbf{k} = 0, \quad \text{and} \quad \int \hbar k n_i\,(f - f_N)\,d\mathbf{k} = 0, \tag{1.8}$$

whence follows by (1.4) and (1.5), (1.6)

$$\Lambda_R^1 = \hbar c\left(\frac{a}{e}\right)^{1/4}, \qquad \Lambda_N^1 \simeq \hbar c\left(\frac{a}{e}\right)^{1/4}, \qquad \Lambda_{iN}^1 \simeq -\hbar c\left(\frac{a}{e}\right)^{1/4}\frac{3}{4}\frac{cp_i}{e}. \tag{1.9}$$

[1]There is also a type of phonon-phonon interaction in which momentum is not conserved. These are called Umklapp-processes and they do not concern us here, since we are considering phonons of small momentum only.

[2]To a certain extent phonons are like photons; both are Bosons. The only difference here is the value of y, which is equal to $3/(2\pi)^3$ now, rather than $2/(2\pi)^3$ for photons. This is because there are no "longitudinal photons."

The approximation implied by $(1.9)_{2,3}$ is due to a linearization: the results $(1.9)_{2,3}$ are valid to within linear terms in p_i. The identification of Lagrange multipliers in (1.9) is the same as for photons in Paragraph 13.2.2—apart from the linearization.

In the phonon field it is assumed that the temperature T of the body determines the energy density of the phonons so that we have[3]

$$e = aT^4 \qquad \text{where} \qquad a = \frac{4\pi^5}{5}\frac{k_B^4}{\hbar^3 c^3} \tag{1.10}$$

In summary we may thus write the phase densities f_R and f_N in the forms

$$f_R = \frac{y}{\exp\left(\frac{\hbar ck}{k_B T}\right) - 1}, \qquad f_N = \frac{y}{\exp\left\{\frac{\hbar ck}{k_B T}\left(1 - \frac{3}{4}\frac{cp_i n_i}{aT^4}\right)\right\} - 1}. \tag{1.11}$$

2 Moments and Moment Equations

2.1 Moments and their Equilibrium Values

We introduce the relevant moments of the phase density by the equation

$$u_{\langle i_1 i_2 \cdots i_n \rangle} = \int k n_{\langle i_1} \ldots n_{i_n \rangle} f \, d\mathbf{k}. \tag{2.1}$$

In comparison with the photon moments in—Paragraph 13.1.3—we remark that (2.1) contains only the $R = 1$ hierarchy. This is sufficient here, because the moments of all other hierarchies are uncoupled from the $R = 1$ hierarchy as long as we assume the relaxation times to be independent of the wave number k.

The first few moments are energy density, momentum density, energy flux, and pressure deviator or, in that order

$$e = \hbar c u, \qquad p_i = \hbar u_i, \qquad Q_i = \hbar c^2 u_i, \qquad N_{\langle ij \rangle} = \hbar c u_{\langle ij \rangle}. \tag{2.2}$$

2.2 Moment Equations and Conservation Laws

Multiplication of the phonon transfer equation ((1.3)), ((1.7)) with $k n_{\langle i_1} \ldots n_{i_n \rangle}$, and integration over $\mathbf{k}$ provides the moment equations

$$\frac{\partial u_{\langle i_1 i_2 \cdots i_n \rangle}}{\partial t} + \frac{n}{2n+1} c \frac{\partial u_{\langle\langle i_1 i_2 \cdots i_{n-1} \rangle}}{\partial x_{i_n \rangle}} + c \frac{\partial u_{\langle i_1 i_2 \cdots i_n k \rangle}}{\partial x_k} = -\int k n_{\langle i_1} \ldots n_{i_n \rangle} \left[\frac{1}{\tau_R}(f - f_R) + \frac{1}{\tau_N}(f - f_N)\right] d\mathbf{k}. \tag{2.3}$$

One of these equations is a conservation law, namely the one for $n = 0$, the energy equation. This follows from the conservation laws $(1.8)_{1,2}$. Unlike energy, momentum is not conserved. Indeed

[3]The constant a is different between phonons and photons by the factor 3/2, because of the difference in y.

we have, because of $(1.9)_2$, $(1.8)_3$,

$$-\int kn_i\left[\frac{1}{\tau_R}(f-f_R)+\frac{1}{\tau_N}(f-f_N)\right]d\mathbf{k} = -\frac{1}{\tau_R}\int kn_i(f-f_R)\,d\mathbf{k} \tag{2.4}$$

$$= -\frac{1}{\tau_R}\int kn_i f d\mathbf{k} \tag{2.5}$$

$$= -\frac{1}{\tau_R}u_i \tag{2.6}$$

$$= -\frac{1}{\tau_R}\frac{p_i}{\hbar} \tag{2.7}$$

due to the definition (2.1) and the isotropy of f_R.

Concerning the productions of the other moments it is obvious from (2.1) and (1.5) that we have

$$u_{\langle i_1 i_2\cdots i_n\rangle|R} = \int kn_{\langle i_1}\dots n_{i_n\rangle} f_R d\mathbf{k} = 0 \qquad \text{for} \qquad n \geq 1, \tag{2.8}$$

since f_R is isotropic. In contrast to this the moments $u_{\langle i_1 i_2\cdots i_n\rangle|N}$ for $f = f_N$ and $n \geq 2$ do not vanish but they are nonlinear in p_i. Therefore we have

$$u_{\langle i_1 i_2\cdots i_n\rangle|N} = \int kn_{\langle i_1}\dots n_{i_n\rangle} f_N d\mathbf{k} \simeq 0 \qquad \text{for} \qquad n \geq 2. \tag{2.9}$$

In summary we may write

$$\frac{\partial u_{\langle i_1 i_2\cdots i_n\rangle}}{\partial t} + \frac{n}{2n+1}c\frac{\partial u_{\langle\langle i_1 i_2\cdots i_{n-1}\rangle}}{\partial x_{i_n\rangle}} + c\frac{\partial u_{\langle i_1 i_2\cdots i_n k\rangle}}{\partial x_k} = \begin{cases} 0 & n=0 \\ -\dfrac{1}{\tau_R}u_i & n=1 \\ -\left(\dfrac{1}{\tau_R}+\dfrac{1}{\tau_N}\right)u_{\langle i_1 i_2\cdots i_n\rangle}. & n\geq 2 \end{cases} \tag{2.10}$$

Obviously the system is not closed at any value n.

2.3 Closure Problem

To achieve closure of the system (2.10) we might use the maximization of entropy as we did for photons. This is unnecessary, however, since we also assume that $u_{\langle i_1 i_2\cdots i_N k\rangle}$ is a constitutive quantity, dependent on the lower moments $u_{\langle i_1 i_2\cdots i_n\rangle}$ $(n = 0, 1, \dots, N)$. Indeed, all of these variables—dependent and independent ones—are traceless, so that there is no way to form linear combinations between them. Therefore we obtain immediately in the present linear theory

$$u_{\langle i_1 i_2\cdots i_N k\rangle} = 0. \tag{2.11}$$

Thus the system (2.10) is closed.

3 The Heat Pulse Experiment

3.1 Experimental Results and One-Dimensional Equations

The early theory of transport by phonons by Ward & Wilks [163] predicted a second sound in solids, namely a pressure wave in the phonon gas; and the first people to look for it were Gutfeld & Nethercot [164]. Although they did find ballistic phonons, these authors did not see the predicted second sound. More theory was needed to identify more closely the propagation conditions and that theory was provided by Guyer & Krumhansl [165], [166], [167]. After their work, observations followed rapidly for various materials, viz. observations of solid He by Ackermann & Bertman [168] and Ackermann & Guyer [169]. McNelly et al. [170], [171] observed second sound in NaF and Narayanamurti & Dynes [172] saw it in bismuth.

The experimental setup is easy: Thin metallic films are deposited from the vapor phase on opposite sides of a crystal. These serve for electric heating on one side and as resistance thermometers on the other. A very short electric pulse produces phonons at the heating end and these will be detected at the other end. In order to minimize the effect of R-processes the experiments are performed with single crystals of high purity. The result of two such experiments is shown in Figure 14.1. They are taken from articles by Jackson, Walker and McNelly [173], [171]. A temperature scale is not given, but the height of the peaks must be considered to be considerably smaller than the temperatures listed on the curves.

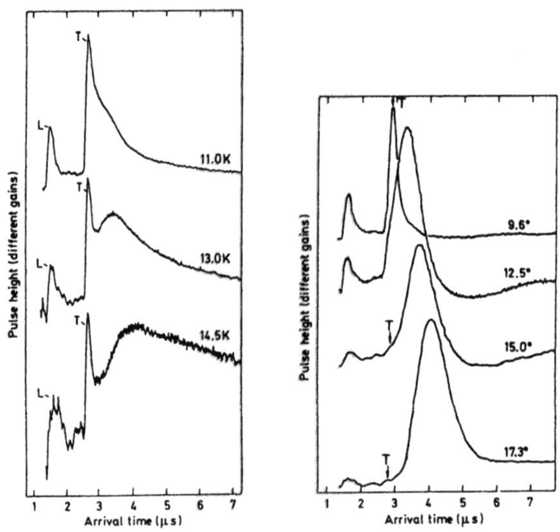

Figure 14.1 Phonon propagation in NaF
Left: Pure NaF. Jackson & Walker [173]
Right: Very pure NaF. Jackson, Walker & McNelly [171]
(T marks the arrival of the ballistic transversal phonons).

In all curves on the left-hand side of Figure 14.1 the first signal to arrive at the thermometer is the peak L. This peak is due to ballistic phonons that have crossed the crystal without any interaction at the speed of the longitudinal sound. A second peak T follows a little later and this is due to transversal ballistic phonons.

The third arrival is the wide peak—clearly seen only for $13K$ and $14.5K$—which is interpreted as partially developed second sound. In all curves we also see a diffusive wake which stems from phonons scattered by impurities of the crystal and other R-processes.

On the right-hand side of Figure 14.1—which shows measurements in a much purer crystal[4]—the wake can hardly be noticed. The ballistic peaks are still there at $9.6K$, but they become weaker at higher crystal temperatures. Starting with $12.5K$ the peak attributed to the second sound is well pronounced and persists up to $17.3K$. The speed of these peaks seems to decrease with increasing temperature.

We shall now proceed to show that extended thermodynamics of phonons is capable of describing the salient features of these experiments. For that purpose we rewrite the field equations (2.10), (2.11) in one-dimensional form as is appropriate for the one-dimensional heat pulse experiment. With $u_{\langle n\rangle} = u_{\langle 11\cdots 1\rangle}$ as the 1-component of tensors of rank n we obtain

$$\frac{\partial u_{\langle n\rangle}}{\partial t} + \frac{n^2}{4n^2-1} c \frac{\partial u_{\langle n-1\rangle}}{\partial x} + c \frac{\partial u_{\langle n+1\rangle}}{\partial x} = \begin{cases} 0 & n = 0 \\ -\dfrac{1}{\tau_R} u_{\langle 1\rangle} & n = 1 \\ -\left(\dfrac{1}{\tau_R} + \dfrac{1}{\tau_N}\right) u_{\langle n\rangle}. & 2 \leq n \leq N \end{cases} \tag{3.1}$$

And, of course $u_{\langle N+1\rangle} = 0$.

3.2 Ballistic Phonons

Ballistic phonons are those that do not interact, neither among themselves nor with the crystal. Thus the system of equations (3.1) is homogeneous in this case. And in fact, it is formally the same system that was used in Paragraph 13.7.2 to calculate characteristic speeds for radiation.

And when in Chapter 13 we had to require that the largest characteristic speed should be equal to the speed of light, we must require here that the largest speed is the sound speed c. This in fact is a criterion for the number N of equations that must be taken into account. We have indicated in Paragraph 13.7.2 that $N = 30$ is the proper number; see also Figure 13.4.

In order to avoid misunderstanding we must emphasize that in the present theory there is only one type of ballistic phonons. We have not taken longitudinal and transversal phonons into account, each endowed with a speed of their own. Therefore the calculated curves in subsequent figures have only one ballistic peak – the sharp one wherever it is visible.

[4]The results are not easy to compare, since the two crystals in Figure 14.1 have different size. Furthermore the duration of the heat pulse affects the detected signal considerably.

3.3 Second Sound in Its Purest Form

Second sound can only occur when there is frequent normal interaction between the phonons. That is to say that τ_N—the mean time of free flight between N-processes—must be small. This leads to a dramatic reduction in the equations (3.1), where the term $\frac{1}{\tau_R}$ may be neglected in comparison with $\frac{1}{\tau_N}$.

We shall also neglect the time derivative $\partial u_{\langle n \rangle}/\partial t$ in comparison with $-u_{\langle n \rangle}/\tau_N$ for $n \geq 2$, where τ_N occurs. Ignoring higher-order terms in τ_N we thus obtain

$$\begin{aligned} u_{\langle N \rangle} &= -\tau_N \frac{N^2}{4N^2-1} c \frac{\partial u_{\langle N-1 \rangle}}{\partial x} \\ u_{\langle N-1 \rangle} &= -\tau_N \frac{(N-1)^2}{4(N-1)^2-1} c \frac{\partial u_{\langle N-2 \rangle}}{\partial x} \\ &\vdots \\ u_{\langle 3 \rangle} &= -\tau_N \frac{3^2}{4 \cdot 3^2 - 1} c \frac{\partial u_{\langle 2 \rangle}}{\partial x} \end{aligned} \tag{3.2}$$

so that, once $u_{\langle 2 \rangle}$ is known, all higher moments may be determined by differentiation. What remains is the simple system consisting of the first three equations in (3.1) with the third one approximated in the same manner as the equations (3.2). We obtain

$$\begin{aligned} \frac{\partial e}{\partial t} \qquad\qquad + c^2 \frac{\partial p}{\partial x} &= 0 \\ \frac{\partial p}{\partial t} + \frac{1}{3}\frac{\partial e}{\partial x} + \frac{\partial N}{\partial x} &= -\frac{1}{\tau_R} p \\ \frac{4}{15} c^2 \frac{\partial p}{\partial x} &= -\frac{1}{\tau_N} N. \end{aligned} \tag{3.3}$$

p and N stand for the 1-components of p_i and $N_{\langle ij \rangle}$ respectively.

This is the system of equations proposed for the second sound by Guyer & Krumhansl [165] through [167], albeit with a different motivation.

We consider the case $\tau_R \to \infty$ and $\tau_N \to 0$, which is ideal for the propagation of the second sound. In this case we must obviously require $N = 0$ in (3.3) and obtain two equations, viz.

$$\frac{\partial e}{\partial t} + c^2 \frac{\partial p}{\partial x} = 0 \quad \text{and} \quad \frac{\partial p}{\partial t} + \frac{1}{3}\frac{\partial e}{\partial x} = 0. \tag{3.4}$$

These imply the wave equation

$$\frac{\partial^2 e}{\partial t^2} - \frac{c^2}{3}\frac{\partial^2 e}{\partial x^2} = 0 \tag{3.5}$$

and predict the speed of the ideal second sound to be equal to $c/\sqrt{3}$.

3.4 Damped Second Sound and Pure Diffusion

If we do not go quite so far in reducing the system (3.3) but allow τ_R to be finite, yet still assume $\tau_N \to 0$, we obtain

$$\frac{\partial e}{\partial t} + c^2 \frac{\partial p}{\partial x} = 0 \qquad \text{and} \qquad \frac{\partial p}{\partial t} + \frac{1}{3}\frac{\partial e}{\partial x} = -\frac{1}{\tau_R} p. \tag{3.6}$$

Equation $(3.6)_2$ is the Cattaneo equation—in essence—and the two equations together imply the telegraph equation

$$\frac{\partial^2 e}{\partial t^2} + \frac{1}{\tau_R}\frac{\partial e}{\partial t} - \frac{c^2}{3}\frac{\partial^2 e}{\partial x^2} = 0. \tag{3.7}$$

This equation describes damped waves, still propagating with the speed $c/\sqrt{3}$.

On the other hand, if τ_R is very small—along with τ_N—we may neglect $\partial p/\partial t$ in comparison with p/τ_R and obtain from (3.6)

$$\frac{\partial e}{\partial t} + c^2 \frac{\partial p}{\partial x} = 0 \qquad \text{and} \qquad \frac{1}{3}\frac{\partial e}{\partial x} = -\frac{1}{\tau_R} p \tag{3.8}$$

which leads to the diffusion equation, or heat conduction equation, viz.

$$\frac{\partial e}{\partial t} = \frac{\tau_R c^2}{3}\frac{\partial^2 e}{\partial x^2}. \tag{3.9}$$

In this case there is so much interaction, both among the photons and between photons and crystal, that no wavelike propagation can occur. The propagation which does occur is diffusive.

3.5 The 9-Field Theory of Extended Thermodynamics

So far we have considered special cases in which either the ballistic phonons, or the second sound, or pure diffusion occurred. This was to show that the theory contains all these phenomena. In reality, however, these phenomena all occur at the same time, and—in order to illustrate that—we investigate the 9-field theory in which the state of the phonon gas is fully characterized by the fields

energy density e, momentum density p_i, pressure deviator $N_{\langle ij \rangle}$ (3.10)

In this case the one-dimensional system (3.1) reduces to the three equations

$$\begin{aligned}
&\frac{\partial e}{\partial t} & & + c^2\frac{\partial p}{\partial x} &&= 0 \\
&\frac{\partial p}{\partial t} & + \frac{1}{3}\frac{\partial e}{\partial x} & + \frac{\partial N}{\partial x} &&= -\frac{1}{\tau_R} p \\
&\frac{\partial N}{\partial t} & + \frac{4}{15} c\frac{\partial p}{\partial x} & &&= -\left(\frac{1}{\tau_R} + \frac{1}{\tau_N}\right) N
\end{aligned} \tag{3.11}$$

We already know from Paragraph 3.2 that this system is not good enough for ballistic phonons, because it would require 30 moments to describe their speed correctly. And yet the system (3.11) is instructive, because it will permit a view on the coupling of phenomena of propagation and diffusion that occurs in crystalline bodies at low temperature – including ballistic phonons, albeit with the wrong speed.

We consider small harmonic waves with the frequency Ω and the wave number q as solutions of the system (3.11). The dispersion relation reads

$$q^2 = \frac{3}{c^2}\frac{2\Omega\tau_N + \Omega\tau_R + i\left(\Omega\tau_R\Omega\tau_N - \left(1+\frac{\tau_N}{\tau_R}\right)\right)}{\Omega\tau_N + \Omega\tau_R + i\frac{9}{5}\Omega\tau_R\Omega\tau_N}\Omega^2. \tag{3.12}$$

We note that $q(\Omega)$ is a complex function with real and imaginary parts both dependent on Ω. Thus there is dispersion and absorption. The dispersion relation (3.12) covers the whole range of frequencies and contains τ_R and τ_N as parameters. So, all aspects of propagation and diffusion are combined in this formula. We proceed to pull out—again—some limiting cases since, after all, these are the most instructive features of (3.12).

i.) (Ballistic Phonons)
For $\tau_R \to \infty$ and $\tau_N \to \infty$ the phonons suffer no interaction, and (3.12) reduces to

$$\frac{\Omega}{q} = \sqrt{\frac{3}{5}}c. \tag{3.13}$$

This is the speed of the ballistic phonons in the 9-field theory which, by rights, should be equal to c. We have already discussed the reason for the wrong value of this speed, which is: too few moments. However, at last *we do have* ballistic phonons, while Guyer & Krumhansl [165] through [167] do not get that propagative mode; the reason is that those authors have not included the term $\partial N/\partial t$ in $(3.11)_3$ so that they arrive at a parabolic system with an infinite characteristic speed.

ii.) (Second Sound)
The pure second sound occurs for $\tau_R \to \infty, \tau_N \to 0$, in which case the dispersion relation reads

$$\frac{\Omega}{q} = \frac{1}{\sqrt{3}}c. \tag{3.14}$$

Once again this means propagation without dispersion and damping. More interesting is the less radical case

$$\tau_R\Omega \gg 1, \qquad \text{and} \qquad \tau_N\Omega \ll 1 \tag{3.15}$$

in which the dispersion relation may be approximated as

$$q^2 = \frac{3}{c^2}\left(1 - \frac{4}{5}\frac{\tau_N}{\tau_R} - i\left[\frac{4}{5}\tau_N\Omega + \frac{1}{\tau_R\Omega}\right]\right)\Omega^2. \tag{3.16}$$

Now, obviously there is dispersion and damping of the second sound because of the imaginary term in (3.16); but both are small, because of the inequalities (3.15). Thus for

$$\frac{1}{\tau_R} \ll \Omega \ll \frac{1}{\tau_N} \tag{3.17}$$

the second sound is weakly damped and weakly dispersive and propagates with the velocity

$$v_{Ph} = \frac{c}{\sqrt{3}}\left(1 + \frac{2}{5}\frac{\tau_N}{\tau_R}\right). \tag{3.18}$$

Condition (3.17) is called the window condition in the literature. It serves as a tool to find crystals which will exhibit second sound. For that purpose it is necessary to know the explicit dependence of τ_R, τ_N on crystal properties and temperature, which we cannot present here; see Jackson & Walker [173]. If the relaxation times are known, (3.17) may be used to find pulse durations which will produce mainly waves with frequencies in the window. The window condition is due to Guyer & Krumhansl [165], and it preceded the experimental proof of the second sound.

iii.) (Diffusion)
In the case of small relaxation times, i.e. $\Omega\tau_R \ll 1, \Omega\tau_N \ll 1$ the dispersion relation reduces to the dispersion relation of the heat equation or diffusion equation, viz.

$$q = \sqrt{-\frac{3i\Omega}{c^2\tau_R}}. \tag{3.19}$$

3.6 Heat Pulses. Numerical Solutions.

The system (3.11) is a linear system of partial differential equations which—for given initial and boundary values—may be solved by use of the Fourier transform. The following figures show results from Dreyer & Struchtrup [160], who took initial and boundary data as

$$e\,(x,t=0) = e_0 = \text{const.} \qquad p\,(x,t=0) = 0, \qquad N\,(x,t=0) = 0$$

$$p\,(x=0,t) = \begin{cases} 0 & t < 0 \\ p_0 = ce_0 & 0 \le t \le \Delta t \\ 0 & t \ge \Delta t \end{cases} \qquad p\,(x \to \infty, t) = 0.$$

Δt is the pulse duration. In Figure 14.2 τ_R was determined from measurements of heat conductivity $\kappa = \frac{c^2}{3}\tau_R c_v$ by Jackson & Walker, while $\tau = 1/\left(1/\tau_R + 1/\tau_N\right)$ was fitted in order to obtain a good qualitative agreement with the experimental results shown on the left-hand side of Figure 14.1. The pulse duration is taken as $\Delta t = 10^{-7}s$ and the relaxation times are

$$\tau = \{3.0, 2.1, 1.9\}\,\Delta t, \qquad \tau_R = \{15.4, 10.4, 7.4\}\,\Delta t.$$

In Figure 14.3 the relaxation time for R-processes was kept constant at $\tau_R = 30\Delta t$ and τ_N was changed. For the large value $\tau_N = 4\Delta t$ we observe the ballistic pulse. If τ_N is decreased to $1.8\Delta t$, a second sound peak appears which absorbs the ballistic pulse completely when τ_N is further reduced to $\tau_N = 1.4\Delta t$ or $\tau_N = \Delta t$. Note that the second sound peak is considerably broadened in comparison with the original signal. Once again we have a good qualitative agreement between theory and experiment. Curves like those in Figure 14.3b have not been observed. According to the 9-field theory such curves should occur in the temperature range between $T = 9.6K$ and $T = 12.5K$.

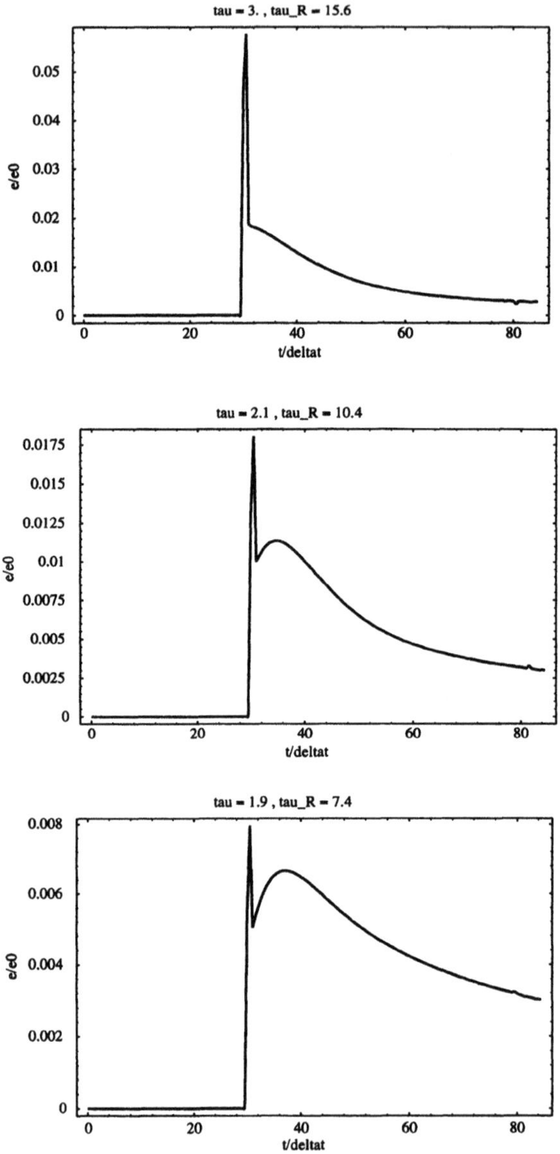

Figure 14.2 Temperature at the detector. Calculated by the 9-field theory. Duration of pulse $\Delta t = 10^{-7}$.

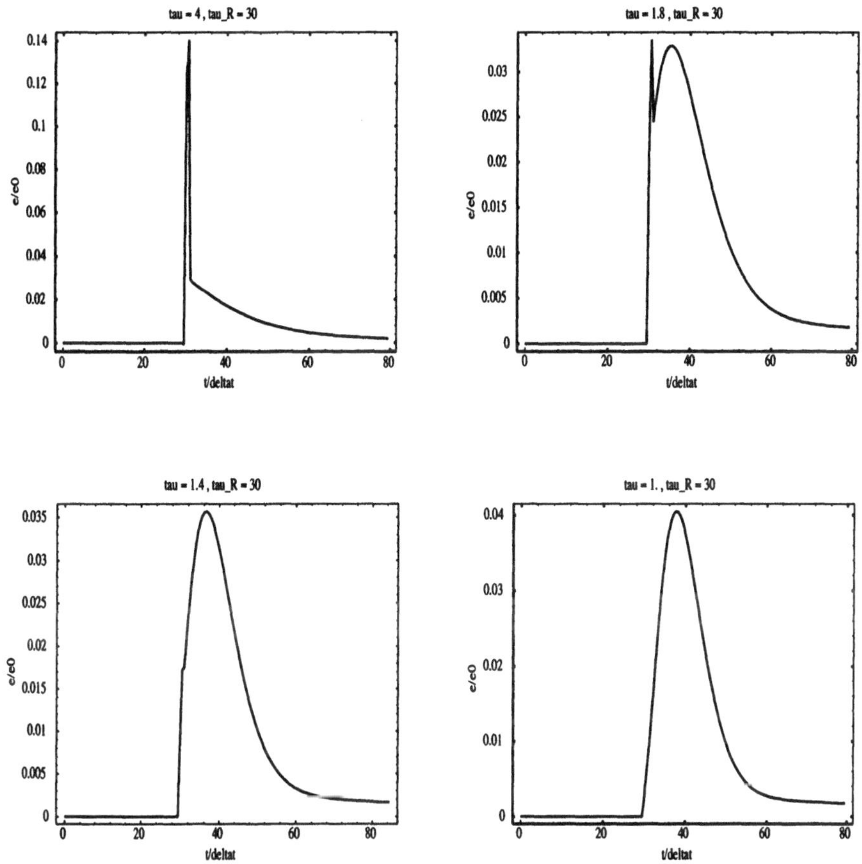

Figure 14.3 Temperature at the detector according to the 9-field theory for different values of τ_N and a big value of τ_R. Duration of pulse $\Delta t = 10^{-7}$.

Chapter 15

Kinetic Theory and Extended Thermodynamics of Metal Electrons

There is one conclusion above all which we may draw from the study of the foregoing chapters: Extended thermodynamics differs from ordinary thermodynamics whenever the constituent particles of the body have large mean free paths, or large mean times of free flight. This situation occurs in dilute gases, and for "gases" of photons and phonons as we have seen, – but also for the electrons in a metallic conductor which carry both the electric current and the heat flux.

The purpose of this section is to call the reader's attention to metal electrons for which an extended thermodynamics has not yet been formulated, although there exists a fairly well-developed kinetic theory.

1 Equations of Balance

1.1 Kinetic Theory of Metal Electrons

When there are many particles in a gas and when they have long mean times of free flight—rarely interrupted by a collision—we may define a phase density; and the phase density will satisfy a transport equation with a collision term, viz.

$$\frac{\partial f}{\partial t} + c_i \frac{\partial f}{\partial x_i} + (f_i + i_i^c) \frac{\partial f}{\partial c_i} = S. \tag{1.1}$$

f is a function of position $\mathbf{x}$, particle velocity $\mathbf{c}$, and time t. f_i and i_i^c are the (specific) external and inertial forces on the particles and S is the collision term.

The "electron gas" in a metal is governed by an equation like this. In this case we consider the external force as an electromagnetic force, so that—with $\mathcal{E}_i$ as the electromotive intensity and B_i as the magnetic flux density—we have

$$f_i = -\frac{q}{m}(\mathcal{E}_i + (\mathbf{c} \times \mathbf{B})_i) \tag{1.2}$$

where $-q$ is the charge of an electron and m is its mass. We neglect the collisions of electrons among themselves so that S is entirely due to collisions which the electrons suffer by interacting with an irregularity in the ionic lattice.[1] Under these circumstances the collision operator reads, e.g. see Sommerfeld [103], or Sommerfeld & Bethe [174]

$$S = \frac{1}{\pi l} \int_0^{2\pi} \int_0^{\pi/2} c \cos\theta (f' - f) \sin\theta d\theta d\varepsilon, \tag{1.3}$$

where l is the mean free path of an electron between interactions—here assumed independent of its speed. $\mathbf{c}$ and $\mathbf{c}'$ are the velocities before and after the collision and f' stands for $f(\mathbf{x}, \mathbf{c}', t)$. The collisions may be assumed to be elastic, because the heavy ions do not move under the impact of the light electron. Therefore we have

$$\mathbf{c}' = \mathbf{c} - 2\mathbf{k}(\mathbf{k} \cdot \mathbf{c}), \tag{1.4}$$

where $\mathbf{k}$ is the collision vector, a unit vector which bisects the angle between $\mathbf{c}$ and $\mathbf{c}'$. The ionic lattice will be considered to be at rest, i.e. $\mathbf{c}$ is the velocity of the electrons with respect to the lattice.

Combining (1.1), (1.2), and (1.3) we obtain

$$\frac{\partial f}{\partial t} + c_i \frac{\partial f}{\partial x_i} + \left[-\frac{q}{m}\ (\mathcal{E}_i + (\mathbf{c} \times \mathbf{B})_i) + i_i^c\right] \frac{\partial f}{\partial c_i} = \frac{1}{\pi l} \int_0^{2\pi} \int_0^{\pi/2} c \cos\theta (f' - f) \sin\theta d\theta d\varepsilon. \tag{1.5}$$

Multiplication by a generic function $\varphi(\mathbf{x}, \mathbf{c}, t)$ and integration over all $\mathbf{c}$ provides after a little calculation

[1] For reasons explained by quantum mechanics the regular periodic lattice does not disturb the free flight of the electrons.

$$\frac{\partial\rho\bar{\varphi}}{\partial t}+\frac{\partial\rho\overline{c_i\varphi}}{\partial x_i}-\rho\overline{\left(\frac{\partial\varphi}{\partial t}+c_i\frac{\partial\varphi}{\partial x_i}+\left[-\frac{q}{m}\ (\mathcal{E}_i+(\mathbf{c}\times\mathbf{B})_i)+i_i^c\right]\frac{\partial\varphi}{\partial c_i}\right)}= \tag{1.6}$$

$$=\frac{1}{\pi l}\int_0^{2\pi}\int_0^{\pi/2}\int_{-\infty}^{\infty} c\cos\theta(\varphi'-\varphi)f\sin\theta d\theta d\varepsilon d\mathbf{c}.$$

$$=\frac{1}{2\pi l}\int_0^{2\pi}\int_0^{\pi}\int_{-\infty}^{\infty} c\cos\theta(\varphi'-\varphi)(f'-f)\sin\theta d\theta d\varepsilon d\mathbf{c}.$$

$\bar{\varphi}$ stands for $\frac{1}{\rho}\int\varphi f d\mathbf{c}$ as in the kinetic theory of gases (see (2.3.12)) and φ' is defined as $\varphi(\mathbf{x},\mathbf{c}',t)$. Equation (1.6) is the generic *equation of transfer for electrons.*

The transfer equation obviously has the character of an equation of balance whose production consists of two terms, which we call

*Eigen*production	$\overline{\frac{\partial\varphi}{\partial t}+c_i\frac{\partial\varphi}{\partial x_i}+\left[-\frac{q}{m}\ (\mathcal{E}_i+(\mathbf{c}\times\mathbf{B})_i)+i_i^c\right]\frac{\partial\varphi}{\partial c_i}}$	and
Collision production	$\frac{1}{\pi l}\int_0^{2\pi}\int_0^{\pi/2}\int_{-\infty}^{\infty} c\cos\theta(\varphi'-\varphi)f\sin\theta d\theta d\varepsilon d\mathbf{c}.$	

Note that the force term is velocity dependent because of the Lorentz force $-q(\mathbf{c}\times\mathbf{B})$ and because of the Coriolis force, which forms part of i_i^c. It is advantageous to render the velocity dependence explicit and write

$$\left[-\frac{q}{m}\ (\mathcal{E}_i+(\mathbf{c}\times\mathbf{B})_i)+i_i^c\right]=-\frac{q}{m}\ \mathcal{E}_i+i_i^0+\left(-\frac{q}{m}\varepsilon_{ijk}B_k+2W_{ij}\right)c_j, \tag{1.8}$$

where i_i^0 is the velocity-independent part of the inertial force.

1.2 Equations of Balance of Mass, Momentum, Energy, and Energy Flux

We set $\varphi=m$, mc_i, $\frac{m}{2}c^2$, and $\frac{m}{2}c^2c_i$ in (1.6) and obtain the conservation laws of mass and energy and the equations of balance of momentum and heat flux. We define

$$\begin{aligned}
\rho &= \int m\, f d\mathbf{c} && - \text{ mass density}\\
J_i &= \int m\, c_i\, f d\mathbf{c} && - \text{ mass flux and momentum density } S_i=-\tfrac{q}{m}J_i \text{ is the electric current density.}\\
P_{ij} &= \int m\, c_i\, c_j\, f d\mathbf{c} && - \text{ momentum flux}\\
e &= \int \tfrac{m}{2}\, c^2\, f d\mathbf{c} && - \text{ energy density}\\
q_i &= \int \tfrac{m}{2}\, c^2 c_i\, f d\mathbf{c} && - \text{ heat flux}\\
\rho_{iijk} &= \int mc^2c_ic_k\, f d\mathbf{c} && - \text{ trace of fourth moment,}
\end{aligned} \tag{1.9}$$

and obtain balance equations of the form

$$\begin{aligned}
&\frac{\partial \rho}{\partial t}+\frac{\partial J_i}{\partial x_i}=0\\
&\frac{\partial J_i}{\partial t}+\frac{\partial P_{ij}}{\partial x_j}-\left[-\frac{q}{m}\,\mathcal{E}_i+i_i^0\right]\rho-\left(-\frac{q}{m}\varepsilon_{ijk}B_k+2W_{ij}\right)J_j=\frac{1}{l}\rho\,\overline{mcc_i}\\
&\frac{\partial e}{\partial t}+\frac{\partial q_i}{\partial x_i}-\left[-\frac{q}{m}\,\mathcal{E}_l+i_l^0\right]J_l=0\\
&\frac{\partial q_i}{\partial t}+\frac{\partial \frac{1}{2}\rho_{llij}}{\partial x_j}-\left[-\frac{q}{m}\,\mathcal{E}_l+i_l^0\right]\left(P_{<li>}+\frac{5}{3}e\delta_{li}\right)-\left(-\frac{q}{m}\varepsilon_{ijk}B_k+2W_{ij}\right)q_j=\frac{1}{l}\rho\,\overline{\frac{m}{2}c^3c_i}.
\end{aligned} \tag{1.10}$$

1.3 Entropy Principle and Phase Density Close to Equilibrium

We know from Paragraph 9.1.3 that the entropy density and the entropy flux of a gas of Fermions[2] is given, in terms of the phase density by

$$\begin{aligned}
h &= -ky\int \left(\frac{f}{y}\ln\frac{f}{y}+\left(1-\frac{f}{y}\right)\ln\left(1-\frac{f}{y}\right)\right)dc,\\
h^i &= -ky\int c_i\left(\frac{f}{y}\ln\frac{f}{y}+\left(1-\frac{f}{y}\right)\ln\left(1-\frac{f}{y}\right)\right)dc.
\end{aligned} \tag{1.11}$$

It is an easy matter to show that the transport equation implies an entropy inequality of the form $\frac{\partial h}{\partial t}+\frac{\partial h_i}{\partial x_i}\geq 0$.

For the present theory we assume that f may be calculated as *the* phase density that maximizes h under the constraints of fixed values of ρ, J_i, e, and q_i. Therefore the phase density assumes the form

$$f=\frac{y}{e^{\frac{1}{k}\chi}+1}\quad\text{where}\quad \chi=\Lambda m+\Lambda_i^J mc+\Lambda_{ii}mc^2+\Lambda_i^q mc^2c_i. \tag{1.12}$$

The Λ's are Lagrange multipliers.

In equilibrium, where J_i and q_i vanish, f must be isotropic and we conclude that $\Lambda_i^J|_E=0$, $\Lambda_i^q|_E=0$ hold. Therefore the equilibrium phase density is given by

$$f_E=\frac{y}{e^{\frac{m}{k}(\Lambda+\Lambda_{ii}c^2)}+1}. \tag{1.13}$$

The Lagrange multipliers Λ and Λ_{ii} follow from the constraints

$$\rho=\int m\,\frac{y}{e^{\frac{m}{k}(\Lambda+\Lambda_{ii}c^2)}+1}dc\quad\text{and}\quad e=\int\frac{m}{2}c^2\,\frac{y}{e^{\frac{m}{k}(\Lambda+\Lambda_{ii}c^2)}+1}dc, \tag{1.14}$$

or else they may be identified in terms of the specific free enthalpy g—or chemical potential—and absolute temperature T by comparison of

$$dh_E=2\Lambda_{ii}de+\Lambda d\rho\quad\text{with the Gibbs equation}\quad dh_E=\frac{1}{T}(de-gd\rho). \tag{1.15}$$

[2]That is what electrons are.

Hence follows

$$\Lambda = -\frac{g}{T} \quad \text{and} \quad \Lambda_{ii} = \frac{1}{2T} \tag{1.16}$$

so that the equilibrium phase density comes out as the Fermi distribution, viz.

$$f_E = \frac{y}{e^{-\frac{mg}{kT}+\frac{mc^2}{2kT}}+1}. \tag{1.17}$$

Close to equilibrium we may expand f to within linear terms in Λ_i^J and Λ_i^q. We obtain

$$f = \frac{y}{e^{-\frac{mg}{kT}+\frac{mc^2}{2kT}}+1} - \frac{ye^{-\frac{mg}{kT}+\frac{mc^2}{2kT}}}{\left(e^{-\frac{mg}{kT}+\frac{mc^2}{2kT}}+1\right)^2}\left(mc_i\Lambda_i^J + mc^2c_i\Lambda_i^q\right). \tag{1.18}$$

2 Extended Thermodynamics and Kinetic Theory

2.1 Toward Extended Thermodynamics of Electrons in Metals

We may define the state of an electron gas by the eight fields

$$\rho, J_i, e, q_i. \tag{2.1}$$

This calls for the formulation of an extended thermodynamics in which the necessary field equations may be based upon the equations of balance (1.10). Obviously the system is not closed. Therefore the

$$\text{fluxes } P_{<ij>}, \rho_{llij} \qquad \text{and productions} \qquad \rho\overline{mcc_i}, \rho\overline{mc^3c_i} \tag{2.2}$$

must be considered as constitutive quantities which we relate to the fields by constitutive functions. In the spirit of extended thermodynamics we write constitutive relations of the generic form

$$C = \hat{C}(\rho, J_i, e, q_i), \tag{2.3}$$

where C stands for any one of the constitutive quantities (2.2); no gradients or time derivatives occur among the variables.

Between the equations of balance (1.10) and the constitutive relations (2.3) we have a system of eight explicit field equations—if the constitutive functions $\hat{C}$ are known explicitly. Every solution of these field equations is called a *thermodynamic process* in an electron gas.

In reality, of course, the constitutive functions are *not known* explicitly, and we may rely on the entropy inequality

$$\frac{\partial h}{\partial t} + \frac{\partial h_i}{\partial x_i} \geq 0 \tag{2.4}$$

to reduce their generality. h and h_i are constitutive quantities which are related to ρ, J_i, e, and q_i, and the inequality (2.4) must hold for all thermodynamic processes.

This strategy has been carried out in many places in this book and we know how to proceed: The entropy inequality will be exploited by use of Lagrange multipliers and these will have to be eliminated or identified. We have seen in several cases that this strategy works well, but it has never been utilized for the electron gas; therefore we pose this problem as homework for the interested reader.

2.2 A Convenient Shortcut via the Kinetic Theory of Electrons

In a linear theory it is obvious, by isotropy, that the constitutive relations must have the forms

$$\begin{aligned} P_{<ij>} &= 0, & \frac{1}{l}\,\rho\overline{mcc_i} &= A_{11}J_i + A_{12}q_i, \\ \rho_{llij} &= a\delta_{ij}, & \frac{1}{l}\,\rho\overline{mc^3c_i} &= A_{21}J_i + A_{22}q_i. \end{aligned} \tag{2.5}$$

$1/A_{11}$ and $1/A_{22}$ are relaxation times for J_i and q_i respectively, and we may assume that they are of the order of magnitude of the mean-time of free flight of the electrons.

The systematic evaluation of the entropy principle would provide us with an explicit expression for the coefficient a that determines ρ_{llij}. We do not pursue that direction; we take a shortcut via the kinetic theory.

First of all we make use of the knowledge that ρ_{llij} is the trace of the fourth moment (cf. $(1.9)_6$) and obviously $a\delta_{ij}$ is its equilibrium value. We calculate that equilibrium value by use of the Fermi distribution and obtain

$$\rho_{llij} = a\delta_{ij} = \frac{1}{3}\int mc^4 \frac{y}{e^{\frac{m}{k}(\Lambda+\Lambda_{ii}c^2)}+1} d\mathbf{c}\delta_{ij} = 4\pi m y \frac{1}{3}\left(2\frac{k}{m}T\right)^{7/2} i_6(\alpha)\delta_{ij}, \tag{2.6}$$

where α stands for -$\frac{mg}{kT}$. For the definition of the special function $i_6(\alpha)$ see (6.4.26), or (4.4.8). In the same places we find ρ and e as functions of T and α as

$$\rho = 4\pi m y\left(2\frac{k}{m}T\right)^{3/2} i_2(\alpha) \quad \text{and} \quad e = 4\pi m y \frac{1}{2}\left(2\frac{k}{m}T\right)^{5/2} i_4(\alpha). \tag{2.7}$$

Next we determine the productions $(2.5)_2$. With (1.18) we have the nonequilibrium phase density and we derive the equations

$$\begin{bmatrix} J_i \\ 2q_i \end{bmatrix} = \begin{bmatrix} K_2 & K_4 \\ K_4 & K_6 \end{bmatrix}\begin{bmatrix} \Lambda_i^J \\ \Lambda_i^q \end{bmatrix} \quad \text{and} \quad \rho m \begin{bmatrix} \overline{cc_i} \\ \overline{c^3c_i} \end{bmatrix} = \begin{bmatrix} K_3 & K_5 \\ K_5 & K_7 \end{bmatrix}\begin{bmatrix} \Lambda_i^J \\ \Lambda_i^q \end{bmatrix}, \tag{2.8}$$

of which the first one results from $(1.9)_{2,5}$. The functions K_A are defined by

$$K_A = -my\int \frac{e^{-\frac{mg}{kT}+\frac{mc^2}{2kT}}}{\left(e^{-\frac{mg}{kT}+\frac{mc^2}{2kT}}+1\right)^2}\, c^A d\mathbf{c}. \tag{2.9}$$

The productions $\rho\overline{mcc_i}$ and $\rho\overline{mc^3c_i}$ follow from (2.8) as functions of J_i and q_i by elimination of the Λ's. We obtain

$$\rho m\begin{bmatrix} \overline{cc_i} \\ \overline{c^3c_i} \end{bmatrix} = K\begin{bmatrix} J_i \\ 2q_i \end{bmatrix} \quad \text{where} \tag{2.10}$$

$$K = \frac{1}{K_2K_6 - K_4^2}\begin{bmatrix} K_3K_6 - K_4K_5 & K_2K_5 - K_3K_4 \\ K_5K_6 - K_4K_7 & K_2K_7 - K_4K_5 \end{bmatrix}.$$

Between (2.6) and (2.10) the constitutive relations (2.5) are now explicit, and so the system is an explicit system of field equation. We proceed to derive some of its properties.

2.3 Characteristic Speeds

We may use the knowledge of the constitutive coefficient a to calculate the characteristic speeds of weak waves propagating into an equilibrium state. The speeds V and jumps [3] $\delta\alpha$, δJ_i, δT, and δq_i follow from the linear algebraic system

$$\begin{bmatrix} -\rho_\alpha V & n_i & -\rho_T V & 0 \\ \frac{2}{3}e_\alpha n_i & -V & \frac{2}{3}e_T n_i & 0 \\ -e_\alpha V & 0 & -e_T V & n_i \\ \frac{1}{2}a_\alpha n_i & 0 & \frac{1}{2}a_T n_i & -V \end{bmatrix} \begin{bmatrix} \delta\alpha \\ \delta J_i \\ \delta T \\ \delta q_i \end{bmatrix} = 0. \tag{2.11}$$

ρ_α, e_T etc. are derivatives of ρ, e etc. with respect to α and T. Note that the matrix is quite explicit in terms of α and T although—obviously—the special functions $i_n(\alpha)$ do occur for which we have no analytic expression.

The characteristic speeds follow from the requirement that the determinant of the matrix (2.11) must vanish and we obtain the biquadratic equation

$$(\rho_\alpha e_T - \rho_T e_\alpha)\, V^4 + \frac{1}{2}\,(\rho_T a_\alpha - \rho_\alpha a_T)\, V^2 + \frac{1}{3}\,(e_\alpha a_T - e_T a_\alpha) = 0, \tag{2.12}$$

or else—by (1.13), (1.14)

$$(5 i_0 i_4 - 9 i_2^2)\left(\frac{V}{\sqrt{2\frac{k}{m}T}}\right)^4 + \left(5 i_2 i_4 - \frac{7}{3} i_0 i_6\right)\left(\frac{V}{\sqrt{2\frac{k}{m}T}}\right)^2 + \left(\frac{7}{3} i_2 i_6 - \frac{25}{9} i_4^2\right) = 0. \tag{2.13}$$

Thus, once again, there are two "sound" speeds in each direction. In the case of complete degeneration—when $i_n(\alpha) \underset{\alpha\to-\infty}{\to} \frac{1}{n+1}\sqrt{-\alpha}^{\,n+1}$ (cf. (4.4.13))—all coefficients in (2.13) vanish, so that no sound can propagate in that state.

2.4 The Laws of Ohm and Fourier

The constitutive equations of Fourier and Ohm of ordinary thermodynamics of metal electrons may be obtained from $(1.10)_{2,4}$ and (2.10) as the first step in the iterative procedure which we have called Maxwell iteration on several previous occasions in this book: In $(1.10)_{2,4}$ we replace the productions $\rho\overline{mcc_i}$ and $\rho\overline{mc^3c_i}$ by the constitutive expressions (2.10) and insert equilibrium

[3] α, J_i, T, and q_i are continuous at the wave. The δ's denote the jumps of the normal derivatives. Thus we have $\delta\alpha = [\alpha_{,i}]\, n_i$ etc.; see Chapter 8 and other places in this book.

values on the left-hand side; in particular $J_i^E = 0$, $q_i^E = 0$. Thus we obtain first iterates $\overset{(1)}{J_i}$ $\overset{(1)}{q_i}$ in the form

$$\begin{bmatrix} \overset{(1)}{J_i} \\ \\ 2\,\overset{(1)}{q_i} \end{bmatrix} = l\,\overset{-1}{K} \begin{bmatrix} \frac{2}{3}e_T & \frac{2}{3}e_\alpha \\ \\ a_T & a_\alpha \end{bmatrix} \begin{bmatrix} \frac{\partial T}{\partial x_i} \\ \\ \left\{ \frac{\partial \alpha}{\partial x_i} + \frac{1}{\frac{k}{m}T}\left[\overset{0}{i_i} - \frac{q}{m}\mathcal{E}_i \right] \right\} \end{bmatrix}. \tag{2.14}$$

We may express this by saying that there are linear relations between thermodynamic forces and fluxes, viz.

thermodynamic forces	temperature gradient	$\frac{\partial T}{\partial x_i}$
	electrochemical potential gradient	$F_i \equiv \left\{ \frac{\partial \alpha}{\partial x_i} + \frac{1}{\frac{k}{m}T}\left[\overset{0}{i_i} - \frac{q}{m}\mathcal{E}_i \right] \right\}$

thermodynamic fluxes	mass flux J_i or electric current	$S_i = -\frac{q}{m} J_i$
	heat flux	q_i.

The two equations (2.14) represent the laws of Ohm and Fourier combined. In particular, in an inertial frame and with ρ and T uniform we obtain Ohm's law in its simplest form

$$S_i = \underbrace{l \left(\overset{-1}{K}_{11} \frac{2}{3} e_\alpha + \overset{-1}{K}_{12}\, a_\alpha \right)}_{\sigma} \mathcal{E}_i \tag{2.15}$$

with σ as the electric conductivity. On the other hand, if the electrochemical potential gradient F_i vanishes, the equations (2.14) imply

$$q_i = \underbrace{\frac{l}{2} \left(\overset{-1}{K}_{21} \frac{2}{3} e_T + \overset{-1}{K}_{22}\, a_T \right)}_{-\kappa} \frac{\partial T}{\partial x_i} \tag{2.16}$$

which is Fourier's law in its simplest form. Note that the mean free path drops out of the quotient κ/σ so that this ratio is an explicit function of ρ and T. Once can prove (see Sommerfeld [103]) that in the limit of complete degeneration we have

$$\kappa/\sigma = \frac{\pi^2}{3} \frac{k^2}{q^2} T,$$

which is the Wiedemann-Franz law, a universal relation between κ and σ.

2.5 Hall and Coriolis Effects

The second iterate in the Maxwell iteration comes out as

$$\begin{bmatrix} -\frac{q}{m} \overset{(2)}{S_i} \\ \\ 2 \overset{(2)}{q_i} \end{bmatrix} = l \overset{-1}{K} \begin{bmatrix} \frac{2}{3}e_T & \frac{2}{3}e_\alpha \\ \\ a_T & a_\alpha \end{bmatrix} \text{x}$$

$$\left\{\delta_{ij} + l \overset{-1}{K} \left(\delta_{ij}\frac{\partial}{\partial t} + \left[\frac{q}{m}\varepsilon_{ijk}B_k - 2W_{ij}\right]\right)\right\} \begin{bmatrix} \frac{\partial T}{\partial x_i} \\ \\ \left\{\frac{\partial \alpha}{\partial x_i} + \frac{1}{\frac{k}{m}T}\left[i_i^0 - \frac{q}{m}\mathcal{E}_i\right]\right\} \end{bmatrix}. \tag{2.17}$$

Inspection shows that S_i and q_i in the second iteration are not linear combinations of temperature gradient and electrochemical potential gradient any more. Indeed there are contributions perpendicular to these gradients and those contributions are due to the magnetic flux density and the Coriolis force.

The effect of the magnetic flux density is known as the Hall effect and it is well understood: The free paths of the electrons are bent into circular arcs by the Lorentz force. We see from (2.17) that the Hall effect is accompanied by a Coriolis effect, because **B** always occurs in combination with **W**, the angular velocity of the noninertial frame. Müller has pointed that out in [56]. The interpretation is much like the one given in Paragraph 4.3.3 for the frame-dependent part of the heat flux, or in Paragraph 5.3.3 for the frame dependent part of the diffusion flux. It is true, of course, that the Coriolis effect is much smaller than the Hall effect, because there are practical limitations for the rate of rotation of the frame.

2.6 Discussion

It has been the intention of this section to demonstrate that the electron theory of metals is a possible area where the systematic procedure of rational extended thermodynamics may be applied in the future. We suspect that the present arguments, based on the simple system (1.10), barely scratch the surface of a systematic theory of the electron gas. Indeed, the experience with ordinary dilute gases and with "gases" of photons and phonons has taught us that *many* variables are needed, and it may well turn out that the present eight fields do not suffice to describe anything out of the ordinary.

Nor is the application of extended thermodynamics to electrons limited to metals. Anile and his coworkers have proposed an extended thermodynamics of semiconductors; see [32]. The carrier transport in semiconductor devices brings a new ingredient into the theory. Indeed, while in the theory of metal electrons the electric field may be considered as an external agent—on a par with gravitational and inertial forces—a satisfactory theory of carrier transport in semiconductors requires coupling between the field $\mathcal{E}_i$ and the charge distribution of electrons and holes. The equation that governs this coupling is the Poisson equation for the electrostatic potential.

Chapter 16

Toward Extended Thermodynamics of Viscoelastic Fluids

There has been a cloud over thermodynamics of second grade non-Newtonian fluids ever since Dunn & Fosdick [33] showed that – in ordinary thermodynamics – the measured sign of the first normal stress coefficient is incompatible with the convexity of free energy. Extended thermodynamics can go a long way to resolve that dilemma, although it provides another one. The new dilemma cocerns the material frame-indifference of rheological equations of state.

Unlike previous chapters this one does not provide finished results. It was written to provoke thought and to encourage work in the field of rheology by exploiting the principles of extended thermodynamics.

1 Viscoelastic Fluids of Second Grade. A Dilemma of Ordinary Thermodynamics.

1.1 The Stress of a Second Grade Fluid

In an incompressible *viscous* fluid the deviatoric stress $\mathbf{S}$ at one time depends on the velocity gradient $\mathbf{L}$ at that time and on temperature. In contrast to that, the stress deviator of a *viscoelastic* fluid may also depend on the history of the velocity gradient. The easiest nontrivial case is a second-grade fluid in which we assume—in ordinary thermodynamics—that the stress deviator $\mathbf{S}$ depends the velocity gradient $\mathbf{L}$ and its time derivative $\dot{\mathbf{L}}$—and on temperature T. Thus we have for the stress $\mathbf{t}$

$$\mathbf{t} = -p\mathbf{1} + \mathbf{S} \text{ and } \mathbf{S} = \hat{\mathbf{S}}(T, \mathbf{L}, \dot{\mathbf{L}}), \quad \text{where} \quad L_{ij} = \frac{\partial v_i}{\partial x_j}. \tag{1.1}$$

$\mathbf{L}$ is traceless in an incompressible fluid, because $\frac{\partial v_i}{\partial x_i} = 0$ holds.

p is the hydrostatic pressure appropriate to an incompressible fluid. For its origin and interpretation we must refer the reader to some book on continuum mechanics or rheology, e.g. see Truesdell & Noll [1], or Coleman, Markowitz & Noll [175]. Some basic knowledge of continuum mechanics is also required for some of the subsequent arguments concerning Rivlin-Ericksen tensors, and representations of isotropic functions.

The constitutive function $\mathbf{S}$ for the stress deviator is subject to the principle of material frame-indifference, which requires that the dependence on $\mathbf{L}$ and $\dot{\mathbf{L}}$ is in fact a dependence on the first two Rivlin-Ericksen tensors

$$\begin{aligned} \mathbf{A}^{(1)} &= \mathbf{L} + \mathbf{L}^T \\ \mathbf{A}^{(2)} &= \dot{\mathbf{A}}^{(1)} + \mathbf{A}^{(1)}\mathbf{L} + \mathbf{L}^T\mathbf{A}^{(1)}. \end{aligned} \tag{1.2}$$

Moreover the function $\hat{\mathbf{S}}$ must be isotropic and this implies a representation which reads

$$\mathbf{S} = \mu\mathbf{A}^{(1)} + \alpha_1\left[\mathbf{A}^{(2)} - tr\mathbf{A}^{(2)}\mathbf{1}\right] + \alpha_2\left[\left(\mathbf{A}^{(1)}\right)^2 - tr\left(\mathbf{A}^{(1)}\right)^2\mathbf{1}\right], \tag{1.3}$$

provided that we are content with linear and quadratic terms in $\mathbf{L}$ and $\dot{\mathbf{L}}$. The coefficients μ, α_1, and α_2 may depend on temperature. μ is the viscosity and we know that it is positive. α_1 and α_2 are the normal stress coefficients. Rheologists measure these coefficients routinely and lay down the results in tables, e.g. see Ginn & Metzner [176]. Invariably, for all substances considered they find negative values for the first normal stress coefficient:

$$\alpha_1 < 0. \tag{1.4}$$

1.2 Ordinary Thermodynamics of Second Grade Fluids

We may say that it is the objective of ordinary thermodynamics of incompressible viscoelastic fluids to determine the velocity and the temperature of the fluid. For simplicity we ignore the heat flux and choose the equations of balance of momentum and internal energy as the basis of the field equations; the mass balance is here reduced to the statement that the divergence of velocity vanishes. Thus we have

$$\frac{\partial v_i}{\partial x_i} = 0, \quad \rho \dot{v}_i + \frac{\partial p}{\partial x_i} - \frac{\partial S_{<ij>}}{\partial x_j} = 0, \quad \rho \dot{\varepsilon} = \mathbf{S} \cdot \mathbf{L}. \tag{1.5}$$

We supplement these equations by constitutive relations for $S_{<ij>}$ and ε, which—in a second grade viscoelastic fluid—we consider to be functions of temperature T, of the velocity gradient $\mathbf{L}$, and of its time derivative $\dot{\mathbf{L}}$

$$S_{<ij>} = \hat{S}_{<ij>}(T, \mathbf{L}, \dot{\mathbf{L}}), \quad \varepsilon = \hat{\varepsilon}(T, \mathbf{L}, \dot{\mathbf{L}}). \tag{1.6}$$

The constitutive functions are reduced by the requirement that the entropy inequality

$$\rho \dot{s} + \frac{\partial \phi_j}{\partial x_j} \geq 0 \quad \text{with} \quad \begin{array}{l} s = \hat{s}\ (T, \mathbf{L}, \dot{\mathbf{L}}) \\ \phi_i = \hat{\phi}_i(T, \mathbf{L}, \dot{\mathbf{L}}) \end{array} \tag{1.7}$$

holds for all solutions of (1.5), (1.6). An equivalent statement results from the use of Lagrange multipliers: The inequality

$$\begin{aligned} \rho \dot{s} + \frac{\partial \phi_j}{\partial x_j} \quad & - \Lambda \frac{\partial v_j}{\partial x_j} \\ & - \Lambda^i \left(\rho \dot{v}_i + \frac{\partial p}{\partial x_i} - \frac{\partial S_{<ij>}}{\partial x_j} \right) \quad - \\ & - \Lambda^{\varepsilon} \left(\rho \dot{\varepsilon} \ - \ \mathbf{S} \cdot \mathbf{L} \right) \qquad \geq 0. \end{aligned} \tag{1.8}$$

must hold for all fields $T(\mathbf{x}, t), v_i(\mathbf{x}, t)$. The Λ's are Lagrange multipliers, which may depend on the variables $\rho, \mathbf{L}, \dot{\mathbf{L}}$.[1]

Among other, less interesting equations we obtain

$$\frac{\partial s}{\partial T} - \Lambda^{\varepsilon} \frac{\partial \varepsilon}{\partial T} = 0, \quad \frac{\partial s}{\partial \dot{\mathbf{L}}} - \Lambda^{\varepsilon} \frac{\partial \varepsilon}{\partial \dot{\mathbf{L}}} = 0 \qquad \text{and} \tag{1.9}$$

$$\left(\frac{\partial s}{\partial \mathbf{L}} - \Lambda^{\varepsilon} \frac{\partial \varepsilon}{\partial \mathbf{L}} \right) \cdot \dot{\mathbf{L}} + \frac{1}{\rho} \Lambda^{\varepsilon} \mathbf{S} \cdot \mathbf{L} \geq 0. \tag{1.10}$$

The left-hand side of the inequality (1.10) represents the entropy production, which assumes its minimum—namely zero—in equilibrium, where $\mathbf{L}$ and $\dot{\mathbf{L}}$ both vanish.

By material frame-indifference s and ε may depend on $\mathbf{L}$ and $\dot{\mathbf{L}}$ only through the Rivlin-Ericksen tensors $\mathbf{A}^{(1)}$ and $\mathbf{A}^{(2)}$; cf. (1.2). We shall neglect terms of third and higher order in $\mathbf{L}$ and $\dot{\mathbf{L}}$ and therefore, since $\mathbf{A}^{(1)}$ is traceless, s and ε can *only* contain second-order terms—apart from

[1] This procedure of exploiting the entropy inequality by use of Lagrange multipliers has been described so often in this book that we may now rush through it.

the equilibrium values which depend on T. By $(1.9)_1$ the same is true for Λ^ε. From equilibrium thermodynamics we know that $ds_E = \frac{1}{T}d\varepsilon_E$ holds, so that $(1.9)_1$ implies

$$\Lambda^\varepsilon = \frac{1}{T} + \text{second order terms in } \mathbf{L}, \dot{\mathbf{L}}. \tag{1.11}$$

With (1.11) it follows from $(1.9)_2$ that in the specific free energy

$$\psi = \varepsilon - Ts$$

all terms with $\dot{\mathbf{L}}$ are of fourth order in $\mathbf{L}$ and $\dot{\mathbf{L}}$. But there may be an additive function $\psi(T, \mathbf{L})$ which is of second order in $\mathbf{L}$.

In the second-order approximation this contribution $\psi(T, \mathbf{L})$ to ψ is the only one that counts in the residual inequality (1.10) and we obtain

$$-\frac{\partial \psi(T, \mathbf{L})}{\partial \mathbf{L}}\dot{\mathbf{L}} + \frac{1}{\rho}\mathbf{S} \cdot \mathbf{L} \geq 0. \tag{1.12}$$

Recalling that $\mathbf{S}$ is given by (1.3), (1.2), we see that both terms on the left-hand side of the inequality are linear in $\dot{\mathbf{L}}$. This linear term must vanish lest the inequality be violated and we conclude

$$\frac{\partial \psi(T, \mathbf{L})}{\partial \mathbf{L}} = \frac{\alpha_1(T)}{\rho}\left(\mathbf{L} + \mathbf{L}^T\right) \quad \text{and} \quad \mu \geq 0. \tag{1.13}$$

Thus the entropy inequality requires a nonnegative viscosity coefficient—as expected—and it furnishes a relation between the $\mathbf{L}$-dependence of $\psi(T, \mathbf{L})$ and the first normal stress coefficient. The latter result is interesting. Indeed, we know that the free energy must assume a minimum in equilibrium. This is one of the thermodynamic stability conditions—on a par with positive specific heats and compressibilities—and in the present case it requires

$$\frac{\partial^2 \psi}{\partial \mathbf{L} \partial \mathbf{L}} = 2\frac{\alpha_1}{\rho} > 0 \tag{1.14}$$

so that we must have

$$\alpha_1 > 0. \tag{1.15}$$

This is in stark contradiction to the experimental result (1.4).

1.3 Discussion

After Dunn & Fosdick [33] published the inequality (1.15), rheologists immediately recognized that there was something wrong with thermodynamics of second grade fluids. The question was what? Dozens of papers appeared and they continue to appear to this day. Some authors resolutely ignore the experimental data, others argue that the entropy inequality is wrong.[2] Joseph [34] cast doubt on the propriety of second-grade fluids—or any nth grade fluids—as a good constitutive model and it seems to us that he is right. Actually the stability problems with n^{th} grad fluids

[2]Defeatists will always find a sympathetic ear when they maintain that the entropy concept cannot be applied outside equilibrium.

were first noticed by Fosdick & Rajagopal [177], and Dunn & Rajagopal [178] recently wrote a review about the matter.

Müller & Wilmanski [179] advocated rate-type relations instead of the differential type n^{th} order relations and they attempted to incorporate their ideas into extended thermodynamics of gases, see also Wilmanski [180]. They did obtain the correct inequality, viz. (1.4), but they failed to formulate a convincing systematic—rational (!)—extended thermodynamics of viscoelastic fluids. Liu [181] went farthest in that direction, but he stood aloof from normal stress coefficients.

So the problem seems to be open as far as rational extended thermodynamics goes. We proceed to offer arguments in order to motivate our belief that the problem is indeed a problem of extended thermodynamics—and we attempt a solution.

2 Rate-Type versus Differential-Type Constitutive Equations

2.1 Cattaneo and Stability

Let us recall Paragraph 2.2.1, where we have discussed the derivation of the Cattaneo equation—the very origin of extended thermodynamics. It will be remembered that Cattaneo used a plausible argument from the elementary kinetic theory—see Figure 1.1—to derive the equation (cf. (2.2.2))

$$q_i = -\kappa \left(\frac{\partial T}{\partial x_i} - \tau \left(\frac{\partial T}{\partial x_i} \right)^{\bullet} \right) \tag{2.1}$$

and the argument was such that τ had to be positive.

And yet, a stability argument based on (2.1) would require that τ is negative: Indeed, if we expect that for $q_i = 0$ the temperature gradient tends to zero, we must have $\tau < 0$.

And now let us recall the true Cattaneo equation (2.2.6), viz.

$$q_i + \tau \dot{q}_i = -\kappa \frac{\partial T}{\partial x_i} \tag{2.2}$$

which was arrived at from (2.1) by a series of seemingly innocuous steps. Stability can now only mean that q_i tends to zero when the temperature gradient vanishes, and this requires that τ be positive as it should be.

So the two arguments lead to opposite results in the two cases, and that indicates that something is badly wrong with the "innocuousness" of the steps from (2.1) to (2.2).

2.2 Viscoelasticity and Stability

An identical argument may be made for viscoelastic fluids. Indeed, the presence of the $\mathbf{A}^{(2)}$-term in (1.3) is motivated by the "memory" of the fluid: The stress deviator $\mathbf{S}$ may depend on the history of $\mathbf{d} = \mathbf{L} + \mathbf{L}^T$, and this means that, in a simple case, it may depend on $\mathbf{d} - \tau\mathbf{d}$ with τ positive. Thus we should have

$$\mathbf{S} = \mu\left(\mathbf{d} - \tau\dot{\mathbf{d}}\right) \tag{2.3}$$

and, if we compare this with the linear version of (1.3), we conclude that $\alpha_1 = -\mu\tau$ should be negative—as in fact it is found to be experimentally; cf. (1.4). However, this sign of α_1 contradicts not only the entropy inequality—recall (1.15)—but also the reasonable stability criterion based on (2.3), by which $\mathbf{d}$ should tend to zero for $\mathbf{S} = 0$.

On the other hand, if we pass from (2.3) to

$$\mathbf{S} + \tau\dot{\mathbf{S}} = \mu\mathbf{d} \tag{2.4}$$

by Cattaneo's steps, we must require $\tau > 0$, and $\alpha_1 < 0$ and everything is as it should be.

2.3 Conclusion

In summary we must conclude that the *differential-type* constitutive equations (2.1), (2.3) are not legitimate, while the *rate-type* constitutive equations (2.2), (2.4) are good—at least with regard to stability.

Once this is agreed on, we need to formulate a thermodynamic constitutive theory for rate-type equations. Extended thermodynamics with its emphasis on balance equations seems to be the proper tool for this research. And, of course, the hope is to find the proper signs—viz. $\tau > 0$, and $\alpha_1 < 0$—as consequences of a thermodynamic stability condition, namely the convexity of the free energy, or the concavity of entropy.

3 Toward Extended Thermodynamics of Viscoelasticity

3.1 Fields and Field Equations

As in Section 1 we consider a fluid without heat conduction, but now we employ extended thermodynamics rather than ordinary thermodynamics. We propose to determine the fields $\rho, \mathbf{v}, T$, and $\mathbf{S}$, where $\mathbf{S}$ is the stress deviator. As a basis for the field equations we choose the following equations of balance

$$\begin{aligned}
\frac{\partial\rho}{\partial t} &+ \frac{\partial\rho v_k}{\partial x_k} &&= 0 \\
\frac{\partial\rho v_i}{\partial t} &+ \frac{\partial F_{ik}}{\partial x_k} &&= \rho f_i \\
\frac{\partial\rho e_{ij}}{\partial t} &+ \frac{\partial F_{ijk}}{\partial x_k} &&= P_{<ij>} + \frac{\rho}{2}\left(v_i f_j + v_j f_i\right).
\end{aligned} \tag{3.1}$$

The first two of these are standard and so is the trace of the third one, which represents the conservation law of energy; $e \equiv e_{ii}$ denotes the specific energy and F_{iik} is the energy flux. The traceless part of $(3.1)_3$ is a new balance law, which we need so as to have the correct number of equations, namely 10 for the 10 variables. e_{ij} may be called the tensor of specific energy, and F_{ijk} is its flux; e_{ij} is symmetric and so is F_{ijk} with respect to i and j. $P_{<ij>}$ is the traceless production of the energy tensor.

The requirement of Galilean invariance determines the velocity dependence of all quantities explicitly, according to Section 3.4, and we obtain

$$\begin{aligned} F_{ij} &= -t_{ij} + \rho v_i v_j \\ e_{ij} &= \varepsilon_{ij} + \tfrac{1}{2} v_i v_j \\ F_{ijk} &= q_{ijk} + \rho \varepsilon_{ij} v_k - \tfrac{1}{2} v_i t_{jk} - \tfrac{1}{2} v_j t_{ik} + \tfrac{\rho}{2} v_i v_j v_k . \end{aligned} \tag{3.2}$$

t_{ij} is the stress, while ε_{ij} and q_{ijk} are the tensor of specific internal energy and the tensor of internal energy flux respectively. These quantities are Galilean tensors. $\varepsilon = \varepsilon_{ii}$ and $q_k = q_{iik}$ are the usual specific internal energy and heat flux.

Insertion of (3.2) into (3.1) provides the equations of balance for the internal quantities, which are equivalent to (3.1), viz.

$$\begin{aligned} \dot\rho + \rho \frac{\partial v_k}{\partial x_k} &= 0 \\ \rho \dot v_i - \frac{\partial t_{ik}}{\partial x_k} &= \rho f_i \\ \rho \dot\varepsilon_{ij} + \frac{\partial q_{ijk}}{\partial x_k} &= P_{<ij>} + t_{k(i} \frac{\partial v_{j)}}{\partial x_k} . \end{aligned} \tag{3.3}$$

Sofar we have tacitly assumed that the equations are referred to an inertial frame. In a non-inertial frame with the specific inertial force i_i and the matrix W_{ik} of angular velocity the equations read

$$\begin{aligned} &\dot\rho + \rho \frac{\partial v_k}{\partial x_k} = 0 \\ &\rho \dot v_i - \frac{\partial t_{ik}}{\partial x_k} = \rho (f_i + i_i) \\ &\rho (\dot\varepsilon_{ij} + 2 W_{p<i} \varepsilon_{j>p}) + \frac{\partial q_{ijk}}{\partial x_k} = P_{<ij>} + t_{k(i} \left(\frac{\partial v_{j)}}{\partial x_k} - W_{j)k} \right) . \end{aligned} \tag{3.4}$$

This follows by transformation of (3.3) to an arbitrary Euclidean frame.

We recall that the stress t_{ij} equals $-p\delta_{ij} + S_{ij}$ and conclude that for closure of the system (3.3) we need constitutive relations for p, ε_{ij} and $P_{<ij>}$. In extended thermodynamics the constitutive quantities must depend on the *present local* state. Therefore a generic constitutive equation reads

$$C = \hat{C}(\rho, T, \mathbf{S}); \tag{3.5}$$

recall that p, ε_{ij}, and $P_{<ij>}$ are internal quantities, independent of $\mathbf{v}$. Note that no history dependence it assumed here, i.e. the constitutive quantities at time t depend only on ρ, T, and $\mathbf{S}$ at that same time. This is a feature on which extended thermodynamics must insist.

3.2 Incompressible Adiabatic Fluid

Incompressibility means that $\rho = const.$ holds and that p is an arbitrary—though positive-valued—hydrostatic pressure to be determined by boundary conditions. Adiabaticity means that the heat flux $q_k = q_{iik}$ vanishes. Therefore in this simple case the balance equations (3.3) reduce to

$$\begin{aligned} &\frac{\partial v_i}{\partial x_i} = 0 \\ &\rho \dot{v}_i + \frac{\partial p}{\partial x_i} - \frac{\partial S_{ik}}{\partial x_k} = f_i \\ &\rho \dot{\varepsilon} = S_{ij} \frac{\partial v_i}{\partial x_j} \\ &\rho \dot{\varepsilon}_{<ij>} + \frac{\partial q_{<ij>k}}{\partial x_k} = P_{<ij>} - p \frac{\partial v_{<i}}{\partial x_{j>}} + S_{k<i} \frac{\partial v_{j>}}{\partial x_k}. \end{aligned} \tag{3.6}$$

and the constitutive equations are given by

$$\varepsilon_{ij} = \hat{\varepsilon}_{ij}(T, \mathbf{S}), \quad q_{<ij>k} = \hat{q}_{<ij>k}(T, \mathbf{S}), \quad P_{<ij>} = \hat{P}_{<ij>}(T, \mathbf{S}). \tag{3.7}$$

Isotropy of the fluids makes tensors of uneven rank vanish if they can only depend on T and $\mathbf{S}$. Thus $q_{<ij>k} = 0$ holds. For the other two constitutive quantities we assume that they are at most of order 2 in $\mathbf{S}$; this will suffice if the fluid is close to equilibrium. Thus we have

$$\begin{aligned} \varepsilon_{ij} &= \tfrac{1}{3}\left(\varepsilon_{00} + \varepsilon_{02} S^2\right)\delta_{ij} + \varepsilon_1 S_{<ij>} + \varepsilon_2\left(S_{<ik>}S_{<kj>} - \tfrac{1}{3}S^2\delta_{ij}\right) \\ q_{<ij>k} &= 0 \\ P_{<ij>} &= P_1 S_{<ij>} + P_2\left(S_{<ik>}S_{<kj>} - \tfrac{1}{3}S^2\delta_{ij}\right). \end{aligned} \tag{3.8}$$

The coefficients ε_{00} through P_2 may depend on T, and S^2 stands for $S_{<ik>}S_{<ki>}$.

3.3 Entropy Inequality

We hope to gain information on the constitutive coefficients from the entropy inequality

$$\rho \dot{s} + \frac{\partial \phi_i}{\partial x_i} \geq 0, \tag{3.9}$$

where s and ϕ_i are constitutive quantities of the type (3.7), (3.8) such that

$$\begin{aligned} s &= s_{00} + s_{02} S^2 \\ \phi_i &= 0. \end{aligned} \tag{3.10}$$

The inequality must hold for all solutions of the field equations (3.6), (3.8).

Equivalently the new inequality with Lagrange multipliers, viz.

$$\begin{aligned}\rho\dot{s}+\phi_{i,i} &- \Lambda\frac{\partial v_k}{\partial x_i}\\ &- \Lambda^i\left(\rho\dot{v}_i+\frac{\partial p}{\partial x_i}-\frac{\partial S_{ik}}{\partial x_k}-f_i\right)-\\ &- \Lambda^\varepsilon\left(\rho\dot{\varepsilon} - S_{ij}\frac{\partial v_i}{\partial x_j}\right)-\\ &- \Lambda^{<ij>}\left(\rho\dot{\varepsilon}_{<ij>}+\frac{\partial q_{<ij>k}}{\partial x_k}-P_{<ij>}+p\frac{\partial v_{<i}}{\partial x_{j>}}-S_{k<i}\frac{\partial v_{j>}}{\partial x_k}\right)\geq 0\end{aligned} \tag{3.11}$$

must hold for all fields T, v_i, and S_{ij}. The Lagrange multipliers may depend on T and $\mathbf{S}$.

It follows that Λ and Λ^i must both vanish and, since ϕ_i and $q_{<ij>k}$ are zero, we obtain

$$\begin{aligned}&\frac{\partial s}{\partial T} - \Lambda^\varepsilon\frac{\partial \varepsilon}{\partial T} - \Lambda^{<rs>}\frac{\partial \varepsilon_{<rs>}}{\partial T}=0\\ &\frac{\partial s}{\partial S_{<ij>}} - \Lambda^\varepsilon\frac{\partial \varepsilon}{\partial S_{<ij>}} - \Lambda^{<rs>}\frac{\partial \varepsilon_{<rs>}}{\partial S_{<ij>}}=0\\ &\Lambda^\varepsilon S_{<ik>} - p\Lambda^{<ik>}+\Lambda^{<i<j>}S_{<k>i>}=0\\ &\Lambda^{<i[\,j>}S_{<k]\,i>}=0;\end{aligned} \tag{3.12}$$

there remains the residual inequality

$$\Lambda^{<ij>}P_{<ij>}\geq 0. \tag{3.13}$$

3.4 Partial Exploitation of the Entropy Inequality

For the exploitation of the conditions (3.12) it is useful to introduce the representations (3.8), (3.10) as well as representations for the Lagrange multipliers, viz.

$$\begin{aligned}\Lambda^\varepsilon &= \lambda_{00}+\lambda_{02}S^2\\ \Lambda^{<ij>} &= \lambda_1 S_{<ij>}+\lambda_2\left(S_{<ik>}S_{<kj>}-\tfrac{1}{3}S^2\delta_{ij}\right).\end{aligned} \tag{3.14}$$

Thus we obtain from (3.12) and (3.13)

$$\lambda_{00}=\frac{1}{T},\quad \varepsilon_{02}-Ts_{02}=-T\lambda_1\varepsilon_1,\quad \lambda_1=\frac{1}{2pT},\quad \lambda_1 P_1\geq 0. \tag{3.15}$$

The identification of λ_{00} as $\frac{1}{T}$ results by comparison with equilibrium thermodynamics, by which the equilibrium values of s and ε are related through $ds_E=\frac{1}{T}\,d\varepsilon_E$. There are more relations to be had from (3.12), but this is all we need for our problem with the normal stress coefficient α_1.

Since the hydrostatic pressure p and the absolute temperature T are positive valued, so are λ_1 by $(3.15)_3$ and P_1 by $(3.15)_4$. On the other hand $\varepsilon_{02}-Ts_{02}$ must be positive for the free energy $\varepsilon-Ts$ to have a minimum in equilibrium, and therefore it follows from $(3.15)_2$ that $\varepsilon_1<0$ holds.

3.5 Evaluation

The inequalities $P_1 \geq 0$ and $\varepsilon_1 < 0$ make perfectly good sense, because they imply that the new balance law $(3.6)_4$—the one for the deviatoric energy tensor—is in fact a rate law for the relaxation of the deviatoric stress $\mathbf{S}$. Indeed, for $\mathbf{v} \equiv 0$ and isothermal conditions this equation reduces to

$$\dot{\mathbf{S}} = \frac{P_1}{\rho\varepsilon_1}\mathbf{S}$$

in the linear case, and obviously $P_1/\rho\varepsilon_1$ must be negative for $\mathbf{S}$ to relax to zero.

But what about the normal stress coefficient α_1? Such a coefficient does not appear in the present theory, nor does the constitutive relation (1.3) for a second-grade fluid appear—at least not in any obvious manner. However, we claim that the constitutive relation (1.3) is a mutilated form of the balance equation $(3.6)_4$. Let us consider:

Ignoring third-order terms in S_{ij} and $\frac{\partial v_j}{\partial x_k}$—and by use of $(3.8)_2$—we may write $(3.6)_4$ in the form

$$-\rho\left|\varepsilon_1\right|\left(\dot{S}_{<ij>} + 2S_{<l<i>}\frac{\partial v_l}{\partial x_{j>}}\right) + (\rho\left|\varepsilon_1\right| - \frac{1}{2})S_{l<i}\frac{\partial v_{<j>}}{\partial x_{l>}} - (\rho\left|\varepsilon_1\right| + \frac{1}{2})S_{l<i}\frac{\partial v_{[j>}}{\partial x_{l]}} +$$
$$+\underline{p\frac{\partial v_{<i}}{\partial x_{j>}} = P_1 S_{<ij>}} + P_2\left(S_{<ik>}S_{<kj>} - \frac{1}{3}S^2\delta_{ij}\right). \quad (3.16)$$

It seems obvious that the underlined terms in (3.16) represent the leading part of the equation and, if we consider *only* that part, we recover the Navier-Stokes equation in the form

$$S_{<ij>} = \frac{p}{P_1}\frac{\partial v_{<i}}{\partial x_{j>}}. \quad (3.17)$$

We regard (3.17) as a first approximation of the stress deviator and proceed to the second approximation by introducing (3.17) into all terms previously neglected. We obtain for isothermal conditions

$$\begin{aligned} S_{<ij>} = \; & \frac{p}{P_1}\frac{\partial v_{<i}}{\partial x_{j>}} - \\ & -\rho\left|\varepsilon_1\right|\frac{p}{P_1}\left[\left(\frac{\partial v_{<i}}{\partial x_{j>}}\right)^{\bullet} + \frac{\partial v_{<l}}{\partial x_{<i>}}\frac{\partial v_l}{\partial x_{j>}} + \frac{\partial v_{<l}}{\partial x_{<j>}}\frac{\partial v_l}{\partial x_{i>}}\right] + \\ & + \left(\frac{P_2}{P_1}\left(\frac{p}{P_1}\right)^2 + (\rho\left|\varepsilon_1\right| - \frac{1}{2})\frac{p}{P_1}\right)\frac{\partial v_{<l}}{\partial x_{<i>}}\frac{\partial v_{<j>}}{\partial x_{l>}} - \\ & -(\rho\left|\varepsilon_1\right| + \frac{1}{2})\frac{p}{P_1}\frac{\partial v_{<l}}{\partial x_{<i>}}\frac{\partial v_{[j>}}{\partial x_{l]}}. \end{aligned} \quad (3.18)$$

Thus we have derived an equation that relates the stress deviator to the velocity gradient and its time derivative. The first three lines of this equation correspond exactly to the constitutive equation (1.3) for the stress in a second-grade fluid. Therefore we may identify the coefficients μ, α_1 and α_2 as

$$\mu = \frac{p}{2P_1}, \qquad \alpha_1 = -\rho\,|\varepsilon_1|\,\frac{p}{P_1}, \qquad \alpha_2 = \frac{P_2}{P_1}\left(\frac{p}{P_1}\right)^2 + (\rho\,|\varepsilon_1| - \frac{1}{2})\frac{p}{P_1}. \tag{3.19}$$

We see that μ is nonnegative, while α_1—the first normal stress coefficient—has the correct negative sign *as a result of the entropy inequality and of the thermodynamic stability condition*, by which the free energy is convex. It is worth noting that the appearance of p in μ, α_1, and α_2 is only apparent; since these quantities are independent of p in an incompressible fluid, we must conclude that P_1 and P_2 are proportional to p.

3.6 Criticism and Outlook

Thus it would seem that we have *derived* the constitutive relation for a second grade fluid within the framework of extended thermodynamics. *And we have found the correct sign of the first normal stress coefficient.*

Not all is perfect, however. Indeed there is also the fourth line in (3.18) and that line depends on the antisymmetric part of the velocity gradient—the spin of the fluid—which is a nonobjective tensor. This must not surprise us, because after all (3.18) is the mutilated form of a balance law, and a balance law generally does contain inertial terms; certainly ours does contain such terms; cf. $(3.4)_3$. But rheologists have firmly embraced the principle of material-frame indifference, according to which such nonobjective terms must not occur in a constitutive equation.

So the question arises whether that term—the fourth line in (3.18)—can be tolerated by the experimentalist. Or is it incompatible with rheological observation? We do not know, and therefore we have to let the question stand.

Apart from experiment a molecular theory may be able to provide an answer. In fact, it may be an interesting proposition to link extended thermodynamics of viscoelastic fluids to the kinetic theory of non-Newtonian fluids as closely as extended thermodynamics of gases is linked to the kinetic theory of gases. It will thus be possible to derive a convincing extended thermodynamics of viscoelastic fluids—not only of 2nd grade fluids, but fluids of arbitrary grade.

The molecular theory of non-Newtonian fluids was pioneered by Bird; see Bird, Armstrong & Hassager [182] and Bird, Warner & Evans [183]. Müller [184] and Reitebuch [185] have made attempts to move these theories closer to thermodynamics, but such efforts are at their beginning.

Bibliography

[1] C. Truesdell, W. Noll. The non-linear field theories of mechanics. *Handbuch der Physik* **III/3**. Springer Heidelberg (1965).

[2] C. Truesdell. *Rational thermodynamics* (2nd ed.). Springer New York (1984).

[3] A. Fisher, D. P. Marsden. The Einstein evolution equations as a first order quasi-linear symmetric hyperbolic system. *Comm. Math. Phys.* **28** (1972).

[4] S. K. Godunov. An interesting class of quasilinear systems. *Sov. Math.* **2** (1961).

[5] K. O. Friedrichs, P. D. Lax. Systems of conservation equations with a convex extension. *Proc. Nat. Acad. Sci. USA* **68** (1971).

[6] G. Boillat. Sur l'éxistence et la recherche d'équations de conservation supplémentaires pour les systèmes hyperboliques. *C. R. Acad. Sc. Paris, Sér A* **278** (1974).

[7] I. Müller. *Zur Ausbreitungsgeschwindigkeit von Störungen in kontinuierlichen Medien.* Dissertation Aachen (1966).

[8] I. Müller. Zum Paradox der Wärmeleitungstheorie. *Zeitschrift für Physik* **198** (1967).

[9] D. Jou, J. Casas-Vázquez, G. Lebon. *Extended irreversible thermodynamics.* Springer Heidelberg (1993).

[10] R. E. Nettleton, S. L. Sobolev. Applications of extended thermodynamics to chemical, rheological and transport processes. A special survey Parts I through III. *J. Non-Equilibrium Thermodynamics* **20** (1995) and **21** (1996).

[11] H. Grad. On the kinetic theory of rarefied gases. *Comm. Appl. Math.* **2**. Wiley New York (1949).

[12] H. Grad. Principles of the kinetic theory of gases. *Handbuch der Physik* **XII**. Springer Heidelberg (1958).

[13] W. Dreyer. Maximization of the entropy in non-equilibrium. *J. Phys. A: Math. Gen.* **20** (1987).

[14] H. Struchtrup. *Zur irreversiblen Thermodynamik der Strahlung.* Dissertation TU Berlin (1996).

[15] H. Struchtrup. An extended moment method in radiative transfer: The matrices of mean absorption and scattering coefficients. *Annals of Physics* **257** (1997).

[16] W. Weiss. Continuous shock structure in extended thermodynamics. *Phys. Review E, Part A* **52** (1995).

[17] B. D. Coleman, W. Noll. The thermomechanics of elastic materials with heat conduction and viscosity. *Arch. Rational Mech. Anal.* **13** (1963).

[18] I. Müller. On the entropy inequality. *Arch. Rational Mech. Anal.* **26** (1967).

[19] I. Müller. A new systematic approach to non-equilibrium thermodynamics. *Pure and Applied Chemistry* **22** (1970).

[20] I-Shih Liu. Method of Lagrange multipliers for exploitation of the entropy principle. *Arch. Rational Mech. Anal.* **46** (1972).

[21] I-Shih Liu, I. Müller. Extended thermodynamics of classical and degenerate gases. *Arch. Rational Mech. Anal.* **83** (1983).

[22] T. Ruggeri, A. Strumia. Main field and convex covariant density for quasi-linear hyperbolic systems. Relativistic fluid dynamics. *Ann. Inst. H. Poincaré* **34** A (1981).

[23] T. Ruggeri. Galilean invariance and entropy principle for systems of balance laws. The structure of extended thermodynamics. *Cont. Mech. Thermodyn.* **1** (1989).

[24] G. Boillat, T. Ruggeri. Hyperbolic principal subsystems: Entropy convexity and subcharacteristic conditions. *Arch. Rational Mech. Anal.* **137** (1997).

[25] G. Boillat, T. Ruggeri. Moment equations in the kinetic theory of gases and wave velocities. *Cont. Mech. Thermodyn.* **9** (1997).

[26] I. Müller. On the frame dependence of stress and heat flux. *Arch. Rational Mech. Anal.* **45** (1972).

[27] W. Weiss. *Zur Hierarchie der Erweiterten Thermodynamik.* Dissertation TU Berlin (1990).

[28] W. Weiss, I. Müller. Light scattering and extended thermodynamics. *Cont. Mech. Thermodyn.* **7** (1995).

[29] W. Israel. Nonstationary irreversible thermodynamics: A causal relativistic theory. *Annals of Physics* **100** (1976).

[30] I-Shih Liu, I. Müller, T. Ruggeri. Relativistic thermodynamics of gases. *Annals of Physics* **169** (1986).

[31] G. M. Kremer, I. Müller. Thermal conductivity and dynamic pressure in extended thermodynamics of chemically reacting mixtures of gases. *Ann. de l'Inst. H. Poincaré* (in press).

[32] A. M. Anile. Mathematical modelling of charge carrier transport in semiconductors: Problems and perspectives. *Suppl. Rendiconti Circ. Mat. di Palermo, Part I, Ser. II* **45** (1996).

[33] J. E. Dunn, R. L. Fosdick. Thermodynamics, stability, and boundedness of fluids of complexity 2 and fluids of second grade. *Arch. Rational Mech. Anal.* **56** (1974).

[34] D. D. Joseph. Instability of the rest state of fluids of arbitrary grade greater than one. *Arch. Rational Mech. Anal.* **15** (1981).

[35] C. Eckart. The thermodynamics of irreversible processes I: The simple fluid. *Phys. Rev.* **58** (1940).

[36] C. Eckart. The thermodynamics of irreversible processes II: Fluid mixtures. *Phys. Rev.* **58** (1940).

[37] C. Eckart. The thermodynamics of irreversible processes III: Relativistic theory of the simple fluid. *Phys. Rev.* **58** (1940).

[38] C. Cattaneo. Sulla conduzione del calore. *Atti sem. Mat. Fis. Univ. Modena* **3** (1948).

[39] M. E. Gurtin, A. C. Pipkin. A general theory of heat conduction with finite wave speeds. *Arch. Rational Mech. Anal.* **31** (1969).

[40] A. E. Green, K. A. Lindsay. Thermoelasticity. *J. Elasticity* **2** (1972).

[41] G. Grioli. Sulla propagazione di onde termomeccaniche nei continui. *Atti Acc. Naz. Lincei* **67** (1979).

[42] D. D. Joseph, L. Preziosi. Heat waves. *Rev. Mod. Phys.* **61** (1989). Addendum to the paper "Heat waves." *Rev. Mod. Phys.* **62** (1990).

[43] S. C. Chapman, T. G. Cowling. *The mathematical theory of nonuniform gases.* Cambridge University Press, London and New York (1961).

[44] C. Truesdell, R. G. Muncaster. *Fundamentals of Maxwell's kinetic theory of a simple monatomic gas.* Academic Press, New York (1980).

[45] I. Müller. *Thermodynamics.* Pitman, New York (1985).

[46] M. Berger, M. Berger. *Perspectives in nonlinearity.* W.A. Benjamin Inc. New York (1968).

[47] G. Boillat. Non-linear hyperbolic fields and waves. In: *Recent mathematical methods in non-linear wave propagation* (T. Ruggeri, ed.). *Lecture Notes in Mathematics* **1640**, Springer Heidelberg (1996).

[48] K. O. Friedrichs. On the laws of relativistic electro-magneto-fluiddynamics. *Comm. Pure Appl. Math.* **27**, Wiley, New York (1974).

[49] T. Ruggeri. Symmetric hyperbolic systems of conservative equations for a viscous, heat-conducting fluid. *Acta Mechanica* **47** (1983).

[50] G. B. Witham. *Linear and non-linear waves*, Wiley, New York (1974).

[51] G. O. Chen, C. D. Levermore, T. P. Liu. Hyperbolic conservation laws with stiff relaxation terms and entropy. *Comm. Pure and Appl. Math.* **67** (1994).

[52] C. Carathéodory. Untersuchungen über die Grundlagen der Thermodynamik. *Math. Annalen* **67** (1909).

[53] B. D. Coleman, V. J. Mizel. Existence of caloric equations of state in thermodynamics. *J. Chem. Physics* **40** (1964).

[54] G. F. Smith. On isotropic integrity bases. *Arch. Rational Mech. Anal.* **18** (1965).

[55] E. Ikenberry, C. Truesdell. On the pressures and the flux of energy in a gas according to Maxwell's kinetic theory. *J. Rational Mech. Anal.* **5** (1956).

[56] I. Müller. On the frame dependence of electric current and heat flux in a metal. *Acta Mechanica* **24** (1976).

[57] M. Abramowitz, I. A. Stegun, eds. *Handbook of mathematical functions.* Dover, New York (1965).

[58] C. Truesdell. Sulle basi della termodinamica. *Acc. Naz. dei Lincei* (8) **22** (1957).

[59] W. Dreyer. *Zur Thermodynamik von Helium II – Superfluides Helium mit und ohne Wirbellinien als binäre Mischung.* Dissertation Technische Universität Berlin (1983).

[60] C. Truesdell. Mechanical basis for diffusion. *J. Chem. Phys.* **37** (1962).

[61] I. Müller. A new approach to thermodynamics of simple mixtures. *Zeitschrift für Naturforschung* **28a** (1973).

[62] S. R. de Groot, P. Mazur. *Non-equilibrium thermodynamics.* North-Holland, Amsterdam (1963).

[63] C. C. Wang. On the concept of frame indifference in continuum mechanics and in the kinetic theory of gases. *Arch. Rational Mech. Anal.* **58** (1975).

[64] C. Truesdell. Correction of two errors in the kinetic theory that have been used to cast unfounded doubt upon the principle of material frame indifference. *Meccanica* **11** (1976).

[65] T. Ruggeri. *Relativistic extended thermodynamics. General assumptions and mathematical procedures.* (A.M. Anile and Y. Choquet-Bruhat, eds.). *Lecture Notes in Mathematics* **1358**, Springer-Verlag, New York (1989).

[66] I. Müller. Relativistic extended thermodynamics II. (A.M. Anile and Y. Choquet-Bruhat, eds.). *Lecture Notes in Mathematics* **1358**, Springer-Verlag, New York (1989).

[67] S. Chandrasekhar. *An introduction to the study of stellar structure.* Dover Publ., New York (1958).

[68] S. Pennisi. Some representation theorems in a 4-dimensional product space. *Fisica Matematica Suppl. B.U.M.I.* **5** (1986).

[69] T. Ruggeri. Convexity and symmetrization in relativistic theories. Privileged time-liked congruence and entropy. *Cont. Mech. Thermodyn.* **2** (1990).

[70] F. Jüttner. Das Maxwell'sche Gesetz der Geschwindigkeitsverteilung in der Relativitätstheorie. *Annalen der Physik* **34** (1911).

[71] F. Jüttner. Die relativistische Quantentheorie des idealen Gases. *Zeitschrift für Physik* **47** (1928).

[72] N. A. Chernikov. The relativistic gas in the gravitational field. *Acta Phys. Polonica* **23** (1964).

[73] N. A. Chernikov. Equilibrium distribution of the relativistic gas. *Acta Phys. Polonica* **26** (1964).

[74] N. A. Chernikov. Microscopic foundation of relativistic hydrodynamics. *Acta Phys. Polonica* **27** (1964).

[75] W. Dreyer, W. Weiss. The classical limit of relativistic extended thermodynamics. *Annales de l'Institut Henri Poincaré* **45** (1986).

[76] G. M. Kremer. Extended thermodynamics of ideal gases with 14 fields. *Annales de l'Institut Henri Poincaré* **45** (1986).

[77] S. R. de Groot, W. A. van Leeuwen, Ch. G. van Weert. *Relativistic kinetic theory.* North-Holland, Amsterdam (1980).

[78] S. Weinberg. Entropy generation and the survival of protagalaxies in an expanding universe. *Astrophys. J.* **168** (1971).

[79] G. M. Kremer, I. Müller. Dynamic pressure in relativistic thermodynamics. *Annales de l'Institut Henri Poincaré* **67** (1997)

[80] d'Ans, Lax. *Taschenbuch für Chemiker und Physiker.* Bd. **1**, Springer Heidelberg (1967).

[81] G. A. Kluitenberg. *Relativistic thermodynamics of irreversible processes.* Dissertation Leiden (1954).

[82] K. Hutter, I. Müller On mixtures of relativistic fluids. *Helvetica physica Acta* **48**, (1975).

[83] A. Muracchini, T. Ruggeri, L. Seccia. Dispersion relation in high frequency limit and non linear wave stability for hyperbolic dissipative system. *Wave Motion* **15**, (2) (1992).

[84] G. Boillat. *La propagation des ondes.* Gauthier-Villars. Paris (1965).

[85] G. Boillat, T. Ruggeri. On the evolution law of the weak discontinuities for hyperbolic quasi-linear systems. *Wave Motion* **1** (1979).

[86] T. Ruggeri. *Stability and discontinuity waves for symmetric hyperbolic systems, Nonlinear wave motion* (A. Jeffrey, ed.). Longman, Harlow (1989).

[87] A. Muracchini, L. Seccia. Thermo-acceleration waves and shock formation in extended thermodynamics of gravitational gases. *Cont. Mech. Thermodyn.* **1** (1989).

[88] P. Chen. Growth and decay of waves in solids. Mechanics of Solids III, *Handbuch der Physik* **6A/3**. Springer-Verlag, Berlin (1973).

[89] M. Kranys. Phase and signal velocities of waves in dissipative media. Special relativistic theory. *Arch. Rational Mech. Anal.* **48** (1972).

[90] J. M. Stewart. On transient relativistic thermodynamics and kinetic theory. *Proc. Roy. Soc. London Ser. A* **357** (1977).

[91] W. Israel, J. M. Stewart. On transient relativistic thermodynamics and kinetic theory. II. *Proc. Roy. Soc. London* Ser. A **365** (1979).

[92] G. Boillat. Wave velocities in relativistic extended thermodynamics. *Proceedings of the Symposium on Kinetic Theory and Extended Thermodynamics*, Bologna, May 18-20th, 1987 (I. Müller and T. Ruggeri, eds.). Pitagora Tecnoprint, Bologna (1987).

[93] L. Seccia, A. Strumia. Wave propagation in relativistic extended thermodynamics. *Cont. Mech. Thermodynamics* **2** (1990).

[94] T. Ruggeri. Shock waves in hyperbolic dissipative systems: non-equilibrium gases. *Pitman Research Notes in Mathematics* **227**. (D. Fusco, A. Jeffrey, eds.) Longman, Harlow (1991).

[95] L. D. Landau, E. Lifshits. *Mécanique des fluides*, MIR, Moscow (1971).

[96] P. D. Lax. *Shock waves and entropy contribution to non-linear functional analysis.* (Zarantonello, ed.). Academic Press, New York (1971).

[97] G. Boillat. Sur une fonction croissant comme l'entropie et generatrice des chocs dans les systemes hyperboliques. *C. R. Acad. Sci. Paris* Sér. A **283** (1976).

[98] P. D. Lax. Hyperbolic systems of conservation laws. *Comm. Pure Appl. Math.* **10**. Wiley, New York (1957).

[99] H. Grad. The profile of a steady plane shock wave. *Comm. Pure Appl. Math.* **5**. Wiley, New York (1952).

[100] A. M. Anile, A. Majorana. Shock structure for heat conducting and viscid fluids. *Meccanica* **16** (3) (1981).

[101] T. Ruggeri, A. Muracchini, L. Seccia. A continuum approach to phonon gas and shape changes of second sound via shock waves theory. *Nuovo Cimento D* **16**, (1) (1994).

[102] T. Ruggeri, A. Muracchini, L. Seccia. Second sound and characteristic temperature in solids. *Phys. Rev. B* **54** (1) (1996).

[103] A. Sommerfeld. Vorlesungen über theoretische Physik. Bd **5**, *Thermodynamik und Statistik. Dietrich'sche Verlagsbuchhandlung*, Wiesbaden (1952)

[104] L. Sirovich, J. K. Thurber. *Rarefied gas dynamics.* (J. A. Laurman, ed.), vol. **1**, Academic Press, New York (1963).

[105] C. S. Wang Chang, G. E. Uhlenbeck. *Studies in statistical mechanics.* (I. de Boer and G.E. Uhlenbeck, eds.), vol. **5**, North-Holland, Amsterdam (1970).

[106] C. Cercignani, A. Majorana. Analysis of thermal and shear waves according the B.G.K. kinetic model. *Z. Angew. Math. Phys.* **36** (1985).

[107] C. Cercignani. Speed of propagation of infinitesimal disturbances in a relativistic gas. *Phys. Rev. Lett.* **50** (1983).

[108] L. Waldmann. Transporterscheinungen in Gasen von mittlerem Druck. *Handbuch der Physik* **XII**. Springer-Verlag, Berlin (1958).

[109] M. N. Kogan. *Proceedings 5th Symp. on Rarefied Gas Dynamics* **1**, suppl. 4, Academic Press, New York (1967).

[110] E. T. Jaynes. Information theory and statistical mechanics. *Phys. Rev.* **106** (1957).

[111] J. N. Kapur. *Maximum entropy models in science and engineering.* John Wiley, New York (1989).

[112] B. J. Berne, R. Pecora. *Dynamic light scattering.* Wiley-Interscience, New York (1976).

[113] P. A. Fleury, J. P. Boon. Laser light scattering in fluid systems. *Adv. Chem. Phys.* **24** (1974).

[114] S. R. de Groot, P. Mazur. *Non equilibrium thermodynamics.* North-Holland Publishing Company, Amsterdam (1962).

[115] G. Simonsohn. The role of the first-order autocorrelation function in conventional grating spectroscopy. *Optics Comm.* **5** (1972).

[116] L. R. Wehr. *Vergleich von Lichtstreuexperimenten an verdünnten Gasen -speziell Xenon - mit Berechnungen des Streuspektrums im kinetischen Grenzfall.* Diploma thesis, Technical University Berlin (1996).

[117] J. Au, L. R. Wehr. Light scattering experiment and extended thermodynamics. *Cont. Mech. Thermodynamics* **9** (1997).

[118] N. A. Clark. *Inelastic light scattering from thermal fluctuations in gases.* Ph. D. Thesis, Massachusetts Institute of Technology (1970).

[119] N. A. Clark. Inelastic light scattering from density fluctuations in dilute gases. The kinetic-hydrodynamic transition in a monatomic gas. *Physical Review A* **12** (1975).

[120] A. Sugawara, S. Yip, L. Sirovich. Spectrum of density fluctuations in gases. *Phys. Fluids* **11** (1968).

[121] B. Kamgar-Parsi, E. G. D. Cohen. Eigenmodes of a dilute simple gas in equilibrium. *Physica* **138** A (1986).

[122] J. Kestin, K. Knierim, E. A Mason, B. Najafi, S. T. Ro, M. Waldmann. Equilibrium and transport properties of the noble gases and their mixtures at low density. *J. Phys. Chem. Ref. Data* **13** (1984).

[123] P. L. Bhatnagar, E. P. Gross, M. Krook. A model for collision processes in gases. I. Small amplitude processes in charged and neutral one-component systems. *Phys. Rev.* **94** (1954).

[124] J. R. D. Copley, S. W. Lovesey. The dynamic properties of monatomic liquids. *Rep. Prog. Phys.* **38** (1975).

[125] S. Yip, M. Nelkin. Application of a kinetic model to time-dependent density correlations in fluids. *Phys. Rev. A* **135** (1964).

[126] E. Meyer, G. Sessler. Schallausbreitung in Gasen bei hohen Frequenzen und niedrigen Drücken. *Zeitschr. f. Physik* **149** (1957).

[127] W. Israel. Covariant fluid mechanics and thermodynamics: An introduction. (A.M. Anile and Y. Choquet-Bruhat, eds.). *Lecture Notes in Mathematics* **1358**, Springer-Verlag Heidelberg (1989).

[128] W. Weiss. *Die Berechnung von kontinuierlichen Stoßstrukturen in der Kinetischen Gastheorie.* Habilitation thesis TU Berlin (1997).

[129] H. Alsmeyer. Density profiles in argon and nitrogen shock waves measured by the absorption of an electron beam. *J. Fluid. Mech.* **74** (1976).

[130] T. Ruggeri. On the shock structure problem in non-equilibrium thermodynamics of gases. *Transport Theory and Statistical Physics* **25** (3-5) (1996).

[131] R. Becker. *Stoßwelle und Detonation.* Z. f. Physik **8** (1922).

[132] D. Gilbarg, D. Paolucci. The structure of shock waves in the continuum theory of fluids. *J. Rat. Mech. Anal.* **2** (1953).

[133] G. M. Kremer, I. Müller. Shock thickness in monatomic gases. *Meccanica* (in press).

[134] D. Jou, D. Pavón. Nonlocal and nonlinear effects in shock waves. *Phys. Rev. A* **44** (1991).

[135] L. H. Holway. Existence of kinetic theory solutions to the shock structure problem. *Phys. Fluids* **7** (1964).

[136] W. Weiss. A comment on "Existence of kinetic theory solutions to the shock structure problem" by L.H. Holway. *Physics of Fluids* **8** (1996)

[137] T. Ruggeri. Breakdown of shock-wave-structure solutions. *Phys. Rev. E* **47** (1993).

[138] K. Zoller. Zur Struktur des Verdichtungsstoßes. *Zeitschr. f. Physik* **130** (1951).

[139] J. D. Foch. On higher order hydrodynamic theories of shock structure. *Acta Physica Austriaca, suppl.* **10** (1973).

[140] H. W. Mott-Smith. The solution of the Boltzmann equation for a shock wave. *Phys. Rev.* **82** (1951).

[141] G. A. Bird. Aspects of the structure of strong shock waves. *Phys. Fluids* **13** (1970).

[142] E. Kamke. *Differentialgleichungen I.* Geest & Portig, Leipzig (1967).

[143] L. Bieberbach. *Theorie der gewöhnlichen Differentialgleichungen.* Springer Berlin (1965).

[144] L. Collatz. *Differentialgleichungen.* B. G. Teubner, Stuttgart (1967).

[145] J. Au. Nonlinear closure in molecular extended thermodynamics with application to the shock wave structure solution. *Suppl. Rendiconti Circ. Mat. Palermo* Part I, Ser. II **45** (1996).

[146] I. Müller. Entropy of radiation - density, flux, production. In: *Comtemporary research in the mechanics and mathematics of materials.* (R. C. Batra, M. F. Beatty, eds.). CIMNE, Barcelona (1996).

[147] K. S. Thorne. Relativistic radiative transfer: Moment formalisms. *Mon. Not. R. Astr. Soc.* **194** (1981).

[148] S. Chandrasekhar. *An introduction to the study of stellar structure.* Dover Pub., London (1957).

[149] D. Mihalas, B. W. Mihalas. *Foundations of radiation hydrodynamics.* Oxford University Press, New York (1984).

[150] J. Oxenius. *Kinetic theory of particles and photons.* Springer, Berlin (1986).

[151] W. Larecki. Symmetric conservative form of low-temperature phonon gas hydrodynamics II. Equations of heat transport and thermal waves in the case of linear isotropic approximation of phonon frequency spectrum. *Il Nuovo Cimento* **1868 D** (1991).

[152] A. M. Anile, S. Pennisi, M. Sammartino. A thermodynamical approach to Eddington factors. *J. Math. Phys.* **31** (1991).

[153] G. M. Kremer, I. Müller. Radiation thermodynamics. *J. Math. Phys.* **33** (1992).

[154] C. D. Levermore. Relating Eddington factors to flux limiters. *J. Quant., Spectrosc. Radiat. Transfer* **31** (1983).

[155] R. Becker, F. Sauter. *Theorie der Elektrizität.* Bd. 1 Teubner Verl. Ges. Stuttgart (1957).

[156] H. A. Bethe, E. E. Salpeter. Quantum mechanics of one- and two-electron systems. *Handbuch der Physik* **XXXV**: Atome I (S. Flügge, eds.). Springer, Berlin (1957).

[157] R. Peierls. Zur kinetischen Theorie der Wärmeleitung in Kristallen. *Ann. Phys.* **3** (1929).

[158] R. Peierls. *Quantum theory of solids.* Oxford University Press (1955).

[159] H. Struchtrup. *Der zweite Schall in der Literatur der Festkörperphysik - eine kritische Durchsicht.* Diploma thesis, TU Berlin (1991).

[160] W. Dreyer, H. Struchtrup. Heat pulse experiments revisited. *Cont. Mech. Thermodyn.* **5** (1993).

[161] P. Debye. Zur Theorie der spezifischen Wärmen. *Ann. Phys.* **39** (1912).

[162] J. Callaway. Model for lattice thermal conductivity at low temperatures. *Phys. Rev.* **113** (1959).

[163] J. C. Ward, J. Wilks. Second sound and the thermo-mechanical effect at very low temperatures. *Phil. Mag.* **42** (1952).

[164] R. J. von Gutfeld, A. H. Nethercot, Jr. Heat pulses in quartz and sapphire at low temperatures. *Phys. Rev. Lett.* **12** (1964).

[165] R. A. Guyer, J. A. Krumhansl. Dispersion relation for second sound in solids. *Phys. Rev.* **133** (1964).

[166] R. A. Guyer, J. A. Krumhansl. Solution of the linearized phonon Boltzmann equation. *Phys. Rev.* **148** (1966).

[167] R. A. Guyer, J. A. Krumhansl. Thermal conductivity, second sound and phonon hydromechanic phenomena in non-metallic crystals. *Phys. Rev.* **148** (1966).

[168] C. C. Ackermann, B. Bertman, H. A. Fairbanks, R. A. Guyer. Second sound in solid helium. *Phys. Rev. Lett.* **16** (1970).

[169] C. C. Ackermann, R. A. Guyer. Temperature pulses in dielectric solids. *Ann. Phys.* **50** (1968).

[170] T. F. McNelly, S. J. Rogers, D. J. Chanin, R. J. Rollefson, W. M. Goubau, G. E. Schmidt, J. A. Krumhansl, R. O. Pohl. Heat pulses in NaF: Onset of second sound. *Phys. Rev. Lett.* **24** (1970).

[171] H. E. Jackson, C. T. Walker, T. F. McNelly. Second sound in NaF. *Phys. Rev. Lett.* **25** (1970).

[172] V. Narayanamurti, D. C. Dynes. Observation of second sound in bismuth. *Phys. Rev. Lett.* **28** (1972).

[173] H. E. Jackson, C. T. Walker. Thermal conductivity, second sound and phonon-phonon interactions in NaF. *Phys. Rev. B* **3** (1968).

[174] A. Sommerfeld, H. Bethe. Elektronentheorie der Metalle. *Handbuch der Physik Bd.* **XIV**, Springer Heidelberg (1956).

[175] B. D. Coleman, H. Markovitz, W. Noll. *Viscometric flows in non-Newtonian fluids.* Springer New York (1966).

[176] R. F. Ginn, A. B. Metzner. Measurement of stresses developed in steady laminar shearing flows of viscoelastic media. *Trans. Soc. Rheol.* **13** (1969).

[177] R. L. Fosdick, K. R. Rajagopal. Anomalous features in the model of second order fluids. *Arch. Rational Mech. Anal.* **70** (1978).

[178] J. E. Dunn, K. R. Rajagopal. Fluids of differential type: Critical review and thermodynamic analysis. *Int. J. Engng. Sci* **33** (1995).

[179] I. Müller, K. Wilmanski. Extended thermodynamics of a non-Newtonian fluid. *Rheologica Acta* **25** (1986).

[180] K. Wilmanski. Thermodynamics of a heat conducting Maxwellian fluid. *Arch. Mech.* **40** (1988).

[181] I-Shih Liu. Extended thermodynamics of viscoelastic materials. *Cont. Mech. Thermodyn.* **1** (1989).

[182] R. B. Bird, R. C. Armstrong, O. Hassager. *Dynamics of polymeric liquids.* John Wiley & Sons New York (1976).

[183] R. B. Bird, H. R. Warner jr., D. C. Evans. Kinetic theory of dumbbell suspensions with Brownian motion. *Adv. of Polym. Sci.* **8**, Springer Heidelberg (1971).

[184] I. Müller. Stress and heat flux in a dumbbell solution. *Arch. Mech.* **31** (1979).

[185] D. Reitebuch. *Zur kinetischen Theorie von nicht-Newton'schen Flüssigkeiten; insbesondere thermodynamische Effekte.* Batchelor thesis, TU Berlin (1995).

Index

Springer Tracts in Natural Philosophy

(Continued from page ii)

MIX
Papier aus verantwortungsvollen Quellen
Paper from responsible sources
FSC® C105338

If you have any concerns about our products,
you can contact us on
ProductSafety@springernature.com

In case Publisher is established outside the EU,
the EU authorized representative is:
Springer Nature Customer Service Center GmbH
Europaplatz 3, 69115 Heidelberg, Germany

Printed by Libri Plureos GmbH
in Hamburg, Germany